The Shell Bitumen Handbook

Sixth edition

The Shell Bitumen Handbook

Sixth edition

Principal authors
Dr Robert N. Hunter, Andy Self and Professor John Read

Published for Shell Bitumen by ICE Publishing, One Great George Street, Westminster, London SW1P 3AA.

Full details of ICE Publishing sales representatives and distributors can be found at: www.icevirtuallibrary.com/info/printbooksales

This title has been previously published as

Mexphalte Handbook, First edition, 1949
Mexphalte Handbook, Second edition, A.W. Jarman (ed), Shell-Mex and BP Ltd, 1955
Mexphalte Handbook, Third edition, 1963
The Shell Bitumen Handbook, Fourth edition, D. Whiteoak, Shell Bitumen UK, 1990
The Shell Bitumen Handbook, Fifth edition, J. Read, D. Whiteoak, Shell Bitumen UK, 2003

www.icevirtuallibrary.com
A catalogue record for this book is available from the British Library

ISBN 978-0-7277-5837-8

Permission to reproduce extracts from British Standards is granted by BSI. British Standards can be obtained in PDF or hard copy formats from the BSI online shop:
www.bsigroup.com/Shop
or by contacting BSI Customer Services for hard copies only:
Tel: +44 (0)20 8996 9001, Email: cservices@bsigroup.com.

Commissioning Editor: Rachel Gerlis
Production Editor: Rebecca Taylor
Market Development Executive: Elizabeth Hobson

Typeset by Academic + Technical, Bristol
Index created by Indexing Specialists (UK) Ltd
Printed and bound in Italy by Rotolito Lombarda SpA

Contents

About the principal authors   xii
Foreword   xiv
Acknowledgements   xv

Chapter 1   **Introduction**   **1**
            1.1   Origins   3
            1.2   Definition of bitumen   4
            1.3   The uses of bitumen   5
            1.4   Health, safety and environment   10
            1.5   Other binders   11
            1.6   Terminology   13
            References   13

Chapter 2   **Manufacture and storage of bitumens**   **15**
            2.1   The manufacture of bitumen   15
            2.2   Fractional distillation of crude oil   15
            2.3   Air blowing of short residues   20
            2.4   Delivery, storage and handling temperatures
                   of bitumen   24
            References   27

Chapter 3   **Handling, health, safety and environmental
            aspects of bitumens**   **29**
            3.1   Handling and safety   29
            3.2   Health aspects of bitumen   36
            3.3   Bitumen and the environment   42
            References   44

Chapter 4   **Constitution and structure of bitumens**   **47**
            4.1   Bitumen constitution   48
            4.2   Bitumen structure   54
            4.3   The relationship between bitumen constitution,
                   structure and rheology   57

4.4 The relationship between broad chemical composition and physical properties during the bitumen life cycle 58

References 62

Chapter 5 **Specifying bitumens and checking their quality 65**

5.1 Penetration grade, hard penetration grade and viscosity grade bitumens 66
5.2 Polymer modified bitumens (PMBs) 66
5.3 Oxidised bitumens 69
5.4 Hard bitumens 70
5.5 Bitumen quality 70
5.6 The CEN bitumen specification 81
5.7 The SHRP/Superpave bitumen specification 82

References 85

Chapter 6 **Routine testing and mechanical properties of bitumens 87**

6.1 Standard specification tests for bitumens 87
6.2 Temperature susceptibility – the penetration index (PI) 92
6.3 Viscosity 93
6.4 The bitumen test data chart 99
6.5 Engineering properties of bitumen 105
6.6 Cohesion 114

References 116

Chapter 7 **Rheology of bitumens 119**

7.1 Rheology, deformation and flow 120
7.2 The rheology of bitumen 134

References 145

Chapter 8 **Polymer modified bitumens and other modified binders 149**

8.1 The role of bitumen modifiers in asphalt pavements 150
8.2 The modification of bitumen 153
8.3 Other modified products used for road and industrial applications 166
8.4 Properties of PMBs and related test methods 174
8.5 Manufacture, storage and handling of PMBs 177
8.6 Performance of PMBs for road application: Shell Cariphalte® example 179

References 180

Chapter 9	**Bitumen emulsions**	**185**
	9.1 A brief history of the development of bitumen emulsions	186
	9.2 The components of emulsions	187
	9.3 The manufacture of bitumen emulsions	193
	9.4 Storage and handling of bitumen emulsions	194
	9.5 Properties of bitumen emulsions	196
	9.6 Modification of bitumen emulsion properties	203
	9.7 Classification and specification of bitumen emulsions	208
	9.8 Uses of bitumen emulsions	209
	9.9 Trends and new developments in the use of bitumen emulsions	214
	9.10 Further information on bitumen emulsions	214
	References	214

Chapter 10	**Aggregates in asphalts**	**217**
	10.1 Introduction	217
	10.2 Origin and type	218
	10.3 Aggregate processing	224
	10.4 Coarse and fine aggregates	231
	10.5 Filler aggregate	250
	10.6 Assuring aggregate quality	255
	References	256

Chapter 11	**Choosing asphalts for use in flexible pavement layers**	**261**
	11.1 Asphalts used in flexible pavements in Europe	261
	11.2 UK pavements	264
	11.3 Asphalts used in flexible pavements in the UK	267
	11.4 Asphalts used in flexible pavements in France	277
	11.5 Asphalts used in flexible pavements in Germany	282
	11.6 Asphalts used in flexible pavements in the USA	284
	11.7 Asphalts used in flexible pavements in India	287
	11.8 Asphalts used in flexible pavements in China	289
	References	290

Chapter 12	**Design of asphalt mixtures**	**295**
	12.1 A very brief history of asphalt mixture design methods	295
	12.2 Mixture volumetrics	297

12.3	Grading	304
12.4	Mixture design methods	311
References		342

Chapter 13	**Design of flexible pavements**	**347**
13.1	Introduction	347
13.2	Background to the design of pavements	351
13.3	The importance of stiffness	353
13.4	The elements of a flexible pavement	356
13.5	Stages involved in pavement design	360
13.6	Pavement designs in the USA	366
13.7	Empirical and analytical pavement design	368
13.8	Analytical pavement design	368
13.9	The contribution of Shell Bitumen to the development of pavement design	369
13.10	Example pavement designs using SPDM-PC	372
13.11	A glimpse of the future for pavement design?	374
References		375

Chapter 14	**Asphalt production plants**	**379**
14.1	Introduction	379
14.2	Types of mixing plants	380
14.3	Plant capacities	402
14.4	Additive systems	403
14.5	Cold mix plants	404
14.6	Warm mix plants	406
14.7	Asphalt plant control	407
14.8	Sampling	410
References		412

Chapter 15	**Transport, laying and compaction of asphalts**	**413**
15.1	Transport	413
15.2	Preparatory works	413
15.3	Application of bond coats	415
15.4	Delivery and discharge	417
15.5	Pavers	418
15.6	Laying in adverse weather conditions	423
15.7	Machine laying	424
15.8	Hot-on-hot paving	427
15.9	Hand laying	427
15.10	Layer thickness and surface regularity	429
15.11	Joints	430

15.12 Chipping of hot rolled asphalt surface course 434
15.13 Installing gussasphalt 435
15.14 Compaction 437
15.15 Application of grit to surface courses 443
15.16 Opening to traffic 443
15.17 Specification and quality control 444
References 444

Chapter 16 **Laboratory testing of asphalts** **447**
16.1 Fundamental tests 449
16.2 Simulative tests 459
16.3 Empirical 467
16.4 Determination of recovered bitumen properties 470
References 474

Chapter 17 **Properties of asphalts** **479**
17.1 Stiffness of asphalts 479
17.2 Deformation of asphalts 486
17.3 Fatigue characteristics of asphalts 491
17.4 Failure theories for cracking in asphalt layers 496
References 499

Chapter 18 **Influence of binder properties on the performance of asphalts** **503**
18.1 Introduction 503
18.2 The influence of binder properties during construction 508
18.3 The influence of binder properties on the performance of asphalts in service 511
References 542

Chapter 19 **Adhesion of bitumen and moisture damage in asphalts** **549**
19.1 Thermodynamic principles of adhesion, cohesion and surface free energy 550
19.2 Factors affecting bitumen–aggregate adhesion 552
19.3 Disbonding mechanisms in asphalts 553
19.4 Methods of measuring and assessing adhesion between bitumen and aggregates and moisture damage 557
19.5 Improving bitumen–aggregate adhesion 567
References 569

Chapter 20 **Durability of bitumens and asphalts** **573**
20.1 Bitumen hardening 574
20.2 Hardening of bitumen during storage,
mixing and in service 576
20.3 Bitumen ageing tests 586
References 588

Chapter 21 **Surface treatments** **591**
21.1 Surface dressing 591
21.2 Slurry surfacing/microsurfacing 621
21.3 High friction surfaces 635
21.4 Grouted macadams 636
References 640

Chapter 22 **Alternative asphalts** **643**
22.1 Asphalts produced using foamed bitumen 643
22.2 Warm mix asphalts 651
22.3 Shell Thiopave®, a sulfur-based asphalt modifier
for enhanced asphalt road applications 654
22.4 Recycling of asphalt pavements 666
22.5 Modification of bitumen by the addition of
rubber 671
References 680

Chapter 23 **Certification of bitumens and asphalts** **687**
23.1 The regulatory framework 688
23.2 Standards 691
23.3 Certification for products and systems outside
the scope of CEN harmonised standards 694
23.4 Requirements for product certification 697
References 700

Chapter 24 **Other important uses of bitumens and
asphalts** **703**
24.1 Bitumen in roofing applications 704
24.2 Bituminous adhesive for use with pavement
markers 714
24.3 Airfield pavements 716
24.4 Conventional and high speed railway
applications 721
24.5 Bridges 726
24.6 Recreational areas 728
24.7 Motor racing tracks, including Formula 1 tracks 729

24.8 Vehicle testing circuits 730
24.9 Hydraulic applications 730
24.10 Coloured surface courses and surface
treatments 732
24.11 Asphalt kerbs 735
References 735

Appendix 1 **Physical constants of bitumens** **739**

Appendix 2 **Conversion factors for viscosities** **743**

Appendix 3 **Blending charts and formulae** **744**

Appendix 4 **Calculation of bitumen film thickness in an asphalt** **748**

Appendix 5 **Bitumen product standards used across the world** **750**

Appendix 6 **Asphalt product standards used across the world** **754**

Appendix 7 **Reporting asphalt compositional analysis results** **756**

Appendix 8 **Volumetrics symbols** **759**

Index **761**

About the principal authors

Dr Robert N. Hunter BSc MSc PhD CEng FICE FIAT MCIArb FCInstCES

Robert has always had an intense interest in asphalt technology and flexible pavements, spending many years on a range of sites supervising highway construction contracts. He undertook two postgraduate research degrees in which he examined various aspects of the behaviour of asphalts. He wrote a computer program that calculates the temperatures in an unchipped asphalt after laying, and wrote the only computer program that calculates the temperatures in a chipped asphalt after laying. In addition, he produced the algorithm on which the only computer program that designs UK compliant pavements and foundations is based.

For much of his career Robert worked for local authorities, and was general manager of a trunk road unit in east and central Scotland. He now runs his own consultancy specialising in providing an expert witness service related to asphalts, pavements and defects in highways. He also advises on contract related issues, and produces contract documents invariably based on NEC3, which he adores. He is also a registered adjudicator with the Institution of Civil Engineers and the Chartered Institute of Arbitrators, and has been appointed as adjudicator in many disputes.

Robert edited and contributed to *Bituminous Mixtures in Road Construction* (1994) and *Asphalts in Road Construction* (2000). He was the technical editor on the 5th edition of *The Shell Bitumen Handbook* (2003).

Andy Self BSc(Hons) MBA

Andy has worked in the road construction industry for over 30 years. He began his career working as a laboratory technician, and spent almost 20 years working in various technical roles for asphalt and quarrying companies in the UK. During this time he gained extensive experience in asphalt production, laboratory testing of construction materials, asphalt mixture design and product quality management. In 2002, he gained a BSc Honours degree with the Open University.

In January 2003, Andy moved to Shell as asphalt development manager for the bitumen business in UK and Ireland, working closely with customers, suppliers and colleagues on product development. More recently, he worked as bitumen technical manager Europe; managing teams in France, Germany, Norway and the UK, and representing Shell on various industry association committees. Andy completed his MBA during his recent role as skillpool manager in the Shell Specialities (bitumen and sulfur) global technical team, and he now works as global learning advisor for the Shell Specialities business.

Professor John Read PhD BEng(Hons) CEng MIM MIAT MIHT MAAPT

Professor John Read has worked in the bitumen and sulfur industry for nearly 30 years, initially for a consultant testing house, then for an asphalt supplier and more recently for Shell in various roles. He is currently the general manager technology for Shell Specialities, where his principal accountabilities are leading a world class team of technical specialists to deliver research and development and technical services to customers in more than 25 countries around the world.

During his career he has delivered more than 100 public presentations at conferences, industry associations, universities and industry training days, and has published more than 80 papers, articles and technical brochures, including co-authoring the 5th edition of the internationally renowned *The Shell Bitumen Handbook*.

John has recently finished two terms as the vice-president of Eurobitume, and serves as Shell's primary director on the Asphalt Institute.

John has two children and lives and works from his home in Derbyshire, UK.

Foreword

It is more than 10 years since I authored the 5th edition of *The Shell Bitumen Handbook* and it gives me great pleasure to write this foreword for the 6th edition. The whole book has been completely reviewed by international experts from Shell, complemented by experts from across the industry, resulting in a much more global approach than the 5th edition, with contributions from Argentina, Australia, China, France, Germany, India, Indonesia, the Netherlands, Malaysia, Thailand, the United Arab Emirates, the USA and the UK. Each chapter has been either edited or completely rewritten. There are several new sections and chapters on recycling, warm asphalt mixtures, foam mixtures, bitumen additives and certification schemes for bitumen and asphalt. The result is a lasting testament to the continued contribution that Shell makes to asphalt technology, as well as confirmation to the wealth of knowledge, expertise and experience that all the authors possess.

Many people have contributed to this edition of the handbook, and I sincerely thank them all, but it would be only fitting for me to give special mention to two individuals in particular: Andy Self and Dr Robert Hunter. As well as being an asphalt expert in his own right, Andy Self has been nursemaid to all the authors, a tyrannical timekeeper and a truly good friend. Without his organisational skills this 6th edition would never have been completed. Dr Robert Hunter agreed to be the technical editor for the second time, and his eye for detail has ensured the quality publication that this book has become. I also need to mention that it was Robert who encouraged me to write the book again, and it is his tenacity that has ensured we have all delivered on time – without him you would not be reading this book. I give my heartfelt thanks to both Andy and Robert.

In closing, I would like to mention all members of the Shell bitumen technology team, who work tirelessly in support of pursuing the advancement of asphalt technology, as well as their families, who are always so understanding and supportive of the amount of time we give – thank you all. You continue to ensure we remain a technology leader, and individually you inspire me each and every day. Lastly, I would like to thank you, the reader, for acquiring this book and taking the time to read it. I hope that you find it a useful resource and are able to benefit from the huge amount of knowledge and experience that has gone into this encyclopaedia of bitumen and asphalt technology.

Professor John Read, General Manager Technology (Bitumen & Sulfur)
May 2014

Acknowledgements

Robert Hunter is delighted to acknowledge the immense support he has enjoyed during the production of this book. First, he would like to thank: the wonderful Debra Francis, librarian at the Institution of Civil Engineers; Professor Ian Walsh, for his immense technical assistance and Bob Allen who, as well as writing a superb chapter dealing with aggregates in asphalts, freely gave enormous assistance on many occasions. Other people he is delighted to thank are Jack Edgar, Lewis Hunter, Russell Hunter, Peter Dick, Eddie Lord, Jeff Farrington, Gordon Steel, Dr Behrooz Saghafi, Andrew Scorer of Miles Macadam Ltd, Rory O'Connor of Tarstone Surfacing Limited, David Merritt of The Transtec Group, Inc, Duncan Tharme of the Curtiss-Wright Corporation, Stephen Collins representing the Road Emulsion Association (REA), Thomas Jennings of The Phoenix Engineering Co Ltd, Peter Wallace of PTS International Ltd and, last but certainly not least, his very good friend Tony Sewell of PTS International Ltd.

Finally, Robert would like to express his gratitude to: Shell, for asking him to work on this book; the authors, many of whom wrote on very complex subjects in a language that is not their native tongue; Andy Self, who is one of the most efficient gentlemen I have ever met and an absolute joy to work with; and Professor John Read, as ever a great friend and a source of substantial support.

Andy Self would like to thank his fellow principal authors and the project team – Professor John Read, Dr Robert Hunter, Diny Rovers, Jayne Davies, Jenny Marsden, Rachel Gerlis and Rebecca Taylor. A very special thanks to Dr Robert Hunter for his tireless and uncompromising commitment to the new edition.

Andy and John would like to acknowledge formally their colleagues at Shell: Dr Richard Taylor and Dr Jia Lu for excellent and timely contributions to several chapters. Others thanked for their support and contributions are: Carlos Maurer, Sonia Hauguel and the team at the Strasbourg Solution Centre, Indra Maizir, Yan Hui, Gary Fitts, Steve Sturridge and John Barlow.

Richard Taylor wishes to thank Dr Mark Bouldin of Oil Re-Refining Specialty Products at Safety-Kleen.

Nilanjan Sarker wishes to thank Dr Robert Hunter.

Bob Allen wishes to thank Dr Paul Phillips, Dr Ignacio Artamendi, Graeme Richards, Dr Ruth Richards, Phil Sabin, Matthew Allen and Benjamin Allen.

John Moore wishes to thank Paul Bolley of Benninghoven UK, Pirmin Hanggi of Ammann Switzerland, David Weeks of Hanson UK, Andrea Barnes Bate of Astec Inc, Chattanooga, USA, Paul Tipper of Control Net Solutions, UK, and Jeannie Moore for proof reading.

Jonathan Core wishes to thank Knut Johannsen of EUROVIA Services GmbH, Phil Higginson of Wirtgen Limited and Melanie Hull of Wirtgen Limited.

Frank Beer wishes to thank Jacques Colange, Dr Martin Vondenhof, Jayne Davies, Dr Andreas Opel, Catherine Noireaux and Punith V Shivaprasad, all of Shell.

The authors of Chapter 18 would like to recognise formally the original contribution of Dr Mike Nunn.

An emulsion manufacturing unit located in Bangkok, Thailand

A polymer modified bitumen manufacturing plant in Buenos Aires, Argentina

Chapter 1

Introduction

Andy Self
Global Learning Advisor, Shell International Petroleum Co. Ltd. UK

This book is about refined bitumen, one of the world's most widely used construction materials. The *Shell Bitumen Handbook* is one of the most well known and highly respected publications on the subject of bitumen and asphalt technology. Thousands of copies have been sold since it was first published. Many people within the construction industry have a copy somewhere in their office or library. Most importantly, it is widely referenced in academic papers.

This new sixth edition covers the same subject matter as previous editions, but a slightly different approach has been taken as to how it has been compiled. For the first time, each chapter has been written by a different subject matter expert. The authors come from many different countries. The intention is to bring a wider, more global focus to the subjects and reflect the fact that growth economies in countries such as China and India are adapting and developing new practices and specifications and driving research in this area.

Shell has a distinguished track record in bitumen research and development, and continues to invest significantly in fundamental research (Table 1.1). Investment in regional technical service laboratories and the recruitment and development of local expertise in many markets has produced a truly global team of subject matter experts. In this edition, this expertise has been used to create something that looks and feels different to previous publications, and is unique in its scope and depth. The aim is to capture all the main new developments in the field since the last edition.

The book is not a review of current academic research and theory, although some chapters refer to recent studies. It is intended to reflect today's state of the art in terms of practice, knowledge and trends. The intention has been to make it relevant for a wide range of readers and to cover the subject in as much detail as possible.

The Shell Bitumen Handbook, Sixth Edition
ISBN 978-0-7277-5837-8
Shell International Petroleum Company Ltd: All rights reserved
http://dx.doi.org/10.1680/tsbh.58378.001

Table 1.1 History of bitumen research at Shell

Research theme	Output
1921–1930 Understanding fundamentals	The effects of the crude oil origin Investigations into the blowing process
1930–1940 Understanding manufacturing	The chemistry of bitumen Colloidal structures, blending, blowing and phase separation R-Grades
1940–1960 Understanding performance	The visco-elastic properties of bitumen Performance and specifications Shell Flintkote® emulsions series
1960–1970 Translating fundamentals into tools	Van der Poel's stiffness nomograph Heukelom's ductility chart Bitumen test data chart Friction reduction (e.g. Compound SL)
1970–1980 Bitumen and asphalt performance	*Shell Pavement Design Manual* (*SPDM*) First polymer modified bitumen (PMB) patents Modelling of the blowing process (Oxytell) New processing routes
1980–1990 Specialities and quality frameworks	Commercialisation of PMB (roads, industrial applications) Coloured binders (Shell Mexphalte C®) Joint sealants, glues Feedstock suitability modelling Qualagon Health and safety aspects of bitumen handling Improved design tools (new modules in *SPDM*)
1990–2000 High performance solutions	Chemically enhanced high performance binder (Shell Multiphalte®) Consistently performing emulsion grades Shell Flintkote Ultra® technology Pipe coating technology Open graded asphalt (low noise, low splash) Warm mix technology Near-infrared quality monitoring Understanding and contributing to the Strategic Highway Research Program (SHRP) Non-bitumen waterproofing systems
2000–2010 Sustainable solutions	Understanding performance warranties Engineered waterproofing solutions Pavement management systems Environmental solutions (WAM Foam, Active Asphalt) Bio-binders
2010–today Moving with the times	New manufacturing solutions Improving asphalt workability Low temperature solutions Crumb rubber modified bitumen (Shell Mexphalte RM®) Odour reduction (Shell Bitufresh®)

Readers can use the book in different ways. If you are new to the subject, it makes sense to start at the beginning and read a few chapters before delving into the more complex technical content. Others may like to read specific chapters, or simply use the book as a reference to dip into when required.

The book is structured as follows: Chapters 2 to 9 focus on bitumen technology, covering manufacturing, handling and the main engineering properties and specifications used around the world. The following six chapters (10 to 15) focus on asphalt and its use in flexible pavements – this includes: the main ingredients, design methods, manufacturing and construction. Chapters 16 to 20 cover the testing and performance characteristics of asphalts. The final four chapters cover some other uses of bitumen and asphalt, including two new chapters (for this edition) on alternative approaches to the design and manufacture of asphalts and product certification schemes.

1.1 Origins

Bitumen is defined in the *Oxford English Dictionary* as 'a tarlike mixture of hydrocarbons derived from petroleum naturally or by distillation, and used for road surfacing and roofing' (Oxford University Press, 1996).

It is widely believed that the term bitumen originated in the ancient and sacred language of Hindus in India, Sanskrit; in which *jatu* means 'pitch' and *jatu-krit* means 'pitch creating'. These terms referred to the pitch produced by some resinous trees. The Latin equivalent is claimed by some to be originally *gwitu-men* ('pertaining to pitch') and by others to be *pixtu-men* ('bubbling pitch'), which was subsequently shortened to bitumen before passing via French into English.

There are several references to bitumen in the Bible, although the terminology used can be confusing. In Genesis, Noah's ark is 'pitched within and without with pitch', and Moses' juvenile adventure is in 'an ark of bulrushes, daubed with slime and with pitch'. Even more perplexing are the descriptions of the building of the Tower of Babel. The Authorised Version of the Bible says 'they had brick for stone, and slime had they for mortar'; the New International Version states that 'they used bricks instead of stone and tar instead of mortar'; Moffat's 1935 translation says 'they had bricks for stone and asphalt for mortar'; but the *New English Bible* states that 'they used bricks for stone and bitumen for mortar'.

The ancient uses of natural bitumens or pitch continued in those inhabited parts of the world where deposits were readily available. However, there seems to have been little development of usage elsewhere. In many countries, none of the present major uses of bitumen were introduced until the end of the nineteenth century. However, there would appear to have been some wider knowledge of large sources of natural bitumen such as lake asphalt in Trinidad. In the middle of the nineteenth century, attempts were made to utilise rock asphalt from European deposits for road surfacing and, from this, there was a slow development of the use of natural products for this purpose, followed by the advent of coal tar and, later, of refined bitumen manufactured from crude oil.

1.2 Definition of bitumen

The term 'bitumen' is used in this book to describe refined bitumen, a hydrocarbon product produced by removing the lighter fractions (such as liquid petroleum gas, petrol and diesel) from crude oil during the refining process. In North America, bitumen is commonly known as asphalt binder or asphalt. For the purpose of this book, the term 'bitumen' is used.

A comprehensive definition of refined bitumen is used in the industry document *The Bitumen Industry – A Global Perspective* (Eurobitume and the Asphalt Institute, 2011) and is reproduced here verbatim:

> Bitumen is an engineering material and is produced to meet a variety of specifications based upon physical properties. Bitumen is the residual product from the distillation of crude oil in petroleum refining. The basic product is sometimes referred to as 'straight run' bitumen and is characterised by CAS# 8052-42-4 or 64741-56-6 which also includes residues obtained by further separation in a deasphalting process. Bitumen can be further processed by blowing air through it at elevated temperatures to alter its physical properties for commercial applications. The general characteristics of oxidized bitumen are described by CAS# 64742-93-4. The vast majority of petroleum bitumens produced conform to the characteristics of these two materials as described in their corresponding CAS definitions.

> Bitumen is produced to grade specification either directly by refining or by blending.

> Bitumen should not be confused with coal derived products such as coal tar or coal tar pitches. These are manufactured by the high temperature pyrolysis (>800°C) of bituminous coals and differ from bitumen substantially in comparison and physical characteristics. The differences between bitumen and coal-tar products are well defined in the literature.

> Similarly, bitumen should not be confused with petroleum pitches (CAS# 68187-58-6), which are often aromatic residues, produced by thermal cracking, coking or oxidation from selected petroleum fractions. The composition of petroleum pitches differs significantly from bitumen.

> Bitumen also should not be confused with natural or lake asphalt such as Trinidad Lake Asphalt, Gilsonite, rock asphalt and Selenice. These products are unrefined and not produced by refining of crude oil. They often contain a high proportion of mineral matter (up to 37% by weight) and light components, leading to a higher loss of mass when heated.

Bitumen is manufactured during the distillation of crude oil. It is generally agreed that crude oil originates from the remains of marine organisms and vegetable matter deposited with mud and fragments of rock on the ocean bed. Over millions of years, organic material and mud accumulated into layers some hundreds of metres thick, the substantial weight of the upper layers compressing the lower layers into sedimentary rock. Conversion of the organisms and vegetable matter into the hydrocarbons of crude oil is thought to be the result of the application of heat from within the Earth's crust and pressure applied by the upper layers of sediments, possibly aided by the effects of bacterial action and radioactive bombardment. As further layers were deposited on the sedimentary rock where the oil had formed, the additional pressure squeezed the oil sideways and upwards through porous rock. Where the porous rock extended to the Earth's surface, oil seeped through to the surface. Fortunately, the majority of the oil and gas was trapped in porous rock, which was overlaid by impermeable rock, thus forming gas and oil reservoirs. The oil remains here until its presence is detected by seismic surveys and recovered by drilling through the impermeable rock.

1.3 The uses of bitumen

The vast majority of bitumen is used by the construction industry, as a constituent of products used in paving and roofing. Excellent waterproofing characteristics and thermoplastic behaviour make it ideal for a wide range of applications. At elevated temperatures (typically between 100 and 200°C) it acts like a viscous liquid, and can be mixed with other components and manipulated and formed as required. Once cooled, it is an inert solid that is durable and hydrophobic (repels water).

Various terms are used to describe conventional bitumen such as straight run, paving grade and penetration grade (or 'pen grade'). When people use these terms they normally mean grades of bitumen that can be produced at a conventional refinery in a relatively simple way. The refinery processes used to manufacture bitumen and the effect these processes have on the final characteristics of the product are described in detail in Chapter 2.

The vast majority of bitumen used in asphalt for road construction is conventional bitumen; that is why it is often known as paving grade. The term 'pen grade' is short for penetration grade, and reflects the fact that this type of product is often classified (in Europe and parts of Asia) using the penetration test. The term straight run refers to the fact that this type of bitumen is often produced direct from the vacuum distillation process, without any further modification.

Current estimates put the world use of bitumen at approximately 102 million tonnes per year (Eurobitume and the Asphalt Institute, 2011), and about

85% of all the bitumen produced is used in asphalt for the construction of roads and other paved areas. Typically, asphalt will contain approximately 5% by mass of bitumen, with the remaining 95% consisting of a mixture of mineral aggregates and much finer materials such as limestone filler (see Chapters 10, 11 and 12 for more information on asphalt mixtures).

A note on terminology is worth making at this point – 'asphalt' is a generic term used to describe a range of road surfacing products containing primarily bitumen and mineral aggregates. A few alternative terms exist including hot mix asphalt (HMA) and asphalt concrete (AC). In this book the term 'asphalt' will be used throughout. Asphalt is often referred to incorrectly in the media and in common parlance as tarmac (short for tarmacadam). Tarmacadam is a road surfacing product using coal tar as a binder and has not been used in road construction for over 30 years.

A further 10% of global bitumen production is used in roofing applications, and the remaining 5% is used mainly for sealing and insulating purposes in a variety of building materials, such as pipe coatings, carpet backing, joint sealants and paint. Chapter 24 contains further information on these types of uses.

The extremely wide range of uses for bitumen is demonstrated by the number of registered uses in Europe under the requirements of the Registration, Evaluation, Authorisation and Restriction of Chemicals (REACH) regulations, which require all chemical substances and associated uses to be registered. Table 1.2 provides an overview (Eurobitume, 2013).

Bitumen is available in a variety of grades. Specifications are used across the world to define these grades to meet the needs of the applications, climate, loading conditions and end use. They are usually based on a series of standard test methods that define the properties of each grade such as hardness, viscosity, solubility and durability. Further information about testing and specification for bitumens can be found in Chapters 5, 6 and 7 of this book.

Bitumens are also used to manufacture mixtures or preparations. In these products, bitumen is often the principal component, but they can contain significant proportions of other materials to meet end use requirements. These mixtures are chemically classified as bitumen preparations.

The most commonly used are as follows (Eurobitume and the Asphalt Institute, 2011):

- *Cut-back bitumen and fluxed bitumen*. Cut-back and fluxed bitumen products are preparations in which the viscosity of the bitumen has been reduced by the addition of a solvent, normally derived from petroleum. Typically the solvents used are white spirit, kerosene and

Table 1.2 REACH registered bitumen uses in Europe (Eurobitume, 2013)

Type	Application
Agriculture	Disinfectants
	Fence post coating
	Mulches
	Mulching paper
	Paved barn floors, barnyards, feed platforms
	Protecting tanks, vats
	Protection for concrete structures
	Tree paints (protective)
Buildings and industrial paving	Water and moisture barriers (above and below ground)
	Floor compositions, tiles, coverings
	Insulating fabrics, papers
	Step treads
	Building papers
	Caulking compounds
	Cement waterproofing compounds
	Glass wool compositions
	Insulating fabrics, felts, papers
	Joint filler compounds
	Laminated roofing shingles
	Liquid roof coatings
	Plastic cements
	Shingles
	Acoustical blocks, compositions, felts
	Bricks
	Damp-proofing coatings, compositions
	Insulating board, fabrics, felts, paper
	Masonry coatings
	Plasterboards
	Putty
	Soundproofing
	Stucco base
	Wallboard
	Air-drying paints, varnishes
	Artificial timber
	Ebonised timber
	Insulating paints
	Plumbing, pipes
	Treated awnings
	Canal linings, sealants
Hydraulics and erosion control	Catchment areas, basins
	Dam groutings
	Dam linings, protection
	Dyke protection
	Ditch linings
	Drainage gutters, structures
	Embankment protection
	Groynes
	Jetties
	Levee protection
	Mattresses for levee and bank protection
	Membrane linings, waterproofing

Table 1.2 Continued	
Type	Application
	Reservoir linings
	Revetments
	Sand dune stabilisation
	Sewage lagoons, oxidation ponds
	Swimming pools
	Waste ponds
	Water barriers
	Backed felts
Industrial	Conduit insulation, lamination
	Insulating boards
	Paint compositions
	Papers
	Pipe wrapping
	Insulating felts
	Panel boards
	Underseal
	Battery boxes, carbons
	Electrical insulating compounds, papers, tapes, wire coatings
	Junction box compound
	Moulded conduits
	Black grease
	Buffing compounds
	Cable splicing compound
	Embalming
	Etching compositions
	Extenders
	Explosives
	Lap cement
	Plasticisers
	Preservatives
	Printing inks
	Well drilling fluid
	Armoured bituminised fabrics
	Burlap impregnation
	Mildew prevention
	Sawdust, cork, asphalt composition
	Acid-proof enamels, mastics, varnishes
	Acid-resistant coatings
	Air-drying paints, varnishes
	Anti-corrosive and anti-fouling paints
	Anti-oxidants and solvents
	Base for solvent compositions
	Baking and heat-resistant enamels
	Boat deck sealing compound
	Lacquers, japans
	Marine enamels
	Blasting fuses
	Briquette binders
	Burial vaults
	Casting moulds
	Clay articles

Table 1.2 Continued	
Type	Application
	Clay pigeons
	Expansion joints
	Flowerpots
	Foundry cores
	Friction tape
	Gaskets
	Mirror backing
	Rubber, moulded compositions
	Shoe fillers, soles
Paving (see also agriculture, hydraulics, railways, recreation)	Airport runways, taxiways, aprons
	Asphalt blocks
	Brick fillers
	Bridge deck, surfacing
	Crack fillers
	Floors for buildings, warehouses, garages
	Highways, roads, streets, shoulders
	Kerbs, gutters, drainage ditches
	Parking lots, driveways
	Portland cement concrete underseal
	Roof-deck parking
	Pavements, footpaths
	Soil stabilisation
Railways	Ballast treatment
	Dust laying
	Paved ballast, sub-ballast
	Paved crossings, freight yards, station platforms
Recreation	Dance pavilions
	Drive-in movies
	Gymnasiums, sport arenas
	Playgrounds, school yards
	Race tracks
	Running tracks
	Skating rinks
	Swimming and wading pools
	Tennis courts, handball courts
	Synthetic playing fields and running track surfaces

gas oil. Cut-back products are typically used for spraying and some mixing applications.

- *Modified bitumen.* Polymer modified bitumens (PMBs) are designed to change the performance properties of straight run bitumen. Properties such as elasticity, adhesive or cohesive strength can be modified by the use of one or more chemical agents. These agents may be polymers, waxes, crumb rubber, sulfur and acids, among other materials. Modified bitumens are now widely used in a variety of applications. More detailed information about PMB is available in Chapter 8.

■ *Bitumen emulsions.* Bitumen emulsions are products in which droplets of bitumen or bitumen preparation are dispersed in an aqueous medium. An emulsifier is used to stabilise the mixture. Bitumen emulsions permit the handling, transport and application of bitumen at lower temperatures, and are used mainly in road surfacing applications. Further information on bitumen emulsions and how they are used can be found in Chapters 9 and 21.

1.4 Health, safety and environment
1.4.1 *Safety*

In service (once applied as an ingredient in a construction product, waterproofing solution, etc.) bitumen is an inert material and represents no hazard to human health. During its storage, transportation and application it is heated to high temperatures, so hazards do exist for those working with the product, although these can be managed by taking appropriate measures to reduce the risk of exposure. Burns remain an immediate hazard, and suitable procedures and safety measures must be used when handling hot bitumen. Hardware such as tanks, pipes and delivery tankers must be specially designed and approved for use. A wealth of industry guidance is available on how to handle hot bitumen. Industry associations such as Eurobitume and the Asphalt Institute are particularly active in this area, and provide user guides that build on many years of improving practice (Eurobitume, 2013; Asphalt Institute, 2013).

Bitumen is not classified as a hazardous substance. However, information on known health, safety and environmental hazards and appropriate measures to reduce any identified risks are captured on safety data sheets (commonly known as SDSs or MSDSs – material safety data sheets). Different country regulators and authorities require certain information to be available, so the format of these can vary, but material safety data sheets are provided to the customer by each bitumen supplier.

1.4.2 *Health*

In 2013, the International Agency for Research on Cancer (IARC) published a monograph entitled *Bitumens and Bitumen Emissions, and Some N- and S-Heterocyclic Polycyclic Aromatic Hydrocarbons* (IARC, 2013). The report represents a scientific evaluation of the hazards associated with occupational exposure to bitumen based on available evidence. More details about this report are available in Chapter 3, but the final evaluation is reproduced below:

■ Occupational exposures to oxidized bitumens and their emissions during roofing are probably carcinogenic to humans (Group 2A).
■ Occupational exposures to hard bitumens and their emissions during mastic-asphalt work are possibly carcinogenic to humans (Group 2B).

■ Occupational exposures to straight run bitumens and their emissions during road paving are possibly carcinogenic to humans (Group 2B).

The report provides details of the studies reviewed during the official scientific evaluation carried out by the monograph team. It details the evidence used, and explains the reasoning behind the final evaluations for occupational exposure to 'bitumen and bitumen emissions during roofing, mastic, and paving applications'.

1.4.3 Environment

Many studies are now available concerning the environmental impact and the carbon footprints of oil products, including bitumen. The most widely referenced life cycle inventory report for bitumen was produced by Eurobitume (Eurobitume, 2012). This report was compiled by experts in accordance with the relevant international standards. It was peer reviewed by an independent expert in the field and provides detail on emissions and resource use of producing bitumen at a typical European refinery.

Many other studies have been commissioned and are publicly available concerning the life cycle and carbon footprint of asphalt (EAPA, 2013).

Further information on the handling, safety and environmental aspects of bitumen can be found in Chapter 3.

1.5 Other binders
1.5.1 Lake asphalt

This is the most extensively used and best known form of 'natural' asphalt binder. It is found in well-defined surface deposits, the most important of which is located in Trinidad. It is generally believed that this deposit was discovered in 1595 by Sir Walter Raleigh.

There are several small deposits of asphalt on the island of Trinidad, but it is the lake in the southern part of the island that constitutes one of the largest deposits in the world. The lake occupies an area of approximately 35 m deep, containing well in excess of 10 million tonnes of material. The surface of the lake is such that it can support the weight of the crawler tractors and dumper trucks that transport excavated material from the surface of the lake to railway trucks that run along the edge of the lake.

Excavated material is refined by heating the material to 160°C, vaporising the water. The molten material is passed through fine screens to remove the coarse, foreign and vegetable matter. This residue is usually termed Trinidad épuré or *refined TLA* ('Trinidad lake asphalt'), and, typically, has the following composition (in percentage by weight): binder 54%, mineral matter 36%, organic matter 10%. This épuré is too hard to be used in asphalts, but is sometimes used as an additive in mastic asphalt production.

1.5.2 Rock asphalt

Natural or rock asphalt (i.e. bitumen-impregnated rock) has been used since the early seventeenth century. Its main uses were for waterproofing, caulking ships and the protection of wood against rot and vermin. Today, rock asphalt enjoys only minimal use. Applications include waterproofing and mastic asphalts where ageing characteristics are particularly important.

Rock asphalt is extracted from mines or quarries, depending on the type of deposit. It occurs when bitumen, formed by the same concentration processes as occur during the refining of oil, becomes trapped in impervious rock formations.

Many deposits have been successfully mined, processed and used in Europe (Italy, France and Switzerland) and in the USA (Utah, Kentucky).

1.5.3 Buton asphalt

Buton asphalt (asbuton) is a source of natural rock asphalt located in Indonesia. It has been used since 1926 (Kramer, 1989), and is located in South Sulawesi province. The deposit of asbuton is considered large, estimated at more than half a billion tonnes, and has been made available in varying forms.

Asbuton originates from crude oil in the depths of Earth layers in Buton Island that migrated upwards along deep-seated faults and deposited heavier residues in the upper beds after lighter fractions evaporated (Wallace, 1989). Asbuton consists of a mixture of natural asphalt and minerals, where the residues have fused and entered into the pores of the minerals. As a consequence, natural asphalt from asbuton cannot effectively be used in asphalt mixtures (Affandi, 2012).

1.5.4 Gilsonite

The state of Utah in the Midwest of the USA holds a sizeable deposit of natural asphalt. Discovered in vertical deposits in the 1860s, it was first exploited by Samuel H. Gilson in 1880 as a waterproofing agent for timber. The material is very hard, and is sometimes blended with petroleum bitumen. It is sometimes used in bridge and roof waterproofing materials as a means of altering the stiffness of mastic asphalt.

1.5.5 Tar

Tar is a generic word for the liquid obtained when natural organic materials such as coal or wood are carbonised or destructively distilled in the absence of air. It is customary to prefix the word 'tar' with the name of the material from which it is derived. Thus, the products of this initial carbonisation process are referred to as crude coal tar, crude wood tar and so on. Two types of crude coal tar are produced as a by-product of the carbonisation of coal–coke oven tar and low temperature tar.

In the mid-1960s, over 2 million tonnes of crude coal tar was produced per annum, of which around half was manufactured as a by-product of the operation of carbonisation ovens that were used to produce town gas. However, the introduction of North Sea gas in the late 1960s resulted in a rapid reduction of tar production from this source, and by 1975 it had disappeared completely.

1.6 Terminology
Some key terminology introduced in this chapter will be used consistently throughout the book.

Bitumen: Hydrocarbon product produced from the refining of crude oil.

Asphalt: Road surfacing material consisting of bitumen, mineral aggregates and fillers and may contain other additives, and includes what are described in some areas as hot mix asphalt (HMA).

Bitumen preparation: A mixture of bitumen and another ingredient (oil, additive, etc.).

Bitumen emulsion: A preparation in which droplets of bitumen or other preparation are dispersed in an aqueous medium.

Polymer modified bitumen (PMB): Widely used preparation in which the primary components are bitumen and polymer(s).

Straight run bitumen: Bitumen produced primarily by distillation processes.

Hard bitumen: A subset of straight run bitumens that have low penetration values commonly specified using a softening point range.

Oxidised bitumen: Bitumen produced by passing air through hot bitumen under controlled temperature and pressure conditions, thus producing a product with specific characteristics.

References
Affandi F (2012) The performance of bituminous mixes using Indonesia natural asphalt. *25th ARRB Conference*, Perth, Australia.
Asphalt Institute (2013) Health & Environmental Documents. http://www.asphaltinstitute. org/public/engineering/health-environmental-documents.dot (accessed 18/02/2014).

EAPA (European Asphalt Pavement Association) (2013) EAPA Publications. http://www.eapa.org/publications.php (accessed 18/02/2014).

Eurobitume and the Asphalt Institute (2011) *The Bitumen Industry – A Global Perspective: Production, Chemistry, Use, Specification and Occupational Exposure*, 2nd edn. Eurobitume, Brussels, Belgium; Asphalt Institute, Lexington, KY, USA.

Eurobitume (2012) http://www.eurobitume.eu/system/files/LCIEurobitumeMarch 2011_0.xlsx (accessed 18/2/2014).

Eurobitume (2013) Publications. http://www.eurobitume.eu/publications (accessed 18/02/2014).

IARC (2013) *Bitumens and Bitumen Emissions, and Some N- and S-Heterocyclic Polycyclic Aromatic Hydrocarbons*. World Health Organization, Lyon, France, IARC Monograph 103. http://monographs.iarc.fr/ENG/Monographs/vol103/index.php (accessed 18/02/2014).

Kramer JW (1989) *Feasibility Study for Refining Asbuton*. Alberta Research Council, Edmonton, AB, Canada.

Oxford University Press (1996) *The Oxford English Reference Dictionary*, 2nd edn. Oxford University Press, Oxford, UK.

Wallace D (1989) *Physical and Chemical Characteristics of Asbuton*. Alberta Research Council, Edmonton, AB, Canada.

Chapter 2

Manufacture and storage of bitumens

Derek Petrauskas
Technical Lead,
Bitumen Manufacturing,
Shell Global Solutions
International. The Netherlands

Saleem Ullah
Consultant,
Bitumen Manufacturing,
Shell Global Solutions US Inc.
USA

2.1 The manufacture of bitumen

Crude oil is the term for 'unprocessed' oil, the viscous liquid that comes out of the ground. It is also known as petroleum. Crude oil is a fossil fuel, meaning that it was made naturally from decaying plants and animals living in ancient seas millions of years ago – most places you can find crude oil were once sea beds. Crude oils vary in colour, from clear to jet black, and in viscosity, from water to almost solid.

Crude oils are such a useful starting point for so many different substances because they contain hydrocarbons. Hydrocarbons are molecules that contain hydrogen and carbon, and come in various lengths and structures, from straight chains to branching chains to rings.

Crude oil is a complex mixture of many hydrocarbons, all differing in molecular weight and, consequently, in boiling range. Before it can be used, crude oil has to be separated into component streams, some of which are further chemically or physically changed. These are often then reblended to final products. Bitumen is one of the components produced from this process.

2.2 Fractional distillation of crude oil

The first process in the refining of crude oil is fractional distillation, which physically separates the crude oil into streams varying in boiling point. This is carried out in tall steel towers known as fractionating or distillation columns. The inside of the column is divided at intervals by horizontal steel trays punctured with holes to allow vapour to rise up the column. Over these holes are small domes called bubble caps that deflect the vapour downwards so that it bubbles through liquid condensed on the tray. This improves the efficiency of separation and also has the advantage of reducing the height of the column.

The Shell Bitumen Handbook, Sixth Edition
ISBN 978-0-7277-5837-8
Shell International Petroleum Company Ltd: All rights reserved
http://dx.doi.org/10.1680/tsbh.58378.015

On entering the distillation plant, the crude oil is heated in a furnace to temperatures between 350 and 380°C before being passed into the lower part of the column operating at a pressure slightly above atmospheric. The material entering the column is a mixture of liquid and vapour: the liquid comprises the higher boiling point fractions of the crude oil, and the vapour consists of the lower boiling point fractions. The vapours rise up the column through the holes in the trays, losing heat as they rise. When each fraction reaches the tray where the temperature is just below its own boiling point, it condenses and changes back into a liquid. As the fractions condense on the trays with the required qualities, they are continuously drawn off by means of pipes.

The lightest fractions of the crude oil remain as vapour and are taken from the top of the distillation column; heavier fractions are taken off the column as side-streams; the heaviest fractions remain as liquids, which, therefore, leave at the base of the column. The lightest fractions produced by the crude distillation process include propane and butane, both of which are gases under atmospheric conditions. Moving down the column, naphtha – a slightly heavier material – is produced. Naphtha is used as a feedstock for gasoline production and the chemical industry. Further down the column, kerosene is produced. Kerosene is used primarily for aviation fuel and, to a lesser extent, domestic fuel. Heavier again is gas oil, which is used as a fuel for diesel engines and central heating. The heaviest fraction taken from the crude oil distillation process is known as the long residue, a complex mixture of high molecular weight hydrocarbons, which requires further processing before it can be used as a feedstock for the manufacture of bitumen. Fractional distillation that is carried out under atmospheric pressure is known as atmospheric distillation (Figure 2.1).

The long residue is further distilled at reduced pressure in a vacuum distillation column (Figure 2.2). This is carried out under a vacuum of 10–100 mmHg at a temperature of between 350 and 425°C, to produce gas oil and distillates fractions, plus a bottom residue known as short residue. If this second distillation were carried out by simply increasing the temperature, cracking or thermal decomposition of the long residue would occur, hence the need to lower the pressure. The short residue is the feedstock used in the manufacture of over 20 different grades of bitumen. The viscosity and yield of short residue is a function of both the origin of the crude oil and the temperature and pressure in the vacuum column during processing, and these vary significantly from crude oil to crude oil. The blend of crude oil processed as well as the conditions in the column are adjusted to produce a short residue with a penetration in the decimillimetre range of 35–300 dmm. Figure 2.3 shows the relationship of the distillation process with other fundamental refining processes such as reforming and cracking for the manufacture of saleable

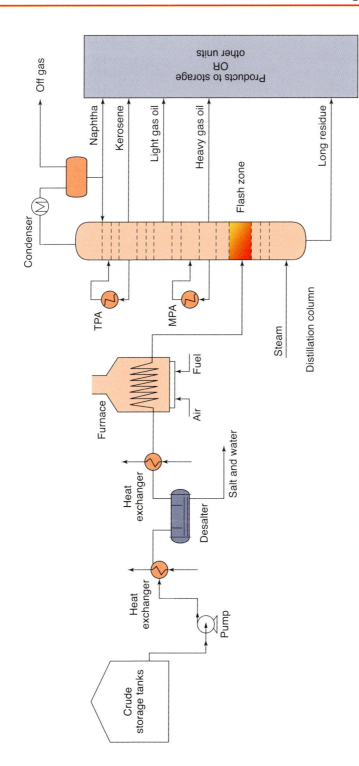

Figure 2.1 Atmospheric distillation (MPA, middle pump-around; TPA, top pump-around)

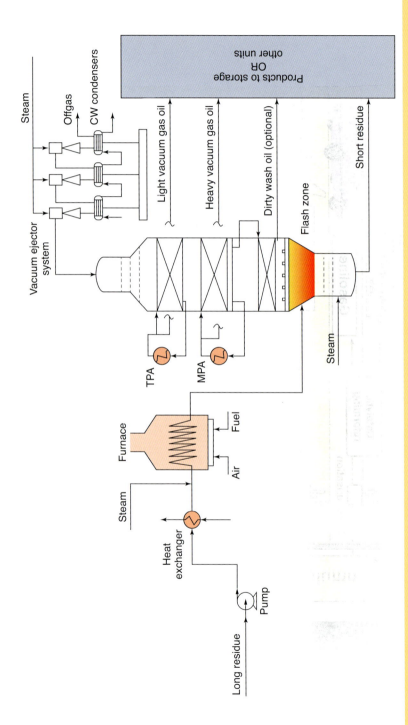

Figure 2.2 Vacuum distillation (CW, cooling water; MPA, middle pump-around; TPA, top pump-around)

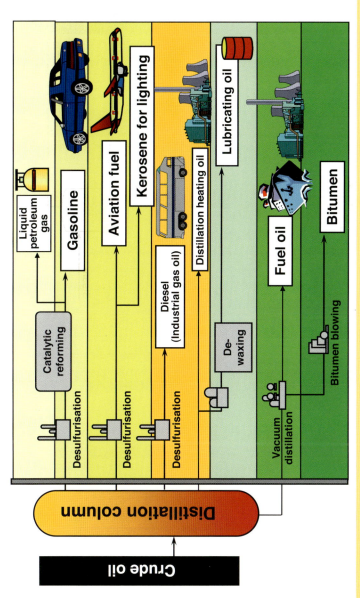

Figure 2.3 Schematic representation of the crude oil distillation process (Ex, from; nnf, normally no flow; SR, short residue; TC, temperature control; OM, oxygen meter)

products, for example aviation, gasoline and diesel fuels, lube oils, heavy fuel oil and chemical feedstocks.

2.3 Air blowing of short residues

Often by the correct selection of the crude oil feed and operating conditions, bitumen complying with a particular specification can be manufactured directly from the distillation process. If required, the physical properties of the short residue can be further modified by 'air blowing'. This is an oxidation process that involves passing air through the short residue, either on a batch or a continuous basis, with the short residue raised to a temperature between 240 and 320°C (Figure 2.4). The main effect of blowing is that it converts some of the relatively low molecular weight 'maltenes' into relatively higher molecular weight 'asphaltenes'. The result is a reduction in the penetration of the bitumen with a comparatively greater increase in the softening point, which has the effect of improving the lower temperature susceptibility of the air-blown bitumen.

2.3.1 The continuous blowing process

After preheating, the short residue is introduced into the blowing column just below the normal liquid level. Air is blown through the bitumen by means of an air distributor located at the bottom of the column. The air acts not only as a reactant but also serves to agitate and mix the bitumen, thereby increasing both the surface area and the rate of reaction. The bitumen absorbs oxygen as air ascends through the material. Steam and water are sprayed into the vapour space above the bitumen level, the former to suppress foaming and dilute the oxygen content of waste gases and the latter to cool the vapours in order to prevent after-burning and the resulting formation of coke.

The blown product passes through heat exchangers to achieve the desired 'rundown' temperatures (the rundown temperature is the temperature at which bitumen can be taken from the tank) and to provide an economical means of preheating the original short residue feed, before pumping the product to storage. The final penetration and softening point of the blown bitumen are affected by many factors that include the viscosity of the feedstock, the temperature in the blowing column, the residence time in the blowing column, the origin of the crude oil used to manufacture the feedstock and the ratio of air to bitumen fed into the process.

Figure 2.5 shows blowing 'curves' for a bitumen feedstock and how these processes can be applied to enable the bitumen to meet the softening point and penetration specification ranges indicated by the boxes. In the blowing processes, the softening point increases and the penetration reduces. However, in the distillation process, the temperature susceptibility (or

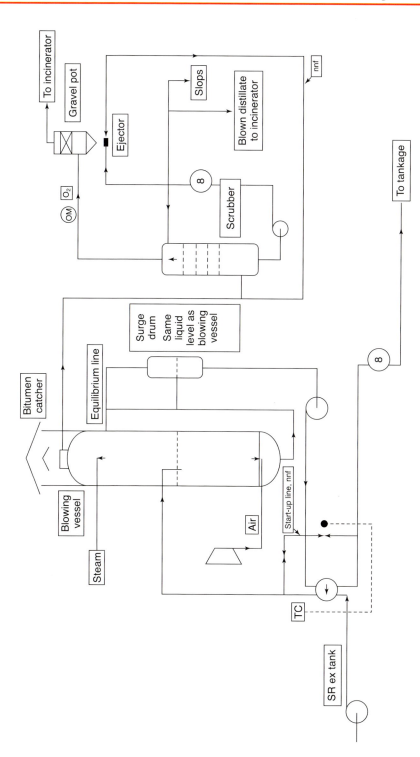

Figure 2.4 Simplified diagram of a bitumen blowing unit (Ex, from; nnf, normally no flow; SR, short residue; TC, temperature control; OM, oxygen meter)

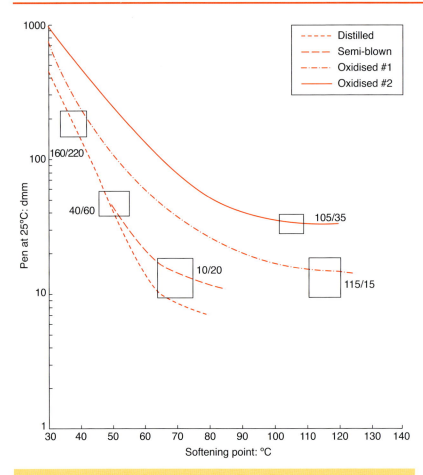

Figure 2.5 Semi-blowing and blowing curves

penetration index) of the material is largely unchanged. Thus, the distillation is a relatively straight line whereas the curves for the blown bitumen flatten substantially as the softening point of the bitumen increases. This demonstrates that the temperature susceptibility of the material is substantially reduced (i.e. the penetration index is increased).

2.3.1.1 Air-rectified or semi-blown bitumens

Many crude oils produce bitumen that requires a limited amount of, ideally, no air blowing in order to generate penetration grade bitumen suitable for road construction. This process is termed semi-blowing or air-rectification. Used judiciously, semi-blowing can be used to reduce the temperature susceptibility of the bitumen (i.e. to increase its penetration index).

2.3.1.2 Fully blown bitumens

Fully blown or oxidised bitumens are often produced by more extensive blowing of a blend of a short residue with a relatively low viscosity flux. The position of the blowing curve in Figure 2.5 is primarily dependent on the viscosity and chemical nature of the feed. The softer, or lower viscosity, the feed, the higher the curve. The amount of blowing that is required depends on the temperature in the column and the air-to-feed ratio. Thus, by selection of a suitable bitumen feedstock, control of the viscosity of the feed and the conditions in the column, a range of blown grades of bitumen can be manufactured.

2.3.1.3 Chemistry of the blowing process

The bitumen blowing process can be described as a conversion process in which oxidation, dehydrogenation and polymerisation take place. Normally, it is effected with the oxygen present in air, and it is this type of process that is considered here.

By subjecting the materials entering and leaving the reaction vessel to a combination of analytical methods, it is possible to draw up an oxygen balance. It has been found that all the oxygen taken up by the bitumen can be accounted for by the formation of hydroxyl, carbonyl, acid and ester groups; no ether oxygen has been detected. The main side products are carbon dioxide, water and some light hydrocarbons. Besides carbon–oxygen bonds, carbon–carbon bonds are also formed.

Of the functional groups mentioned, the esters are particularly important because they serve as a link up of two different molecules and thus contribute to the formation of material of higher molecular weight. This mechanism results (together with the direct formation of carbon–carbon bonds) in an increase in the asphaltene content. In other words, the formation of esters and the direct formation of carbon–carbon bounds are desired reactions; all others are less desirable or even undesirable reactions.

As a typical trend, the amount of oxygen used for condensation reactions decreases from 60 to 20%, whereas the amount of oxygen used for side reactions increases from 40 to 80% with an increase in temperature from 150 to 350°C. In other words, increasing temperature decreases the contribution of the desirable reactions at the cost of the undesirable chemical reactions.

As the formation of asphaltenes – the main aim of the blowing process – is a function of the number of bonds obtained, the number of bonds per mole of reacting oxygen is of paramount importance. The amount of oxygen required for one ester bond is much higher than that required for a carbon–carbon bond, and the amount of ester bonds formed sharply decreases with temperature, while the amount of carbon–carbon bonds increases with temperature. Consequently, at a certain temperature, an optimum in bond formation may be expected. It has been found that this optimum is reached at a blowing temperature of 250°C.

However, at this preferred temperature the transfer of oxygen from the gaseous to the liquid phase is often not very satisfactory because, due to the relatively high viscosity of the liquid bitumen, the distribution of air bubbles may be poor. An increase in the blowing temperature results in a reduced viscosity and in an increased gas–liquid interface, thus promoting the oxygen uptake (efficiency) by the bitumen. However, this is at the cost of a shift towards a regime where the total number of bonds per mole of reacted oxygen is lower. On balance, the total number of bonds formed per mole of oxygen intake is more favourable at higher temperatures. A shift in the optimum occurs, which explains the trend towards higher blowing temperatures (275–285°C) frequently encountered in practice.

2.4 Delivery, storage and handling temperatures of bitumen

As most bitumen grades are solid at ambient temperature, in order to enable them to be moved through the distribution system as a liquid, they must be heated to temperatures in the range of 140°C to over 200°C, depending on the grade; higher temperatures up to 230°C are applied with an inerted atmosphere for highly oxidised grades. When handled properly, bitumen can be reheated or maintained at elevated temperatures for a considerable time without adversely affecting its properties. However, mistreatment of bitumen by overheating or by permitting the material to be exposed to conditions that promote oxidation can adversely affect the properties of the bitumen and may influence the long term performance of mixtures that contain bitumen. The degree of hardening (or, under certain circumstances, softening) that is produced as a result of mishandling is a function of a number of parameters such as temperature, the presence of air, the surface-to-volume ratio of the bitumen, the method of heating and the duration of exposure to these conditions.

2.4.1 Safe delivery of bitumen

Over 50% of bitumen related accidents and incidents that result in lost working days occur during the delivery of bitumen. In Europe, Eurobitume has produced a detailed code of practice (Eurobitume and the Asphalt Institute, 2011) designed to assist in reducing the frequency of incidents and accidents by raising awareness of their causes. Personnel involved in the delivery and receipt of bitumen are strongly advised to refer to this document.

2.4.2 Bitumen tanks

All bitumens should be stored in tanks specifically designed for the purpose (Eurobitume and the Asphalt Institute, 2011). In order to minimise the possible hardening of the bitumen during storage, certain aspects of the design of the tank should be considered. In order to minimise the risk of over-heating the bitumen, the tank should be fitted with accurate temperature sensors and gauges. These should be positioned in the region of the heaters and preferably be removable to facilitate regular cleaning and maintenance. Oxidation and the loss of volatile fractions from bitumen are both related to the exposed surface-to-volume ratio of the storage tank, which, for a cylindrical vessel, equals the reciprocal height of the filled part of the tank and is given by

$$\frac{\text{surface area}}{\text{volume}} = \frac{\pi r^2}{\pi r^2 h} = \frac{1}{h}$$

where h is the height of the bitumen and r is the radius of the tank.

Thus, the dimensions of the bulk storage tank should be such that the surface-to-volume ratio is minimised. Accordingly, vertical storage tanks with a large height-to-radius ratio are preferable to horizontal tanks.

It is common practice for bitumen in bulk storage tanks at mixing plants to be recirculated around a ring main in order to heat the pipework that carries the bitumen to the processing point. Return lines in a recirculation system should re-enter the storage tank below the bitumen surface to prevent hot bitumen cascading through the air. Often, the bitumen is returned to the bulk storage tank through a pipe fitted into the upper part of the tank, flush with the side or roof or protruding just into the air space at the top of the tank. If the bitumen enters the tank above the bitumen, all the factors that promote oxidation are present

- high temperature
- access to oxygen
- high exposed surface-to-volume ratio.

Fortunately, the residence time of bitumen in the tank is usually sufficiently low for any hardening to be insignificant. However, if material is stored for a prolonged period, recirculation should be used only intermittently, and the bitumen should be tested before use to ensure its continued suitability for the proposed application. A recommended layout for a bitumen storage tank is shown in Figure 2.6.

Bitumen storage tanks should be fitted with automatic level indicators together with low and high level alarms to avoid having to dip manually for a reading. Such an approach avoids exposing the hot heater tubes to a potentially combustible or explosive atmosphere should the bitumen level fall below that of the heater tubes. Automatic level control also ensures that the tank is not overfilled. Regardless of whether a high level alarm is fitted, a

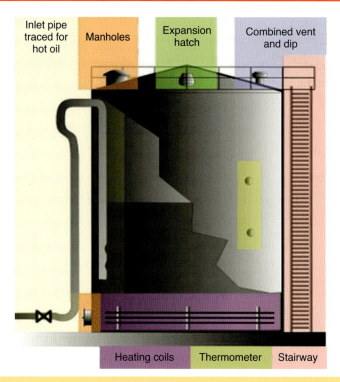

Inlet pipe traced for hot oil Manholes Expansion hatch Combined vent and dip

Heating coils Thermometer Stairway

Figure 2.6 Recommended layout for a bitumen storage tank

maximum safe filled level for the tank should be predetermined, taking into account the effects of thermal expansion of the bitumen in the tank.

Before ordering additional bitumen, it is essential to check that the ullage in the tank is capable of taking the delivery without exceeding the maximum safe working level.

Every tank should be clearly labelled with the grade of bitumen it contains. When the grade of bitumen in a tank is changed, it is important to ensure that the tank is empty and relabelled before the new grade is delivered.

2.4.3 Bitumen storage and pumping temperatures

Bitumen should always be stored and handled at the lowest temperature possible, consistent with efficient use. As a guide, working temperatures for specific operations are given in the Energy Institute's bitumen safety code (Energy Institute, 2005). These temperatures have been calculated on the basis of viscosity measurements, and are supported by operational experience. For normal operations (i.e. the blending and transferring of liquid bitumen), temperatures of 10–50°C above the minimum pumping temperature are recommended, but the maximum safe handling temperature of 230°C must never be exceeded.

The period during which bitumen resides in a storage tank at elevated temperatures and is recirculated should be minimised, to prevent hardening of the bitumen. If bitumen must be stored for an extended period, say for a period exceeding 1 week without the addition of fresh material, the temperature should be reduced to approximately 20–25°C above the softening point of the bitumen and, if possible, recirculation stopped.

When bitumen is being reheated in bulk storage, care must be taken to heat the bitumen intermittently over an extended period to prevent localised overheating of the product around the heating pipes or coils. This is particularly important where direct flame tube heating is used because surface temperatures in excess of 300°C may be reached. In such installations, the amount of heat that is applied should be limited, sufficient only to raise the temperature of the product to just above its softening point. This will allow the material to soften, after which further heat can be applied to raise the temperature of the product to the required working value. This technique is beneficial because when the bitumen is a fluid, albeit a viscous fluid, convection currents dissipate the heat throughout the material, and localised overheating is thus less of a problem. Circulation of the tank contents should begin as soon as the product is sufficiently fluid, thereby further reducing the likelihood of local overheating. With hot oil, steam or electric heaters that are designed properly, reheating from cold should not cause these problems.

References

Energy Institute (2005) *Bitumen Safety Code. Model Code of Safe Practice in the Petroleum Industry*, part 11, 4th edn. Energy Institute, London, UK.

Eurobitume and the Asphalt Institute (2011) *The Bitumen Industry – A Global Perspective: Production, Chemistry, Use, Specification and Occupational Exposure*, 2nd edn. Eurobitume, Brussels, Belgium; Asphalt Institute, Lexington, KY, USA.

Chapter 3

Handling, health, safety and environmental aspects of bitumens

Prof. John Read
General Manager Specialities Technology, Shell Global Solutions UK. UK

Bitumen has a long history of being used safely in a wide range of applications. Although it is primarily used in road construction, it is also employed in roofing felts, reservoir linings and as an internal lining for potable water pipes as well as having a myriad of other uses (many of which are given in Table 1.2). Bitumen presents a low order of potential hazard, and provided that good handling practices are observed, the risks of the hazard ever being released are very small. These are described in detail in part of the Model Code of Safe Practice (Energy Institute, 2005), and are also described in the material safety data sheets provided by each supplier. A substantial amount of health, safety and environmental data on bitumen and its derivatives are detailed in CONCAWE Product Dossier No. 92/104 (CONCAWE, 1992). In addition, with the implementation of REACH (Registration, Evaluation, Authorisation and Restriction of Chemicals – the base legislation to which there have been numerous amendments and corrigenda; EC, 2007) a large amount of work around the manufacturing, safe handling and safe delivery of bitumen, focusing on hazard and risk has been documented by Eurobitume (Eurobitume, 2011a) – Eurobitume is the European Bitumen Association, and comprises a number of bitumen companies and bitumen associations in Europe. Notwithstanding the safe use indicated by these various publications, bitumens are generally applied at elevated temperatures, and this brings with it a number of hazards that are considered below.

3.1 Handling and safety
3.1.1 Personal protective equipment as related to occupational exposure limits
The principal hazard from handling hot bitumen is thermal burns resulting from contact with the product. Thus, it is essential to wear clothing that

The Shell Bitumen Handbook, Sixth Edition
ISBN 978-0-7277-5837-8
Shell International Petroleum Company Ltd: All rights reserved
http://dx.doi.org/10.1680/tsbh.58378.029

provides adequate protection (PPE, personal protective equipment), including

- helmet and neck apron to provide head protection
- visor to protect the face (goggles only protect eyes)
- heat-resistant gloves (with cuffs worn inside coverall sleeves)
- safety boots
- coveralls (with coverall legs worn over boots).

Table 3.1 provides an overview of the occupational exposure limits (OELs) for various countries. Further discussion on these, as evaluated in the International Agency for Research on Cancer (IARC) monograph *Bitumens and Bitumen Emissions, and Some N- and S-Heterocyclic Polycyclic Aromatic Hydrocarbons* (IARC, 2013), is given in section 3.2.1.

Exposure to bitumen fumes can result in irritation to the eyes, nose and respiratory tract, headaches and nausea. The symptoms are usually mild and temporary. Removal of the affected personnel from the source results in rapid recovery. Even though the irritation is usually mild, exposure to bitumen fumes should be minimised and, where there is any doubt, tests (e.g. Drægar tube analysis or personal exposure monitoring) should be undertaken to determine the concentration of bitumen fumes or hydrogen sulfide in the working atmosphere.

Persons affected by the inhalation of bitumen fumes should be removed to fresh air as soon as possible. If the symptoms are severe or if the symptoms persist, medical help should be sought without delay.

3.1.2 Personal hygiene
Garments soiled with bitumen should either be replaced or dry cleaned in order to avoid permeation of bitumen into the underclothing. Soiled rags or tools should not be placed in the pockets of overalls, as contamination of the lining of the pocket will result.

Personnel handling bitumen and asphalts should be provided with and use barrier creams to protect exposed skin, particularly hands and fingers. Skin should be thoroughly washed after any contamination, and always before going to the toilet, eating or drinking.

The application of barrier creams prior to handling bitumen assists in subsequent cleaning, should accidental contact occur. However, barrier creams are no substitute for gloves or other impermeable clothing. Consequently, they should not be used as the sole form of protection. Solvents such as petrol, diesel oil, white spirit or similar should not be used for removing bitumen from the skin, as they may spread the contamination as well as being skin sensitisers or carcinogenic or both. An approved skin cleanser together with warm water should be used.

Table 3.1 A selection of OELs

Country/ Advisory body	OEL long term (8 h/day): mg/m^3	OEL short term (15 min): mg/m^3	Authority
Argentina	0.5		Decree 351 of 1979, amended by Resolution 295/2003 – Annex IV, 21/11/2003
Australia	5		Adopted National Exposure Standards for Atmospheric Contaminants in the Occupational Environment (NOHSC:1003)
Bahrain	5		Resolution No. 4 Regarding the Management of Hazardous Chemicals, Annex 3 (2006)
Belgium	5		Moniteur Belge No. 187, 30/06/2011
Brazil	0.5^a		Ordinance No. 3214, 6/8/78, NR-15, Annex 11, as amended through 10/03/1994 and NR-09, as amended through 29/12/1994
Bulgaria	5	10	Regulation No. 13 on protection of workers from exposure to chemical agents at work (D.V.8/2004, as amended through 06/01/2012, D.V. 2/2012)
Canada	0.5		Ontario & British Columbia Provincial 2013
Chile	4		Reg. 594/1999, arts. 61 and 66, as of 08/11/2012
China	5		Occupational Exposure Limits for Hazardous Agents in the Workplace, Part 1, Chemical Hazardous Agents (GBZ 2.1-2007, 03/2008)
Colombia	0.5^c		Resolución Número 02400, 22/05/1979. Normas sobre vivienda, higiene y seguridad en los establecimientos de trabajo) (as updated through ACGIH publication from 20/03/2013)
Costa Rica	0.5^c		Regulation for the Registration of Hazardous Products, Executive Decree No. 28113S, as amended by Executive Decree No. 30718, 02/10/2002) updated with 20/03/2013 ACGIH data
Croatia	5	10	Dangerous Substance Exposure Limit Values in the Workplace (OELs), Annexes 1 and 2, as amended through 20/06/2013 (NN 75/2013)
Cyprus	5		OELs. Control of factory atmosphere and dangerous substances in factories regulation, PI 311/73, as amended through PI 41/86 (1986)
Denmark	1	1	Grænseværdier for stoffer og materialer. Arbejdstilsynet, An. 2 and 3, amended through Order No. 986, 12/10/2012
Dominican Republic	0.5^c		Regulation of Safety and Health in the Workplace – Decree No. 522–06 and Resolution No. 04-2007 of 30/01/2007) updated with ACGIH data published on 20/03/2013
Ecuador	0.5^c		INEN 2266: 2013, 2013-01 2nd rev.: Transport, storage and handling of hazardous materials. Requirements. 1st edition, 29/01/2013

Table 3.1 Continued			
Country/ Advisory body	OEL long term (8 h/day): mg/m^3	OEL short term (15 min): mg/m^3	Authority
Egypt	0.5		Threshold limits of air pollutants in the workplace (Decree No. 338, 1995, as amended through Decree No. 710, Annex 8, Table 1, 2012)
Estonia	5		Occupational Exposure Limits of Hazardous Substances (Annex of Regulation No. 293 of 18/09/2001), as amended 11/2011
Finland	5	10	OEL limits referenced in several publications
France	No mandatory threshold[d]	No mandatory threshold[d]	
Germany	No mandatory threshold[e]	No mandatory threshold[e]	
Greece	5		Decree No. 307/1986, last amended by Decree No. 12/2012, 09/02/2012
Hong Kong	0.5		Code of Practice on Control of Air Impurities (Chemical Substances) in the Workplace (First Edition, 04/2002)
Iceland	5		Iceland. OELs. Regulation 390/2009 on Pollution Limits and Measures to Reduce Pollution at the Workplace, 2 April 2009; as amended by Regulation 1296/2012 of 21/12/2012
Indonesia	5		Minister of Manpower and Transmigration Regulation No. Per.13/MEN/X/2011 concerning Threshold Limit Values, Annex II, 11/2011
Ireland	0.5	10	2011 Code of Practice for the Safety, Health and Welfare at Work [Chemical Agents] Regulations 2001 (SI No. 619 of 2001)
Israel	0.5		Environmental Monitoring and Biological Monitoring of Workers with Harmful Agents – 5771, 2011; updated through ACGIH, 20/03/2013
Italy	0.5		Legislative Decree No. 81, 9 April 2008, as last amended by Decree 06/08/2012
Jordan	5		Resolution No. 43 (1998) Safety and Protection from Industrial Equipment, Machinery and Workplaces
Kenya	5	10	The Factories and Other Places of Work Rules in 2007 of the Factories and Other Places of Work Act (CAP. 514)
Korea, Republic of	0.5		ISHL Article 42; MOL Public Notice No. 1986-45, as amended through MOEL Notice 2013-38, 14/08/2013
Kuwait		5	Decision No. 210/2001 Appendix (3-1): Amendment 2011

Table 3.1 Continued

Country/ Advisory body	OEL long term (8 h/day): mg/m^3	OEL short term (15 min): mg/m^3	Authority
Malaysia	5		Occupational Safety and Health (Use and Standards of Exposure of Chemicals Hazardous to Health) Regulations 2000: Schedule 1
Mexico	5	10	NOM-010-STPS-1999, Condiciones de Seguridad e Higiene en los Centros de Trabajo. Diario Oficial de la Federación, 13/03/2000
New Zealand	5		Workplace Exposure Standards and Biological Exposure Indices, 7th edition, 02/2013
Nicaragua	0.5^c		General Law of Workplace Hygiene and Safety. Law No. 618, published in Official Gazette No. 133, 13/07/2007
Norway		5	Administrative normer for forurensning i arbeidsatmosfære 2003, No. 361, as amended through 11/01/2013
Paraguay	0.5		Decree No. 14.390/92 that approves the General Technical Regulation of Safety, Hygiene and Medicine in the Workplace, 28/07/1992
Peru	0.5		Decreto Supremo 015-2005-SA (Reglamento sobre Valores Límites Permisibles para Agentes Químicos en el Ambiente de Trabajo, 07/2005)
Poland	5	10	Regulation of 29 November 2002 regarding maximum permissible concentrations and intensities in working environment, as amended by DzU, No. 274, item 1621, 21/12/2011
Portugal	0.5		NP 1796-2007, Valores limite de exposição (VLEs) profissional a agentes químicos, 4th edition, 09/2007
Romania	5		Protection of workers from exposure to chemical agents at the workplace (Hotarârea, No. 1218, 6 September 2006; Monitorul Oficial, Partea I, No. 845, 13/10/2006)
Serbia and Montenegro	5		Yugoslav Standard JUS Z.B0.001, 1991; Reg. No. 15/01-149/52 of 23/05/1991 on maximum allowable concentration of airborne toxic gases, vapors and aerosols in working premises, 17/09/1991
Singapore	5		Workplace Safety and Health (General Provisions) Regulations 2006 (S 134/2006), First Schedule: Permissible Exposure Limits of Toxic Substances, 28/02/2006
Slovenia	10^f		Regulation on Protection of Workers from Risks due to Exposure to Chemicals at Work (OGRS Nos. 100/01, 39/05 and 53/07; as amended by OGRS No. 102/2010, item 5233, 17/12/2010)

Table 3.1 Continued

Country/ Advisory body	OEL long term (8 h/day): mg/m^3	OEL short term (15 min): mg/m^3	Authority
South Africa	5	10	Regulations for Hazardous Chemical Substances, Table 2, 1995
Spain	0.5		Valores Límites Ambientales (VLAs), Table 1, Límites de Exposición Profesional para Agentes Químicos 2013
Switzerland	10		Limit Values at the Workplace 2013, as per SUVA
UAE[g]	0.5		AD EHSMS RF – Occupational Standards and Guideline Values, Schedule A, 02/2012
Uruguay	0.5[c]		Decreto Supremo 015-2005-SA (Reglamento sobre Valores Límites Permisibles para Agentes Químicos en el Ambiente de Trabajo, 07/2005)
UK	5	10	Workplace Exposure Limits (WELs) (EH40/2005), as amended through 12/2011
Venezuela	0.5		Concentraciones ambientales permisibles (CAPs), Table 1, COVENIN 2253:2001
Vietnam	5	10	Allowable Limit in the Work Atmosphere Standard (Ministry of Health Decision Number 3733/2002/ QD-BYT Re: The issuance of 21 Labor Health Standards, 10/10/2002)
ACGIH[g]	0.5		ACGIH Threshold Limit Values (as amended by the Annual Report of the Committees on TLVs and BEIs, 03/02/2014)
NIOSH[g]		5	*Pocket Guide to Chemical Hazards*, 2010

[a] Sourced from ACGIH limit
[b] Quebec uses an 8 hour limit of 5 mg/m^3
[c] Updated with and follows ACGIH limit from 2013
[d] Ministry of Labour advises use of the ACGIH/NIOSH thresholds for OEL
[e] Former threshold of 10 mg/kg removed by the employer's mutual insurance association (Berufsgenossenschaft)
[f] Units are ppm
[g] More than one limit listed for UAE; most stringent one recorded in this table
[h] Advisory organisations based in USA. Limits stated are not regulated unless formally adopted by a country into their OEL legislation

3.1.3 *Fire prevention and fire-fighting*

The adoption of safe handling procedures will substantially reduce the risk of fire. However, if a fire occurs, it is essential that personnel are properly trained and well equipped to extinguish the fire, thereby ensuring that the risk of injury to personnel and damage to plant is minimised. Detailed advice on fire prevention and fire-fighting is given in the Model Code of Safe Practice (Energy Institute, 2005) and in supplier material safety data sheets.

Small bitumen fires can be extinguished using dry chemical powder, foam, vaporising liquid or inert gas extinguishers, fog nozzle spray hoses and steam lances. Direct water jets must not be used because frothing may occur, which tends to spread the hot bitumen and, therefore, the fire.

Injection of steam or a 'fog' of water into the vapour space can extinguish internal tank fires where the roof of the tank is largely intact. However, only trained operatives should use this method, as the water vaporises instantly on contact with the hot bitumen. This initiates foaming, which may result in the tank overflowing, creating an additional hazard. Alternatively, foam extinguishers may be used. The foam ensures that the water is well dispersed, thereby reducing the risk of froth-over. The disadvantage of using this type of extinguisher is that the foam breaks down rapidly when applied to hot bitumen.

Portable extinguishers containing either aqueous film-forming foam or dry chemical powder are suitable for dealing with small bitumen fires, at least initially. In bitumen handling areas, these should be placed at strategic, permanent and conspicuous locations. The type and location of equipment to be used if initial attempts fail should be discussed with the local fire brigade before installation.

3.1.4 Sampling
Sampling of hot bitumen is particularly hazardous because of the risk of heat burns from spills and splashes of the material. It is therefore essential that appropriate protective clothing is worn. The area should be well lit, and safe access to and egress from the sample point should be provided. Gantry access should be provided where samples are required from the tanks of vehicles, as climbing on top of vehicles should be avoided.

3.1.4.1 Dip sampling
In this process, a sample of bitumen is obtained by dipping a weighted can or 'thief' on the end of a rope or rod through the access lid into bitumen stored in a bulk tank. The sample is then transferred to a suitable permanent container. The method is simple, but is only appropriate for small samples. Dip sampling from cut-back tanks should be avoided because of the presence of flammable atmospheres in tank vapour spaces. Additionally, where possible, dip sampling should be avoided altogether, as it is also possible to be exposed to hydrogen sulfide when the access lid is opened.

3.1.4.2 Sample valves
Properly designed sample valves are very useful for sampling from pipelines or from tanks. Their design should ensure that they are kept hot by the product in the pipeline or tank in order to avoid blockage when in the closed position, or have a means of flushing them clean.

Sample valves should preferably be the screw-driven plunger type. When closed, the plunger of this type of valve extends into the fresh product.

Thus, when the valve is opened, a representative sample of product is obtained without 'fore-runnings'. With ball and plug type valves, fore-runnings have to be collected and disposed of before a representative sample can be obtained. Designs for bitumen sample valves are described in detail in the European Norm EN 58:2012 (BSI, 2012).

3.2 Health aspects of bitumen
3.2.1 Hazards associated with bitumen
Some of the key hazards are given below. A more comprehensive review can be found in the joint Eurobitume/Asphalt Institute publication *The Bitumen Industry – A Global Perspective* (Eurobitume, 2011b).

3.2.1.1 Elevated temperatures
The main hazard associated with bitumen is that the product is held at elevated temperatures during transportation, storage and processing. Thus, it is critical that appropriate PPE is worn and any skin contact with hot bitumen is avoided. Some guidance is given below, but detailed advice is available in a number of publications (CONCAWE, 1992; Energy Institute, 2005; Eurobitume, 2011a) and the relevant supplier material safety data sheets.

3.2.1.2 Vapour emissions
Bitumens are complex mixtures of hydrocarbons that do not have well defined boiling points because their components boil over a wide temperature range. Visible emissions or fumes normally start to develop at approximately 150°C. The amount of fumes generated doubles for each 10–12°C increase in temperature. Fumes are mainly composed of hydrocarbons (Brandt and De Groot, 1996) and small quantities of hydrogen sulfide. Bitumen fumes also contain small quantities of polycyclic aromatic compounds (PACs) and more specifically polycyclic aromatic hydrocarbons (PAHs). These chemicals consist of a number of benzene rings that are grouped together. Some of these with three to seven (usually four to six) fused rings are known to cause or are suspected of causing cancer in humans. However, the concentrations of these carcinogens in bitumen are extremely low (CONCAWE, 1992). The most common PAHs included in this category by the US Environmental Protection Agency (2008) by name and CAS number, are

- benzo[*a*]anthracene, 56-55-3
- benzo[*a*]phenanthrene (chrysene), 218-01-9
- benzo[*a*]pyrene, 50-32-8
- benzo[*b*]fluoranthene, 205-99-2
- benzo[*j*]fluoranthene, 205-82-3
- benzo[*k*]fluoranthene, 207-08-9

- benzo[j,k]fluorene (fluoranthene), 206-44-0
- benzo[r,s,t]pentaphene, 189-55-9
- dibenz[a,h]acridine, 226-36-8
- dibenz[a,j]acridine, 224-42-0
- dibenzo[a,h]anthracene, 53-70-3
- dibenzo[a,e]fluoranthene, 5385-75-1
- dibenzo[a,e]pyrene, 192-65-4
- dibenzo[a,h]pyrene, 189-64-0
- dibenzo[a,l]pyrene, 191-30-0
- 7H-dibenzo[c,g]carbazole, 194-59-2
- 7,12-dimethylbenz[a]anthracene, 57-97-6
- indeno[1,2,3-cd]pyrene, 193-39-5
- 3-methylcholanthrene, 56-49-5
- 5-methylchrysene, 3697-24-3
- 1-nitropyrene, 5522-43-0

A public statement on PAHs by the US Agency for Toxic Substances and Disease Registry states that

> One of the most common ways PAHs can enter the body is through breathing contaminated air. PAHs get into your lungs when you breathe them. If you live near a hazardous waste site where PAHs are disposed, you are likely to breathe PAHs. If you eat or drink food and water contaminated with PAHs, you could be exposed. Exposure to PAHs can also occur if your skin contacts PAH-contaminated soil or products like heavy oils, coal tar, roofing tar, or creosote. Creosote is an oily liquid found in coal tar and is used to preserve wood. Once in your body, PAHs can spread and target fat tissues. Target organs include the kidneys and liver. However, PAHs will leave your body through urine and faeces in a matter of days.

Therefore, it is exposure to PAHs over an extended period that causes concerns to health, and measures such as reducing temperature and avoiding enclosed spaces should be practised as a matter of good product stewardship.

In 2013, a comprehensive review was carried out in IARC Monograph 103 that concluded that there was an increased probability of carcinogenicity when exposed to fumes from fully oxidised bitumen versus straight run paving and hard bitumens (IARC, 2013). The IARC Monograph found that that there was limited evidence in humans for the carcinogenicity of occupational exposures to bitumens and bitumen emissions during roofing and mastic asphalt work and that there was sufficient evidence in experimental animals for carcinogenicity of fume condensates generated from oxidised bitumens, and on this basis they classified bitumen into three groups as follows.

- Occupational exposures to oxidised bitumens and their emissions during roofing are probably carcinogenic to humans (Group 2A).
- Occupational exposures to hard bitumens and their emissions during mastic–asphalt work are possibly carcinogenic to humans (Group 2B).
- Occupational exposures to straight run bitumens and their emissions during road paving are possibly carcinogenic to humans (Group 2B).

Hence, in contrast to the previous monograph published in 1985 (IARC, 1985), oxidised bitumen moved from a classification of possibly carcinogenic (Group 2B) to probably carcinogenic (Group 2A) and this is now reflected in supplier materials safety data sheets. It must be noted that IARC monographs only look at whether the hazard is present, and take no account of the risk of exposure: that is, whether the hazard will ever be released.

3.2.1.3 Hydrogen sulfide

Hydrogen sulfide is of particular concern, as it can accumulate in enclosed spaces, such as the tops of storage tanks, and exposure to this gas at concentrations of as little as 500 ppm can be fatal, so it is essential that any space where hydrogen sulfide may be present is tested and approved as being gas free before anyone enters the area (Table 3.2). In addition, with the increase in the use of sulfur as a cross-linking agent in the production of stable polymer modified bitumens, it is ever more important to actively manage hydrogen sulfide, which potentially can be generated throughout the supply chain.

3.2.1.4 Combustion

Very high temperatures are required to make bitumen burn. Certain materials, if hot enough, will ignite when exposed to air, and are sometimes described as being 'pyrophoric'. In the case of bitumens, this auto-ignition temperature is generally around 400°C. However, despite storage and use of temperatures well below the auto-ignition temperature, fires have very occasionally occurred. Under conditions of low oxygen content, hydrogen sulfide from bitumen can react with rust (iron oxide) on the roof and walls of storage tanks, to form 'pyrophoric iron oxide'. This material reacts readily with oxygen and can self-ignite if the oxygen content of the tank increases suddenly, which, in turn, can ignite coke deposits on the roof and walls of the tank. Coke deposits are the result of condensate from the bitumen that has been deposited on the roof and walls of a tank degrading over a period of time, forming carbonaceous material. Under conditions of high temperature and in the presence of oxygen or a sudden increase in available oxygen, an exothermic reaction, leading to the risk of fire or explosion, can occur. Accordingly, manholes in bitumen tanks should be kept closed and access to tank roofs should be restricted at all times.

3.2.1.5 Contact with water

It is essential that water does not come into contact with hot bitumen. If this does happen, the water is converted into steam. In the process, its volume increases by a factor of 1673 at atmospheric pressure, resulting in spitting, foaming and, depending on the amount of water present, possible boil-over of the hot bitumen.

3.2.1.6 Potential hazards through skin contact

Other than thermal burns, the hazards associated with skin contact of most bitumens are negligible. Studies (Bofetta, 2001) concluded that there was no direct evidence to associate bitumen with long term skin disorders in humans despite bitumens having been widely used for many years. Nevertheless, it is prudent to avoid intimate and prolonged skin contact with bitumen.

Cut-back bitumens and bitumen emulsions are handled at lower temperature, which increases the chance of skin contact. If personal hygiene is poor, regular skin contact may occur. However, studies carried out by Shell (Brandt *et al.*, 1999; Deygout, 2011; Potter *et al.*, 1999) have demonstrated that bitumens are unlikely to be bio-available (skin penetration and body uptake) and bitumens diluted with solvents are unlikely to present a carcinogenic risk. Nevertheless, bitumen emulsions can cause irritation to the skin and eyes, and can produce allergic responses in some individuals.

3.2.2 First aid for skin burns

The following is taken from the Eurobitume notes for the guidance of first aid and medical personnel (Eurobitume, 2011a). This advice is produced in the form of an A5-sized card, copies of which are available from Eurobitume in a number of different languages. It is intended that the card should accompany a burns victim to the hospital, to provide immediate advice on proper treatment.

3.2.2.1 First aid

Bitumen burns should be cooled for at least 15 min, first with cool water to reduce pain, then with warm water to prevent hypothermia if the burned surface is larger than the size of a hand. Burns to the eyes should be irrigated for at least 5 min.

NO ATTEMPT SHOULD BE MADE TO REMOVE THE BITUMEN AT THE WORK SITE.

3.2.2.2 Medical care

(If in doubt do not hesitate to contact a burns centre.)

Measures to remove the bitumen layer from the skin should be taken as soon as possible under the supervision of a doctor, or at a hospital. However, this

Table 3.2 Various hydrogen sulfide OELs in force (March 2014)

Country/Advisory body	OEL long term (8 h/day): ppm	OEL short term (15 min): ppm	OEL long term (8 h/day): mg/m³	OEL short term (15 min): mg/m³	Authority
Austria	5	5	7	5	Grenzwerteverordnung 2011 – GKV 2011
Belgium	10	15	14	21	Belgisch Staatsblad, 30/06/2011; No. 2011-1687
Bulgaria	10	15	14	21	РБ МТСП и М³ Наредба No. 13/2003
Cyprus	5	10	7	14	
Czech Republic	5	10	7.2	14.4	178/2001 (12/2007)
Denmark	10	15	14	21	Arbejdstilsynet; Grænseværdier for stoffer og materialer, augustus 2007 (publicatie C.0.1)
Estonia	5	10	7	14	Sotsiaalminister, 10/2007
Finland	5	10	15	21	Työterveyslaitos, Sosiaali- ja terveysministeriö, 07/2009
France	5	10	7	14	Valeurs limites d'exposition professionnelle aux agents chimiques en France; INRS ED 984; 06/2008 (mandatory based on the decree 05/2012)
Germany	5	10	7.1	14.2	TRGS 900; version 03/2011
Greece	5	10	7	14	
Hungary	5	10	7	14	EüM-SxCsM, 12/2007
Iceland	10	15		21	
Ireland	5	10	7	14	Health and Safety Authority
Italy	1	5	14	21	EU OEL/list of indicateve OEL values, 12/2009
Latvia	5	10	7	14	LV National Standardisation and Meteorological Centre, 2007
Lithuania	5	10	7	14	Del Lietuvos Higienos Normos 10/2007

Luxembourg	5	10	NR	NR	EU OEL; List of Indicative OEL values 12/2009
Malta	5	10	7	14	
The Netherlands	1.6	NR	2.3	NR	DECOS 2010/06OSH
Norway	5	10	7	14	Nye administrative normer for forurensning I arbeidsatmosfaere; version 12/2011
Poland	5	10	7	14	Ministra Pracy i Polityki Spolecznej (Poland, 07/2009)
Portugal	10	15	NR	NR	Insituto Português da Qualidade
Romania	5	10	7	14	
Slovakia	5	10	7	14	Nariadenie Vlady Slovenskej Republiky
Slovenia	5	10	7	14	
Spain	5	10	10	20	Límites de Exposición Profesional para Agentes Químicos en España, 2012; Ministerio de Trabajo e Inmigración, INSHT
Sweden	10	15	14	20	AFS 2005:17
Switzerland	5	10	7.1	14.2	SuvaPro Grenzwerte am Arbeitsplatz 2009
UK	5	10	7	14	Health and Safety Executive EH40/2005
OSHA[a]	b	b	b	b	US Department of Labor
ACGIH[a]	1	5	1.5	21	ACGIH
NIOSH[a]	10	5	14		NIOSH
SCOEL	5	10	7	14	Scientific Committee on Occupational Exposure Limits

a North American value

b OSHA PEL (permissible exposure limit) (General Industry): 'Exposures shall not exceed 20 ppm (ceiling) with the following exception: if no other measurable exposure occurs during the 8-hour work shift, exposures may exceed 20 ppm, but not more than 50 ppm (peak), for a single time period up to 10 minutes'

treatment should be carried out with caution because careless removal of the bitumen may result in the skin being damaged further, bringing with it the risk of infection and the possibility of complications.

Initially, it is not important to know whether the burn is superficial or deep. The priority should be to remove the bitumen without causing further damage.

3.2.2.3 Removal of bitumen adhering to the burned areas
Different methods can be recommended.

- The bitumen layer should be left in place and covered with thick gauze containing paraffin or a paraffin-based antibiotic cream such as Flammazine (silver sulfadiazine). Such treatment will have the effect of softening the bitumen, enabling it to be gently removed after a few days.
- Alternatively, olive oil (new bottle) should be applied and left to soak the affected areas for a few hours. Thereafter, the bitumen can be removed by rubbing gently with some gauze. Any remaining bitumen can be removed by wrapping the affected areas with gauze soaked in olive oil. The dressing should be changed every 4 h. After 24 h, any remaining bitumen can be removed, and the burn may be disinfected and treated conventionally.

3.2.2.4 Circumferential burns with tourniquet effect
When bitumen completely encircles a limb, or other body part, the cooled and hardened bitumen may cause a tourniquet effect due to oedema (swelling) in the burn. In the event of this occurring, the bitumen must be softened as soon as possible and/or split, to prevent restriction of blood flow.

3.2.2.5 Eye burns
No attempt should be made to remove the bitumen by unqualified personnel. The patient should be referred urgently to an ophthalmologist or hospital with an ophthalmology unit for diagnosis and appropriate treatment.

3.3 Bitumen and the environment
3.3.1 Life cycle assessment of bitumen
Life cycle assessment (LCA) is a tool to investigate the environmental aspects and potential impact of a product, process or activity by identifying and quantifying energy and material flows. LCA covers the entire life cycle, including extraction of the raw material, manufacturing, transport and distribution, product use, service and maintenance, and disposal (recycling, incineration or landfill). It is a complete cradle-to-grave analysis focusing on the environmental input (based on ecological effects) and resource use.

LCA can be divided into two distinct parts – life cycle inventory (LCI) and life cycle impact. Eurobitume has carried out a life cycle inventory of bitumen (Eurobitume, 2012) to generate inventory data on the production of paving grade bitumen for future LCI studies where bitumen is used. In addition there are many other bitumen, asphalt and construction material LCAs, but to be comparable they need to meet the standard of the ISO 14040 series (ISO, 2006).

3.3.2 Use of bitumen as a potable water lining

Bitumen and asphalt have been used for many years for applications in contact with water, such as reservoir linings, dams and dykes (Schönian, 1999). A number of studies have been carried out both in the USA and Europe to determine if components are leached out of asphalt and bitumen when in prolonged contact with water.

Shell has carried out laboratory studies of leaching on a range of bitumens and asphalts (Brandt and De Groot, 2001). These concluded that although prolonged contact with water will result in PACs being leached into water, the levels rapidly reach an equilibrium level that is well below the surface water limits that exist in a number of EU countries and more than an order of magnitude below the EU limits given in the Drinking Water Directive and listed in Table 3.3 (Council Directive 98/83/EC) (EC, 1998).

Table 3.3 EU chemical limits for drinking water

Parameter	Parametric value	Parameter	Parametric value
Acrylamide	0.10 μg/l	Fluoride	1.5 mg/l
Antimony	5.0 μg/l	Lead	10 μg/l
Arsenic	10 μg/l	Mercury	1.0 μg/l
Benzene	1.0 μg/l	Nickel	20 μg/l
Benzo[a]pyrene	0.010 μg/l	Nitrate	50 mg/l
Boron	1.0 mg/l	Nitrite	0.50 mg/l
Bromate	10 μg/l	Pesticides	0.10 μg/l
Cadmium	5.0 μg/l	Pesticides (total)	0.50 μg/l
Chromium	50 μg/l	PAHs	
		Sum of concentrations of specified compounds	0.10 μg/l
Copper	2.0 mg/l	Selenium	10 μg/l
Cyanide	5 μg/l	Tetrachloroethene and trichloroethene	
		Sum of concentrations of specified parameters	10 μg/l
1,2-Dichloroethane	3.0 μg/l	Trihalomethanes (total)	
		Sum of concentrations of specified compounds	100 μg/l
Epichlorohydrin	0.10 μg/l	Vinyl chloride	0.50 μg/l

The objective of the Drinking Water Directive is to protect human health from the adverse effects of any contamination of water intended for human consumption by ensuring that it is wholesome and clean. The directive applies to

- all distribution systems serving more than 50 people or supplying more than 10 m^3/day, but also distribution systems serving less than 50 people or supplying less than 10 m^3/day if the water is supplied as part of an economic activity
- drinking water from tankers
- drinking water in bottles or containers
- water used in the food processing industry, unless the competent national authorities are satisfied that the quality of the water cannot affect the wholesomeness of the foodstuff in its finished form.

References

Boffetta PT (2001) Cancer risk in asphalt workers and roofers: review and meta-analysis of epidemiologic studies. *American Journal of Industrial Medicine* **26(6)**: 721–740.

Brandt HCA and De Groot PC (1996) Emission and composition of fumes from current bitumen types. *Proceedings of the 1st Eurasphalt/Eurobitume Congress*. EAPA, Brussels, Belgium.

Brandt HCA and De Groot PC (2001) Aqueous leaching of polycyclic aromatic hydrocarbons from bitumen and asphalt. *Journal of Water Research* **35(17)**: 4200–4207.

Brandt HCA, Booth ED, De Groot PC and Watson WP (1999) Development of a carcinogenic potency index for dermal exposure to viscous oil products. *Archives of Toxicology* **73(3)**: 180–188.

BSI (British Standards Institution) (2012) BS EN 58:2012. Bitumen and bituminous binders – Sampling bituminous binders. BSI, London, UK.

CONCAWE (Conservation of Clean Air and Water in Europe) (1992) *Bitumens and Bitumen Derivatives*. CONCAWE, The Hague, The Netherlands, Product Dossier No. 92/104.

Deygout F (2011) Personal exposure to PAHs in the refinery during truck loading of bitumen. *Journal of Occupational and Environmental Hygiene* **8**: D97–D100.

EC (European Community) (1998) Council Directive 98/83/EC. Quality of water intended for human consumption. *Official Journal of the European Union* **L330/32**.

EC (2007) Regulation (EC) No 1907/2006 [1]; Corrigendum version, 29 May 2007. *Official Journal of the European Union* **L136/3**.

Energy Institute (2005) *Bitumen Safety Code. Model Code of Safe Practice in the Petroleum Industry*, part 11, 4th edn. Energy Institute, London, UK.

Eurobitume (2011a) *Guide to Safe Delivery of Bitumen*, 2nd edn. Eurobitume, Brussels, Belgium.

Eurobitume (2011b) *The Bitumen Industry – A Global Perspective: Production, Chemistry, Use, Specification and Occupational Exposure*, 2nd edn. Eurobitume, Brussels, Belgium; Asphalt Institute, Lexington, KY, USA.

Eurobitume (2012) *Life Cycle Inventory: Bitumen*, 2nd edn. Eurobitume, Brussels, Belgium.

IARC (1985) *Polynuclear Aromatic Compounds*, part 4. *Bitumens, Coal-tars and Derived Products, Shale-oils and Soots*. World Health Organization, Lyon, France, IARC Monograph 35.

IARC (2013) *Bitumens and Bitumen Emissions, and Some* N- *and* S-*Heterocyclic Polycyclic Aromatic Hydrocarbons*. World Health Organization, Lyon, France, IARC Monograph 103.

ISO (International Standards Organisation) (2006) ISO 14040: 2006. Environmental management – Life cycle assessment – Principles and framework. ISO, Geneva.

Potter D, Booth ED, Brandt HCA *et al*. (1999) Studies on the dermal and systemic bioavailability of polycyclic aromatic compounds in high viscosity oil products. *Archives of Toxicology* **73(3)**: 129–140.

Schönian E (1999) *The Shell Bitumen Hydraulic Engineering Handbook*. Shell International Petroleum Company, London, UK.

US Environmental Protection Agency (2008) *Polycyclic Aromatic Hydrocarbons (PAHs)*. Office of Solid Waste, Washington, DC, USA.

Chapter 4

Constitution and structure of bitumens

Dr Dawid D'Melo
Chemist,
Shell India Markets Pvt. Ltd.
India

Dr Richard Taylor
Global Technical
Development Manager,
Shell International Petroleum
Co. Ltd. UK

Bitumen is a mixture of organic compounds composed of linear aliphatic, cycloaliphatic and aromatic derivatives. Linear aliphatic compounds are saturated linear carbon compounds (e.g. n-heptane). Cycloaliphatic compounds are ring structures that are composed of saturated carbon atoms (e.g. cyclohexane), or cyclic structures that have a low number of unsaturated groups (e.g. cyclohexene), and aromatic compounds are those compounds that have at least one aromatic ring (e.g. benzene). The interaction between molecules and the dilution effect of other molecules in the mixture are reflected in the properties of bitumen. The wide range of compounds and the varying nature of these compounds from bitumen to bitumen make their individual isolation and identification a challenging proposition. The constituents of bitumen are usually separated into classes based on parameters such as solubility, polarity or hydrodynamic size. The hydrodynamic size refers to the volume that a molecule or molecular aggregate (e.g. a colloid particle) occupies when in a solvent. This volume includes the adherence of solvent molecules onto the molecule or aggregate. This volume will be dependent on the molecular structure as well as the solvent used.

The most common method to characterise bitumen is by means of its rheology. Rheology is the science that deals with the flow and deformation of matter. The rheological characteristics of a bitumen at a particular temperature are determined by both the constitution (chemical composition) and the structure (physical arrangement) of the molecules in the material. Changes to the constitution, structure or both will result in a change to the rheology.

Thus, to understand changes in bitumen rheology, it is essential to understand how the structure and constitution of a bitumen interact to influence its rheology. Rheology is considered in detail in Chapter 7.

The Shell Bitumen Handbook, Sixth Edition
ISBN 978-0-7277-5837-8
Shell International Petroleum Company Ltd: All rights reserved
http://dx.doi.org/10.1680/tsbh.58378.047

4.1 Bitumen constitution

4.1.1 Elemental analysis of bitumen

Bitumen is largely composed of hydrocarbon molecules, with some hetero-cyclic species and functional groups containing sulfur, nitrogen and oxygen atoms (Romberg et al., 1959; Traxler, 1936; Traxler and Coombs, 1936). Bitumen also contains trace amounts of metals such as nickel, vanadium, iron, calcium and magnesium, which occur in the form of metallic salts, oxides or in porphyrin structures. Porphyrins are complex organic compounds that occur naturally: for example, haemoglobin, found in blood, and chlorophyll, found in green plants, are examples of porphyrins associated with metal atoms. Porphyrins contain four nitrogen atoms, each of which can bond with a metal atom to result in a metalloporphyrin. Elemental analysis of bitumen manufactured from a variety of crude oils shows that most bitumen contains

carbon	82–88%
hydrogen	8–11%
oxygen	0–1.5%
sulfur	0–6%
nitrogen	0–1%

It has been known since the 1930s that bitumens contain metallopor-phyrins, from analytical work that identified iron and vanadium and, consequently, established the link between marine plant chlorophyll and petroleum genesis. Ash from fuel oil on examination by ultraviolet emission spectrography has been shown to contain the following metallic elements (Crump, 1981):

aluminium	lead	sodium
barium	magnesium	strontium
calcium	manganese	tantalum
chromium	molybdenum	tin
copper	nickel	uranium
gallium	potassium	vanadium
iron	silicon	zinc
lanthanum	silver	zirconium

These elements occur principally in the heavier or involatile components of the oils, some as inorganic contaminants, possibly in colloidal form, but also as salts (e.g. of carboxylic acids), transition metal complexes and porphyrin-type complexes. An analysis of bitumen from various sources is shown in Table 4.1.

48

Table 4.1 Elemental analysis of bitumen from various sources (Crump, 1981)

	Carbon: % w	Hydrogen: % w	Nitrogen: % w	Sulfur: % w	Oxygen: % w	Nickel: ppm	Vanadium: ppm
Range	80.2–84.3	9.8–10.8	0.2–1.2	0.9–6.6	0.4–1.0	10–139	7–1590
Average	82.8	10.2	0.7	3.8	0.7	83	254

	Iron: ppm	Manganese: ppm	Calcium: ppm	Magnesium: ppm	Sodium: ppm	Atomic ratio: H/C	
Range	5–147	0.1–3.7	1–335	1–134	6–159	1.42–1.50	
Average	67	1.1	118	26	63	1.47	

The predominant metals present in most fuel oils are sodium, vanadium, iron, nickel and chromium, with most of the sodium present as sodium chloride. Vanadium and nickel are largely present as porphyrin structures, which also represent large numbers of different molecules depending on ring substituents (various organic groups that have been added to the core structure of a molecule) and structural isomerism (repositioning of organic groups in different positions within a molecule, with the same molecular formula but with a different chemical or physical behaviour). Bitumen and fuel oil are related products, and Goodrich surmises that similar elements will be present in bituminous compounds (Goodrich *et al.*, 1986).

The precise composition varies according to the source of the crude oil from which a bitumen originates and chemical modification induced during the manufacturing process and ageing in service.

The chemical composition of bitumen is extremely complex. Thus, a complete analysis of bitumen (if it was possible) would be extremely laborious and would produce such a large quantity of data that correlation with the rheological properties would be impractical, if not impossible. In addition, the dataset obtained would be valid for a particular bitumen sample only and not for bitumen in general.

Traditionally, bitumen has been divided into two broad chemical groups called *asphaltenes* and *maltenes*. The maltenes can be further subdivided into *saturates*, *aromatics* and *resins*. The four groups are not well defined, and there is some overlap between them. However, this classification does enable bitumen rheology to be set against broad chemical composition.

Solvent extraction is attractive as it is a relatively rapid technique (Traxler and Schweyer, 1953), but the separation obtained is generally poorer than that which results from using chromatography, where a solvent effect is combined with selective adsorption. Similarly, simple adsorption methods (Marcusson, 1916) are not as effective as column chromatography, in which the eluting solution is constantly re-exposed to fresh adsorbent and different equilibrium conditions as it progresses down the column. (An eluting solution is one that is used to remove an adsorbed substance by washing.)

In 1987, the US Congress authorised the Strategic Highway Research Program (SHRP) – a 5-year applied research initiative – to develop and evaluate techniques and technologies to combat the deteriorating conditions of US highways and to improve their performance, durability, safety and efficiency (Halladay, 1998). As part of the SHRP study to investigate the chemistry of bitumen, methods were developed to separate the constituents based on the nature of their functional groups and molecular weight.

4.1.2 Bitumen classification based on solubility

Chromatographic techniques are the most widely used methods to define bitumen constitution (Corbett and Swarbick, 1958; Schweyer *et al.*, 1955). The basis of these methods is initially to precipitate asphaltenes using n-heptane followed by chromatographic separation of the remaining constituents/fractions. Figure 4.1 shows a schematic representation of the chromatographic method. Using this technique, bitumen can be separated into the four groups: saturates, aromatics, resins and asphaltenes (sometimes described by the acronym SARA). The main characteristics of these four broad component groups and the metallic constituents are now discussed.

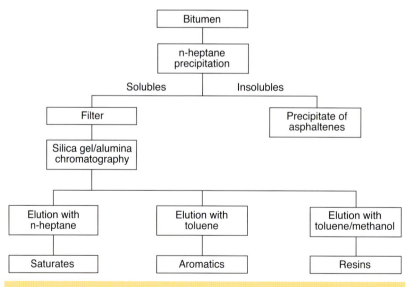

Figure 4.1 Schematic of the saturates, aromatics, resins and asphaltenes (SARA) chromatographic method

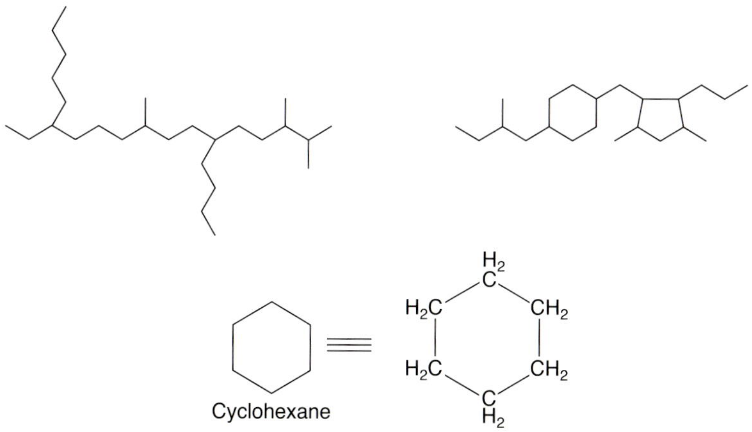

Figure 4.2 Schematic representations of saturate structures

4.1.2.1 Saturates

Saturates consist of straight and branched-chain aliphatic hydrocarbons together with cycloaliphatic compounds. They are non-polar viscous oils that are straw coloured or colourless. The hydrogen-to-carbon molar ratio (H/C) of the saturate fraction is approximately 2 (Lesueur, 2009). The average molecular weight range has been reported to vary between 470 and 880 g/mol (Lesueur, 2009). This fraction forms 5–20% of bitumen. Figure 4.2 shows two different saturate structures and the structure of cyclohexane.

4.1.2.2 Aromatics

Aromatics comprise the lowest molecular weight naphthenic aromatic compounds in bitumen, and represent the major proportion of the dispersion medium for the peptised asphaltenes. They constitute 40–65% of the total bitumen, and are dark-brown viscous liquids. The H/C ratio of the aromatic fraction could be said to range from 1.4 to 1.6. The average molecular weight range is in the region of 570–980 g/mol (Lesueur, 2009). They consist of non-polar carbon chains attached to unsaturated ring systems (aromatics) (Figure 4.3).

4.1.2.3 Resins

Resins are soluble in n-heptane. They are largely composed of hydrogen and carbon, and contain a small number of oxygen, sulfur and nitrogen atoms. They are dark brown in colour, solid or semi-solid and, being polar in nature, strongly adhesive. Resins are dispersing agents or peptisers for the asphaltenes. The proportion of resins to asphaltenes governs, to a degree, the solution (sol) or gelatinous (gel) character of the bitumen. Resins separated from bitumen are found to have molecular weights ranging from 780 to 1400 g/mol and an H/C atomic ratio of 1.4–1.7.

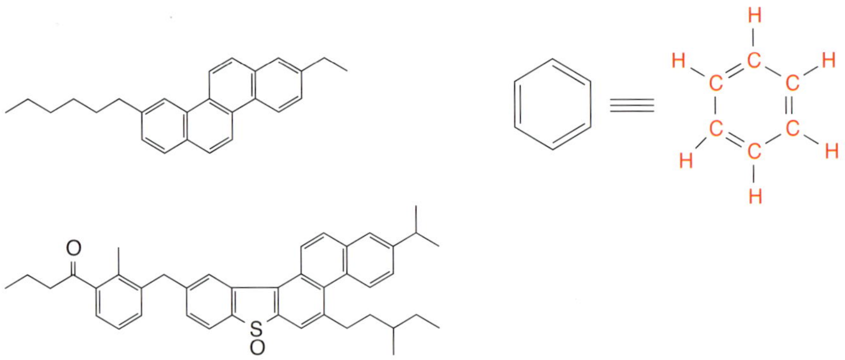

Figure 4.3 Schematic representations of aromatics in bitumen

4.1.2.4 Asphaltenes

These are n-heptane insoluble black or brown amorphous solids containing, in addition to carbon and hydrogen, some nitrogen, sulfur and oxygen atoms. Asphaltenes are generally considered to be highly polar and complex aromatic materials of fairly high molecular weight. Different methods of determining molecular weights have led to various values, ranging widely from 800 to 3500 g/mol, based on vapour pressure osmometric measurements (a method used to measure the molecular weight of compounds based on determining the vapour pressure of solutions using Raoult's law), and the H/C ratio of asphaltenes ranges from 0.98 to 1.6 (Lesueur, 2009). They have a particle size of 2–5 nm. The asphaltene content has a significant effect on the rheological characteristics of a bitumen. Increasing the asphaltene content produces a harder, more viscous bitumen with a lower penetration, higher softening point and, consequently, higher viscosity. Asphaltenes constitute 5–25% of the bitumen. Figure 4.4 shows a typical chemical structure of an asphaltene.

Recent studies have shown that the asphaltenes can be further separated, based on their solubility in solvents with an increasing solubility parameter,

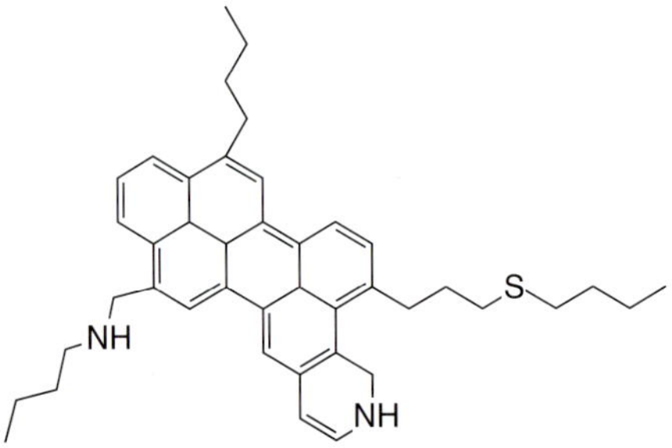

Figure 4.4 Schematic representation of an asphaltene in bitumen

Table 4.2 Typical elemental analysis of the four groups of a bitumen with a penetration value of 100 dmm (Chipperfield, 1984)

	Yield on bitumen: % w	Carbon: % w	Hydrogen: % w	Nitrogen: % w	Sulfur: % w	Oxygen: % w	Atomic ratio: H/C
Asphaltenes (n-heptane)	5.7	82.0	7.3	1.0	7.8	0.8	1.1
Resins	19.8	81.6	9.1	1.0	5.2	–	1.4
Aromatics	62.4	83.3	10.4	0.1	5.6	–	1.5
Saturates	9.6	85.6	13.2	0.05	0.3	–	1.8

using a preparative or chromatographic method (Schabron *et al.*, 2010). The method separates n-heptane insoluble asphaltenes into three sub-fractions. It was seen that the relative amounts of aromatic fused-ring structures increased in the asphaltene sub-fractions as the solubility parameter of the extracting solvent increased. This method could be of interest to study further the structure and type of asphaltenes present in bitumen.

The elemental analysis of the above four groups from a bitumen with a penetration value of 100 dmm is detailed in Table 4.2.

4.1.3 Other classification methods

Bitumen constituents can also be divided into groups based on their polarity. This is accomplished by using an ion exchange chromatography (IEC) technique (Branthaver *et al.*, 1993). This technique separated bitumen into five components, namely strong acid, weak acid, strong base, weak base and the neutral fraction. The structural analysis of these components was carried out by means of elemental analysis and Fourier transform infrared spectroscopy, while the molecular weight of the fractions was determined by vapour pressure osmometry. The speciation and distribution of functional groups within the fractions and the rheology of the individual fractions obtained from a bitumen sample were also investigated.

The IEC technique has been reported to be sensitive to experimental variables, and the recovery of the sample is never complete (i.e. some of the material remains permanently adsorbed in the columns). When analysing aged bitumen samples, the amount of sample permanently adsorbed on the resin was higher than that observed for fresh bitumen samples.

The separation of bitumen components based on their hydrodynamic volumes using size exclusion chromatography (SEC) was also carried out (Branthaver *et al.*, 1993). Methods to separate bitumen into its constituents, based on molecular size or the size of associated molecules, were used. Compared with IEC, the recovery of the sample using the SEC technique was higher. Viscosity and rheological studies were carried out on the fractions obtained. Interestingly, it was stated that fractions that contained sulfoxides and ketones, which are oxidation products, were found to be present in the low molecular weight fractions. This indicated that these molecules did not readily associate with other molecules with polar groups.

4.2 Bitumen structure
4.2.1 Simple colloidal models

Bitumen is traditionally regarded as a colloidal system consisting of high molecular weight asphaltene micelles dispersed or dissolved in a lower molecular weight oily medium (maltenes) (Girdler, 1965). The micelles are considered to be asphaltenes together with an absorbed sheath of aromatic resins that act as a stabilising solvating layer. Away from the centre of the micelle, there is a gradual transition to less polar aromatic resins, these layers extending outwards to the less aromatic oily dispersion medium.

In the presence of sufficient quantities of resins and aromatics of adequate solvating power, the asphaltenes are fully peptised, and the resulting micelles have good mobility within the bitumen. These are known as 'sol'-type bitumens, illustrated in Figure 4.5.

If the aromatic/resin fraction is not present in sufficient quantities to peptise the micelles, or has insufficient solvating power, the asphaltenes can associate together further. This can lead to an irregular, open packed structure of linked micelles in which the internal voids are filled with an intermicellar fluid of mixed constitution. These bitumens are known as 'gel' types, as depicted

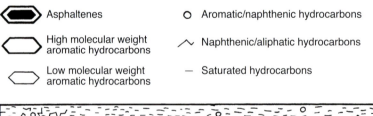

Asphaltenes	○	Aromatic/naphthenic hydrocarbons
High molecular weight aromatic hydrocarbons	∿	Naphthenic/aliphatic hydrocarbons
Low molecular weight aromatic hydrocarbons	—	Saturated hydrocarbons

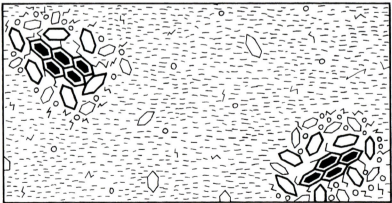

Figure 4.5 Schematic representation of a sol-type bitumen

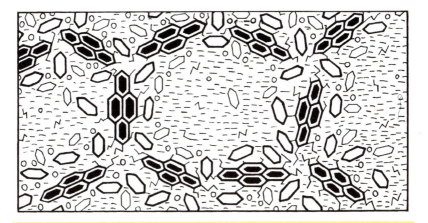

◆ Asphaltenes   ○ Aromatic/naphthenic hydrocarbons

⬡ High molecular weight aromatic hydrocarbons   ∿ Naphthenic/aliphatic hydrocarbons

▭ Low molecular weight aromatic hydrocarbons   — Saturated hydrocarbons

Figure 4.6 Schematic representation of a gel-type bitumen

in Figure 4.6, the best examples being the oxidised grades used for roofing purposes. In practice, most bitumens are of an intermediate character.

The colloidal behaviour of the asphaltenes in bitumen results from their aggregation and solvation. The degree to which the asphaltenes are peptised will have a considerable influence on the resultant viscosity of the system. Such effects decrease with increasing temperature, and the gel character of certain bitumens may be lost when they are heated to high temperatures. The viscosities of the saturates, aromatics and resins depend on the molecular weight distribution and interaction between molecules. The higher the molecular weight and the greater the degree of interaction, the higher the viscosity. The viscosity of the continuous phase (i.e. the maltenes) imparts an inherent viscosity to the bitumen that is increased by the presence of the dispersed phase (i.e. the asphaltenes). The saturates fraction decreases the ability of the maltenes to solvate the asphaltenes because high saturate contents can lead to marked agglomeration of the asphaltenes. Accordingly, an increase in gel character and a lower temperature dependence for bitumen results not only from the asphaltene content but also from the saturates content.

4.2.2 More complex models
Dickie and Yen (1967) proposed a model of a colloidal system formed by asphaltenes stacking over each other. The asphaltenes in this model conform

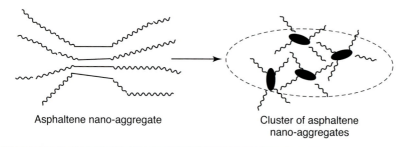

Asphaltene nano-aggregate

Cluster of asphaltene
nano-aggregates

Figure 4.7 Schematic representation of asphaltene nano-aggregates and clusters

to what is known as the 'island' structure (i.e. where there is a central poly-aromatic nucleus with aliphatic side-chains). The polyaromatic cores of the asphaltenes result in the proposed stacking behaviour. Further studies and a refinement of the model by Mullins (2010) suggested that the asphaltene stacks form what is termed a 'nano-aggregate'. These nano-aggregates form clusters (Figure 4.7).

It was suggested that the nano-aggregates are formed by approximately six asphaltene molecules. The polyaromatic core of the asphaltenes was reported to contain, on average, seven fused aromatic rings. The asphaltene clusters of the nano-aggregates form the colloidal portion of the bitumen.

An alternative assembly architecture dealing with asphaltene nano-aggrega-tion was proposed by Gray *et al.* (2005). The asphaltenes proposed also conform to the 'island' configuration of the asphaltene molecule. Here, the nano-aggregation is based on hydrogen bonding, Brøndsted acid–base interactions, metal complex formation and hydrophobic group interactions as well as aromatic core stacking. The asphaltene nano-aggregate structure suggested by Gray *et al.* results in a more random distribution of molecules compared with the nano-aggregate structure proposed by Dickie and Yen–Mullins.

Other studies have shown that, in addition to asphaltenes that conform to the 'island' type of asphaltene, 'archipelago'-type asphaltenes are also present. It has been suggested that both of these asphaltene types occur (Long *et al.*, 2006). The 'archipelago' type of asphaltene consists of a few fused aromatic groups linked by aliphatic chains, as opposed to the central large fused core with aliphatic chains attached to it proposed for the 'island' type of asphaltene.

Regardless of the internal structure of the asphaltene nano-aggregates, the colloidal dispersion is composed of the asphaltenes as the colloidal particles dispersed in the maltene phase.

4.3 The relationship between bitumen constitution, structure and rheology

4.3.1 Influence of bitumen components on properties

The influence of bitumen components (i.e. the SARA fractions) on the properties of bitumen has been studied previously (Griffin and Miles, 1961; Griffin *et al.*, 1959; McKay *et al.*, 1978; Reerink, 1973). It was observed, by keeping the concentration of asphaltenes constant in a sample while varying the other components, that

- increasing the aromatics content at a constant saturates-to-resins ratio has little effect on the rheology other than a marginal reduction in the shear susceptibility
- maintaining a constant ratio of resins to aromatics and increasing the saturates content softens the bitumen
- the addition of resins hardens the bitumen and reduces the penetration index and shear susceptibility, but increases the viscosity.

It has also been shown that the rheological properties of bitumen depend strongly on the asphaltene content (Lin *et al.*, 1996). At a constant temperature, the viscosity of a bitumen increases as the concentration of the asphaltenes blended into the parent maltenes is increased. However, the increase in viscosity is substantially greater than would be expected if the asphaltenes were spherical, non-solvated entities. This suggests that the asphaltenes can interact with each other and/or the solvating medium. Even in a dilute toluene solution, the viscosity increase observed with increasing asphaltene concentrations corresponds to a concentration of non-solvated spheres some five times higher than the amount of asphaltenes used. Bitumen asphaltenes are believed to be stacks of plate-like sheets formed of aromatic/naphthenic ring structures. The viscosity of a solution, in particular a dilute solution, depends on the shape of the asphaltene particles. Size is important only if the shape changes significantly as the size increases. At high temperatures, the bonds holding the asphaltene aggregates together are broken, resulting in a change in both the size and shape of the aggregates or clusters. Consequently, the viscosity falls as the temperature increases. However, as a hot bitumen cools, associations between asphaltene nano-aggregates occur, to produce clusters.

The marked increase in non-Newtonian behaviour as bitumen cools is a consequence of the inter-molecular and intra-molecular attractions between asphaltenes and other entities. Under shear, these extended associations will deform or even dissociate in a way that is not adequately described by classical Newtonian concepts. Consequently, at ambient and intermediate temperatures, it is reasonable to conclude that the rheology of bitumen is dominated by the degree of association of asphaltene particles and the relative amount of other species present in the system to stabilise these associations.

The question regarding the interaction between resins and asphaltenes was considered in the review by Mullins (2010). There has been some evidence to indicate that there is limited interaction between the asphaltene nano-aggregates and the resin fraction (Mullins, 2010). When considering the separation of asphaltenes from bitumen or even crude oils, it must be remembered that asphaltenes are a solubility class; that is, that fraction that is insoluble in n-heptane (although other paraffinic solvents, such as pentane, have also been used). This would mean that what would be considered as the resin fraction of the maltenes separated using n-heptane could be included in the asphaltene fraction, when pentane is used as the asphaltene precipitating solvent.

The argument put forward to support the theory that the resins do not interact with the asphaltenes to stabilise them is that the asphaltene nano-aggregates or dispersion in toluene are stable in the absence of any resin, and the nano-aggregates are similar to those observed in crude oils (Mullins, 2010). Experiments carried out where the separation of asphaltenes from crude oils showed that there was lower than expected amounts of resins present have also been cited as evidence of the limited stabilising effect of resins on asphaltene nano-aggregates.

However, other studies using microcalorimetric measurements (i.e. measurements of extremely small changes in heat content) have shown that there is indeed interaction between the resin and asphaltene phases. These measurements were seen to be comparable to those predicted by simulations. Other studies involving small angle neutron scattering measurements showed that the size of the asphaltene nano-aggregate was smaller in the presence of resins, indicating that there was an interaction between the two components.

The question regarding the interaction or extent of interaction between the resins and asphaltene components has not been addressed by the Dickie and Yen–Mullins model.

4.4 The relationship between broad chemical composition and physical properties during the bitumen life cycle

Atmospheric and vacuum distillation removes the lighter components from the bitumen feedstock. The loss of distillates leads to the preferential removal of saturates and the concentration of asphaltenes. Air blowing is carried out to meet certain bitumen application requirements, such as that used for roofing grades. These grades require a higher softening point than paving grade bitumen samples, at similar penetration values. Air is blown through bitumen, maintained at a temperature ranging from 240 to 320°C, until the specific

Table 4.3 Comparison of the broad chemical compositions of distilled and blown bitumen manufactured from a single short residue (Chipperfield, 1984)[a]

	Vacuum residue	Distillation				Blowing		
Penetration at 25°C: dmm	285	185	99	44	12	84	46	9
Asphaltenes: % w	9.1	9.9	10.5	11.3	12.5	15.2	17.3	22.9
Resins: % w	18.6	16.7	18.2	17.7	21.3	21.0	22.1	21.5
Aromatics: % w	51.2	53.0	52.4	58.4	53.8	47.6	45.0	40.5
Saturates: % w	16.2	15.1	14.1	11.2	9.4	16.2	15.6	15.1

[a] The recovery of components is incomplete because of the techniques used

bitumen properties are obtained. Goppel and Knoterus (1955) determined that, during the bitumen blowing process, carbonyl, carboxylic acid, ester and hydroxyl groups are introduced into the bitumen. Goppel and Knoterus also stated that the molecular weight of the asphaltenes increased during the blowing process.

Air blowing of bitumen from a given vacuum residue or fluxed vacuum residue results in a considerable increase in the asphaltene content and a decrease in the aromatics content. Saturate and resin contents remain substantially of the same order as before blowing. Table 4.3 shows a comparison of the chemical composition of bitumen derived by distillation and that from blowing of the same feedstock, at similar bitumen penetration values.

4.4.1 Changes during bitumen processing and asphalt manufacture

Asphalts are manufactured at elevated temperatures. During this process, the properties of the bitumen change compared with those exhibited by the bitumen prior to the manufacturing process. These alterations occur because the bitumen is oxidised, a process often described as 'ageing'. It has been established that either the rolling thin-film oven test (RTFOT) or the thin-film oven test (TFOT) simulates this ageing of bitumen (Lewis and Welborn, 1940). However, subsequent research found that the RTFOT and TFOT are not good predictors for in-service ageing (Schmidt and Santucci, 1969).

The analysis of bitumen samples after subjecting them to standard RTFOT and TFOT conditions has shown an increase in the concentrations of sulfoxide and ketone groups (Lu and Isacsson, 2002). It was also found that on oxidation, the aromatic fraction of the bitumen usually decreased and the resin and asphaltene fractions increased.

4.4.2 In-service changes to bitumen constitution

The oxidation of bitumen during its service life takes place at lower temperatures than those that occur during the manufacture of asphalts. These lower temperatures have an impact on the kinetics of reactions that occur (Lu and Isacsson, 2002). The effect of temperature on bitumen oxidation kinetics has been investigated using various oxidation methods, and the results are extensively discussed in the review of bitumen ageing by Petersen (2009).

4.4.3 Chemical changes observed during the oxidation of bitumen

Bitumen contains limited concentrations of polar functional groups, which are normally present or formed during oxidation (Petersen, 2009). It has been suggested that bitumen naturally contains some functional groups: phenolic, 2-quinolone type, pyrrolic, pyridinic, sulfide and acidic moieties or functional groups. It has been further suggested that ketonic, sulfoxide and anhydride functional groups are formed due to the oxidation reactions.

Polar groups in bitumen and its fractions impact the properties of bitumen. It has been shown that the presence and concentration of polar groups (in this case the phenolic group) affects the performance of pavement (i.e. its susceptibility to cracking) (Davis and Petersen, 1967). This was also shown to be true in the case of roofing bitumen. This trend has also been shown to be valid in relation to the concentration of ketones (Branthaver *et al.*, 1993; Martin *et al.*, 1990; Petersen *et al.*, 1993). The mechanism of the chemical reactions resulting in an increase in the polar content of oxidised bitumen has also been discussed (Petersen, 2009).

Studies have shown that the nature of the bitumen, as well as the void content of the asphalt pavement, plays a large role in the rate at which bitumen oxidises (Petersen, 2009). The influence of metals and aggregates was also investigated as part of SHRP (Branthaver *et al.*, 1993). It is interesting to note that studies using anti-oxidants and additives, which have been proved to be effective when used in polymeric systems, showed that these were not universally effective when added to bitumen (Martin, 1968).

A study of the change in the composition, in relation to the SARA fractions, was carried out for bitumen while in service in asphalt pavements (Chipperfield *et al.*, 1970) (Figure 4.8). The concept of the 'ageing index' has been used to study bitumen when in service: the ageing index is the ratio of the bitumen viscosity after ageing to that before ageing. It is commonly used to normalise the ageing behaviour of samples with different viscosities. The major changes in viscosity were observed to be associated with the mixing and laying process. The recovered bitumen samples were obtained from the top 3 mm of cores extracted from the test sections. Changes in the viscosity of the binder were found to be small, over time. With regard to the chemical composition, the asphaltene content increased with mixing, and showed a gradual increase with time, while the resin and aromatic content

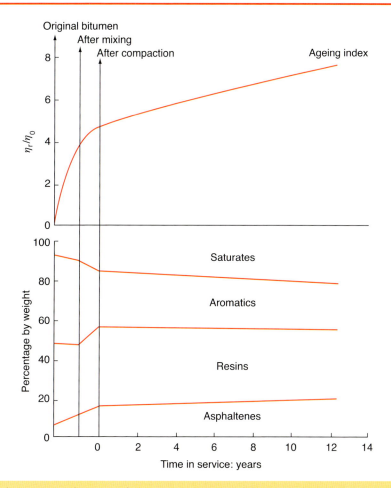

Figure 4.8 Changes to bitumen composition while in service in pavements

decreased with time. Unexpectedly, the saturates content was observed to increase with time. Overall, it was observed that changes after mixing were very small, even though the initial void content of the mixtures studied was relatively high (5–8%). The ageing index is given by:

$$\text{The ageing index} = \frac{\text{Viscosity of recovered bitumen}}{\text{Viscosity of original bitumen}} = \frac{\eta_r}{\eta_0}$$

As stated above, bitumen is a mixture of complex compounds, and their interaction directly impacts the performance of bitumen and its products. Although a large amount of research has been carried out to study the composition of bitumen, based on solubility, molecular weight and polarity, at present no precise relationship between the composition and performance of bitumen has been established. Further research will help to elucidate the influence of bitumen components on its properties.

References

Branthaver JF, Petersen JC, Robertson RE *et al.* (1993) *Binder Characterization and Evaluation*, vol. 2. *Chemistry*. Strategic Highway Research Program, National Research Council, Washington, DC, USA, SHRP-A-368.

Chipperfield EH (1984) *IARC Review on Bitumen Carcinogenicity: Bitumen Production Properties and Uses in Relation to Occupational Exposures*. Institute of Petroleum, London, UK, Report IP 84-006.

Chipperfield EH, Duthie JL and Gridler RB (1970) Asphalt characteristics in relation to road performance. *Proceedings of the American Asphalt Paving Technologists* **39**: 575–613.

Corbett LW and Swarbick RE (1958) Clues to asphalt composition. *Proceedings of the American Asphalt Paving Technologists* **27**: 107–115.

Crump GB (1981) Black but such as in esteem – the analytical chemistry of bitumen. *Proceedings of the Chairman's Retiring Address to the NW Region Analytical Division of the Royal Society of Chemistry*, London, UK.

Davis TC and Petersen JC (1967) An inverse GLC study of the asphalts used in the Zaca–Wigmore experimental test road. *Proceedings of the Association of Asphalt Paving Technologists* **36**: 1–10.

Dickie JP and Yen TF (1967) Macrostructures of asphaltic fractions by various instrumental methods. *Analytical Chemistry* **39(14)**: 1847–1852.

Girdler RB (1965) Constitution of asphaltenes and related studies. *Proceedings of the Association of Asphalt Paving Technologists* **34**: 45–79.

Goodrich JL, Goodrich JE and Kari WJ (1986) Asphalt composition tests: their application and relation to field performance. *Transportation Research Record* **1096**: 146–167.

Goppel JM and Knoterus J (1955) Fundamentals of bitumen blowing. *4th World Petroleum Congress*, Rome, Italy.

Gray MR, Tywinski RR, Stryker JM and Tan X (2005) Supramolecular assembly model for aggregation of petroleum asphaltenes. *Energy and Fuels* **25(7)**: 3125–3134.

Griffin RL and Miles TK (1961) Relationship of asphalt properties to chemical constitution. *Journal of Chemical and Engineering Data* **6(3)**: 426–429.

Griffin RL, Simson WC and Miles TK (1959) Influence of composition of paving asphalt on viscosity, viscosity–temperature, susceptibility and durability. *Journal of Chemical and Engineering Data* **4(4)**: 349–354.

Halladay M (1998) The strategic highway research program: An investment that has paid off. *Public Roads* **1(5)**: 11–17.

Lesueur D (2009) The colloidal structure of bitumen: consequences on the rheology and on the mechanisms of bitumen modification. *Advances in Colloid Interface Science* **145(1-2)**: 42–82.

Lewis RH and Welborn JY (1940) Report on the physical and chemical properties of petroleum asphalts on the 50-60 and 85-100 penetration grades. *Proceedings of the Association of Asphalt Paving Technologists* **11**: 86–92.

Lin MS, Chaffin JM, Liu M *et al.* (1996) The effect of asphalt composition on the formation of asphaltenes and their contribution to asphalt viscosity. *Fuel Science and Technology International* **14(1–2)**: 139–162.

Long J, Xu Z and Masliyah JH (2006) Role of colloidal interactions in oil sand tailings treatment. *Langmuir* **52(1)**: 371–383.

Lu X and Isacsson U (2002) Effect of ageing on bitumen chemistry and rheology. *Construction and Building Materials* **16(1)**: 15–22.

McKay JR, Amend PJ, Cogswell TG *et al.* (1978) Petroleum asphaltenes: chemistry and composition. In: *Analytical Chemistry of Liquid Fuel Sources* (Uden PC, Siggia S and Jensen HB (eds)). American Chemical Society, Washington, DC, USA, pp. 128–142.

Marcusson J (1916) Der chemische aufbau der naturasphalte. *Angewandte Chemie* **29(73)**: 346–351 (in German).

Martin KG (1968) Laboratory evaluation of antioxidants in bitumen. *Proceedings of the 4th Conference of the Australian Road Research Board*, Melbourne, Australia, vol. 4, pp. 1–14.

Martin KL, Davidson KK, Glover CJ and Bullin JA (1990) Asphalt ageing in Texas roads and test section. *Transportation Research Record* **1269**: 9–18.

Mullins OC (2010) The modified Yen model. *Energy and Fuels* **24(4)**: 2179–2207.

Petersen JC (2009) *A Review of the Fundamentals of Asphalt Oxidation: Chemical, Physicochemical, Physical Property, and Durability Relationships*. Transportation Research Board, Washington, DC, USA, Transport Research Circular E-C140.

Petersen JC, Branthaver JF, Robertson RE *et al.* (1993) Effect of physicochemical factors on asphalt oxidation kinetics. *Transportation Research Record* **1391**: 1–10.

Reerink H (1973) Size and shape of asphaltene particles in relationship to high temperature viscosity. *Industrial and Engineering Chemistry Product Research and Development* **12(1)**: 82–88.

Romberg JW, Nesmitts SD and Traxler RN (1959) Some chemical aspects of the components of asphalt. *Journal of Chemical Engineering Data* **4(2)**: 159–161.

Schabron JF, Rovani JF and Sanderson MM (2010) Asphaltene determinator method for automated on-column precipitation and redissolution of pericon-densed aromatic asphaltene components. *Energy and Fuels* **24**: 5984–5996.

Schmidt RJ and Santucci LE (1969) The effect of asphalt properties on the fatigue cracking of asphalt concrete on the Zaga–Wigmore test project. *Proceedings of the Association of Asphalt Paving Technologists* **38**: 39–45.

Schweyer HE, Chelton H and Brenner HH (1955) A chromatographic study of asphalt. *Proceedings Association of Asphalt Paving Technologists* **24**: 3–8.

Traxler RN (1936) The physical chemistry of asphaltic bitumen. *Chemical Reviews* **19(2)**: 119–143.

Traxler RN and Coombs CE (1936) The colloidal nature of asphalt as shown by its flow properties. *Journal of Physical Chemistry* **40(9)**: 1133–1147.

Traxler RN and Schweyer HE (1953) How to make component analysis. Increase of viscosity of asphalts with time. *Oil and Gas Journal* **52**: 133–158.

Chapter 5

Specifying bitumens and checking their quality

Dr Martin Vondenhof
Senior Application Specialist,
Shell Deutschland Oil GmbH.
Germany

Norbert Clavel
Application Advisor,
Société des Pétroles Shell.
France

In Europe, bitumens and polymer modified bitumens for asphalts are manufactured to three standards

- EN 12591:2009 (BSI, 2009a), covering penetration and viscosity grade bitumens
- EN 13924:2006 (BSI, 2006), covering hard paving grade bitumens
- EN 14023:2010 (BSI, 2010a), covering polymer modified bitumens (PMBs).

Bitumens for industrial uses are specified by

- EN 13304:2009 (BSI, 2009b), covering oxidised bitumens
- EN 13305:2009 (BSI, 2009c), covering hard industrial bitumens.

The key feature of all of these specifications is a combination of penetration and softening point and, in the case of the soft bitumen grades, viscosity. These properties are generally used to designate the bitumen grade required by the purchaser.

Cut-back and fluxed bitumens are bitumens blended with more or less volatile hydrocarbon components (fuels), mainly kerosene. They are characterised by a viscosity specification, EN 15322:2013 (BSI, 2013) providing the framework for such products. While such products remain popular in some parts of the world (e.g. Australia), their use has decreased markedly in the first decade of the twenty-first century in Europe.

For penetration grade bitumen, simple test methods such as those mentioned in EN 12591 are considered appropriate. Modified bitumens may require more sophisticated test methods to describe their performance adequately.

The Shell Bitumen Handbook, Sixth Edition
ISBN 978-0-7277-5837-8
Shell International Petroleum Company Ltd: All rights reserved
http://dx.doi.org/10.1680/tsbh.58378.065

In addition to the tests specified in the standards cited above, some bitumen suppliers carry out a further range of laboratory tests to ensure that bitumens are manufactured to a consistent standard, maintaining suitability for purpose.

5.1 Penetration grade, hard penetration grade and viscosity grade bitumens

Penetration grade bitumens are specified by the penetration (BSI, 2007a) and softening point (BSI, 2007b) tests. Designation is by penetration range only: for example, 40/60 pen bitumen has a penetration that ranges from 40 to 60 inclusive. The unit of penetration is the decimillimetre (0.1 mm). This is the unit that is measured in the penetration test (discussed, along with the softening point test, in detail in Chapter 6). However, penetration grade bitumens are usually referred to without stating units. Tables 1A, 1B, 2A and 2B of EN 12591 give details of key performance parameters. They are summarised here as Tables 5.1 and 5.2.

In addition to the harder penetration grades, softer bitumens are covered by EN 12591. Because the penetration and softening point tests are not applicable to soft bitumens, the viscosity measurement at 135°C is used as the main differentiating property. Such grades are also called viscosity grade bitumens, and their specifications according to EN 12591 are summarised in Table 5.3.

Hard penetration grade bitumens represent the other end of the viscosity scale: they are generally characterised and designated by penetration range only (e.g. 10/20). Over recent years, the tendency has been to use harder bitumens in asphalts.

5.2 Polymer modified bitumens (PMBs)

During the last three decades, PMBs have become increasingly popular as a replacement for penetration grade bitumens in the upper layers of asphalt pavements. Clients usually do so because of the superior properties of PMBs. Although they represent advanced technology binders, they are still designated both by the penetration range and a minimum softening point: for example, 45/80-55 is a PMB with a penetration of 45–80 dmm and a softening point of at least 55°C. Apart from the test methods that apply to these traditional methods of nomenclature, the specification framework EN 14023:2010 (BSI, 2010a) includes in its Annex B several more sophisticated tests as key performance parameters that need to be validated:

- bending beam rheometer based on test method EN 14771:2012 (BSI, 2012a)

Table 5.1 Specifications for paving grade bitumens with penetrations from 20 × 0.1 mm to 220 × 0.1 mm: Tables 1A and 1B of EN 12591 combined, including examples of specific regional requirements (BSI, 2009a)

Property	Test method	Unit	20/30	30/45	35/50	40/60	50/70	70/100	100/150	160/220
Penetration at 25°C	EN 1426	0.1 mm	20–30	30–45	35–50	40–60	50–70	70–100	100–150	160–220
Softening point	EN 1427	°C	55–63	52–60	50–58	48–56	46–54	43–51	39–47	35–43
Resistance to hardening at 163°C	EN 12607-1									
Retained penetration		%	≥55	≥53	≥53	≥50	≥50	≥46	≥43	≥37
Change of mass (absolute value)		%	≤0.5	≤0.5	≤0.5	≤0.5	≤0.5	≤0.8	≤0.8	≤1.0
Increase in softening point – severity 1		°C	≤8	≤8	≤8	≤9	≤9	≤9	≤10	≤11
or			or	or	or	or	or	or	or	or
Increase in softening point – severity 2 a		°C	≤10	≤11	≤11	≤11	≤11	≤11	≤12	≤12
Flash point	EN ISO 2592	°C	≥240	≥240	≥240	≥230	≥230	≥230	≥230	≥220
Solubility	EN 12592	%	≥99.0	≥99.0	≥99.0	≥99.0	≥99.0	≥99.0	≥99.0	≥99.0
Penetration index	Annex A b	–	–1.5 to +0.7							
Dynamic viscosity at 60°C	EN 12596	Pa·s	≥440	≥260	≥225	≥175	≥145	≥90	≥55	≥30
Breaking point (Fraass)	EN 12593	°C	≤–5	≤–5	≤–5	≤–7	≤–8	≤–10	≤–12	≤–15
Kinematic viscosity at 135°C	EN 12595	mm²/s	≥530	≥400	≥370	≥325	≥295	≥230	≥175	≥135
France			x		x				x	x
Belgium			x		x		x	x	x	x
The Netherlands			x					x		x
Germany				x			x	x	x	x
UK			x	x		x		x	x	x
Switzerland					x		x	x	x	x
Czech Republic			x	x	x		x	x	x	x
Poland			x		x		x	x	x	x

Table 5.2 Specifications for paving grade bitumens with penetrations from 250 × 0.1 mm to 900 × 0.1 mm: Tables 2A and 2B of EN 12591 combined (BSI, 2009a)

Property	Test method	Unit	250/330	330/430	500/650	650/900
Penetration at 25°C	EN 1426	0.1 mm	250–330	–	–	–
or						
Penetration at 15°C	EN 1426	0.1 mm	70–130	90–170	140–260	180–360
Dynamic viscosity at 60°C	EN 12596	Pa·s	≥18	≥12	≥7	≥4.5
or						
Softening point	EN 1427	°C	30–38	–	–	–
Resistance to hardening at 163°C	EN 12607-1					
Ratio of viscosities at 60°C		–	≤4.0	≤4.0	≤4.0	≤4.0
or						
Increase in softening point		°C	≤11	–	–	–
Change of mass (absolute value)		%	≤1.0	≤1.0	≤1.5	≤1.5
Flash point	EN ISO 2719	°C	≥180	≥180	≥180	≥180
Solubility	EN 12592	%	≥99.0	≥99.0	≥99.0	≥99.0
Breaking point (Fraass)	EN 12593	°C	≤−16	≤−18	≤−20	≤−20
or			NRᵃ	NRᵃ	NRᵃ	NRᵃ
Kinematic viscosity at 135°C	EN 12595	mm²/s	≥100	≥85	≥65	≥50
or			NRᵃ	NRᵃ	NRᵃ	NRᵃ

ᵃ NR: no requirement may be used, if there is no legal or other regional requirement for this property

Table 5.3 Specifications for soft paving grade bitumens with viscosity grading: Tables 3A and 3B of EN 12591 combined (BSI, 2009a)

Property	Test method	Unit	V1500	V3000	V6000	V12000
Kinematic viscosity at 60°C	EN 12595	mm²/s	1000–2000	2000–4000	4000–8000	8000–16 000
Resistance to hardening at 120°C	EN 12595					
Change of mass (absolute value)		%	≤2.0	≤1.7	≤1.4	≤1.0
Flash point	EN ISO 2719	°C	≥160	≥160	≥180	≥180
Solubility	EN 12592	%	≥99.0	≥99.0	≥99.0	≥99.0
Resistance to hardening at 120°C (thin-film oven test)	EN 12607-2					
Ratio of viscosities at 60°C			≤3.0 or NR[a]	≤3.0 or NR[a]	≤2.5 or NR[a]	≤2.0 or NR[a]

[a] NR: no requirement may be used, if there is no legal or other regional requirement for this property

- dynamic shear rheometer based on test method EN 14770:2012 (BSI, 2012b)
- deformation energy by force ductility based on test methods EN 13589:2008 (BSI, 2008) and EN 13703:2003 (BSI, 2003)
- tensile properties by the tensile test EN 13587:2010 (BSI, 2010b)
- elastic properties by the elastic recovery test EN 13398:2010 (BSI, 2010c).

It has been shown that the softening point of PMBs has a different meaning as in the case of penetration grade bitumens, as it is strongly influenced by the type and amount of polymer added to the bitumen. This also explains why the penetration index (when determined from the penetration and softening point) does not apply to PMBs.

While styrene–butadiene copolymers were the most widely used bitumen modifiers in the first decade of the twenty-first century, in recent years other modifiers have become increasingly popular in the quest to enhance bitumen performance. Examples include paraffin, amide waxes and recycled crumb rubber. Specifications for such modified bitumens are beginning to be developed, and the same test methods used for PMBs are likely to describe the performance of such binders adequately.

5.3 Oxidised bitumens

Oxidised bitumens are used almost entirely for industrial applications (roofing, flooring, mastics, pipe coatings, paints, etc.). They are specified and designated by reference to both the softening point and penetration tests: for example, 85/40 is an oxidised grade bitumen with a softening point of 85 ± 5°C and a penetration of 40 ± 5 dmm. Oxidised bitumens also have to comply with solubility, loss on heating and flash point criteria (BSI, 2009b).

The softening points of oxidised grades of bitumen are considerably higher than those of the corresponding penetration grade bitumens, and therefore the temperature susceptibility (i.e. the penetration index is high) is much lower, from +2 to +8.

5.4 Hard bitumens

Hard bitumens are used solely for industrial applications (coal briquetting, paints, etc.). They are specified by reference to both the softening point and penetration tests, but are designated by a softening point range only and the prefix H: for example, H80/90 is a hard grade bitumen with a softening point between 80 and 90°C. Hard bitumens also have to comply with solubility, loss on heating and flash point criteria (BSI, 2009c).

5.5 Bitumen quality

Over many years, Shell has investigated the relationship between laboratory measured properties of penetration grade bitumens and their performance in asphalts on the road. The ability to predict the long term behaviour of asphalts becomes more important as traffic loading has increased and performance requirements have become ever more demanding. Performance on the road depends on many factors, including the design, the nature of the application and the quality of the individual components. In volumetric terms, bitumen is a relatively minor component of an asphalt, but it has a crucial role – acting both as a durable binder and conferring visco-elastic properties on the asphalt.

Essentially, satisfactory performance of a bitumen on the road can be ensured if four properties are controlled:

- rheology
- cohesion
- adhesion
- durability.

5.5.1 Rheology

Although more sophisticated test methods have been developed in the last two decades, the rheology of penetration grade bitumen at service temperatures is still adequately characterised by the values of penetration and the penetration index (Appendix A in BSI 2009a). However, for assessing the rheology of more complex modified bitumens, testing with the dynamic shear rheometer (BSI, 2012b) has found its way into the standards (BSI, 2010a). It allows users to assess bitumen stiffness and its viscous and elastic

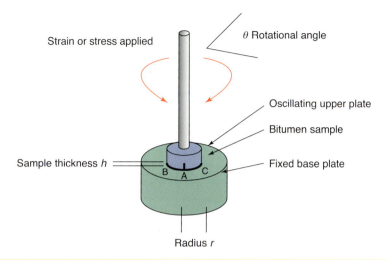

Strain or stress applied

θ Rotational angle

Oscillating upper plate

Bitumen sample

Sample thickness h

B A C

Fixed base plate

Radius r

Figure 5.1 The dynamic shear rheometer test

behaviour in a wide range of service temperatures. Figure 5.1 shows the core of the dynamic shear rheometer.

Rheology also plays a part when it comes to assessing the tendency of the bitumen to become brittle at low temperature. While the breaking point Fraass test (BSI, 2007c) has traditionally been used for this purpose, a more advanced test, the bending beam rheometer (BBR), is now available (BSI, 2012a).

The BBR (Bahia *et al.*, 1991) is a simple device that measures how much a beam of bitumen will deflect under a constant load at temperatures corresponding to its lowest pavement service temperature when bitumen behaves like an elastic solid. The creep load is intended to simulate the stresses that gradually increase in a pavement as the temperature falls.

Two parameters are determined in this test: the creep stiffness is a measure of the resistance of the bitumen to constant loading, and the creep rate is a measure of how the bitumen stiffness changes as loads are applied.

If the creep stiffness is too high, the asphalt will behave in a brittle manner, and cracking will be more likely. A high creep rate (sometimes denoted the *m* value) is desirable because, as the temperature changes and thermal stresses accumulate, the stiffness will change relatively quickly. A high value for the creep rate indicates that the bitumen will tend to disperse stresses that would otherwise accumulate to a level where low temperature cracking could occur.

Figure 5.2 shows a schematic diagram of the BBR test equipment.

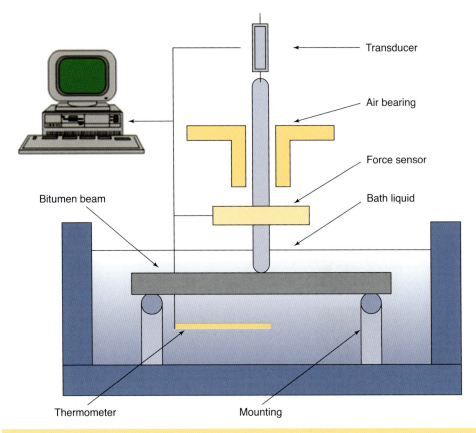

Transducer

Air bearing

Force sensor

Bitumen beam

Bath liquid

Thermometer   Mounting

Figure 5.2 The BBR test

5.5.2 *Cohesion*

The cohesive strength of bitumen is characterised either by the force ductility at low temperature (BSI, 2008) or by the tensile test (BSI, 2010b).

In the original ductility test (Figure 5.3), dumb-bell shaped specimens of bitumen are immersed in a water bath and stretched at a constant rate of 50 mm/min until fracture occurs. The distance the specimen is stretched before failure is reported as the ductility. The test temperature is adjusted, depending on the penetration of the bitumen under test (e.g. 10°C for 80–100 pen, 13°C for 60–70 pen and 17°C for 40–50 pen). Under these conditions, the test has been found to discriminate between bitumens of different cohesive strengths.

However, by recording not only the strain at fracture but also the energy required to stretch the sample during the entire test, as described in the force ductility test method (BSI, 2008), a much better discrimination of tensile and therefore cohesive behaviour of different bitumens can be obtained.

(a)

(b)

Figure 5.3 (a) Ductility test; (b) elastic recovery of 50/70 (top) and Shell Cariphalte® 45/80-50 (bottom)

For example, the energy recording of unmodified paving grade bitumens generally only shows one maximum at low strain, which is characteristic of the stiffness of the bitumen. Thereafter, the force required to stretch the sample further drops quickly to values close to zero. On the other hand, PMBs are often characterised in this test by two maxima, the first again showing the stiffness of the bitumen compound while the second, occurring at larger strains, typically 200–400 mm, points to the effect of a polymer network in the blend, providing additional resilience. Figure 5.4 shows examples of curves obtained in this test.

5.5.3 Adhesion
In contrast to rheology and cohesion, the adhesion of bitumen can be measured only in combination with a substrate. Numerous laboratory tests were developed in the last century, and many of these determine the surface

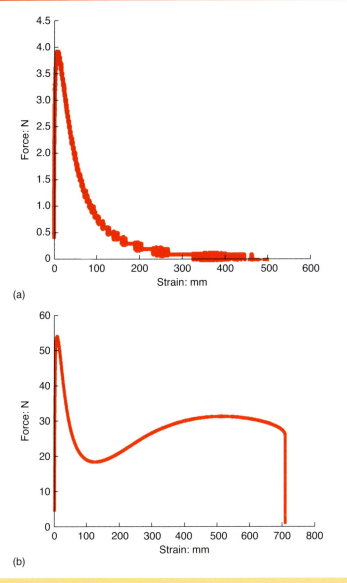

(a)

(b)

Figure 5.4 (a) Characteristic energy curve obtained for unmodified bitumen 30/45 (measured at 25°C); (b) characteristic energy curve obtained for PMB 25/55-65 (measured at 10°C)

coverage of mineral aggregates with bitumen or the stability of Marshall samples before and after storage under water.

The results of these laboratory tests together with observations of performance in practice have identified key links between functional properties and the constitution of the bitumen. This work has indicated that if the molecular weight distribution and chemical constitution of the bitumen is unbalanced, it can exhibit inhomogeneity that may adversely affect both the cohesive and adhesive properties of the bitumen (Van Gooswilligen *et al.*, 1989).

5.5.4 *Durability*

Durability can be defined as the ability to maintain satisfactory rheology, cohesion and adhesion in service. The term 'ageing' is often used to describe the change in bitumen properties during storage, mixing, laying and in service. The following four mechanisms have been identified as the prime durability factors

- oxidation
- evaporative hardening
- structural ageing
- exudation.

While, for penetration grade bitumens, ageing is synonymous with hardening, some PMBs may become softer under certain conditions.

5.5.4.1 *Oxidation and evaporative hardening*

Oxidation – reaction with oxygen from the air – is regarded as the main cause of ageing of bitumen. It is strongly accelerated by increasing temperature and increasing surface area. Also, the bitumen composition influences the rate of oxidation. It has been shown that asphaltene-rich bitumens tend to age quicker than bitumens with low asphaltene contents (Neumann, 1990). Finally, the reaction time determines the extent of oxidative ageing.

This is why the conditions during the preparation of the asphalt (i.e. at temperatures exceeding 120°C) need to be controlled, and the minimum mixing time and temperature should always be aimed at in order to minimise oxidation.

Looking at the mixture, high void contents or even interconnected voids (as in an uncompacted mix or in a porous asphalt) favour the access of oxygen. Therefore, surface courses are more prone to ageing in situ by oxidation than the lower layers of a road pavement.

It is difficult to distinguish between oxidation and evaporation in practice. Evaporation of volatile components, leaving behind a higher viscosity bitumen, can be significant at high temperatures, and may also play a role during the service life.

Tests on the bitumen

The rolling thin-film oven test (RTFOT) (BSI, 2007d) is an ageing test, and measures changes by both oxidation and evaporation. The apparatus for this test is shown in Figure 5.5. In this test, a thin film of bitumen is continuously

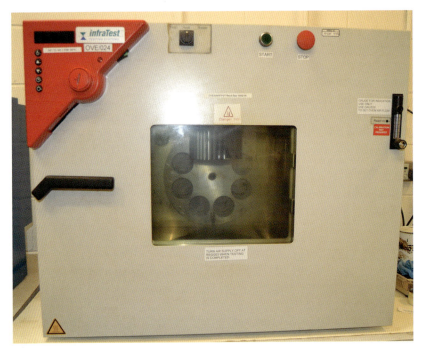

(a)

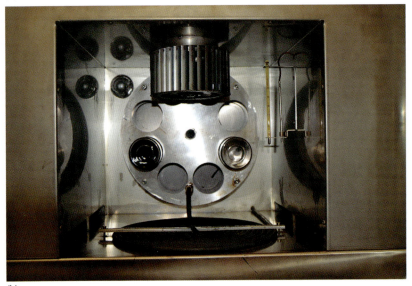

(b)

Figure 5.5 The rolling thin-film oven test: (a) external view of RTFOT equipment with control panel on door; (b) internal view of RTFOT equipment with partially loaded carousel with one empty glass jar and another with binder. (Photographs courtesy of PTS International Ltd.)

rotated around the inner surface of a glass jar at 163°C for 75 min, with an injection of hot air into the jar every 3–4 s. The amount of bitumen hardening during the test (i.e. the drop in penetration and the increase in the softening point) correlates strongly with that observed during the manufacture of an asphalt. As mentioned before, PMBs may show a decrease in the softening point, which is caused by the destruction of large polymer chains – these may then react with bitumen molecules, leading to a different structure (Vonk *et al.*, 1994).

Usually a change in mass of the sample before and after the test will be observed. A loss of mass always indicates volatile components in bitumen; however, due to reaction with oxygen, quite often the mass of the sample increases during the test.

After its introduction in the USA, the pressure ageing vessel test (PAV) has also become popular in Europe, being used to predict the changes in bitumen properties during the service life of the pavement (see Figure 5.6 and section 20.3). Again, the softening point and penetration are measured before and after the test, and changes are reported. The test is now standardised as a European Norm (BSI, 2012c).

Evaporative hardening without the interference of oxidation can be assessed by operating the RTFOT with nitrogen instead of hot air or, more directly, by measuring the volatility of the bitumen by true boiling point gas–liquid chromatography (TBP-GLC). In this test, a small sample of bitumen (150 mg) is dissolved in carbon disulfide and is separated on two chromatographic columns. The first column separates the heavy components such as asphaltenes and heavy polar aromatics from the bitumen. The hydrocarbons eluted from this column are then separated on a second column.

TBP-GLC is a rapid and accurate method of front-end volatility analysis, and is plotted on the Qualagon (described in section 5.5.5) as the percentage by mass recovered at 450 and 500°C, thus taking account of the shape of the volatility curve. The loss of 0.2% of the mass in the RTFOT correlates well with the TBP-GLC recovery limits at 450 and 500°C (Van Gooswilligen *et al.*, 1989).

Tests on the mixture
The hardening due to the oxidation and evaporation of a thin film of bitumen in contact with aggregate is assessed by two mixture tests: the hot mixture storage test (Van Gooswilligen *et al.*, 1989) and the change in the softening point of the bitumen during commercial asphalt manufacture. The hot mixture storage test simulates ageing conditions during mixing and hot storage. A prescribed mixture is manufactured in the laboratory, and a specified

(a)

(b)

(c)

(d)

Figure 5.6 The PAV test. (a) Front view of the equipment; (b) plan view; (c) the pressure vessel with cover open; and (d) the samples rack placed inside the pressure vessel binder. (Photographs courtesy of PTS International Ltd.)

quantity of this mixture is stored for 16 h at 160°C in a sealed tin. Thus, the volume of air entrained in the sample is known and is constant from test to test. The bitumen is recovered from both the mixed and stored material, and the penetration and softening points are determined from samples of bitumen recovered from these two materials. The ageing of the bitumen during mixing and storage is expressed as the difference between the softening point of the bitumen after storage and the softening point of the original bitumen.

At laboratory scale, this is a very severe test, and the change in the softening point is very much larger than that which would be found during actual bulk storage. Nevertheless, the test correlates with the hardening tendency of a mixture at high temperature when in prolonged contact with air. Again, the limited meaning of the softening point change of PMBs must be taken into account.

5.5.4.2 Structural ageing

There are hints that the structure of bitumen may slowly change during extended service life, leading to a system that is no longer a gel type but a sol. The effect is sometimes also called physical hardening, and is believed to be the result of the slow approach of the thermodynamic equilibrium of the bitumen. Asphaltenes are formed and successively no longer held in solution by the remaining maltenes. However, the results do not unambiguously prove that physical hardening is the sole cause of this process: oxidative ageing is also a suspect.

5.5.4.3 Exudative hardening

If the constitution of a bitumen is unbalanced, it may, when in contact with a porous aggregate, exude oily, less viscous components into the surface pores of the aggregate. Such constitutions can, for instance, appear when highly oxidised bitumens – either blown or strongly aged – are blended with light components, such as flux oils, to produce the desired viscosity of a fresh bitumen. This exudation process may result in a hardening of the bitumen film remaining on the surface of the aggregate. Exudation is primarily a function of the amount of low molecular weight components present in the bitumen relative to the amount and type of asphaltenes.

Shell Research has developed the exudation droplet test (Van Gooswilligen *et al.*, 1989) to measure quantitatively the exudation tendency of a bitumen. In this test, bitumen droplets are applied to the recesses in custom-made white marble plates. The plates are stored at 60°C for 4 days under a nitrogen blanket. During this period, oily rings develop around the bitumen droplet, which can be measured under ultraviolet light using a microscope. Ring widths vary from a few tenths of a millimetre for a balanced bitumen to several millimetres for an unbalanced bitumen.

Hardening in service as a result of exudation can be substantial, and depends not only on the exudation tendency of the bitumen but also on the porosity of the aggregate. If the aggregate possesses low porosity, the quantity of exudate absorbed is negligible, irrespective of the exudation tendency of the bitumen. Similarly, if the exudation tendency of the bitumen is low, the quantity of exudate absorbed will be negligible, irrespective of the porosity of the aggregate. However, highly porous aggregate and bitumen with a strong exudation tendency may lead to quicker hardening and ensuing crack formation of the asphalt pavement.

5.5.5 Qualagon tests and test criteria

By studying the correlation between field performance and the measured properties of experimental and commercial bitumens, Shell Research has developed a set of laboratory tests to assess the quality of a bitumen. The set

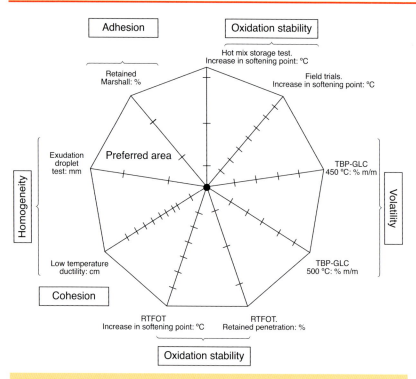

Figure 5.7 The penetration grade Qualagon (Van Gooswilligen *et al.*, 1989)

of tests includes six on the bitumen and three on the asphalt. For ease of assessment, these nine results are presented in the form of a regular polygon, called the bitumen Qualagon, depicted in Figure 5.7. There are tests within the Qualagon that cover the three remaining key performance elements, apart from rheology

- cohesion – low temperature ductility (see section 5.5.2)
- adhesion – the retained Marshall test
- durability
  - the RTFOT (oxidation stability)
  - the true boiling point – gas–liquid chromatography (volatility)
  - the exudation droplet test (homogeneity)
  - field trials (oxidation stability).

In the Qualagon, the adhesion characteristics of a bitumen are assessed by a retained Marshall test (Marshall Consulting and Testing Laboratory, 1949). In the test, eight Marshall specimens are manufactured using a prescribed aggregate, aggregate grading and bitumen content. The eight specimens are divided into two groups of four such that the average void content of the two groups is equal. One group of four is tested immediately using the standard Marshall test, and the remaining four specimens are vacuum treated under water at a temperature of between 0 and 1°C to saturate, as far as possible, the pore volume of the mixture with water. Subsequently, the specimens are stored in a water bath at 60°C for 48 h. The Marshall stability of these four specimens is then determined. The ratio of the Marshall stability of the treated specimens to the initial Marshall stability is termed the 'retained Marshall stability'.

Although a relatively wide scatter in absolute stability values and poor reproducibility are inherent in Marshall testing, it has been found that consideration of the percentage relative to the result from the standard procedure reduces the differences between results from different laboratories.

The oxidation stability of bitumens according to the field test is plotted on the Qualagon as the percentage retained penetration and the increase in softening point.

Many years' experience using the Qualagon in conjunction with rheological data have confirmed that the set of criteria on which it is based provides a satisfactory quantitative measure of bitumen quality and its performance in service.

Good quality bitumens generally have properties within the 'preferred' area of the Qualagon. However, this does not mean that a bitumen that is partially outside the preferred area is necessarily a poor bitumen or one of low quality. The Qualagon limits are not pass/fail criteria, and the whole figure must be interpreted with care and judgement. For this reason, the Qualagon is not suitable for specification purposes, but it is an invaluable assessment tool providing an excellent guide to performance in service.

5.6 The CEN bitumen specification

Harmonisation of European standards for petroleum products was set as a target by the European Committee for Standardisation (Comité Européen de Normalisation – CEN) during the mid-1980s to eliminate barriers to trade within the member states of the EU. One of the first steps in achieving this aim was the publication of the Construction Products Directive (CPD) in December 1988. The CPD was replaced by the Construction Products Regulation (CPR) in 2011. It requires that construction products used in member states must be fit for their intended use, satisfying certain essential requirements

■ mechanical resistance and stability
■ safety in case of fire
■ hygiene, compliance with health regulations and the environment
■ safety and accessibility in use
■ protection against noise

- energy, economy and heat retention
- sustainable use of natural resources.

These essential requirements have to be taken into account when drafting European standards (often described as 'ENs' – Euro Norms). The task of producing ENs has been entrusted to the BSI and, in turn, to the BSI technical committees (TCs) and working groups (WGs). It was the BSI working group dealing with paving grade bitumens, TC227 WG1, which proposed a pan-European specification that has resulted in the publication of EN 12591:2009 (BSI, 2009a). This is based on the traditional tests that are used for characterising bitumens. The specification, shown in Tables 5.1–5.3, includes a series of mandatory tests and properties that must be adopted by all members of the EU. It also includes optional properties that can be adopted or rejected by each national body.

EN specifications are routinely resubmitted every four years, thus the next version of EN 12591 is in preparation.

5.6.1 CE marking

The European Community decided to implement harmonisation of products for construction. Accordingly, asphalt binders were considered. The process began with the original Council Directive (89/106/CEE), and has now moved to Regulation 305/2011. The effect of this harmonisation is to define the minimum specifications that will be applied throughout the EU.

The direct consequence is the mandatory CE marking certification by manufacturers of binders.

In effect, this is a legally enforceable commitment to deliver a product with declared performance, such performance being defined in the framework of associated ENs.

In the case of bitumens, CE marking is obligatory for

- paving grade bitumens complying with EN 12591:2009 (BSI, 2009a)
- hard grade paving bitumen complying with EN 13924:2006 (BSI, 2006)
- PMB complying with EN 14023:2010 (BSI, 2010a).

For more detail on the CE marking of bitumens, refer to section 23.1.4.

5.7 The SHRP/Superpave bitumen specification

The Strategic Highway Research Program (SHRP), initiated in the USA in 1987, was a coordinated effort to produce rational specifications for bitumens and asphalts based on performance parameters. The motivation was to produce pavements that performed well in service. These pavements were subsequently called 'Superpave' (*superior performing pavements*).

One of the results of this work is the 'Superpave asphalt binder specification', which categorises grades of bitumen according to their performance characteristics in different environmental conditions. The specification was intended to limit the potential of a bitumen to contribute to deformation, fatigue failure and low temperature cracking of asphalt pavements. The specifications do not distinguish between unmodified and modified bitumens, although some grades can only be achieved through polymer modification of bitumen.

Table 5.4 shows the complete Superpave binder specification. It is intended to control deformation, low temperature cracking and fatigue in asphalt pavements. This is achieved by controlling various physical properties measured with the equipment described in this chapter. In this specification, the physical properties remain constant for all grades but the temperature at which these properties must be achieved varies, depending on the climate in which the binder is to be used. For example, a PG 52-40 grade is designed to be used in an environment where the average 7 day maximum pavement temperature is 52°C and the minimum pavement design temperature is −40°C. Maps of the USA have been prepared showing these upper and lower temperature regimes, thus facilitating binder selection.

5.7.1 Rotational viscometer
The rotational viscometer is specified to ensure that the viscosity of the bitumen at normal application temperatures is capable of being pumped and coating the aggregate, and will enable satisfactory compaction of the asphalt.

5.7.2 Direct tension test
The dynamic shear rheometer and the bending beam rheometer (BBR) (see section 5.5.1) provide information relating to the (creep) stiffness behaviour of bitumen over a wide range of temperatures. Although stiffness can be used to estimate failure properties, for modified binders the relation between stiffness and failure is less well defined. Accordingly, an additional test to measure the strength and strain at break has been included in the Superpave specifications: the direct tension test (ASTM, 2012).

Bitumens that can be stretched long distances before failure are termed 'ductile', and those that break before significant stretching has occurred are termed 'brittle'. Unfortunately, the BBR is unable to characterise fully the ability of some bitumens to stretch before failure. The direct tension test measures the ultimate tensile strain of a bitumen at low temperatures (between 0 and −36°C). A dumb-bell specimen is loaded in tension at a

Table 5.4 The complete Superpave binder specification; where G^*, Complex Shear Modulus; δ, phase angle; S, Stiffness modulus (Asphalt Institute, 1997)

	Performance grade																	
	PG 52						PG 58						PG 64					
	−10	−16	−22	−28	−34	−40	−46	−16	−22	−28	−34	−40	−10	−16	−22	−28	−34	−40
Average 7 day maximum pavement design temperature: °C	<52						<58						<64					
Minimum pavement design temperature: °C	>−10	>−16	>−22	>−28	>−34	>−40	>−46	>−16	>−22	>−28	>−34	>−40	>−10	>−16	>−22	>−28	>−34	>−40
Original binder																		
Minimum flash point: °C	230																	
Viscosity, ASTM D 4402																		
Maximum 3 Pa·s, test temperature: °C	135																	
Dynamic shear TP5:																		
$G^*/\sin\delta$, minimum, 1.00 kPa																		
Test temperature at 10 rad/s: °C	52						58						64					
Binder after the RTFOT																		
Mass loss, maximum, %	1.00																	
Dynamic shear TP5:																		
$G^*/\sin\delta$, minimum, 2.20 kPa																		
Test temperature at 10 rad/s: °C	52						58						64					
Binder after the PAV test																		
PAV ageing temperature, °C	90						100						100					
Dynamic shear TP5:																		
$G^*/\sin\delta$, maximum, 5000 kPa																		
Test temperature at 10 rad/s: °C	25	22	19	16	13	10	7	25	22	19	16	13	31	28	25	22	19	16
Physical hardening	Report																	
Creep stiffness, TP1:																		
S, maximum, 300 MPa or m-value, maximum, 0.30																		
Test temperature at 60 s: °C	0	−6	−12	−18	−24	−30	−36	−6	−12	−18	−24	−30	0	−6	−12	−18	−24	−30
Direct tension, TP3:																		
Failure strain, minimum, 1.0%																		
Test temperature at 1.0 mm/min: °C	0	−6	−12	−18	−24	−30	−36	−6	−12	−18	−24	−30	0	−6	−12	−18	−24	−30

constant rate, and the failure strain (the change in length divided by the original length) is determined. In this test, failure is defined by the stress where the load on the specimen reaches its maximum value. At this failure stress, the minimum strain at failure must be 1%.

References

Asphalt Institute (1997) *Superpave Performance Graded Asphalt Binder Specifica-tion and testing. Superpave Series No 1.* Asphalt Institute, Lexington, KY, USA.

ASTM (American Society for Testing and Materials) (2012) D6723-12. Standard test method for determining the fracture properties of asphalt binder in direct tension (DT). ASTM, West Conshohocken, PA, USA.

Bahia HU, Anderson DA and Christensen DW (1991) The bending beam rheometer: a simple device for measuring low-temperature rheology of asphalt binders. *Proceedings of the Association of Asphalt Paving Technologists.* Associ-ation of Asphalt Paving Technologists, Seattle, WA, USA, pp. 117–135.

BSI (British Standards Institution) (2003) BS EN 13703:2003. Methods of test for petroleum and its products. BS 2000-515. Bitumen and bituminous binders. Determination of deformation energy. BSI, London, UK.

BSI (2006) BS EN 13924:2006. Bitumen and bituminous binders. Specifications for hard paving grade bitumens. BSI, London, UK.

BSI (2007a) BS EN 1426:2007. Bitumen and bituminous binders. Determination of needle penetration. BSI, London, UK.

BSI (2007b) BS EN 1427:2007. Bitumen and bituminous binders. Determination of the softening point. Ring and Ball method. BSI, London, UK.

BSI (2007c) BS EN 12593:2007. Bitumen and bituminous binders. Determination of the Fraass breaking point. BSI, London, UK.

BSI (2007d) BS EN 12607-1. Bitumen and bituminous binders. Determination of the resistance to hardening under influence of heat and air. RTFOT method. BSI, London, UK.

BSI (2008) BS EN 13589:2008. Bitumen and bituminous binders. Determination of the tensile properties of modified bitumen by the force ductility method. BSI, London, UK.

BSI (2009a) BS EN 12591:2009. Bitumen and bituminous binders. Specifications for paving grade bitumens. BSI, London, UK.

BSI (2009b) BS EN 13304:2009. Bitumen and bituminous binders. Framework for specification of oxidised bitumen. BSI, London, UK.

BSI (2009c) BS EN 13305:2009. Bitumen and bituminous binders. Framework for specification of hard industrial bitumens. BSI, London, UK.

BSI (2010a) BS EN 14023:2010. Bitumen and bituminous binders. Specification framework for polymer modified bitumens. BSI, London, UK.

BSI (2010b) BS EN 13587:2010. Bitumen and bituminous binders. Determination of the tensile properties of bituminous binders by the tensile test method. BSI, London, UK.

BSI (2010c) BS EN 13398:2010. Bitumen and bituminous binders. Determination of the elastic recovery of modified bitumen. BSI, London, UK.

BSI (2012a) BS EN 14771:2012. Bitumen and bituminous binders. Determination of the flexural creep stiffness. Bending beam rheometer (BBR). BSI, London, UK.

BSI (2012b) BS EN 14770:2012. Bitumen and bituminous binders. Determination of complex shear modulus and phase angle. Dynamic shear rheometer (DSR). BSI, London, UK.

BSI (2012c) BS EN 14769:2012. Bitumen and bituminous binders. Accelerated long-term ageing conditioning by a pressure ageing vessel (PAV). BSI, London, UK.

BSI (2013) BS EN 15322:2013. Bitumen and bituminous binders. Framework for specifying cut-back and fluxed bituminous binders. BSI, London, UK.

Marshall Consulting and Testing Laboratory (1949) *The Marshall Method for the Design and Control of Bituminous Paving Mixtures*. Marshall Consulting and Testing Laboratory, Jackson, MS, USA.

Neumann HJ (1990) *Bitumen und seine Anwendung. Bitumen, Asphalt, Industriebitumen*. Expert Verlag, Renningen, Germany (in German).

Van Gooswilligen G, De Bats FTh and Harrison T (1989) Quality of paving grade bitumen – a practical approach in terms of functional tests. *Proceedings of the 4th Eurobitume Symposium*, Madrid, pp. 290–297.

Vonk W, Phillips MC and Roele M (1994) Alterungsbeständigkeit von Strassenbaubitumen – Vorteile der Modifizierung mit SBS. *Bitumen* 3/94: 98–104 (in German).

Chapter 6

Routine testing and mechanical properties of bitumens

Dr Burgard Koenders
Resource Manager, Shell Global Solutions International. The Netherlands

Bitumen is a complex material. Its response to stress is equally complex. The response of a bitumen to stress is dependent on both the temperature and the loading time. Thus, the nature of any bitumen test and what it indicates about the properties of a bitumen must be interpreted in relation to the nature of the material. A wide range of tests is performed on bitumens, from specification tests to more fundamental rheological and mechanical tests.

6.1 Standard specification tests for bitumens

As a wide variety of bitumens is manufactured, it is necessary to have tests to characterise different grades. In several countries, the two tests often used to specify different grades of bitumen are the needle penetration test and the softening point test. Although they are both empirical tests, it is possible to estimate important engineering properties from the results, including high temperature viscosity and the stiffness modulus. The use of the penetration test for characterising the consistency of bitumen dates from the late nineteenth century (Bowen, 1889).

As the tests have to be carried out under well defined conditions, standard methods of testing bitumen exist, as published by the Energy Institute (IP test methods), the American Society for Testing and Materials (ASTM) and the British Standards Institution (BSI), for example. In many cases, the methods are identical. However, some methods differ in detail, for example the BS EN (European standard) and ASTM softening point methods, and in these cases a factor is provided to relate the test results obtained using the different methods.

The majority of the methods quote limits for assessing the acceptability of test results. Limits of variability for results obtained by a single operator

The Shell Bitumen Handbook, Sixth Edition
ISBN 978-0-7277-5837-8
Shell International Petroleum Company Ltd: All rights reserved
http://dx.doi.org/10.1680/tsbh.58378.087

(repeatability) and by different operators in different laboratories (reproducibility) are specified. Thus, tolerance is given to allow for differences between operators and equipment at different locations.

6.1.1 The penetration test

The consistency of a penetration grade or oxidised bitumen is measured by the penetration test (e.g. according to ASTM D5 (ASTM, 2013a) or EN 1426 (BSI, 2007a)). In this test, a needle of specified dimensions is allowed to penetrate a sample of bitumen, under a known load at a fixed temperature for a known time. The test apparatus is shown in Figure 6.1.

The penetration is expressed as the distance in decimillimetres (1 dmm = 0.1 mm) that a standard needle will penetrate vertically into a sample of bitumen under specified conditions of temperature, load and load duration. Usually, the applied load is 100 g, the duration of loading 5 s and the test temperature 25°C. For each test, after specified conditioning at the test temperature, three individual measurements of penetration are taken. The average of the three values is recorded to the nearest integer. The recorded penetration is reported if the difference between the individual three measurements does not exceed a specified limit.

Figure 6.1 The needle penetration test

The lower the value of penetration, the harder the bitumen. Conversely, the higher the value of penetration, the softer the bitumen. This test is the basis on which penetration grade bitumens are classified into standard penetration ranges.

The specifications for paving grade bitumens and hard paving grade bitumens are provided in EN 12591 (BSI, 2009) and EN 13924 (BSI, 2006), and are discussed in detail in Chapter 5.

It is essential that the test methods are followed precisely, as even a slight variation can cause large differences in the result. The most common errors are

- poor sampling and sample preparation
- badly maintained apparatus and needles
- incorrect temperature and timing.

Temperature control is critical, to within $\pm 0.1°C$. Needles must be checked regularly for straightness, correctness of profile and cleanliness. Automatic timing devices are also necessary for accuracy, and these must be checked regularly. Penetration values less than 2 dmm and greater than 500 dmm cannot be determined with accuracy with this equipment. Soft bitumens require longer needles and deeper cups. Very soft bitumens are often better characterised in terms of viscosity. If the penetration value is less than 50 dmm, the repeatability for the penetration test is 2 dmm.

6.1.2 The softening point test

The softening point is another property commonly used to determine the consistency of a penetration grade or oxidised bitumen (e.g. according to ASTM D36 (ASTM, 2012a) or EN 1427 (BSI, 2007b) or IP 58 (EI, 2007a)). In this test, a small steel ball is placed on a sample of bitumen contained in a brass ring, and the set up is then suspended in a bath (in the form of a glass beaker) containing water or glycerine. The apparatus is shown in Figure 6.2.

Water is used for bitumen with a softening point of 80°C and below, and glycerine is used for bitumen with a softening point above 80°C. When water is used, the initial temperature of the bath liquid is 5°C. The initial temperature of the bath with glycerine is 30°C. The bath temperature is raised at 5°C per minute, and the bitumen softens and eventually deforms slowly with the ball through the ring. The softening point temperature is the temperature indicated by the thermometer at the instant the bitumen surrounding the ball touches the bottom plate 25 mm below the ring. The test is performed twice, and the average of the two measured temperatures is reported to the nearest 0.2°C for softening points below or equal to 80°C and 0.5°C for softening

Figure 6.2 The softening point test

points above 80°C. If the difference between the two results exceeds 1°C for softening points below 80°C or exceeds 2°C for softening points above 80°C, the test must be repeated. The reported temperature is designated the softening point of the bitumen. In the ASTM D36 procedure for the softening point test, the liquid in the bath is not stirred whereas in the IP 58 and EN 1427 procedure the water or glycerine is stirred. Consequently, the softening point determined by these two methods differs. The ASTM results are generally around 1.5°C higher than those obtained with the IP and EN methods (Krom, 1950).

As with the penetration test, the procedure for carrying out the softening point test must be followed precisely in order to obtain accurate results (Pfeiffer, 1950). Dimensions of the ring and the ball, the sample preparation, the rate of heating and the accuracy of temperature measurement are all critical (Pfeiffer, 1950). Automatic softening point instruments are available, and these ensure close temperature control and automatically record the result at the end of the test.

The consistency of bitumen at the softening point temperature was determined in terms of penetration value by Pfeiffer and Van Doormaal (1936). Using a specifically prepared extra long penetration needle, they found a value of 800 dmm for many, but not all, bitumens. The exact value was found to vary with the penetration index (PI) and the wax content. It has also been demonstrated by direct measurement that the viscosity at the softening point temperature of the majority of bitumens is about 1300 Pa·s (13 000 poise) (Heukelom, 1973).

6.1.3 The Fraass breaking point test

The Fraass breaking point test (Fraass, 1937) is one of very few tests that can be used to describe the behaviour of bitumens at very low temperatures (as low as $-30°C$). It is essentially a research tool that determines the temperature at which bitumen reaches a critical stiffness, and cracks. A number of countries with very low winter temperatures have specified maximum allowable Fraass temperatures for individual grades of bitumen.

In the Fraass test, shown in Figure 6.3, a steel plate 41 mm long, 20 mm wide and 0.15 mm thick coated with bitumen at a uniform thickness is subjected to a constant cooling rate and flexed repeatedly until the bitumen layer breaks (IP 80 (EI, 2007b) and EN 12593 (BSI, 2007c)). The temperature is reduced at a rate of 1°C per minute until the bitumen reaches a critical stiffness, and cracks. The temperature at which the sample cracks is termed the Fraass breaking point, and represents an equi-stiffness temperature. It has been shown that, at fracture, the bitumen has a stiffness modulus of 2.1×10^9 Pa, which is approaching the maximum stiffness modulus of 2.7×10^9 Pa (Thenoux et al., 1987). The Fraass temperature can be predicted from the penetration and the softening point for penetration grade bitumens because it is equivalent to the temperature at which the bitumen has a penetration of 1.25 dmm (Heukelom, 1969).

Figure 6.3 The Fraass breaking point test

6.2 Temperature susceptibility – the penetration index (PI)

All bitumens become softer when heated, and harden when cooled. One of the best known equations is that describing the temperature susceptibility of the penetration of a bitumen (Pfeiffer and Van Doormaal, 1936). If the logarithm of penetration, log pen, is plotted against temperature T, a straight line is obtained such that

$$\log \text{pen} = AT + K$$

where A is the temperature susceptibility of the logarithm of the penetration and K is a constant.

The value of A varies from about 0.015 to 0.06, showing that there may be a considerable difference in temperature susceptibility. Pfeiffer and Van Doormaal developed an equation for the temperature susceptibility that assumes a value of about zero for road bitumens (Pfeiffer and Van Doormaal, 1936; Van der Poel, 1954). For this reason, they defined the PI as

$$\frac{20 - \text{PI}}{10 + \text{PI}} = 50A$$

or, explicitly,

$$\text{PI} = \frac{20(1 - 25A)}{1 + 50A}$$

The value of the PI ranges from around -3 for highly temperature-susceptible bitumens to around $+7$ for highly blown low temperature susceptible (high PI) bitumens. The PI is an unequivocal function of A, and hence it may be used for the same purpose. The values of A and the PI can be derived from penetration measurements at two temperatures, T_1 and T_2, using the equation

$$A = \frac{\log \text{pen at } T_1 - \log \text{pen at } T_2}{T_1 - T_2}$$

The consistency at the softening point can be expressed in terms of penetration, both by linear extrapolation of the logarithm of the penetration versus temperature and by direct measurement with an extra long penetration needle at the ASTM softening point temperature. Pfeiffer and Van Doormaal found that most bitumens had a penetration of about 800 dmm at the ASTM softening point temperature (Pfeiffer and Van Doormaal, 1936). Replacing T_2 in the above equation by the ASTM softening point temperature (SP) and the penetration at T_2 by 800, they obtained the equation

$$A = \frac{\log \text{pen at } T_1 - \log 800}{T_1 - \text{SP}}$$

Substituting this equation in the equation for the PI and assuming a penetration test temperature of 25°C gives

$$PI = \frac{1952 - 500 \log pen - 20SP}{50 \log pen - SP - 120}$$

The assumption of a penetration value of 800 dmm at the softening point temperature is not valid for all bitumens. It is therefore advisable to calculate the temperature susceptibility using the penetration at two temperatures, T_1 and T_2.

The nomographs shown in Figures 6.4 and 6.5 enable the approximate value of the PI to be deduced from either the penetration at 25°C and the softening point temperature, or the penetration of the bitumen at two different temperatures. Due to the spread of the actual value of penetration at the softening point temperature, the value of the PI calculated from one penetration and one softening point may vary from the precise value calculated from two penetration values. However, since the penetration at 25°C and softening point are generally determined for bitumen specification control, it is normally the case that these properties are used.

One drawback of the PI system is that it uses the change in bitumen properties over a relatively small range of temperatures to characterise bitumen. Extrapolations to extremes of temperature can sometimes be misleading. The PI can be used to give a good approximation of the behaviour to be expected, but confirmation using stiffness or viscosity measurements is desirable.

6.3 Viscosity

Viscosity is a physical property of a fluid, and is a measure of its resistance to flow. Viscosity is defined as the ratio between the applied shear stress and the rate of shear strain. Various types of viscometers and rheometers are used to determine the flow behaviour under different conditions.

Viscometers often used for bituminous products are the cup viscometer, the capillary viscometer and the rotational viscometer. These are discussed below.

A fundamental method of measuring viscosity is using the sliding plate viscometer in which the shear stress (Pa) and the rate of strain (s^{-1}) on a thin film of bitumen between two parallel flat plates are determined. Dynamic shear rheometers are used to characterise the visco-elastic behaviour of bitumens. The fundamentals of viscosity and rheology are considered in more detail in Chapter 7.

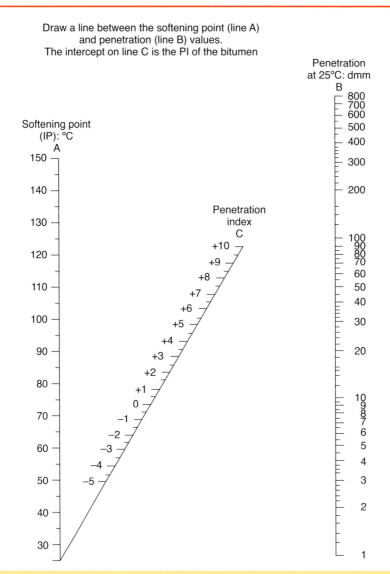

Draw a line between the softening point (line A)
and penetration (line B) values.
The intercept on line C is the PI of the bitumen

Figure 6.4 Nomograph for the PI (softening point/pen)

6.3.1 Cup viscometers

The use of a cup viscometer is a simple method for determining the viscosity at a given temperature. A metal cup is filled with the material at a standard temperature, and the time is recorded in seconds for a standard volume of material to flow out through the orifice in the bottom of the cup. There are several cup viscometers available, which differ mainly in the size of opening through which the material is drained. Such viscosity tests are often used for

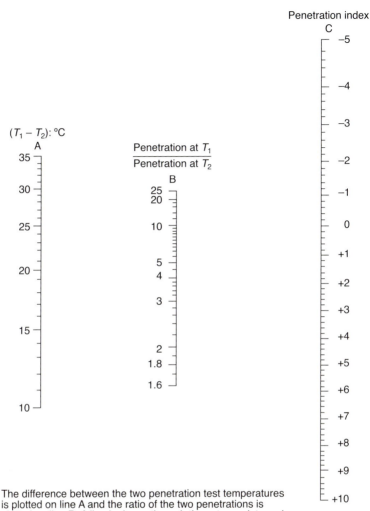

The difference between the two penetration test temperatures is plotted on line A and the ratio of the two penetrations is plotted on line B. A line is drawn through these two points and where it intercepts line C is the PI of the bitumen.

Figure 6.5 Nomograph for the PI (pen/pen)

emulsions (EN 12846-1 (BSI, 2011a)) and cut-back bitumens (EN 12846-2 (BSI, 2011b)).

The standard tar viscometer is shown in Figure 6.6.

The dynamic (absolute) viscosity (Pa·s) is given by

$$\eta = \text{flow time (in s)} \times \text{density} \times \text{constant}$$

The test results may also be expressed as the kinematic viscosity, ν (mm^2/s)

$$\nu = \text{flow time} \times \text{constant}$$

Figure 6.6 Standard tar viscometer

In the above equations, the value of the constant depends on the instrument used. Conversion factors for several cup viscometers are provided in Appendix 2.

6.3.2 Capillary viscometers

The flow conditions are much better defined if the opening or orifice of the cup viscometer is replaced by a long narrow tube or capillary. Capillary viscometers are essentially narrow glass tubes through which the bitumen flows. The tube has narrow and wide sections, and is provided with two or more marks to indicate a particular volume or flow, as shown in Figures 6.7 and 6.8.

The value of the kinematic viscosity is measured by timing the flow of bitumen through a glass capillary viscometer at a given temperature. Each viscometer is calibrated, and the product of the flow time and the viscometer calibration factor gives the kinematic viscosity (in mm^2/s).

ASTM D2170 (ASTM, 2010a) and EN 12595 (BSI, 2007d) specify procedures to ascertain the kinematic viscosity, which is often determined at a temperature of 135°C. Measurements of kinematic viscosity at a number of different temperatures are used to obtain temperature/viscosity curves.

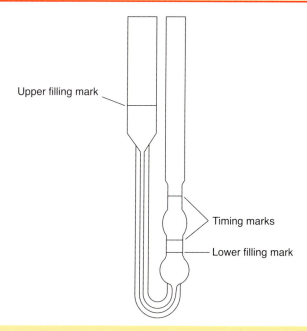

Figure 6.7 U tube reverse flow viscometer

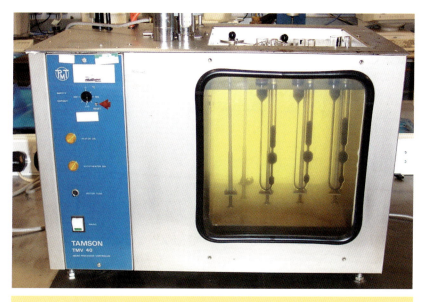

Figure 6.8 Kinematic viscosity test equipment

For specifications, the dynamic viscosity at a temperature of 60°C is often required, and the test is carried out using a vacuum capillary viscometer in accordance with ASTM D2171 (ASTM, 2010b) and EN 12596 (BSI, 2007e).

6.3.3 Rotational viscometers

Rotational viscometers are based on the concept that the torque required to rotate an object in a fluid at a certain speed is a function of the viscosity of that fluid. Rotational viscometers are normally used to determine the viscosity of bitumens at application temperatures (ASTM D4402 (ASTM, 2013b) and EN 13302 (BSI, 2010)). An example of a Brookfield rotational viscometer with a Thermosel system is shown in Figure 6.9. Essentially, the device consists of a thermostatically controlled chamber containing a sample of hot bitumen. The spindle is lowered into the bitumen and rotated. The torque

Figure 6.9 Rotational viscometer

required to rotate the spindle is measured, and converted into the viscosity of the bitumen (in Pa·s).

The rotational viscosity of bitumen is usually determined at 135 or 150°C, but with this type of apparatus the viscosity can be determined over a relatively wide range of temperatures (i.e. between 120 and 180°C).

6.4 The bitumen test data chart

In the late 1960s, Heukelom developed a system that permitted penetration, softening point, Fraass breaking point and viscosity data to be described as a function of temperature on one chart (Heukelom, 1969). This is known as the 'bitumen test data chart' (BTDC). The chart consists of one horizontal scale for the temperature and two vertical scales for the penetration and viscosity. The temperature scale is linear, the penetration scale is logarithmic and the viscosity scale has been devised so that penetration grade bitumens with 'normal' temperature susceptibility or penetration indices give straight line relationships. A typical BTDC is shown in Figure 6.10.

The BTDC shows how the viscosity of a bitumen depends on temperature, but it does not take account of the loading time. However, as the loading times for penetration, softening point and Fraass breaking point tests are similar, these test data can be plotted with viscosity test data on the BTDC, as shown in Figure 6.10. As the test results on this chart form a straight line relationship, it is possible to predict the temperature/viscosity characteristics of a penetration grade bitumen over a wide range of temperatures using only the penetration and softening point.

During the production and compaction of an asphalt, it is important to know the bitumen viscosities. This is illustrated in Figure 6.11 for a dense asphalt concrete manufactured using 200 pen bitumen. If the viscosity of the bitumen is too high during mixing, the aggregate will not be properly coated, whereas if the viscosity is too low, the bitumen will coat the aggregate easily but may drain off the aggregate during storage or transportation. For satisfactory coating, the viscosity should be approximately 0.2 Pa·s.

During compaction, if the viscosity is too low, the mixture will be excessively mobile, resulting in pushing of the material in front of the roller. High viscosities will significantly reduce the workability of the mixture, and little additional compaction will be achieved. It is widely recognised that the optimum bitumen viscosity for compaction is between 2 and 20 Pa·s.

Thus, the BTDC is a useful tool for ensuring that the correct operating temperatures are selected to achieve the appropriate viscosity for any grade of bitumen. Consideration of the viscosity requirements during asphalt production and laying has led to the operating temperatures given in

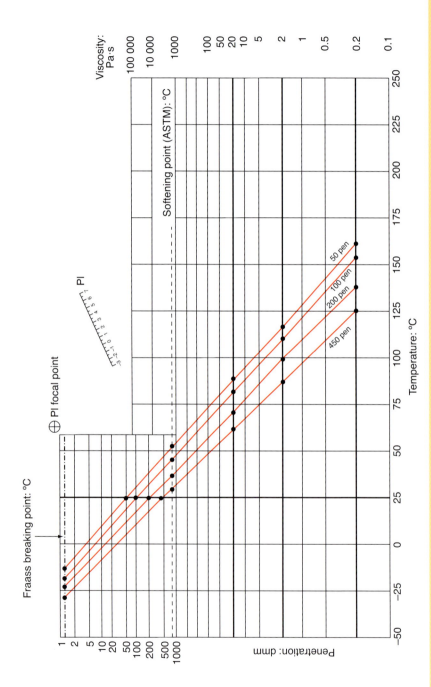

Figure 6.10 BTDC comparing penetration grade bitumens manufactured from one crude

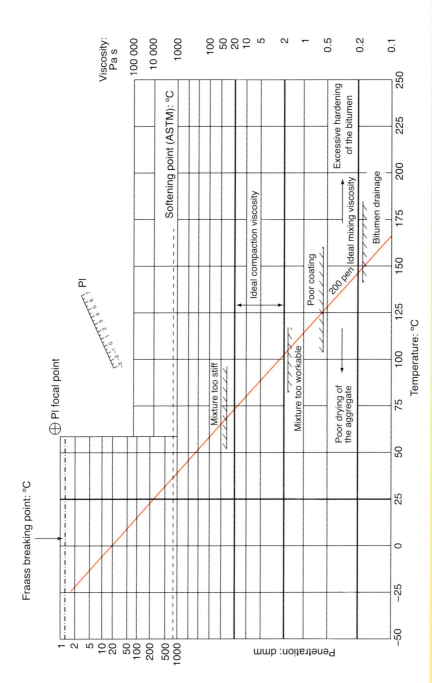

Figure 6.11 BTDC showing bitumen viscosities for optimal mixing and compaction of a dense asphalt concrete

Figure 6.11. Particular mixtures or circumstances may dictate other operating temperatures.

The chart can also be used for comparing the temperature/viscosity characteristics of different types of bitumen. Three classes of bitumen can be distinguished using the BTDC: class S, class B and class W bitumens.

6.4.1 Class S bitumens

The test data for a large group of bitumens can be represented by straight lines on the BTDC, within the repeatability of the test. This group, which has been designated class S ('straight line'), comprises penetration bitumens of different origins with limited wax content. Figure 6.10 shows a chart with straight lines for a number of different penetration grade bitumens manufactured from one base crude. The lines move towards the left of the chart as the bitumens become softer. However, their slopes are equal, indicating that their temperature susceptibilities are similar. Bitumens with the same penetration at 25°C but having different origins are shown in Figure 6.12. The origin may influence the temperature susceptibility, which is reflected by the slope of the line. Accordingly, the temperature/viscosity characteristics of S-type bitumens may be determined from their penetration and softening point only.

6.4.2 Class B bitumens

The test data of class B ('blown') bitumens give curves on the chart as shown in Figure 6.13. The curves can be represented by two intersecting straight lines. The slope of the line in the high temperature range is about equal to that of an unblown bitumen of the same origin, but the line in the lower temperature range is less steep. Physically, there is no transition point, but it is very convenient that they are still straight lines in the penetration and viscosity regions. Each of them can be characterised with two test values. Thus, in all, four tests are required for a complete description: penetration, softening point and two high temperature viscosity measurements.

6.4.3 Class W bitumens

Class W ('waxy') bitumens also give curves consisting of two straight lines; however, they are different from those of blown bitumens. The two branches of the curve give slopes that are similar but are not aligned. Figure 6.13 shows an example of an S-type bitumen together with a curve for a similar bitumen with a wax content of 12%. At low temperatures, when the wax is crystalline, there is hardly any difference between the two curves. At higher temperatures, where the wax is molten, the curve for the waxy bitumen is significantly lower down on the chart. Between the two straight branches, there is a transition range in which the test data are scattered because the

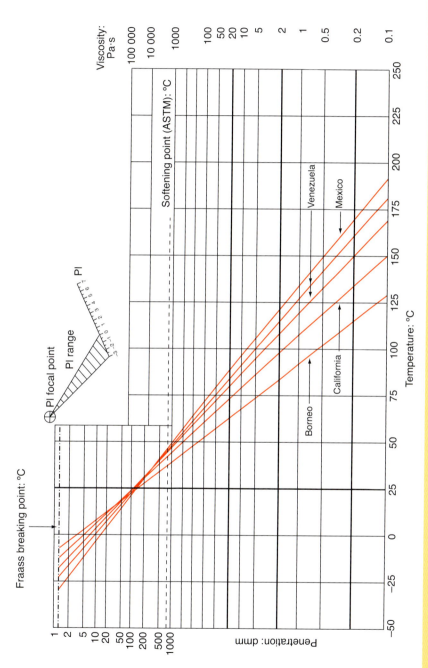

Figure 6.12 BTDC comparing several 100 pen bitumens manufactured from different crudes

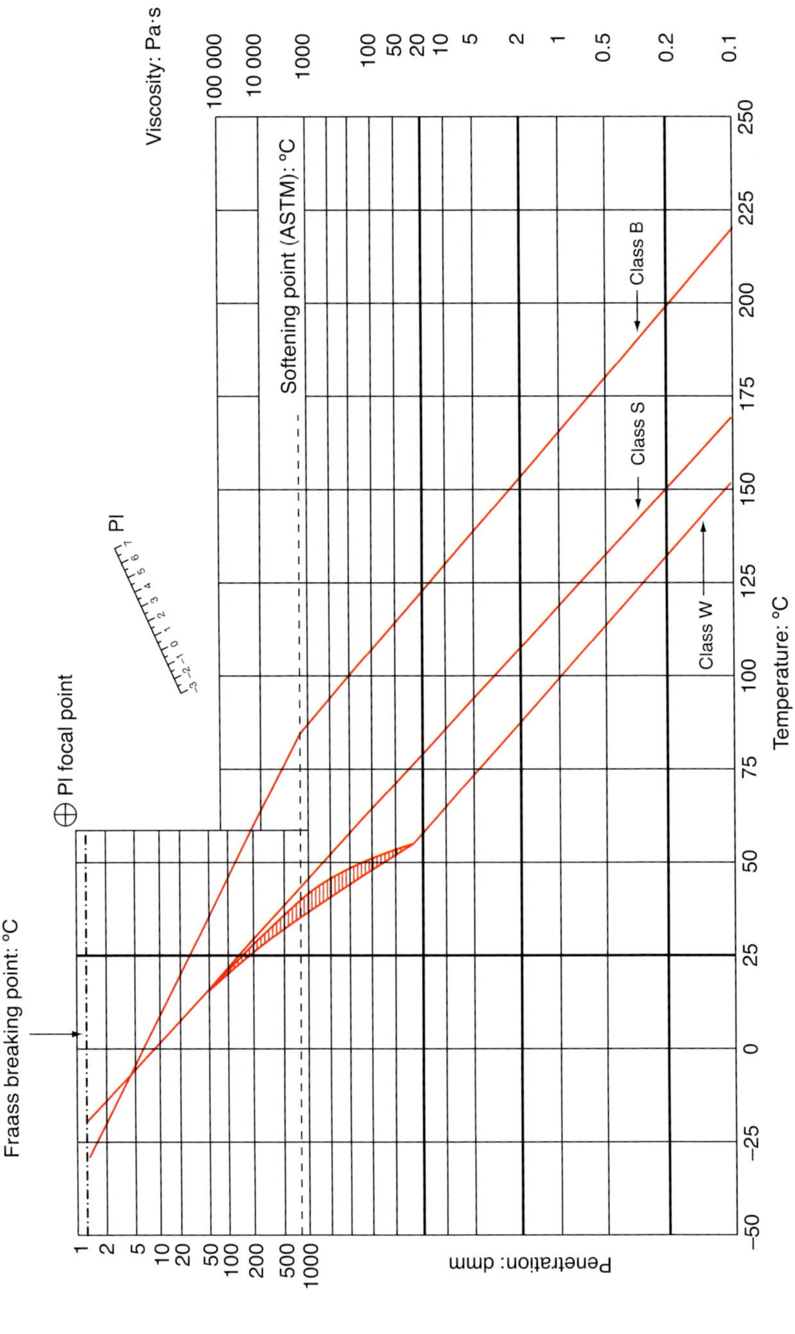

Figure 6.13 BTDC comparing class S, B and W bitumens

thermal history of the sample influences the viscosity result obtained over this range of temperatures.

6.5 Engineering properties of bitumen

The use of bitumen as such, or as a binder in diverse composite materials, is based on its mechanical properties and on the way these properties depend on the loading time and temperature. Bitumens are manufactured in a variety of grades with different mechanical properties to meet the demands of road and industrial applications. For the civil engineer and the industrial bitumen user, it is of considerable value to define the mechanical properties of bitumen in terms that are analogous to the elastic moduli of rigid materials.

For the structural design of pavements, it is essential to be able to measure and predict the fundamental mechanical characteristics of paving materials (Bonnaure *et al.*, 1977). The stiffness modulus of an asphalt depends on the stiffness modulus of the bitumen and the volume fractions of the different mixture components (aggregates, bitumen, air). The stiffness modulus of bitumen is also a suitable parameter for the breaking properties of the bitumen itself and of the asphalts prepared with it (Heukelom, 1966).

6.5.1 The concept of stiffness modulus

A viscous material is one that is semi-fluid in nature. When stressed, it will deform or tend to deform, any deformation being permanent because it is not recovered when the loading is removed. Elastic materials also deform or tend to deform when stressed, but, when the loading is removed, any deformation is fully recovered. Bitumens are visco-elastic materials. The degree to which their behaviour is viscous and elastic is a function of both temperature and the period of loading (usually referred to as the 'loading time'). At high temperatures or long loading times they behave as viscous liquids, whereas at very low temperatures or short loading times they behave as elastic (brittle) solids. The intermediate range of temperature and loading times, more typical of conditions in service, results in visco-elastic behaviour.

In order to define the visco-elastic properties, the concept of the stiffness modulus as a fundamental parameter to describe the mechanical properties of bitumens by analogy to the elastic modulus of solids was introduced (Van der Poel, 1954). If a tensile stress σ is applied at a loading time $t = 0$, a strain ε is instantly attained that does not increase with the loading time. The elastic modulus E of the material is expressed by Hooke's law as stress divided by strain.

In the case of visco-elastic materials such as bitumen, a tensile stress σ applied at a loading time $t = 0$ causes a strain ε that increases, but not proportionately, with the loading time. The stiffness modulus S_t at a loading time t is defined as the ratio between the applied stress and the resulting strain at the loading time t. It follows from the above that the value of the stiffness modulus is dependent on the temperature and the loading time that is due to the special nature of bitumen. Consequently, it is necessary to state both the temperature T and the loading time t of any stiffness modulus measurement

$$S_{t,T} = \frac{\sigma}{\varepsilon_{t,T}}$$

The methods used to measure the stiffness modulus of bitumen are often based on shear deformations. The resistance to shear is expressed in terms of the shear modulus G, which is defined as

$$G = \frac{\text{shear stress}}{\text{shear strain}}$$

The elastic modulus and shear modulus are related by

$$E = 2(1 + \mu)G$$

where μ is Poisson's ratio. The value of μ depends on the compressibility of the material, and may be assumed to be 0.5 for almost incompressible pure bitumens, while values of <0.5 have to be considered for asphalts. Thus,

$$E \sim 3G$$

In the static creep test, a constant load is applied, and the resulting deformation is then measured as a function of the loading time. The deformations at loading times from about 1 to 10^5 s or longer can be measured.

In dynamic tests, the shear stress is usually applied as a sinusoidally varying stress of constant amplitude and fixed frequency. The deformation of the material under test also varies sinusoidally with the same frequency as the applied stress. Tests are carried out at various values of frequency, and the ratio of the stress to strain can be plotted against the inverse of the angular frequency. The degree of elasticity of the bitumen under the test conditions is given by how much the deformation response is out of phase from the applied shear stress, and this is referred to as the phase angle. For purely elastic materials the phase angle is zero. For purely viscous materials the phase angle is 90°.

In analogy to the elastic modulus E and the shear modulus G, the stiffness modulus of bitumen can be given as (Van der Poel, 1954)

$$S_E \sim 3S_G$$

The rheology aspects and the experimental determination of stress–strain relationships for bitumen are considered in more detail in Chapter 7.

By combining creep tests with dynamic tests, a considerable range of stiffness moduli and loading times can be covered. It appears that the stiffness–loading time curves obtained at different temperatures for one grade of bitumen all have the same shape, and, if shifted along the loading time axis, would coincide.

The effect of changes in the temperature and the loading time on the stiffness modulus of three different bitumens is shown in Figures 6.14, 6.15 and 6.16.

Figure 6.14 shows a bitumen of low PI, −2.3. At very short loading times, the stiffness modulus is virtually constant, asymptotic towards 2.5×10^9 to 3.0×10^9 Pa, and is, in this region, largely independent of the temperature and the loading time (i.e. $S = E$). The effect of the PI is clearly illustrated by comparing Figures 6.14 and 6.15. The bitumen with the higher PI (+5) (Figure 6.15) is considerably stiffer at higher temperatures and longer loading times (i.e. it is less temperature susceptible).

Figure 6.16 shows the relationship for a 100 pen bitumen with a PI of −1.0. At a loading time of 0.02 s (which equates to a vehicle speed of around 50 km/h) the stiffness modulus is approximately 10^7 Pa at 25°C, but falls to approximately 5×10^4 Pa at 60°C. At low temperatures, the stiffness modulus is high, and therefore permanent deformation does not occur. However, at higher temperature or longer loading times (slow moving or

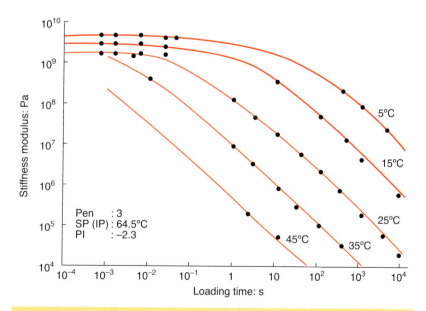

Figure 6.14 The effect of the temperature and the loading time on the stiffness modulus of a low-PI bitumen

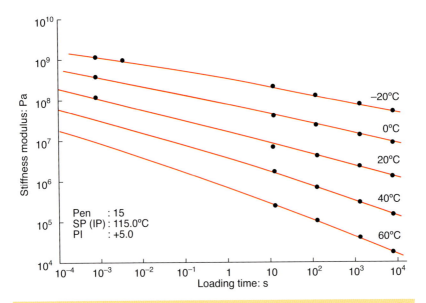

Figure 6.15 The effect of the temperature and the loading time on the stiffness modulus of a 115/15 bitumen

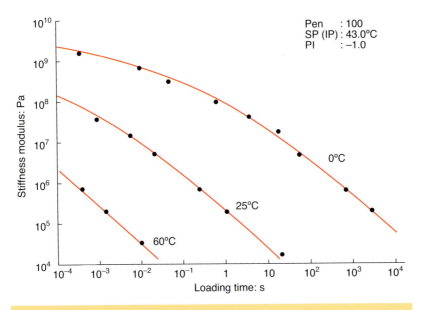

Figure 6.16 The effect of the temperature and the loading time on the stiffness modulus of a 100 pen bitumen

stationary traffic), the stiffness modulus is substantially reduced and, under these conditions, permanent deformation of the road surface is much more likely to occur.

In order to appreciate fully the significance of the stiffness modulus and its measurement, it is necessary to consider the deformation under stress of simple solids and liquids. The deformation behaviour of visco-elastic materials can then be derived.

6.5.2 Prediction of the stiffness modulus

If direct measurement of the stiffness modulus is not feasible, it can be predicted using the Van der Poel nomograph (Heukelom, 1973; Van der Poel, 1954). Van der Poel showed that two bitumens of the same PI at the same loading time have equal stiffness moduli at temperatures that differ from their respective softening points by the same amount. Over 40 bitumens were tested with PI values varying from +6.3 to −2.3 at many temperatures and frequencies, using both creep tests and dynamic tests. From the test data, Van der Poel produced a nomograph from which, using only penetration and the softening point, it is possible to predict the stiffness modulus of a bitumen over a wide range of conditions of temperature and loading times. The nomograph allows the prediction of the stiffness modulus within a factor of 2, which is considered of practical value in relation to the wide range of stiffness moduli. Figure 6.17 shows a Van der Poel nomograph with the stiffness modulus determined for a 40/60 pen bitumen at a loading time of 0.02 s and a test temperature of 5°C.

6.5.3 Elongation at break

In the characterisation of bitumen (and asphalt) as a construction material, essential factors are the resistance to deformation (rheology) and the permissible deformation (fracture). The stiffness modulus, relating stress and strain, was also used in explaining the breaking properties of bitumen (Heukelom, 1966). Two tests often used in this context are the Fraass breaking point test, to determine the temperature at which a bitumen obtains a certain degree of brittleness, and the ductility test, to determine the elongation at break at a fixed temperature. However, the results obtained in these tests cannot be compared directly because they are measures of different parameters (i.e. temperature and distance). The fracture of bitumen occurs after a certain deformation, the magnitude of which depends on the temperature and the loading time or rate of deformation. It was assumed that there is a relationship between fracture and rheological behaviour at the instant of failure. The rheological behaviour can be expressed in terms of the amounts of elastic strain, delayed elastic strain and viscous strain. Analysing the data from ductility tests, tensile tests and bending tests, it was shown that the

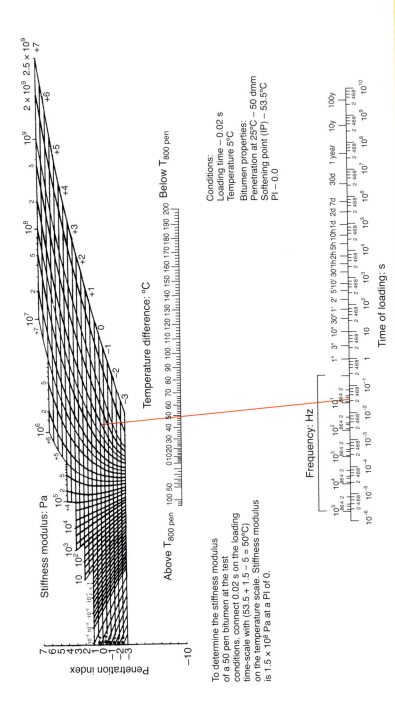

To determine the stiffness modulus of a 50 pen bitumen at the test conditions, connect 0.02 s on the loading time-scale with (53.5 + 1.5 − 5 = 50°C) on the temperature scale. Stiffness modulus is 1.5 × 10⁸ Pa at a PI of 0.

Conditions:
Loading time − 0.02 s
Temperature 5°C
Bitumen properties:
Penetration at 25°C − 50 dmm
Softening point (IP) − 53.5°C
PI − 0.0

Figure 6.17 Nomograph for determining the stiffness modulus of bitumens

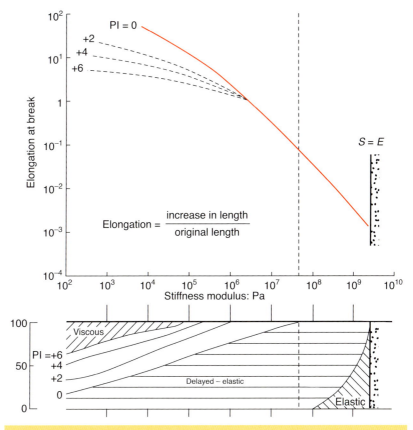

Figure 6.18 Elongation at break

stiffness modulus is the main parameter, and the PI is only significant at low stiffness modulus when the viscous deformation is large. At high stiffness modulus, $S = E$. As the stiffness modulus was not measured in the ductility test, it was estimated from Van der Poel's nomograph.

Figure 6.18 shows the elongation at break as a function of the stiffness modulus and the PI (Heukelom and Wijga, 1973).

When all the results were plotted in Van der Poel's nomograph, a new nomograph was obtained that allows the elongation at break to be estimated, starting from the same input parameters (Heukelom 1966; Heukelom and Wijga, 1973). This nomograph is shown in Figure 6.19.

The elongation at break is closely related to the strain at break. At low levels of elongation and strain, they are equal. As the strain at break multiplied by the stiffness modulus at break is equal to the breaking strength, the latter is also a function of S and the PI. The elongation being measured in tension,

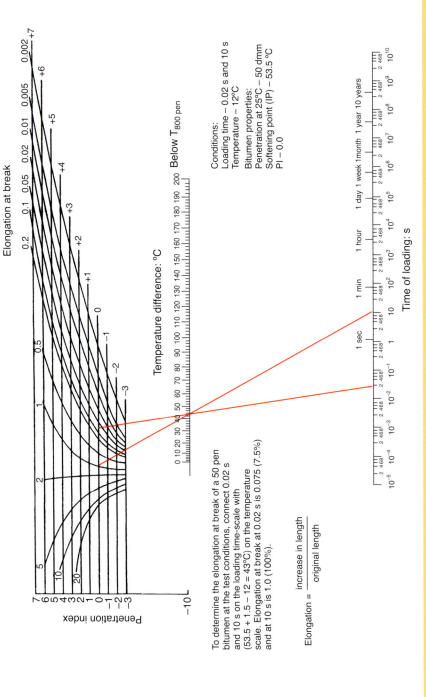

Figure 6.19 Nomograph for determining the elongation at break of bitumens

these are the tensile strain at break and the tensile strength. It was also found that the strain at break in tension is influenced by the presence of a hard carrier in sample testing, as is used, for example, in the Fraass test (Heukelom, 1966; Heukelom and Wijga, 1973).

6.5.4 Fatigue strength

Like many other materials, the strength of bitumen decreases as a result of repeated loading (fatigue). The fatigue phenomenon has to do with fracture produced by repeated applications of stresses smaller than the breaking strength. Fatigue tests were carried out in bending at constant stress amplitude. The fatigue strength is the stress that causes failure after N load cycles. Figure 6.20 shows the fatigue strength and the breaking strength as a

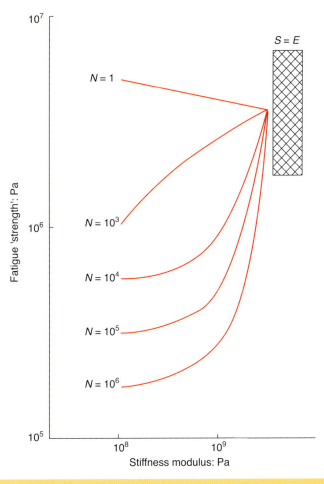

Figure 6.20 The breaking strength and fatigue strength of bitumen as a function of the stiffness modulus

function of the stiffness modulus (Heukelom, 1966). The breaking strength corresponds to one cycle. The fatigue strength is reduced as the number of loading cycles increases. At higher stiffness moduli, the effect of load repetitions is considerably reduced, and there is a tendency for the lines to converge at $S = E$ (the maximum stiffness modulus of 2.7×10^9 Pa).

6.6 Cohesion
6.6.1 Ductility and force ductility
The cohesiveness of bitumen can be assessed by its ductility at a given temperature, as mentioned in Chapter 5. In this test, ASTM D113 (ASTM, 2007), a dumb bell shaped specimen of bitumen (a briquette with a specified form of the mould) is immersed in a water bath, and the two ends are pulled apart at a specified temperature and at a specified speed (usually a constant rate of 50 mm/min) until fracture occurs. The distance to which the bituminous material will elongate before breaking is reported as the ductility. The test temperature is adjusted, depending on the penetration of the bitumen in the test. In these conditions, the test has been found to discriminate between bitumens of different cohesive strengths (see Chapter 5).

Under conditions where bitumen forms a very thin thread at large values of elongation, the cohesive strength is close to zero. When bitumen is modified with elastomers, much thicker threads are formed, and the force required to stretch the material indicates a better cohesion. In the force ductility test, EN 13589 (BSI, 2008a), a moulded test specimen is extended in the ductilometer at the specified test temperature and at constant speed until fracture or an elongation of at least 1333% (400 mm) is achieved. The force–elongation curves for an unmodified bitumen and a polymer modified bitumen are presented in Figure 5.4. From the area under the curve, the deformation energy can be calculated, as described in EN 13703 (BSI, 2003), and is a measure of the cohesion.

6.6.2 Toughness and tenacity
In addition to the force ductility – making use of the ductility test set up – other tensile tests, which differ in the shape of the sample, the testing device and the applied loading, are available.

The shape of the force–elongation curve can be used not only to distinguish between modified and unmodified bitumens but also to differentiate between the level and type of modification (in particular, elastomeric components).

As is described in EN 13587 (BSI, 2010), a specimen, held by its ends between two jaws, is extended in a chamber, regulated at the test temperature, at constant speed until fracture or a given percentage elongation is achieved. The calculations for the different parts of the area under the force–elongation curve are detailed in EN 13703 (BSI, 2003).

ASTM D5801 (ASTM, 2012b) describes a procedure to determine the toughness and tenacity of asphalts. At a specified temperature, a tension head of specified size and shape is pulled from a sample of asphalt at a specified rate, and the force is measured as a function of elongation. The toughness of the sample is defined as the work required to separate the tension head from the sample under the specified test conditions. It is calculated as the total area under the force versus elongation curve. The tenacity of the sample is defined as the work required to stretch the sample after the initial resistance has been overcome. A tangent line is placed against the force versus elongation curve as the force decreases from the initial maximum value so that the tangent line intersects the elongation axis (zero force). The area under the curve to the right of the tangent line is the tenacity of the sample.

6.6.3 Vialit cohesion
The Vialit cohesion pendulum is a test device used to assess the cohesion of bitumen for road construction. The method is described in EN 13588 (BSI, 2008b). The apparatus for carrying out the test is shown in Figure 6.21.

Figure 6.21 The Vialit pendulum test

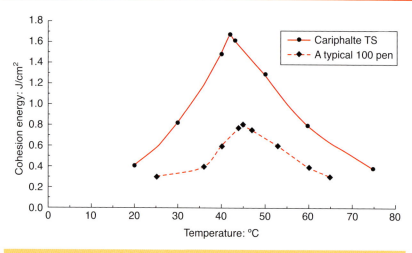

Figure 6.22 Cohesion curves for a penetration grade bitumen and a polymer modified bitumen

The procedure is to place a thin film of a bitumen or a polymer modified bitumen between two cubes and to determine the energy required to remove, by pendulum impact, the upper cube. This test is of interest in situations where aggregate is placed in direct contact with traffic stresses, for example in surface dressing. It is important to use binders that have a sufficient level of cohesion for the level of traffic and the site conditions to be supported. The cohesion energy can be significantly increased by polymer modification across a wide temperature range. An example is given in Figure 6.22.

References

ASTM (2007) D113-07. Standard test method for ductility of bituminous materials. ASTM, Philadelphia, PA, USA.

ASTM (2010a) D2170/D2170M-10. Standard test method for kinematic viscosity of asphalts (bitumens). ASTM, Philadelphia, PA, USA.

ASTM (2010b) D2171/D2171M-10. Standard test method for viscosity of asphalts by vacuum capillary viscometer. ASTM, Philadelphia, PA, USA.

ASTM (2012a) D36/D36M-12. Standard test method for softening point of bitumen (ring-and-ball apparatus). ASTM, Philadelphia, PA, USA.

ASTM (2012b) D5801-12. Standard test method for toughness and tenacity of bituminous materials. ASTM, Philadelphia, PA, USA.

ASTM (2013a) D5/D5M-13. Standard test method for penetration of bituminous materials. ASTM, Philadelphia, PA, USA.

ASTM (2013b) D4402/D4402M-13. Standard test method for viscosity determination of asphalt at elevated temperatures using a rotational viscometer. ASTM, Philadelphia, PA, USA.

Bonnaure F, Gest G, Gravois and Uge P (1977) A new method of predicting the stiffness of asphalt paving mixtures. *Proceedings of the Association of Asphalt Paving Technologists* **46**: 64–104.

Bowen HC (1889) An apparatus for determining the relative degree of cohesion of a semi-liquid body. *The School of Mines Quarterly* **10**: 297.

BSI (British Standards Institution) (2003) BS EN 13703: 2003. Bitumen and bituminous binders. Determination of deformation energy. BSI, London, UK.

BSI (2006) BS EN 13924:2006. Bitumen and bituminous binders. Specifications for hard paving grade bitumens. BSI, London, UK.

BSI (2007a) BS EN 1426:2007. Bitumen and bituminous binders. Determination of needle penetration. BSI, London, UK.

BSI (2007b) BS EN 1427:2007. Bitumen and bituminous binders. Determination of the softening point. Ring and Ball method. BSI, London, UK.

BSI (2007c) BS EN 12593:2007. Bitumen and bituminous binders. Determination of the Fraass breaking point. BSI, London, UK.

BSI (2007d) BS EN 12595:2007. Bitumen and bituminous binders. Determination of kinematic viscosity. BSI, London, UK.

BSI (2007e) BS EN 12596:2007. Methods of test for petroleum and its products. Bitumen and bituminous binders. Determination of dynamic viscosity by vacuum capillary. BSI, London, UK.

BSI (2008a) BS EN 13589:2008. Bitumen and bituminous binders. Determination of the tensile properties of modified bitumen by the force ductility method. BSI, London, UK.

BSI (2008b) BS EN 13588:2008. Bitumen and bituminous binders. Determination of cohesion of bituminous binders with pendulum test. BSI, London, UK.

BSI (2009) BS EN 12591:2009. Bitumen and bituminous binders. Specifications for paving grade bitumen. BSI, London, UK.

BSI (2010) BS EN 13302:2010. Bitumen and bituminous binders. Determination of dynamic viscosity of bituminous binder using a rotating spindle apparatus. BSI, London, UK.

BSI (2011a) BS EN 12846-1:2011. Bitumen and bituminous binders. Determination of efflux time by the efflux viscometer. Bituminous emulsions. BSI, London, UK.

BSI (2011b) BS EN 12846-2:2011. Bitumen and bituminous binders. Determination of efflux time by the efflux viscometer. Cut-back and fluxed bituminous binders. BSI, London, UK.

EI (Energy Institute) (2007a) IP 58. Bitumen and bituminous binders. Determination of softening point – Ring and ball method. EI, London, UK.

EI (2007b) IP 80. Bitumen and bituminous binders. Determination of the Fraass breaking point. EI, London, UK.

Fraass A (1937) Test methods for bitumen and bituminous mixture with specific reference to low temperature. *Bitumen* **7**: 152–155.

Heukelom W (1966) Observations on the rheology and fracture of bitumens and asphalt mixes. *Proceedings of the Association of Asphalt Paving Technologists* **35**: 358–399.

Heukelom W (1969) A bitumen test data chart for showing the effect of temperature on the mechanical behaviour of asphaltic bitumens. *Journal of the Institute of Petroleum* **55**: 404–417.

Heukelom W (1973) An improved method of characterizing asphaltic bitumens with the aid of their mechanical properties. *Proceedings of the Association of Asphalt Paving Technologists* **42**: 67–98.

Heukelom W and Wijga PWO (1973) *Bitumen Testing*. Koninklijke Shell Laboratorium, Amsterdam, The Netherlands.

Krom CJ (1950) Determination of the ring and ball softening point of asphaltic bitumens with and without stirring. *Journal of the Institute of Petroleum* **36**: 36.

Pfeiffer JPH (1950) *The Properties of Asphaltic Bitumen: With Reference to its Technical Applications*. Elsevier Science, Amsterdam, The Netherlands.

Pfeiffer JPH and Van Doormaal PM (1936) The rheological properties of asphaltic bitumens. *Journal of the Institute of Petroleum* **22**: 414–440.

Thenoux G, Lees G and Bell CA (1987) Laboratory investigation of the Fraass brittle test. *Asphalt Technology* **39**: 34–46.

Van der Poel C (1954) A general system describing the visco-elastic properties of bitumen and its relation to routine test data. *Journal of Applied Chemistry* **4**: 221–236.

Chapter 7

Rheology of bitumens

Dr Richard Taylor
Global Technical
Development Manager,
Shell International Petroleum
Co. Ltd. UK

Prof. Gordon Airey
Director,
Nottingham Transportation
Engineering Centre (NTEC),
Faculty of Engineering,
University of Nottingham

Rheology is the fundamental interdisciplinary science that is concerned with the study of the internal response of real materials to stresses. The word 'rheology' is derived from the Greek, and translates literally as *rheos* (stream or flow) and *logy* (science); therefore, 'rheology' literally means 'the study of flow'. Bitumen rheology can thus broadly be defined as the fundamental measurements associated with the flow and deformation characteristics of bitumen.

Bitumen is a thermoplastic, visco-elastic liquid that behaves as a glass-like elastic solid at low temperatures and/or during short loading times and as a viscous fluid at high temperatures and/or during slow loading. The response of bitumen to stress is therefore dependent on both the temperature and the loading time, and consequently the rheology of bitumen is defined by its stress/strain/time/temperature response.

Traditionally, bitumen has been predominantly characterised and specified using empirical tests. A classic example is the penetration test, which has historically been the backbone for most of the specification regimes globally. The penetration test conditions neither relate to the actual stresses or strains encountered in service nor do they capture any fundamental engineering properties of the material. Instead, the procedure derives a relative measure of the 'stiffness' related to the material's viscosity, and researchers have been successful in using penetration values to formulate an array of extremely useful measures that have greatly facilitated grading and predicting bitumen and asphalt performance.

This simplistic approach has worked rather well for conventional bitumen; however, the growth in use of polymer modified bitumens has necessitated a review of this thinking. Concurrently, there emerged recognition that to characterise a bituminous binder adequately over the entire performance spectrum, it is necessary to capture not only the properties at elevated temperatures but also at intermediate and low temperatures.

The Shell Bitumen Handbook, Sixth Edition
ISBN 978-0-7277-5837-8
Shell International Petroleum Company Ltd: All rights reserved
http://dx.doi.org/10.1680/tsbh.58378.119

Early work, particularly at Shell (Bouldin *et al.*, 1991; Collins *et al.*, 1991), was important in establishing the importance of rheometric testing. Not only does rheometric testing provide an avenue to quantify the temperature dependence of a given property appropriately but it also provides a window to understand the rate (or frequency) dependency of the rheological properties. The US Strategic Highway Research Program (SHRP) built on these insights, and in 1993 published the initial version of the American Association of State and Highway Transportation Officials (AASHTO) performance graded (PG) binder specification, the first bitumen specification to be entirely based on fundamental material properties, derived either from rheological or strength testing (AASHTO, 2009).

This chapter provides an introduction to key rheological concepts and their relevance for bitumen, introduces some of the main test methods employed to make a rheological characterisation of bitumen, and discusses the rheological properties of unmodified and polymer modified bitumens.

7.1 Rheology, deformation and flow

Rheology is concerned with the study of the internal response of real materials to stresses. Eugene C. Bingham, a professor at Lafayette College in the USA, who was researching the properties of materials that showed both elastic properties, akin to solids, as well as flow characteristics typical of liquids, coined the term 'rheology' in 1929 from the Greek, as explained above. There is a body of different tests, commonly referred to as rheometry, which has been devised to quantify these properties. More often than not, they have been customised to elicit a response from a material to a specific mechanical stimulus. Rheology can be considered to be the study of materials that exhibit both solid and liquid characteristics.

At rest, a solid or a liquid retains its form; under load the material deforms, and when the load is removed this deformation can be either irrecoverable (viscous) or the deformation can be recoverable (elastic). In the simplest of terms, ideal solid materials behave in an elastic manner, and the deformation is recovered when the force or load is removed, whereas liquids behave in a viscous manner, and the load is not recovered but results in a permanent deformation of the material. Between these two ideal states, materials can exhibit what is referred to as visco-elastic behaviour, in which the response to stress is partially viscous and partially elastic.

Central to the discipline of rheology is the concept of deformation and flow. Deformation is associated with solid behaviour, and can be defined as an alteration in shape or size as a response to an applied load. Flow is associated with liquids (and gases), and is a deformation that continues for as long as the load is applied. Importantly, the effects of temperature and loading time during the measuring process can give rise to different observations of the behaviour, and a material will exhibit a range of behaviours in response to a force, depending on the temperature and the loading time.

Although flow is typically associated with liquids, over long time-scales all materials can exhibit flow. In the Old Testament (Book of Judges, Chapter 5, Verse 5), the biblical prophetess Deborah captured this phenomenon in the phrase 'The mountains flow before the Lord'. Implicit in this quotation is an understanding that the way a body behaves, or is perceived to behave, depends intrinsically on the time-scale of the observation and the nature of the material in question. Given sufficient time, even a mountain will flow. A famous example of long term flow behaviour is the windows of Renaissance cathedrals that have gradually become thinner at the top than at the bottom – the viscosity of glass in ambient conditions is too high to observe any cold flow during the brevity of our human lifespan, but over centuries the glass has flowed (Zanotto and Gupta, 1999).

7.1.1 Elasticity, plasticity and the behaviour of solids
The ability of a deformed material body to return to its original shape and size when the forces causing the deformation are removed is referred to as elasticity, and materials exhibiting this behaviour are referred to as elastic. Most solid materials exhibit elasticity, and the response to relatively small stresses results in a directly proportional strain: this relationship was observed by Robert Hooke in 1660, and is referred to as Hooke's law. Hooke's law states that the applied force $F =$ a constant k multiplied by the displacement or change in length: $F = kx$. The value of k depends not only on the kind of elastic material under consideration but also on its dimensions and shape. Bitumen exhibits elastic behaviour at low temperatures.

The limit of proportionality is defined as the point beyond which Hooke's law is no longer true when deforming a material, and the elastic limit is the point beyond which the material under load becomes permanently deformed so that the material does not return to its original length when the force is removed. It marks the onset of plastic behaviour. For most brittle materials, stresses beyond the elastic limit result in fracture with almost no plastic deformation.

7.1.2 Viscosity and the behaviour of liquids
There are two basic forms of relative movement (flow) of adjacent particles of liquid: shear flow, where adjacent particles move over and past each other, and extensional flow, where particles move away from or towards each other (Barnes, 2000). Viscosity (η) is the measure of the resistance to flow of a liquid, and is defined as the ratio between the applied shear stress (σ) and the rate of shear strain ($\dot{\gamma}$). The SI unit for viscosity is the Pascal

second (Pa·s). The viscosity of all simple liquids, including bitumen, decreases with increasing temperature.

7.1.2.1 Newtonian behaviour and non-Newtonian behaviour

Many everyday fluids exhibit what is referred to as Newtonian behaviour (i.e. where stresses arising are linearly proportional to the local strain rate), and the viscosity is independent of the rate of deformation. The reason that the bitumen viscosity-grading system in the USA and other regions has endured for standard refined bitumens is that, at temperatures close to or exceeding 60°C and very low shear rates (which generally prevail within the standard capillary viscometers used for bitumen), bitumen tends to exhibit Newtonian behaviour.

At a high enough shear rate, all liquids will exhibit 'non-Newtonian' behaviour. Some liquids display what is known as shear thickening, and are termed 'dilatant'. Shear thickening is typically associated with suspensions, and occurs as a result of restructuring of the particles in suspension as the shear rate increases. More typically, liquids display what is referred to as shear thinning, and are termed 'pseudoplastic', and their viscosity decreases as the shear rate increases. Most polymer modified bitumens can be considered to be shear-thinning liquids.

The behaviour of liquids under different shear conditions can be described using models. The simplest model is known as the power law model, and is widely used in engineering calculations (Hieber and Chiang, 1992)

$$\eta = m\dot{\gamma}^{n-1} \tag{7.1}$$

where η is the viscosity, $\dot{\gamma}$ is the shear rate, and m and n are positive constants for any given material.

In the case of $n < 1$ the viscosity decreases, and these liquids are generally referred to as pseudoplastic, when $n = 1$ they are Newtonian and when $n > 1$ the fluid is considered dilatant (i.e. the viscosity increases as a function of the shear rate). The limitation of this model is that there is no time constant (i.e. relaxation time) associated with the function, and, for example in the case of a pseudoplastic material, the viscosity monotonically increases to infinity as the shear rate decreases while tending towards zero as the shear rate increases. The most simplistic fix is known as the truncated power law model. Although this model provides a zero-shear viscosity (see section 7.2.3.1) and a characteristic time, it does not address the issue of the monotonically decreasing viscosity, and, hence, does not provide a comprehensive solution for high shear rates. In addition, this approach does not generate a realistic transition from Newtonian to non-Newtonian behaviour.

The Cross model offers, with a simple modification to the truncated power law, an approach that describes the entire range from the Newtonian region to the power law region in one equation (Cross, 1969). This model works relatively well in describing the behaviour for most polymer modified bitumens over a wide range of temperatures and shear rates.

7.1.3 Visco-elasticity

In the previous sections, the idealised behaviour of solids and liquids was introduced. The term visco-elasticity refers to the mechanical properties of a material that, in two limiting extremes, can result in the material behaving either as an elastic solid or a viscous fluid, depending on the temperature and the time of loading. Visco-elasticity differs from plasticity in that visco-elastic materials exhibit a time-related recovery when a load is removed: this is often referred to as a delayed elastic response. A plastic material does not return to its original form after the load is removed; plasticity is referred to as ductility in metals. Under load, a material can exhibit a mixed visco-elastic and plastic response in which a proportion of deformation occurs; this is an important characteristic of bitumen and asphalts where accumulated non-recovered deformation manifests itself in the development of ruts in asphalt pavements. Reducing this phenomenon has resulted in the adoption of ways to reduce the permanent deformation, and includes the modification of bitumen using polymers.

Bitumen can be considered a visco-elastic material where, typically, at low temperatures the elastic properties dominate, while at high temperatures the material behaves as a viscous fluid. A well known material that can be used to describe visco-elastic behaviour, particularly with respect to loading time, is Silly Putty^{TM}. If it is kneaded into a ball and thrown against the floor (a high rate of deformation), it will bounce back just like a rubber ball. If, however, it is placed over an orifice it will, over time and under the force of gravity, flow in a viscous manner. Bitumen, and in particular polymer modified bitumen, would exhibit a similar response to loading time.

The degree to which a material displays visco-elastic behaviour can be quantified. In deference to the biblical prophetess Deborah, Marcus Reiner, a professor at the Israel Institute of Technology, proposed a dimensionless number, the Deborah number (De) as a measure of a material's visco-elasticity (Reiner, 1964)

$$De = \lambda/t_0 \qquad\qquad (7.2)$$

where λ is the relaxation time and t_0 is the observation time.

If the relaxation time is very long relative to the observation time, then the Deborah number is large, and the material will exhibit elastic behaviour because it cannot relax the stresses imposed on it. Conversely, if the time of the observation is very long relative to the relaxation time, then the material will behave in a viscous manner – the energy exerted on it will be dissipated in the form of viscous flow.

As the loading periods become longer, the elastic contribution will steadily decrease, and, in terms of the Deborah number, De $\ll$ 1. If the time that the stress is applied is much shorter than the relaxation time, then De $\gg$ 1, and the material will behave more elastically than viscously.

In transient (steady) flow, the Weissenberg number (Wi) – rather than the Deborah number which applies to dynamic flows – is used to determine the nature of the visco-elastic response (Rohn, 1995). If the relaxation time is long and the rate of deformation high, then a material will display non-linear visco-elastic behaviour. Nevertheless, the underlying principle is the same as in the case of the Deborah number

$$Wi = \lambda \dot{\gamma} \qquad (7.3)$$

where λ is the relaxation time and $\dot{\gamma}$ is the shear rate.

An important non-linear visco-elastic effect that may accompany high Weissenberg numbers is referred to as the Weissenberg effect. This refers to a build-up of normal stresses perpendicular to the shear plane. It manifests itself in polymer modified bitumen 'climbing' up the shaft of a mixer during production. These effects can, particularly in asphalts, be of significant importance because they will generate confining pressures and can con-siderably enhance the material's performance (Sousa et al., 1991).

7.1.4 Measuring visco-elasticity properties
There are two fundamentally different modes (and an infinite number of vari-ations) of strain or stress application during tests to determine visco-elastic properties of materials: transient flow and dynamic flow.

7.1.4.1 Transient flow: creep and relaxation tests
Creep and relaxation are two methods used to characterise the response of a material under transient loading. Creep is a physical phenomenon that causes non-reversible deformation of a material exposed to constant stress over a given length of time. Creep compliance ($J(t)$) is defined as the inverse of stiffness as a function of time.

For ideal elastic solids there would be an immediate response to stress during a creep test, and this would result in a constant strain. For a Newtonian liquid the same applied stress would result in an ever increasing strain, which results in a straight line if strain is plotted against time, the slope of which gives the shear rate (Barnes, 2000).

The behaviour of a visco-elastic material in creep tests varies with time, and in a generalised example consists of three regions

- an instantaneous elastic response
- a delayed elastic response
- a steady state viscous response.

The elastic response of the bitumen dominates at short loading times and/or low temperatures, while the viscous response dominates at long loading times and/or high temperatures. The delayed elastic response dominates at intermediate loading times and temperatures. The purely viscous component and the delayed elastic component constitute the time-dependent deformation of the visco-elastic material. Although none of the viscous deformation is recovered once the load is removed, the delayed elastic deformation is recovered, but not immediately as with purely elastic deformation. Because the relative magnitudes of the three components change with loading time and temperature, both the magnitude and the shape of the creep curve will change with the loading time and the temperature (Anderson *et al.*, 1991).

Linear visco-elastic models are used to model deformation in creep tests. A linear spring and a linear dashpot are used to represent elastic (recoverable) and viscous (irrecoverable) behaviour, respectively, and a linear spring connected in parallel to a linear dashpot is used to represent delayed (recoverable) elastic behaviour. A spring and a dashpot in series are referred to as a Maxwell element, and a spring and a dashpot in parallel are known as a Kelvin element. Combining one Maxwell element and one Kelvin element gives one of the best known models for visco-elastic behaviour: the Burgers model.

Visco-elastic creep data are typically presented in one of two ways. Total strain can be plotted as a function of time for a given temperature or temperatures. Below a critical value of applied stress, a material may exhibit linear visco-elasticity. Above this critical stress, the creep rate grows disproportionately faster. The second way of graphically presenting visco-elastic creep in a material is by plotting the creep compliance as a function of time. Below its critical stress, the visco-elastic creep modulus is independent of the stress applied. A family of curves describing the strain versus time response to various applied stresses may be represented by a single visco-elastic creep modulus versus time curve if the applied stresses are below the material's critical stress value.

If the stress is removed, then the retained 'strain memory' will cause the material to recover, and creep and recovery testing can therefore be used to determine how elastic a material is by measuring the recovered strain

125

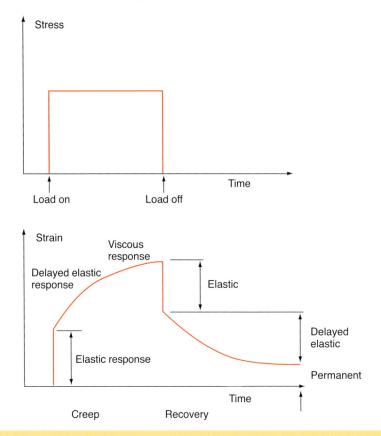

Figure 7.1 Stress–strain response during a creep and recovery test

(Figure 7.1). Data from recovery tests can be treated and represented in the same manner as described above, and the inverse of stiffness as a function of time in recovery is known as the recovery compliance (J_r). The proportion recovered can be treated as either a percentage recovery or the magnitude of the creep compliance not recovered, referred to in the multiple stress creep recovery (MSCR) test for bitumen (see section 7.2.3.1) as J_{nr}, the non-recovered creep compliance – which has been proposed recently as a predictor for the permanent deformation characteristics of bitumen.

A relaxation test is the reverse of a creep test (Figure 7.2). In relaxation tests, the deformation is imposed suddenly and then maintained at a constant level. The total visco-elastic resistance to deformation at a constant stress level can be evaluated by measuring the total accumulated strain, after the material has had sufficient time to relax.

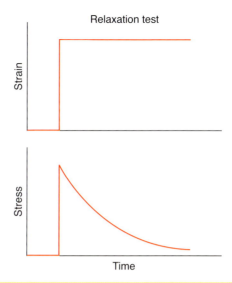

Figure 7.2 Stress–strain response during a relaxation test

7.1.4.2 Dynamic flow: oscillatory tests

A dynamic shear rheometer (DSR) can be used to conduct visco-elastic analysis of a material. This is a widespread technique employed to measure bitumen properties, and is a key technique as part of the PG grading system (M 320-10 (AASHTO, 2010a)). In such tests, a sinusoidal strain is applied to a specimen, and the resulting stress is monitored as a function of frequency. This is termed 'strain controlled testing' and, for bitumen, is more common than the stress controlled approach in which a sinusoidal varying stress is applied and the strain response measured. If a material is exposed to a strain or stress profile as a function of time, and if the strain is sufficiently small, then it will behave in a linear visco-elastic manner, meaning that there is a linear relationship between the shear stress and the strain, as premised by Hooke's law

$$\sigma_{12} \propto \gamma \qquad\qquad (7.4)$$

where σ_{12} is the shear stress and γ is the shear strain.

The operational procedure used in DSR testing for bitumen is to impose sinusoidal strains, as an oscillatory shear, on samples of bitumen sandwiched between the parallel plates (discs) of the rheometer. The amplitude of the responding stress is measured by determining the torque transmitted through the sample in response to the applied strain (Figure 7.3). The strains that are applied during the testing must be kept small to ensure that the test remains in the linear visco-elastic region. Strain sweeps can be used to verify that testing occurs in the linear visco-elastic region. The strain must generally be less than

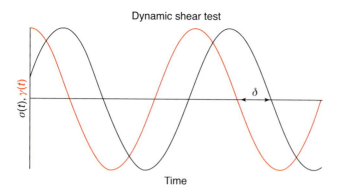

Figure 7.3 Stress–strain response during oscillatory testing

0.5% at low temperatures, but can be increased at higher temperatures (Goodrich, 1988).

The loading frequency (ω) is known as the angular frequency, rotational frequency or radian frequency (Eurobitume, 1996; Van der Poel, 1954), and is defined as

$$\omega = 2\pi f \tag{7.5}$$

where ω is the angular frequency (rad/s) and f is the frequency (Hz).

The response of the applied strain is the development of a stress that, for linear visco-elasticity (small strains), is sinusoidal, and the response of the material is out of phase with the applied strain, and is referred to as the phase angle δ. This is defined as the phase difference between the stress and the strain, and is also called the loss angle or the phase lag. An alternative symbol, ϕ, can also be used for the phase angle. For purely elastic materials, the phase angle will be zero, whereas for purely viscous materials the phase angle will be 90°. The phase angle is therefore important in describing the visco-elastic properties of a material such as bitumen.

The ratio of the resulting stress to the applied strain is called the complex shear modulus, G^*, and is also referred to as the complex modulus, shear modulus or, simply, the stiffness.

The norm of the complex modulus is analogous to the magnitude of a vector (Figure 7.4), as the value is calculated from the square root of the sum of the squares of the components.

The in-phase component of G^* is called the shear storage modulus (G'), or more commonly the storage modulus. The storage modulus equals the stress that is in phase with the strain divided by the strain, or

$$G' = G^* \cos \delta \tag{7.6}$$

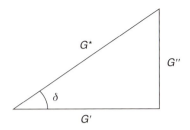

Figure 7.4 Relationship between the shear modulus and the phase angle

The storage modulus describes the amount of energy that is stored and released elastically in each oscillation, and is therefore also known as the elastic modulus, or the elastic component of the complex modulus.

The shear loss modulus G'', or simply the loss modulus, is the out-of-phase component of G^*. The loss modulus equals the stress 90° out of phase with the strain divided by the strain, or

$$G'' = G^* \sin \delta \tag{7.7}$$

The loss modulus describes the average energy dissipation rate in the continuous steady oscillation found in the dynamic test. The loss modulus is also referred to as the viscous modulus or the viscous component of the complex modulus. The storage and loss modulus, G' and G'', are sometimes misinterpreted as the elastic and viscous modulus, respectively. In reality, the elastic component of the response represents only part of the storage modulus, and the viscous response only part of the loss modulus. In addition, visco-elastic materials exhibit a significant amount of delayed elastic response that is time dependent but completely recoverable. The storage and loss modulus both reflect a portion of the delayed elastic response.

The loss tangent is defined as the ratio of the viscous and elastic components of the complex modulus, or simply the tangent of the phase angle

$$\tan \delta = \frac{G''}{G'} \tag{7.8}$$

The complex viscosity, also termed the complex dynamic shear viscosity, is a viscosity value obtained from dynamic oscillatory tests. The complex viscosity (Pa·s) is defined as the ratio of the complex modulus and the angular frequency

$$\eta^* = \frac{G^*}{\omega} \tag{7.9}$$

As the complex viscosity is a function of a complex number pair, a real part and an imaginary part of the complex viscosity can also be defined.

The real part of the complex viscosity is termed the dynamic viscosity (Pa·s), and is defined as

$$\eta' = \frac{G''}{\omega} \qquad (7.10)$$

The imaginary part of the complex viscosity is called the out-of-phase component of η^* (Pa·s), and is defined as

$$\eta'' = \frac{G'}{\omega} \qquad (7.11)$$

7.1.4.3 Relationship between creep recovery and oscillatory measurements

Van der Poel (1954) defined stiffness as either the inverse of the creep compliance at a loading time t, or the uniaxial dynamic modulus at a loading frequency $1/t$. Therefore, a simple approximation between the creep compliance and the dynamic complex modulus can be defined as

$$G^*(\omega) \approx 1/J(t) \quad \text{as} \quad t \rightarrow 1/\omega \qquad (7.12)$$

where G^* is the complex modulus, ω is the angular frequency (rad/s) and $J(t)$ is the creep compliance.

The reciprocal of the shear compliance ($J(t)$) obtained at time t_i is approximately equal numerically to the complex shear modulus ($G^*(\omega)$) at a frequency ω_i, where $t_i = 1/\omega_i$.

The transient and dynamic visco-elastic functions can be related mathematically, although the exact relationships are complicated. However, approximate relationships exist that are frequently accurate enough for practical purposes (Anderson et al., 1994).

7.1.5 Representation of rheological data

In order to study the rheological characteristics of materials (including bitumens and polymer modified bitumens), the data obtained from testing need to be represented in a useful form. The most common data representation diagram and curves used for bitumen (Eurobitume, 1996) are discussed in the following sections.

7.1.5.1 Simple graphical representations

An isochronal plot or isochrone is an equation, or a curve on a graph, representing the behaviour of a system at a constant frequency or loading time. In a dynamic test, such as the DSR test, curves of the complex modulus (G^*) or other visco-elastic functions versus temperature at constant frequencies are isochrones.

An isothermal plot or isotherm is an equation, or a curve on a graph representing the behaviour of a system at a constant temperature. In a dynamic test, curves of the complex modulus (G^*) as a function of frequency at constant temperatures are isotherms.

A Cole–Cole diagram is a graph of the loss (viscous) modulus (G'') as a function of the storage (elastic) modulus (G'). The plot provides a means of representing the visco-elastic balance of the bitumen without incorporating frequency and/or temperature as one of the axes.

A Black diagram is a graph of the magnitude (norm) of the complex modulus ($[G^*]$) versus the phase angle (δ) obtained from a dynamic test. The frequency and the temperature are therefore eliminated from the plot, which allows all the dynamic data to be presented in one plot without the need to perform time–temperature superposition manipulations of the raw data (see the following section). A smooth curve in a Black diagram is a useful indicator of time–temperature equivalency, while a disjointed curve indicates the breakdown of the time–temperature superposition, and potentially the presence of a high asphaltene structured bitumen, high wax content bitumen or a highly polymer modified bitumen (Lesueur et al., 1996; Planche et al., 1996).

7.1.5.2 Master curves

To characterise fully the rheology of materials, it is important to be able to construct so-called master curves. Master curves can be constructed from both dynamic and transient loading tests. These master curves can be constructed either in the time domain or in the temperature domain. In their simplest form, master curves are produced by manually shifting modulus versus frequency plots (isotherms) at different temperatures along the logarithmic frequency axis to produce a smooth master curve (Marasteanu and Anderson, 1996).

The extended time or frequency scale used in a master curve is referred to as the reduced time or reduced frequency scale, where the reduced frequency scale is defined as

$$\log f_r = \log f + \log a(T) \tag{7.13}$$

where f_r is the reduced frequency (Hz), f is the frequency (Hz) and $a(T)$ is the shift factor.

Viscosity–temperature equations are used to characterise the temperature dependency of bitumens, and therefore to determine the shift factors needed for the time–temperature superposition principle. The time–temperature superposition principle can be expressed as (Dobson, 1969)

$$G(\omega, T) = G(\omega a(T), T_r) \tag{7.14}$$

where G is the modulus (and may be G', G'' or $[G^*]$), $a(T)$ is the shift factor, ω is the loading frequency, T is the temperature and T_r is the reference temperature.

The shift factor ($a(T)$) can be defined in several ways, depending on the mathematical equation used for its determination. Temperature dependency, as indicated by shift factors, should not be confused with temperature susceptibility, which is an empirical concept based on the change of consistency or hardness of a bitumen with temperature. The temperature dependency of the visco-elastic behaviour of bitumen is indicated by means of shift factors, and expressed as

$$a_T = a_T(T, T_{ref}) \tag{7.15}$$

and therefore depends, for a given system, only on the temperature.

The Williams, Landel and Ferry (WLF) equation (Williams et al., 1955) has been widely used to describe the relationship between the shift factors and temperature, and thereby determine the shift factors of bitumens. The equation is theoretical, based on the free volume theory (Ferry, 1971), and makes use of temperature differences, which makes it suitable for practical manipulations. The equation has also been found to be applicable to bitumen results.

The WLF equation is

$$\log a(T) = \log \frac{\eta_0(T)}{\eta_0(T_r)} = -\frac{C_1(T - T_r)}{C_2 + (T - T_r)} \tag{7.16}$$

where $a(T)$ is the shift factor at a temperature T, $\eta_0(T)$ is the Newtonian viscosity at a temperature T, $\eta_0(T_r)$ is the Newtonian viscosity at the reference temperature T_r, and C_1 and C_2 are empirically determined coefficients.

The WLF equation requires three constants to be determined, namely C_1, C_2 and T_r. The temperature dependency of bitumens can be described by one parameter, T_r, if universal constants are used for C_1 and C_2 in the WLF equation. Williams et al. (1955) proposed that if T_r is suitably chosen for each material, then C_1 and C_2 could be allotted universal values of 8.86 and 101.6, respectively. Brodnyan et al. (1960) showed that, for bitumens, the universal parameters fitted the data for $T - T_r > -20°C$, but at lower temperatures the predicted shift factors were too great.

Anderson et al. (1991) have found that for aged and unaged bitumen the constants in the WLF equation are all essentially the same value, with $C_1 = 19$ and $C_2 = 92$ based on a defining temperature, T_d, which is bitumen specific. These values have also been obtained by Jongepier and Kuilman (1969).

Unfortunately T_r is difficult to determine. Brodnyan et al. (1960) suggested that T_r is very similar to the softening point. Williams et al. (1955) proposed that T_r was related to the glass transition temperature (T_g) by the relationship

$$T_r - T_g = 50°C \tag{7.17}$$

An alternative equation that can be used to describe the relationship between the shift factors and temperature is the Arrhenius equation:

$$\log a(T) = \frac{-E_a}{2.303R}\left(\frac{1}{T} - \frac{1}{T_r}\right) \tag{7.18}$$

where $a(T)$ is the horizontal shift factor, E_a is the activation energy (typically 250 kJ/mol), R is the universal gas constant (8.314 J/K mol, T is the temperature (K) and T_r is the reference temperature (K).

The Arrhenius expression or function requires only one constant to be determined, namely the activation energy. The reference temperature (T_r) can be arbitrarily chosen (Nielsen, 1995).

Both the Arrhenius and WLF equations are based on theoretical considerations, and therefore their parameters provide some insight into the molecular structure of bitumen (Marasteanu and Anderson, 1996).

Master curves are only valid for the reference temperature used in the master curve plot. Time–temperature superposition must be applied to calculate the rheological properties at other temperatures. It must be remembered that, in interpreting the master curve, both the time dependency, as indicated by the master curve, and the temperature dependency, as indicated by the shift factors, must be considered in evaluating the rheological properties of the visco-elastic material. For this reason, isochronal plots are probably more informative and more easily interpreted when characterising the rheological properties of bitumen (Anderson et al., 1992).

The shape of the master curve of the complex modulus as a function of the reduced frequency, on a log–log scale, resembles the shape of a hyperbola. The curve has a horizontal asymptote (glassy modulus) at high frequencies and an asymptote at an angle of 45° (viscous part) at low frequencies, with a transition range in between.

Modulus curves at low temperatures crowd together at high frequency values, and at very high frequencies they nearly all coincide with one horizontal asymptote. The limiting elastic behaviour is therefore not only independent of frequency but also nearly independent of temperature. The elastic modulus corresponding with this asymptote is called the glassy modulus (G_g). However, under viscous conditions, there is no convergence to a single viscous asymptote, as viscosity strongly depends on temperature, and therefore each temperature gives rise to a separate viscous flow asymptote. The phase angle can also be shifted together with its modulus value, to obtain a curve of the phase angle versus the logarithm of the reduced frequency.

As with Black diagrams, breaks in the smoothness of the master curve indicate the presence of structural changes with temperature within the bitumen, as would be found for waxy bitumens, highly structured 'gel'-type bitumens and polymer modified bitumens.

A plot of log $a(T)$ versus temperature with respect to the reference temperature curve is generally prepared in conjunction with a master curve. These values can be considered to be viscosity changes with respect to the viscosity at the reference temperature, and give a visual indication of how the properties of a visco-elastic material change with temperature. The visco-elastic behaviour of a bitumen is therefore represented by two curves, namely viscosity as a function of temperature (the shift factor versus temperature) and modulus as a function of frequency at a fixed temperature (the master curve).

7.2 The rheology of bitumen

The behaviour of bitumen can be discussed in three regions of differing behaviour, namely

- a low temperature linear elastic region
- a high temperature viscous region
- an intermediate temperature visco-elastic region.

Linear behaviour is fulfilled at low temperatures and short loading times (high frequencies), where the bitumen behaves as an elastic solid. The linearity is also maintained (for unmodified bitumens) at high temperatures and long loading times (low frequencies), where the material behaves almost entirely as a Newtonian fluid. The area where non-linearity is prominent is therefore in the range of moderate temperatures and moderate loading times (Van der Poel, 1954). This range of temperatures and loading times corresponds to the conditions experienced in the field.

7.2.1 Low temperature linear elastic region

The bending beam rheometer (BBR) is the most widely used test device for determining the stiffness of bitumen at low temperatures. In principle, the BBR is a constant stress extensional rheometer that yields a Young's modulus (E). However, generally the stiffness is referred to in the bitumen industry as a flexural stiffness ($S(t)$). The test method has been adopted as a standard test in AASHTO's 'Standard specification for performance-graded asphalt binder' (AASHTO M 320-10) (AASHTO, 2010a). There is also a detailed

American Society for Testing and Materials specification (ASTM D6648) (ASTM, 2008).

During this testing, a beam of bitumen is subjected to a constant stress by means of a loaded piston (100 g) in a three-point bending machine. This beam is suspended in an approximately equi-dense cooling fluid, and the travel of the piston measured as a function of time (Lee, 1997). The relationship between the relevant parameters is

$$S(t) = \frac{pL^3}{4\delta(t)bh^3} \qquad (7.19)$$

where L is the beam length, h is the beam height, p is the applied constant load, b is the beam width and $\delta(t)$ is the displacement at time t.

For specification purposes, $S(t)$ and the slope of the curve, commonly referred to as the m value, are both determined at a load time of 60 s. Alternatively, the stiffness curves can be obtained at different temperatures, and flexural stiffness master curves constructed. The data can also be treated in exactly the same way as creep test data, to compute compliance information. To convert the results from the extensional mode to the shear mode, it is imperative to take Poisson's ratio (ν) into consideration. Poisson's ratio accounts for the fact that, when extending, an object will compress in the perpendicular direction and expand in the axial direction (in which it is being pulled). For the ideal case of an incompressible, isotropic material, ν should be equal to 0.5. However in practice, the value of Poisson's ratio can be significantly smaller.

7.2.2 High temperature viscous region

At very low shear rates or very low stress levels, almost all bitumens exhibit Newtonian behaviour. Non-Newtonian behaviour appears gradually as the shear rate or stress level increases. Generally, unmodified bitumens tend to exhibit Newtonian behaviour at temperatures greater than approximately 60°C. However, polymer modified bitumens tend to be shear susceptible at temperatures above 60°C and sometimes at mixing and compaction temperatures, and therefore exhibit non-Newtonian behaviour at higher temperatures than those associated with unmodified bitumens (Anderson et al., 1991).

Historically, and in the USA particularly, the Asphalt Institute vacuum capillary viscometer has been the standard equipment used to determine the viscosity of conventional bitumens, and more recently of polymer modified bitumens. A detailed description of the equipment is given in *The Asphalt Handbook* (AI, 2007). Generally, the viscosity is measured with this device at 60°C or 135°C.

The drawback of capillary viscometers is that a particular shear rate cannot be selected because the value depends on the viscosity of the material. The lower the viscosity, the faster it flows through the tube and the higher the shear rate. This does not present difficulties for conventional grades of bitumen, which tend to behave in a Newtonian manner at the rates observed in a vacuum capillary viscometer. However, the non-Newtonian behaviour of polymer modified bitumens gives rise to limitations in this approach.

Rotational viscometers consist of one cylinder rotating coaxially inside a second (static) cylinder containing the bitumen sample. The material between the inner cylinder and the outer cylinder is therefore analogous to the thin bitumen film found in the sliding plate viscometer. The torque on the rotating cylinder or spindle is used to measure the relative resistance to rotation of the bitumen at a particular temperature. The torque value is then altered by means of calibration factors to yield the viscosity of the bitumen. Rotational viscometers are also commonly used as a practical means of determining the viscosity and/or the shear rate dependency of bitumen, because such instruments allow the testing of a wide range of bitumens over a wide range of temperatures, more so than most other viscosity measurement systems (Zacharias and Emery, 1994).

The shear susceptibility, or shear rate dependency, of bitumen can be defined as the change in rheological properties of the bitumen as a function of the loading time. The shear susceptibility of bitumen can be represented by modelling the flow properties using a power law model in which the logarithms of the shear stress and shear strain rate are linearly related (Traxler et al., 1944)

$$\log \sigma = c(\log \dot\gamma) + B \qquad (7.20)$$

where σ is the shear stress, $\dot\gamma$ is the $d\gamma(t)/dt$ shear strain rate, c is the degree of complex flow and B is a constant.

A more common representation of the shear rate dependency is the following function

$$\eta = \frac{\sigma}{\dot\gamma^c} \qquad (7.21)$$

where η is the apparent viscosity.

Viscosity testing, although a more fundamental method of determining the rheological performance of bitumen, does not provide information on the time dependence of bitumen. These measurements are consequently not adequate for describing the visco-elastic properties needed for the complete rheological evaluation of bitumen.

7.2.3 Intermediate temperatures and the visco-elastic region

At in-service pavement temperatures, bitumen has properties that are in the visco-elastic region. At these temperatures, the bitumen therefore exhibits both viscous and elastic behaviour and displays a time-dependent relationship between the applied stress or strain and the resultant strain or stress (Goodrich, 1988). The conventional methods of characterising the rheological properties of bitumen cannot therefore completely describe the visco-elastic properties that are needed to relate fundamental physical binder properties to performance. The penetration and softening point tests, although conducted around the temperatures in question, are almost completely empirical, and hence not useful for characterising the visco-elastic behaviour of bitumen.

The visco-elastic characteristics of bitumen are typically determined by means of time-dependent tests. The two most common means of determining the visco-elastic properties of bitumen are by creep and recovery measurements (transient loading) and dynamic oscillatory measurements (alternating stress and strain of constant amplitude and frequency).

7.2.3.1 Creep and recovery tests

Steady state creep measurements can be used to calculate an apparent viscosity at intermediate pavement service temperatures, ranging from 0 to 25°C (Griffen *et al.*, 1956; Moavenzadeh and Stander, 1967; Romberg and Traxler, 1947). To conduct such measurements, it is necessary to apply a shear stress to the bitumen until the strain rate becomes constant. At low temperatures, longer times are required for delayed elasticity to be expended and steady state flow to occur. This results in the bitumen being subjected to very large strains, causing geometric non-linearity. The delayed elasticity and geometric non-linearity can be overcome by assuming that the steady state strain rate has been attained at a series of shear stress levels, and extrapolating the calculated apparent viscosities to a zero shear rate. This apparent viscosity at a zero shear rate is known as the zero shear viscosity (η_0), and is equivalent to the maximum Newtonian viscosity of the bitumen at the test temperature.

The time–temperature superposition principle can be applied to creep test data to produce a master curve from the creep data. The horizontal shift factor $(a(T))$ needed for the production of the master curve can be produced for each creep curve determined at a particular test temperature. These shift factors can be plotted against temperature to form, together with the master curve, a complete characterisation of the linear stress–strain–time–temperature response of the bitumen. The time dependency of the bitumen is reflected in the master curve whereas the temperature dependency is reflected in the temperature shift factors (log $a(T)$).

Recently, creep and recovery testing has been proposed as a standard means of more accurately predicting the rutting behaviour of bitumen and asphalts (AASHTO TP 70-09 (AASHTO, 2009) and AASHTO MP 19-10 (AASHTO, 2010b)). In this approach, a sample of bitumen is subjected to a 1 s load and a recovery time of 9 s using a DSR. The test starts with the application of a low stress (0.1 kPa) for ten creep/recovery cycles, after which the stress is increased to 3.2 kPa and repeated for an additional ten cycles. It is considered that the higher levels of stress applied during the test are more representative of the conditions experienced in the pavement (Federal Highways Administration, 2011). Results from the multiple stress creep recovery (MSCR) test are represented as a percentage recovery and non-recovered creep compliance (J_{nr}).

7.2.3.2 Oscillatory tests: dynamic shear rheometry

The rheology of bitumen is typically measured by using oscillatory type testing, generally conducted within the region of linear visco-elastic response. This approach allows the viscous and elastic (visco-elastic) nature of the bitumen to be determined over a wide range of temperatures and loading times (Anderson *et al.*, 1994; Goodrich 1988; Petersen *et al.*, 1994a). It is important that strain sweeps are carried out to ensure that the dynamic tests are conducted in the linear visco-elastic range so that the material functions, such as the complex modulus (G^*) and the complex viscosity (η^*), are independent of the applied strain levels.

Sliding plate rheometers were one of the first devices developed to characterise the rheology of bitumens. These devices can be used for transient or stress and strain controlled experiments, but they are also appropriate for performing oscillatory shear experiments. The strain amplitude of the oscillations needs to be kept sufficiently small to ensure linearity and angular frequencies, and gaps are adjusted to minimise inertial effects.

The DSR using plate–plate geometry is the most common example of a rheometer (Figure 7.5). In such tests, a sample of bitumen, which is sandwiched between two parallel discs or plates, is subjected to alternating shear stresses and strains. The test can be either stress controlled or strain controlled, depending on which of these variables is controlled by the test apparatus. The test condition usually used to determine the dynamic rheological properties of the bitumen is typically the controlled strain condition (Goodrich 1988; Petersen *et al.*, 1994b). The use of the controlled strain condition ensures that the strains are kept small and therefore within the linear visco-elastic region.

The principle that is used to relate the equivalency between time and temperature and thereby produce the master curve is known as the time–temperature superposition principle or the method of reduced variables (Ferry,

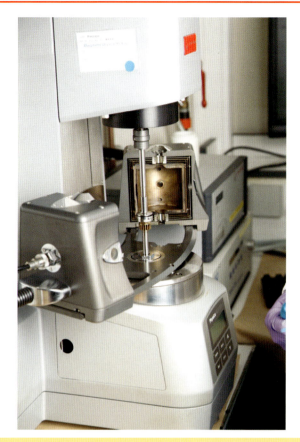

Figure 7.5 A dynamic shear rheometer typically used in bitumen testing

1971). Because bitumens are linear visco-elastic in a wide range of their applications and because they are found to be 'thermo-rheologically simple', the time–temperature superposition principle can be used to determine master curves and shift factors. Typically, dynamic data are first collected over a range of temperatures and frequencies, a standard reference temperature is selected that is usually between 0 and 25°C, and the data at all other temperatures are then shifted with respect to time until the curves merge into a single smooth function.

The shifting may be carried out based on any of the visco-elastic functions, such as the complex modulus (G^*), and if the time–temperature superposition principle is valid, the other visco-elastic functions will all form continuous functions after shifting. This continuous curve represents the binder behaviour at a given temperature for a wide range of frequencies. The complex modulus increases with decreasing penetration values for a given frequency, and over the reduced frequency scale of the master curve the complex modulus of the

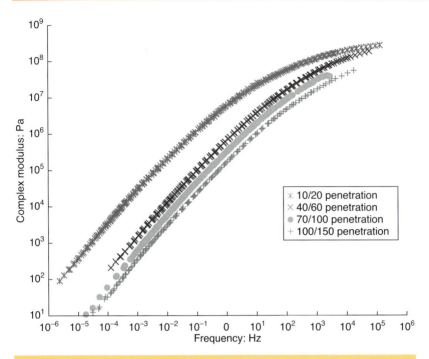

Figure 7.6 Reduced-frequency master curves for various penetration grade bitumens

bitumen varies by over six orders of magnitude (i.e. powers of 10). Master curves of the complex modulus for a range of bitumens are given in Figure 7.6.

Over the last two decades, DSR testing of bitumen has become much more widespread, primarily as a result of the US bitumen research programme of the Strategic Highway Research Program (SHRP). SHRP developed and implemented a number of binder rheological and fracture (failure) tests that measure the fundamental properties of bitumen. These new or improved test methods were used to develop a performance related specification as part of Project A-002A (Anderson and Kennedy, 1993), which introduced DSR testing and rheology into bitumen specifications. Properties, such as the complex modulus (G^*) and the phase angle (δ), are measured across a range of frequencies and temperatures.

Although the rutting of asphalts is influenced primarily by mixture properties, the properties of the binder are also important. This is particularly true for polymer modified bitumens, which are claimed to enhance the rutting resistance of asphalt pavements. A measurement of the non-recoverable deformation of bitumen at high temperatures using the DSR was established as a suitable rutting parameter for bitumen. A loading time of 0.1 s was chosen to represent the loading time within the pavement, which can be attributed to a truck tyre travelling at 80 km/h (Petersen et al., 1994a). This 0.1 s loading time is equivalent to a sinusoidal loading at a frequency of 1.59 Hz. The specification criterion for rutting was taken as the inverse of the loss compliance, $1/J''$, which is numerically equal to the complex modulus divided by the sine of the phase angle, $G^*/\sin \delta$ (Anderson and Kennedy, 1993). The SHRP specification states that, at the maximum pavement design temperature, $G^*/\sin \delta$ for an unaged bitumen must be greater than 1.0 kPa, and for a bitumen aged by rolling thin-film oven test it must be greater than 2.2 kPa (Anderson and Kennedy, 1993; Petersen et al., 1994a).

Fatigue cracking generally occurs late in the life of the pavement, and therefore the bitumen needs to be tested after appropriate long term ageing. The fatigue parameter was chosen to reflect the energy dissipated per load cycle, which can be calculated as $G^* \sin \delta$ in a dynamic shear test (Ferry, 1971). The SHRP specification requires that, at the intermediate pavement design temperature, the value of $G^* \sin \delta$ at a frequency of 1.59 Hz must be less than 5.0 MPa (Anderson and Kennedy, 1993; Petersen et al., 1994b).

The selection of the basically empirical parameters of the complex modulus divided by the sine of the phase angle ($G^*/\sin \delta$) and the complex modulus multiplied by the sine of the phase angle ($G^* \sin \delta$) to represent the binder characteristics from the available DSR rheological data to describe the performance of bitumens presents some limitations. This is particularly relevant for modified binders, such as polymer modified bitumens, which, due to their complex rheological behaviour, require greater quantities of data for their complete characterisation. Additionally, the small strain measurements associated with the DSR are not considered to be wholly representative of pavement conditions, and this has given rise to alternative approaches such as the MSCR test outlined in section 7.2.3.1.

7.2.4 Relationships between fundamental and empirical bitumen testing

Many studies link between index properties taken from rheological test data and empirical tests such as penetration and the softening point (Eurobitume, 2012; Nicholls et al., 1999). In general, unmodified bitumen relationships can be successfully developed between fundamental and empirical tests. For example, the penetration value of bitumen can be estimated using the following formula (Gershkoff, 1995)

$$\log(G^*_{T=25°C, \ f=0.4 \ Hz}) = 2.923 - 1.9 \log(\text{Pen}) \qquad (7.22)$$

where $G^*_{T=25°C, \, f=0.4 \, Hz}$ is the complex modulus at 25°C and 0.4 Hz and Pen is the penetration.

Such approaches work reasonably well for unmodified bitumens, but when relating the rheological properties of polymer modified bitumens to empirical tests, the more complex rheology of the latter gives rise to difficulties, and a more detailed approach is often required to characterise them.

7.2.5 Rheology of polymer modified bitumen

The use of synthetic polymers to modify the performance of conventional bituminous binders dates back to the early 1970s (Taylor and Airey, 2008), with these binders subsequently having decreased temperature susceptibility, increased cohesion and modified rheological characteristics. Polymers have been primarily used to improve the high temperature properties of bitumen and provide asphalts with a better resistance to deformation (Collins et al., 1991; Goodrich, 1988; Isacsson and Lu, 1995; King et al., 1993). In addition, the polymer modified bitumen should possess improved resistance to thermal and fatigue cracking (Brown et al., 1990).

A polymer modified bitumen extends the temperature range, sometimes referred to as the plasticity interval. In simple terms, the plasticity interval can be defined as the temperature range between the measure of high temperature performance (e.g. the softening point or criteria based on the complex modulus in the SHRP PG grading approach) and the low temperature measure of performance (e.g. a brittleness point or limiting stiffness value determined by the BBR). In rheological terms, the plasticity interval broadly describes the temperature range between the bitumen behaving as an elastic solid and as a viscous liquid. The extent to which polymer modification extends the plasticity interval is dependent on the degree of modification, the polymer properties, the polymer content and the nature of the base bitumen.

Plastomers modify bitumen by forming a tough, rigid, three dimensional network to resist deformation, while elastomers have a characteristically high elastic response and, therefore, resist deformation by stretching and recovering their initial shape. Elastomers, such as styrene–butadiene–styrene (SBS) copolymers, derive their strength and elasticity from the physical cross-linking of the molecules into a three dimensional network within the bitumen. The polystyrene end-blocks impart strength to the polymer, while the polybutadiene rubbery matrix mid-blocks give the material its elasticity. The effectiveness of these cross-links diminishes rapidly above the glass transition temperature of polystyrene (approximately 100°C), although the polystyrene domains will reform on cooling, restoring the strength and elasticity of the copolymer (Isacsson and Lu, 1995).

7.2.5.1 Rheological differences between modified and unmodified bitumen

Many polymer modified bitumens tend to behave more as a polymer than as a bitumen. For a cross-linked SBS (or styrene–butadiene) polymer blend, four distinct regions are found

- a flow region, where the properties are primarily driven by the viscous response
- a plateau region, where the storage modulus is more or less independent of the frequency (or, conversely, temperature)
- a transition region at intermediate temperatures
- a glassy region, where the properties are entirely controlled by the physico-chemical make-up of the bitumen at low temperatures.

Conventional bitumens are characterised by a flow region that transitions into the glassy region (i.e. no truly visco-elastic region can be identified). Figure 7.7 shows schematically how cross-linking greatly enhances the elastic response relative to a conventional bitumen.

The effect of SBS modification on the rheological parameters (the complex modulus and the phase angle) has been combined in the form of Black space diagrams, an example of which is shown in Figure 7.8. The morphology, and therefore the rheological characteristics of the polymer modified bitumens, are functions of the mutual effect of the polymer and the bitumen and, consequently, are influenced by the bitumen composition and the polymer nature and content. At low polymer contents (up to 3%), the behaviour of the modified binder remains closer to that of the base bitumen. At

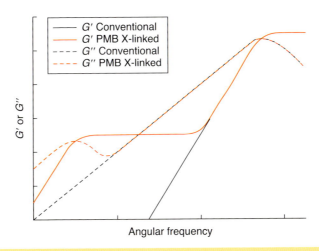

Figure 7.7 Comparison of elastic and viscous components as a function of angular frequency for modified and unmodified bitumens

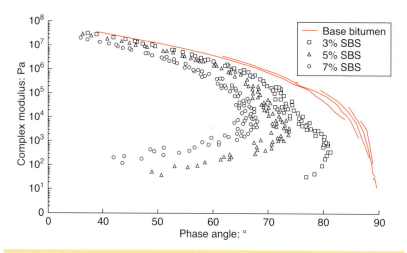

Figure 7.8 Black space diagram showing the effect on the rheology of bitumen with increasing SBS addition

higher polymer contents, the polymer network in the SBS polymer modified bitumens results in a continued reduction in the phase angle (increased elastic response) at low complex modulus values.

Isochronal plots of the complex modulus (G^*) and the phase angle (δ) versus the temperature at 0.02 Hz for a base bitumen with an increasing quantity of SBS are shown in Figures 7.9 and 7.10. Although there are only minor increases in G^* at low temperatures due to SBS modification, there is

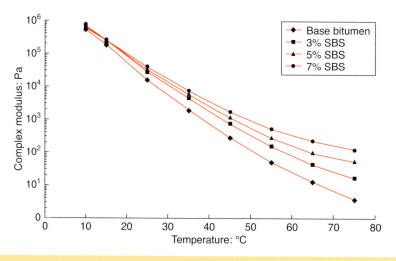

Figure 7.9 Isochronal plots of the complex modulus at 0.02 Hz for SBS modified bitumens

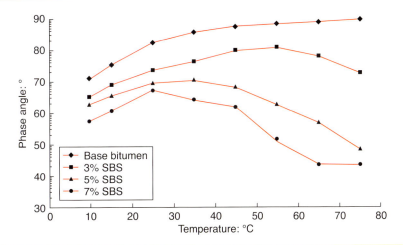

Figure 7.10 Isochronal plots of the phase angle at 0.02 Hz for SBS modified bitumens

considerable evidence of extreme polymeric modification at high temperatures, with the establishment of a plateau region indicative of a dominant polymer network.

The phase angle isochrones in Figure 7.10 illustrate the improved elastic response (reduced phase angles) of the modified binders compared with their respective base bitumen. Whereas the phase angles of the base bitumen approaches 90°, indicating predominantly viscous behaviour, with increasing temperature the SBS polymer significantly improves the elastic response of the modified binders. This increase in elastic response at high temperatures can be attributed to the viscosity of the base bitumen being low enough to allow the elastic network of the polymer to influence the mechanical properties of the modified binders.

References

AASHTO (2009) AASHTO TP 70-09. Standard method of test for multiple stress creep recovery (MSCR) test of asphalt binder using a dynamic shear rheometer (DSR). AASHTO, Washington, DC, USA.

AASHTO (2010a) AASHTO M 320-10. Standard specification for performance-graded asphalt binder. AASHTO, Washington, DC, USA.

AASHTO (2010b) AASHTO MP 19-10. Standard specification for performance-graded asphalt binder using multiple stress creep recovery (MSCR) test. AASHTO, Washington, DC, USA.

AI (Asphalt Institute) (2007) The Asphalt Handbook. AI, Lexington, KY, USA.

Anderson DA and Kennedy TW (1993) Development of SHRP binder specification. Journal of the Association of Asphalt Paving Technologists **62**: 481–507.

Traxler RN, Schweyer HE and Romberg HW (1944) Rheological properties of asphalt. *Industrial and Engineering Chemistry* **36(9)**: 823–829.

Van der Poel C (1954) A general system describing the visco-elastic properties of bitumen and its relation to routine test data. *Journal of Applied Chemistry* **4**: 221–236.

Williams ML, Landel RF and Ferry JD (1955) The temperature-dependence of relaxation mechanisms in amorphous polymers and other glass-forming liquids. *Journal of the American Chemical Society* **77**: 3701–3706.

Zacharias MP and Emery SJ (1994) Application of the Brookfield viscometer test method in South African road grade bitumen specifications. *Proceedings of the 6th Conference on Asphalt Pavements for Southern Africa*, Cape Town, South Africa, vol. 2, pp. 109–123.

Zanotto ED and Gupta PK (1999) Do cathedral glasses flow? – Additional remarks. *American Journal of Physics* **67(3)**: 260–263.

Chapter 8

Polymer modified bitumens and other modified binders

Dr Catherine Rodrigues
Global Technology Development Executive, Société des Pétroles Shell. France

Raghu Hanumanthgari
Product Researcher, Shell India Markets Pvt. Ltd. India

On the majority of roads, conventional bitumen grades possess satisfactory adhesion and mechanical properties for use in asphalt pavements for the vast majority of traffic and climatic conditions that are encountered. However, demands made on roads increase year by year and, in some cases, the limits of performance of conventional bitumens have been reached mainly due to

- increases in rainfall and temperature variations
- increases in axle weight, tyre pressures and freight movement, leading to a higher maintenance requirement and consequent increased costs for road owners
- a tendency to use thinner layers in pavements
- an ongoing demand for financial efficiency and cost savings to reduce the frequency of maintenance required in order to minimise disruption to traffic flow, and to provide an increased service life.

Thus, road failures happen earlier than expected, and maintenance downtime becomes more expensive for road owners. One means of overcoming these challenges is to modify bitumen with polymers or other additives to provide improved pavement performance. Modification technology depends on the specific applications and techniques (surface dressings, thin surface course systems, porous asphalt, surface courses for heavily trafficked roads, bases for heavily trafficked roads, anti-cracking membranes, bridge waterproofing layers, etc.). The degree of improvement required, and hence the cost, will depend on the particular needs of the site where treatment is proposed.

In this chapter, the main additives used for bitumen modification for road applications (Brown *et al.*, 1990) are described. The polymers most commonly used are elastomers, plastomers, reclaimed tyre rubbers and, to a lesser extent, viscosity modifiers and reactive polymers. Special asphalts for road and industrial applications, developed through particular processes such as

The Shell Bitumen Handbook, Sixth Edition
ISBN 978-0-7277-5837-8
Shell International Petroleum Company Ltd: All rights reserved
http://dx.doi.org/10.1680/tsbh.58378.149

the use of clear pigmentable binders, multigrade bitumens, cold application bituminous adhesives and a special bitumen additive for odour neutralisation are also discussed. Finally, basic properties of polymer modified bitumens (PMBs) and the related test methods, the current industry practice for manufacturing and handling PMBs, and some general information on the performance of PMBs for use on road applications are considered.

8.1 The role of bitumen modifiers in asphalt pavements

Traditional bitumens, derived from the distillation and/or blowing of crude oils, possess adequate performance characteristics, but increasing the high temperature performance can sometimes lessen the low temperature performance properties of the bitumen. The development of steps to counter this effect has been the incentive that has driven the early stages of the development of PMBs for use in roads. Modified binders are binders whose properties have been changed by the use of a chemical agent that, when added to the original bitumen, alters its chemical structure and physical and/or mechanical properties. As the bituminous binder is responsible for the visco-elastic behaviour of asphalts, it plays a large part in determining many aspects of road performance, particularly resistance to deformation and cracking, the two most common reasons for the structural failure of thicker pavements. In general, the proportion of any induced strain in asphalt that is attributable to viscous flow (i.e. non-recoverable flow) increases with both the loading time and the temperature. One of the prime roles of many bitumen modifiers is to increase the resistance of the asphalt to deformation at high road temperatures without adversely affecting the properties of the bitumen or asphalt at other temperatures. This is achieved by one of the two following methods, both of which result in a reduction in the permanent strain. The first approach is to stiffen the bitumen so that the total visco-elastic response of the asphalt is reduced. The second approach is to increase the elastic component of the bitumen, thereby reducing the viscous component. Increasing the stiffness of the bitumen is also likely to increase the dynamic stiffness of the asphalt. This will improve the load spreading capability of the material, increase the structural strength and lengthen the expected service life of the pavement. Alternatively, it may be possible to achieve the same structural strength but with a thinner layer. Increasing the elastic component of the bitumen will improve the flexibility of the asphalt. This is important where high tensile strains are induced.

For many years, researchers and chemists have experimented with modified bitumens, mainly for industrial uses, adding asbestos, special fillers, mineral fibres, plastomers, thermoplastic elastomers and rubbers. In the 1970s, the development of PMBs (Jiqing *et al.*, 2014; Taylor and Airey, 2008; Yetkin, 2007) and the use of additives for bitumen by controlled industrial processes in Europe was very much linked to the development of new mixture designs for asphalts such as thin surface course systems, which provide improved performance. For instance, the highly gap graded asphalts containing modified binders or additives, particularly fibres (asbestos, organic or mineral fibres), made it possible to increase the maximum binder content without the danger of the asphalt fatting up (the migration of binder to the surface of the asphalt causing a reduction in surface texture), providing the asphalt with the required level of cohesion and waterproofing the substrate. The addition of fibres helped in the development of stone mastic asphalt in Germany, subsequently adopted in many other countries. Another application of modified binders is in the anti-cracking membranes and waterproofing layers on steel and concrete bridges, because of their ability to withstand the considerable movements that occur in the surfacing of this type of construction. Modified binders have also been considered for structural layers because they improve fatigue strength. In the case of hot rolled asphalt, it is recognised that some polymers can improve the embedment of precoated chippings into the asphalt, particularly when laying takes place in less favourable weather conditions. In the case of open textured asphalts (essentially porous asphalt), the most important objectives for the modified binder are to improve its cohesive strength, which assists the resistance to weathering and its adhesion to aggregate particles, and in turn reduces the risks of the binder being stripped by water. Furthermore, the use of modified binders or fibres (the two additives are often combined for porous asphalts with very high void contents of over 25%) improves the resistance to binder drainage during transportation and laying, by increasing the binder viscosity. It allows higher binder contents to be used, increasing the thickness of the binder film, thus making the asphalt more durable by delaying binder ageing.

The most common modifying agents are polymers. These are macromolecules in which the same group of atoms is repeated many times. These repeated groups can be formed from one or several different molecules (monomers). Most of the additives assessed for modifying bitumen are shown in Table 8.1. Bitumens modified by polymers often prove to be the most cost-effective alternative to conventional bitumen because they improve targeted aspects of the performance of roads, and the polymers employed to modify bitumens are readily available at reasonable cost. This has led to the development of a wide range of proprietary asphalts made with PMBs and a range of PMBs that can be added to generic asphalts. The advantages of using these products have been proved in service for decades, while complying with product standards in different countries.

For the modifier (polymer, additives) to be effective and for its use to be both practicable and economic, it must:

Table 8.1 Examples of additives used to modify bitumen (Jiqing *et al*, 2014; Taylor and Airey, 2008; Yetkin, 2007)

Type of modifier	Examples	Abbreviation
Thermoplastic elastomers	Styrene–butadiene elastomer	SBE
	Styrene–butadiene–styrene elastomer (linear or radial)	SBS
	Styrene–butadiene rubber	SBR
	Styrene–isoprene–styrene elastomer	SIS
	Styrene–ethylene–butadiene–styrene elastomer	SEBS
	Ethylene–propylene–diene terpolymer	EPDM
	Isobutene–isoprene random copolymer	IIR
	Polyisobutene	PIB
	Polybutadiene	PBD
	Polyisoprene	PI
Latex	Natural rubber	NR
Thermoplastic polymers	Ethylene–vinyl acetate	EVA
	Ethylene–methyl acrylate	EMA
	Ethylene–butyl acrylate	EBA
	Atactic polypropylene	APP
	Polyethylene	PE
	Polypropylene	PP
	Polyvinyl chloride	PVC
	Polystyrene	PS
Thermosetting polymers	Epoxy resin	
	Polyurethane resin	PU
	Acrylic resin	
	Phenolic resin	
Chemical modifiers	Organometallic compounds	
	Sulfur	S
	Phosphoric acid, polyphosphoric acid	PA, PPA
	Sulfonic acid, sulfuric acid	
	Carboxylic anhydrides or acid esters	
	Dibenzoyl peroxide	
	Silanes	
	Organic or inorganic sulfides	
	Urea	
Recycled materials	Crumb rubber, plastics	
Fibres	Lignin	
	Cellulose	
	Alumino-magnesium silicate	
	Glass fibres	
	Asbestos	
	Polyester	
	Polypropylene	PP
Adhesion improvers	Organic amines	
	Amides	
Anti-oxidants	Phenols	
	Organo-zinc or organo-lead compounds	
Natural asphalts	Trinidad Lake Asphalt	TLA
	Gilsonite	
	Rock asphalt	

Table 8.1 Continued		
Type of modifier	Examples	Abbreviation
Fillers	Carbon black Hydrated lime Lime Fly ash	C
Reactive polymers	Random terpolymer of ethylene, acrylic ester and glycidyl methacrylate Maleic anhydride-grafted styrene–butadiene–styrene copolymer	
Viscosity modifiers	Flux oils (aromatics, napthenics, parrafinics) Fischer–Tropsch waxes	

- be readily available
- be cost effective
- blend with bitumen
- resist degradation at asphalt mixing temperatures
- improve resistance to flow at high road temperatures without making the bitumen too viscous at mixing and laying temperatures or too stiff or brittle at low road temperatures
- improve binder cohesion or adhesion properties.

In addition, the modifier, when blended with bitumen, should

- be capable of being processed using conventional equipment
- maintain its premium properties during storage, transportation, application and in service
- be physically and chemically stable during storage, transportation, application and in service
- achieve a coating or spraying viscosity at normal application temperatures.

8.2 The modification of bitumen
8.2.1 The modification of bitumen using thermoplastic elastomers

Thermoplastic elastomers include polyurethane, polyether–polyester copolymers and olefinic or styrenic block copolymers, but it is the styrenic block copolymers that have proved to have the greatest potential when blended with bitumen (Bonemazzi et al., 1996; Bull and Vonk, 1984; Lewandowski, 1994). Styrene block copolymers, also termed styrene elastomers, thermoplastic rubbers or thermoplastic elastomers, may be produced by a sequential operation of successive polymerisation of styrene and butadiene (resulting in styrene–butadiene elastomers, SBEs) or styrene and isoprene (resulting in styrene–isoprene elastomers). The linear copolymer (styrene–butadiene–styrene, SBS) consists of a central rubbery backbone chain joined at each

153

end to a polystyrene chain. Alternatively, a di-block precursor can be produced by successive polymerisation of styrene and a mid-block monomer, followed by reaction with a coupling agent. Thus, not only linear copolymers but also multi-armed SBEs can be produced. These polymers are often referred to as radial, star or branched copolymers. When added to bitumen, a radial SBE may result in a much higher viscosity and softening point when compared with a bitumen to which has been added the same level of linear copolymer (e.g. linear SBS or SB). This may make it more difficult to achieve good blending and dispersion of the radial polymer in a bitumen compared with a bitumen made with the linear block copolymers.

In general, SBEs derive their strength and elasticity from a physical cross-linking or entanglements of molecules: a cross-linking tri-dimensional network can be achieved by the association of the polystyrene end-blocks into separate domains to form semi-crystalline zones, providing the physical cross-links with the polybutadiene chains. Cross-links are formed at close to ambient temperatures but are broken at elevated temperatures, only to reform again on cooling, being then a three dimensional reversible network (Figure 8.1). SBEs are characterised by a glass transition point (T_g), and, as elastomers, exhibit increased tensile strength with elongation. They have the ability to revert to their initial condition after an applied load has been removed. It is the polystyrene end-blocks that impart strength to the polymer, and the rubber mid-block that gives the material its exceptional elasticity (Vonk and Gooswilligen, 1989). At temperatures above the glass transition point of polystyrene (100°C), the polystyrene softens as the domains weaken, and will even dissociate under stress, thus allowing easy processing. On cooling, the domains re-associate, and the strength and elasticity are restored (i.e. the material is thermoplastic). Once this has occurred, the level of shear exerted on the swollen particles is critical if a satisfactory dispersion is to be achieved within a realistic blending time at the manufacturing plant.

Ultraviolet (UV) fluorescence microscopy is a technique relying on the use of UV light to cause polymer molecules (but not the bitumen) to fluoresce at a wavelength that can be detected by the eye. Thus, the polymer molecules appear as bright objects against a dark (bitumen) background. Figures 8.2 and 8.3 show examples of typical morphologies of SBE blended into a bituminous matrix (Oliver and Khoo, 2012), the yellow nodules being the SBE particles dispersed in the black bitumen matrix. The addition of a thermoplastic polymer to bitumen makes a significant change to its physical properties, even at fairly low concentrations (<10%). Medium or, preferably, high shear mixers are required to disperse thermoplastic elastomers adequately into the bitumen matrix. The mixing and manufacturing of PMB will be discussed later in this chapter. When the SBE is added to the hot bitumen, the

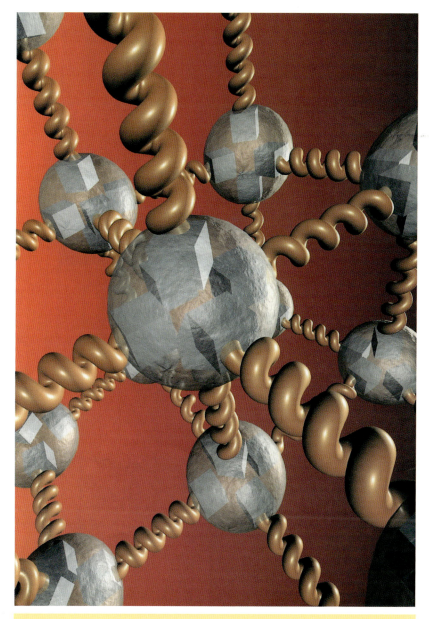

Figure 8.1 Schematic representation of SBS copolymer in bitumen – a three dimensional polymer network at ambient temperatures

bitumen constituents immediately start to penetrate the SBE particles, causing the styrene domains of the polymer to become solvated (the styrene domain of the SBS polymer becomes soluble with specific bitumen species). For instance, SBE, while absorbing the oily maltenes in the bitumen, may swell

Figure 8.2 UV fluorescence microscopy of 4.5% SBS blended with bitumen

by as much as nine times its volume when used at high concentrations (above 4%), and, from a morphology viewpoint, is forming a 'continuous polymer-rich phase' (Figure 8.2) in the bitumen. At lower concentrations (around 2–3%), the SBE will be dispersed in the bitumen matrix, giving a more 'bitumen-rich phase' (Figure 8.3).

The modification of bitumen morphology results in an increase in the softening point of the blend with the incremental addition of SBE, leading to a

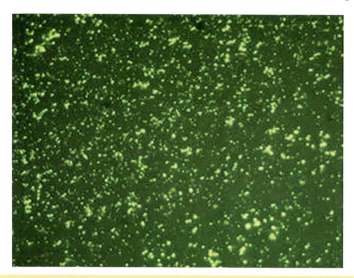

Figure 8.3 UV fluorescence microscopy of 3% SBS blended with bitumen

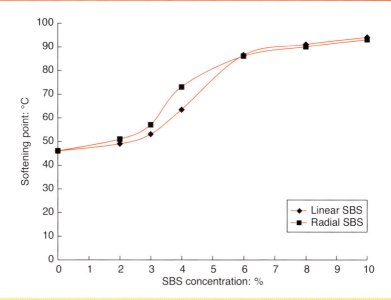

Figure 8.4 Typical 'S-shape' curve of SBS modified bitumen blends (softening point with increasing content of linear and radial SBSs)

typical 'S-shape' curve as shown in Figure 8.4. The rate of increase in the softening point of the modified binder also depends on the bitumen type, the SBE content, the SBE chemical structure and the grades of the SBS polymer.

As discussed in Chapter 4, bitumens are complex systems that can be subdivided into groups of molecules, saturates, aromatics, resins and asphaltenes. Saturates and aromatics can be viewed as carriers for the 'polar' aromatics (i.e. the resins and asphaltenes). The polar aromatics are responsible for the visco-elastic properties of the bitumen at ambient temperatures. This is due to the association of the polar molecules that leads to large structures, which in some cases may result in the generation of three dimensional networks (i.e. 'gel'-type bitumen). The degree to which this association takes place depends on the temperature, the molecular weight distribution, the concentration of the polar aromatics, and on the solvency power of the saturates and aromatics in the maltenes phase. If the concentration and molecular weight of the asphaltenes is relatively low, the result will be a 'sol'-type bitumen. The addition of thermoplastic elastomers with a molecular weight similar to or higher than asphaltenes disturbs the phase equilibrium: the polymer and the asphaltene will then 'compete' for the solvency power of the maltene phase, and, if insufficient maltenes are available, phase separation (also sometimes called de-mixing) between the polymer and the bitumen may occur. There are other parameters that have an effect on the compatibility of polymers with bitumen. These parameters may contribute to the instability of the system, depending on the handling conditions (e.g. the storage conditions of the PMBs such as time, temperature, efficiency of mixing system). The quality of the polymer dispersion in the bitumen phase is influenced by a number of factors

- the difference in solubility parameters of the polymers and the maltene phase of the bitumen
- the amount and type of asphaltenes present in the bitumen
- the type and concentration of the polymer
- the manufacturing processes and handling conditions of the SBE modified binder (Whiteoak, 1990).

The complexities of the inter-relationships between these factors are shown in Figure 8.5. It can be seen that very high aromaticity levels of the bitumen constituents lead to a weakening of the polystyrene domains (of the thermoplastic rubber), causing low softening points and low flow resistance properties (Collins *et al.*, 1991). At low aromaticity levels, however, insufficient polymer will be incorporated into the bitumen, which also leads to low flow resistance properties.

In this context, the molecular weight of the polymer directly relates to the solubility parameters (Nielsen and Charles, 2005; Redelius, 2000). It is thus easier to blend polymers of low molecular weight than those of high molecular weight. Solubility parameters can vary according to the SBS type and within different commercial grades of SBS. This will depend on the ratio

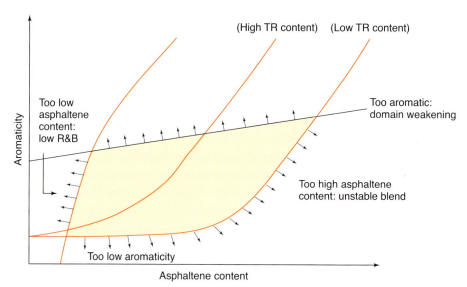

Figure 8.5 Effect of aromaticity and asphaltene content on the stability of thermoplastic rubber (TR) bituminous blends (Whiteoak, 1990)

of polymer segments in the case of copolymer structure and on the chemical structure of the polymer (linear or branched, molecular weight and nature of the double bonds inside the butadiene chains). Suppliers of polymers can provide different grades of SBEs for example, giving different properties to the bitumen (Bulatovic et al., 2013; Jianga and Zhang, 2011; Zhang et al., 2014).

The stability of SBE modified bitumens during handling (storage, transport and delivery to the customer site) can be enhanced by adding chemical co-modifiers such as peroxides, polyphosphoric acid, carboxylic anhydrides or esters, sulfur or any cross-linking materials. Sulfur vulcanisation, as used commonly in the rubber industry since 1839, is currently the predominant technology to improve the stabilisation of SBE modified bitumen products in a variety of handling conditions. The mechanism of vulcanisation by sulfur is a fairly complex chemical reaction (Akiba and Hashim, 1997; Krejsa et al., 1937; Weng et al., 2002). It is strongly dependent on the bitumen type, the polymer structure, the polymer concentration and the processing conditions at the plant. The SBE particles, which are sheared and shattered to a very small size, are homogeneously dispersed in the bitumen matrix. There are competing kinetic reactions involving sulfur compounds, SBE chains and bitumen species. From the literature, it can be concluded that a reaction between bitumen and sulfur (S) occurs predominantly within the aromatic and naphthenic centres (Ar) of the bitumen species, leading to higher asphaltene/resin ratios and to the formation of mono or polysulfide bonding such as Ar–S–Ar, Ar–SS–Ar, Ar–SSS–Ar and Ar–SSSSS–Ar (Weng et al., 2001). Fine SBE particles are wrapped by some aromatic components in the bitumen through sulfide or polysulfide links (Gawel, 2000), creating polymer–$(S)_x$–bitumen chains links (Figure 8.6).

These sulfur bonds contribute to the enhancement of polymer compatibility with bitumen, and may also influence the rheological properties of the binder, depending on the cross-link density and bonding lengths. Sulfur reacts with the carbon atoms next to the carbon–carbon double bonds of polybutadiene chains in SBEs by substituting allylic hydrogen atoms with sulfur by a free radical mechanism (Akiba and Hashim, 1997; Gawel, 2000; Krejsa et al., 1937; Weng et al., 2001, 2002). Thus, storage-stable sulfur cross-linked SBE modified bitumen is generally obtained by creating an irreversible three dimensional network, although the reactions between sulfur and SBE modified bitumens have not been studied extensively in the literature.

8.2.2 The modification of bitumen using plastomers
Some plastomers are used in the bitumen industry as an alternative to elastomers for road paving applications. Polyethylene, polypropylene, polyvinyl

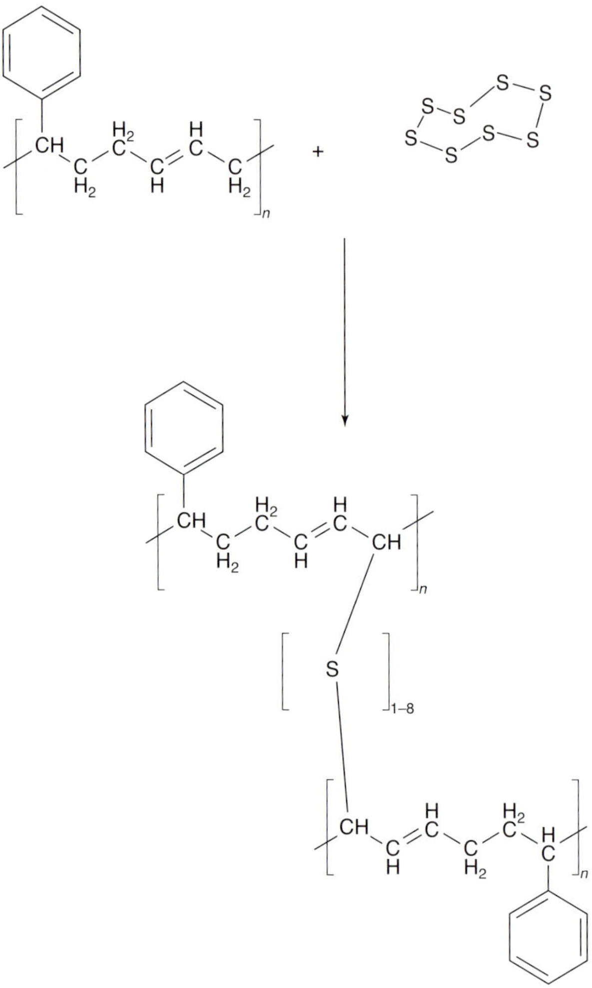

Figure 8.6 Representation of SBS–S–SBS bonding in bitumen

chloride, polystyrene and ethylene–vinyl acetate (EVA) copolymer are the main non-rubber thermoplastic polymers that have been examined in recent decades. As thermoplastic polymers, they are characterised by softening on heating and hardening on cooling. These polymers tend to influence the penetration more than the softening point properties when added to bitumen, which is the opposite tendency of thermoplastic elastomers. Polyolefins have been used for the modification of bitumen due to their relatively low cost and the benefits that they induce. The light components of bitumen are usually absorbed by the polyolefins. A bi-phasic morphological structure is formed (i.e. the polyolefin is dispersed in the bitumen matrix). As the concentration of the polymer increases, phase inversion phenomena occur, leading to a polymer matrix that can be detected in higher concentration polyolefin modified bitumen formulations. In addition to this, the regular and long chains of polyolefin materials have a tendency to pack closely and crystallise, which leads to a lack of interaction between the bitumen and the polyolefin and causes instability of the modified bitumen. The compatibility of polyolefins with bitumen is usually found to be very poor because of the non-polar nature of the polyolefin materials (Fawcett and McNally, 2000; Yousefi, 2003). Adding polyolefins to bitumen does not significantly increase its elastic properties, and, when heated, the polyolefin can separate, which may give rise to a coarse dispersion on cooling. PMBs obtained with olefinic polymers are used, but the storage stability of the system remains an unresolved issue. A possible solution is the use of polyethylene based copolymers, in which the co-monomer is polar, either inert or reactive with respect to the bitumen species. Typical examples are EVA and ethylene–butyl acrylate random copolymers (Gonzalez *et al.*, 2004; Isacsson and Lu, 1999), while maleic anhydride and glycidyl methacrylate groups being grafted to polyethylene chains or being incorporated by a copolymerisation process to polyethylene could interact with the reactive functions of bitumen species, leading to better compatibility of the polyolefin with the bitumen.

EVA copolymers, as thermoplastic materials, are commonly used for modifying bitumen. EVA has a random structure produced by the copolymerisation of ethylene and vinyl acetate. EVA copolymers with a low vinyl acetate content possess properties similar to low density polyethylene. As the level of vinyl acetate increases, the properties of the copolymer alter, which may induce changes in the properties of the bitumen. The properties of EVA copolymers are classified by molecular weight and vinyl acetate content as follows.

- *Molecular weight.* Standard practice for EVAs is to measure the melt flow index (MFI), a viscosity test that is inversely related to the polymer molecular weight: the higher the MFI, the lower the molecular weight and viscosity. This is analogous to the penetration test for bitumen: the higher the penetration, the lower the average molecular weight and viscosity of the bitumen.
- *Vinyl acetate content.* Regular polyethylene segments of the EVA chain pack closely together and form so-called 'crystalline' regions, and can be represented graphically, showing the main effects of vinyl acetate species on the properties of a bitumen (Figure 8.7). At the same time, the bulky vinyl acetate groups disrupt this closely packed arrangement to give 'non-crystalline' or 'amorphous rubbery' regions. The crystalline regions are relatively stiff, and have a considerable reinforcing effect whereas the amorphous regions are rubbery. Obviously, the more vinyl

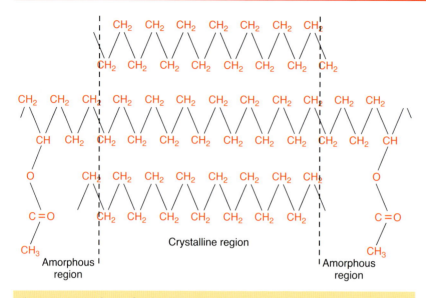

Figure 8.7 Packing of polyethylene in EVA copolymer

acetate groups present (or the higher the vinyl acetate content), the higher the proportion of rubbery regions will be, and, conversely, the lower the proportion of crystalline regions.

A wide range of EVA copolymers is available, specified by both the MFI and the vinyl acetate content. EVA copolymers are easily dispersed and have good compatibility with bitumen. They are thermally stable at the temperatures at which asphalt is normally mixed. However, during static storage, some separation between the polymer and bitumen may occur, and it is therefore recommended that the blended product should be thoroughly mixed before use.

8.2.3 *The modification of bitumen by the addition of rubbers*
Polybutadiene, polyisoprene, natural rubber, butyl rubber, chloroprene and random styrene–butadiene rubber, among other rubbers, have all been used to modify bitumen: their effect is mainly to increase its viscosity. Rubbers are also used in a vulcanised (cross-linked) state. For example, particles of crumb rubber (CR) reclaimed from old scrap tyres, are used to modify and improve the performance of conventional bitumen while providing an outlet for the disposal of waste tyres (Gillen, 2007; Oliver, 1999; WRAP, 2008). This technology began in the 1960s, and there are two main types of process, described as 'wet' and 'dry'. In the 'wet' process, the CR modifier is added to the bitumen while heating to 180–210°C, prior to the manufacture of the hot asphalt. In the 'dry' process, the CR modifier is mixed directly into the

conventionally prepared asphalt at the paddle mixer. The CR modifier acts as an 'elastic mineral aggregate', and replaces part of the aggregate compound. The description of the processes for manufacturing CR modified bitumen (CRMB) and its applications are detailed in section 22.5.

Typical concentrations of CR particles in bitumen are between 8 and 15%. High CR particle content (above 20%) allows the formation of a physical network that partially prevents the sedimentation of the particles. In all cases, when dispersed in bitumen, these crumb particles strongly influence its rheology and increase its viscosity, which may sometimes be detrimental to its workability. A high temperature is necessary (180–195°C), which unfortunately results in the generation of odours and emissions. In addition, storing the CRMB at high temperature may accelerate the process of particle sedimentation to the bottom of the tank or truck, which results in inconsistent supply quality. Accordingly, it is usually recommended that the CRMB blends are continuously agitated during storage to maintain the stability of the product, or even mixed immediately with aggregates. The preparation of these blends directly at the asphalt plant is also said to be advantageous. Chemical compatibility and processing conditions, such as time, temperature and shear, are important in obtaining the desired properties and stable binder product as per specification (Pérez-Lepe et al., 2003). Successful developments were made and patent applications on this were filed by Shell Bitumen (Shell Mexphalte RM®) to improve the homogeneity and the stability of CRMB during transportation to and storage at the asphalt plant. Other advances have resulted in limiting the maximum mixing and laying temperature to 170°C, thus avoiding the generation of odours and emissions from the rubber.

8.2.4 *The modification of bitumen by the addition of viscosity modifiers*

As bitumen is responsible for the visco-elastic behaviour characteristic of asphalt, it plays a large part in determining many aspects of road performance, particularly resistance to deformation (Edwards, 2005) and resistance to cracking (Prowell, 2005; Suleiman, 2011), the two most common structural failure modes in roads. (Structural failure is usually interpreted as requiring the replacement of the surface course, binder course and the base of the pavement.)

A wax additive is used to modify the binder, which itself is used subsequently to lower the mixing and compaction temperatures in asphalts. Several performance properties of asphalts can be improved by adding low molecular weight polyethylene, synthetic wax obtained from a Fischer–Tropsch synthesis or paraffin wax to bitumen. Fischer–Tropsch synthetic waxes have a microcrystalline structure, and consist of particles with a large number of carbon atoms, up to 100. Due to the morphology of waxes, they can be used to lower the viscosity of conventional bitumen at higher temperatures (typically above 120°C) and increase the softening point. In relation to the modification of viscosity, the following points should be noted.

■ At temperatures above the melting point of the wax additive, the viscosity of the bitumen is significantly reduced. This, in turn, allows for a reduction in asphalt mixing and laying temperatures by up to 30°C, thereby saving energy and reducing fume emissions. Also, critical applications, where hand laying may be required or material has to be laid at low ambient temperatures, may be facilitated as a result of the longer time period available between asphalt mixing and the time when compaction is no longer possible.

■ At the crystallisation temperature of the additive, the viscosity of the modified bitumen rises sharply, which, in turn, may lead to higher asphalt stiffness, thereby reducing the permanent deformation characteristics of the carriageway. However, the low temperature properties may be adversely affected.

Figure 8.8 shows typical viscosity–temperature relationships for an unmodified penetration grade bitumen and the same bitumen modified with a selected paraffin wax content. The magnitude and nature of the effect on bitumen rheology depends on the bitumen itself as well as on the type and quantity of additive. Bitumen composition is of critical importance, but there are many additives that can change the visco-elastic properties of bitumen: examples are natural bitumens such as rock asphalt, which harden bitumen, and paraffinic flux oils or aromatics, which soften bitumen.

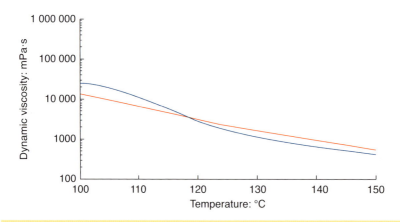

Figure 8.8 Dynamic viscosity profile of binders with wax (blue curve) and without wax modification (red curve)

8.2.5 *The modification of bitumen by reactive chemistry*

In addition to physical blends of bitumens and polymers, another way to improve the binder properties is through chemical modification: reactive ethylene terpolymers, comprising ethylene, ester groups of methyl, ethyl or butyl acrylate and glycidyl methacrylate groups (also known as reactive epoxy functions), can chemically react with bitumen species (carboxylic groups present in asphaltenes) and can be used to enhance asphalt performance and increase the compatibility between the polymer backbone and the bitumen, keeping the product stable during storage and transportation (Dupont, 2014; Kanabar, 2010; Keyf *et al.*, 2007). There are many other reactive polymeric materials described in the literature or in patent publications that can be used to enhance the properties of bitumen, examples being grafted maleic anhydride styrene block copolymers, polymers with high vinyl content, and hydroxyl, carboxylic or silane-grafted polyolefins (Chaverot *et al.*, 2012; Cong *et al.*, 2011; Crews and Kalinowski, 1954). As this involves various chemical reactions between the reactive functions of the polymer and bitumen species, in certain instances some catalysts such as phosphoric acid (PA) can be used. However, in such circumstances, processing conditions (mixing time, temperature, polymer content) should be carefully defined to avoid uncontrolled kinetic reactions and gelling of the reacted product. In the case of reactive ethylene terpolymers, PA can be used to catalyse the reactivity of glycidyl methacrylate (epoxy) groups with bitumens that have carboxylic reactive functions.

Among the different acids that can be used in bitumen, polyphosphoric acid (PPA) added at a low content (<2%) to improve asphalt binder properties through chemical modification occurs in numerous patented technologies described in articles and patent publications that appeared in the USA in the 1990s and 2000s. In the main, these lead to increased asphalt stiffness without deterioration of its low temperature properties (Baumgardner *et al.*, 2005; Masson, 2008), greater adhesion to aggregates and an increase in the performance grade range by one to two classification grades (e.g. an increase in the grade of the binder from PG70 to PG76 for the high temperature range) to meet Superpave – see also section 5.7. (Superpave specifications are commonly used in the USA – see section 12.4.5). PPA is different from orthophosphoric acid, with no free water, and is defined as an inorganic polymer modifier having different distributions of chain lengths, with the number of repeating units varying from one chain to another. The modification of bitumen with PPA appears to be a complex physico-chemical process, and the resulting properties may be strongly dependent on the nature of the bitumen. The investigation of the reactions occurring between the bitumen species and PPA is difficult due to the large number of molecules with different chemical structures and their possible interactions. Many mechanisms have been proposed in the literature to explain the increase in bitumen stiffness by PPA addition, for instance a model is proposed based on the reactivity of asphaltenes with PPA molecules, through an acid–base reaction mechanism and esterification reactions between those molecules (Orange *et al.*, 2004). These reactions may cause a de-agglomeration of the aggregates in the asphaltenes, which improves bitumen rheology. In combination with a polymer or a rubber, a positive effect also seems to occur: adding very small amounts of PPA jointly with SBS or CR to bitumen with a suitable formulation is used to improve the handling and performance of PMBs (TRB, 2009; Zhang and Hu, 2013), achieving higher Superpave performance grades while improving the mixing and compaction characteristics. (Superpave performance grades are commonly used in the USA. They apply to the performance of asphalts, and are based on the proposition that the properties of a bitumen should be related to the conditions under which it is to be used.)

8.3 Other modified products used for road and industrial applications

8.3.1 Multigrade bitumens

If it was possible to create the ideal pavement binder, it would have a uniform designed 'stiffness' and 'flow' behaviour across the operational temperature range to combat both softness/deformation and fatigue/brittleness. Multigrade bitumens are a step in this direction, as they are designed to be less temperature susceptible than penetration grade bitumens, resulting in improved performance at high temperatures and better low temperature characteristics than hard grades, making them ideal for use in surface or binder courses. Multigrade bitumens are generally manufactured through a unique processing route that involves catalytic oxidation using PA. This process makes the product perform as both

- a soft bitumen at low temperatures
- a hard, stiffer bitumen at higher temperatures, but not so stiff that premature cracking occurs at low temperatures.

Multigrade bitumens are specified in national/international bitumen standards such as the Australian AS 2008 standard (AS, 2008) and the European EN 13924-2 standard (BSI, 2014), to provide a framework for specifying the properties and relevant test methods for multigrade bituminous binders.

8.3.1.1 Shell Multiphalte®

In the 1990s, Shell developed and introduced to the market a range of proprietary multigrade bitumens called Multiphalte after more than 10 years of laboratory research and field trial experience (Koole *et al.*, 1991).

These products are part of a well established technology with an improved viscosity–temperature relationship designed to enhance stiffness and viscosity at high service temperatures, with the aim of substantially improving deformation resistance. At low temperatures they have lower stiffness, and are therefore less likely to crack.

This can be seen, for example, when a Multiphalte bitumen is compared with a conventional penetration grade bitumen on the bitumen test data chart (Figure 8.9). The top left corner of the chart shows that at low temperatures Multiphalte is less susceptible to low temperature cracking than conventional bitumen, meaning that as a component in an asphalt it is less brittle at low winter temperatures. The bottom right corner shows that, for the same high temperature, Multiphalte has a higher viscosity than conventional bitumen, meaning that it improves the resistance of pavements to deformation and reduces stripping during high summer road temperatures.

Shell Multiphalte is produced and used in asphalt applications (sealing, surface or binder course applications) worldwide (Europe, Argentina, Australia, etc.), typically in the pen range 20/30, 35/50 and 45/60, although some grades can be tailor-made to suit specific market requirements and countries. In France, the high modulus asphalt version (Multiphalte HM 'High Modulus' with a 20/30 penetration range) is available to satisfy certain French asphalt specifications. There are two Multiphalte grades in Australia: the M1000 grade is designed to resist rutting, and is often used on airport runways, while the M500 grade is more commonly employed in sprayed sealing applications.

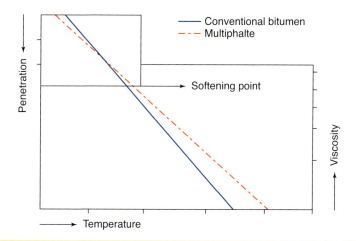

Figure 8.9 Bitumen test data chart illustrating the properties of a Multiphalte bitumen compared with a conventional bitumen

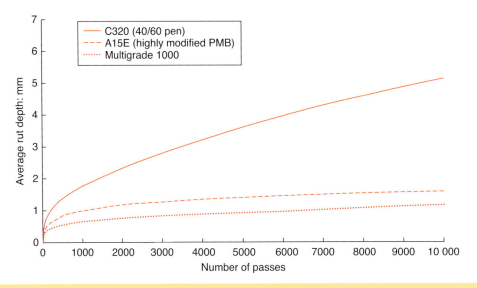

Figure 8.10 Comparison of the deformation resistance of Multiphalte M1000, conventional penetration grade C320 bitumen and highly modified bitumen (A15E)

Figure 8.10 illustrates the concept of Shell's Multiphalte bitumen, showing the improved resistance to deformation of the M1000 grade compared with reference asphalts made with C320 bitumen (40/60 pen) and a highly modified PMB (A15E), using the wheel track tester. The wheel tracking machine measured the rut depth generated by a loaded wheel at 60°C over 10 000 passes, following the Australian reference standard AG:PT/T231 (AS, 2006).

Asphalts manufactured with Multiphalte can be mixed, transported, laid and rolled with minimal changes to traditional and well established procedures. The most critical difference is that it is somewhat more viscous than conventional bitumen, and consequently needs to be mixed and laid at a temperature 10°C higher than normal. All other paving operations remain within the normal range.

8.3.2 Synthetic pigmentable binders

The different methods of achieving a coloured asphalt surfacing are described in the literature (Pierard *et al.*, 2013). One important way of producing a coloured surface is to use a pigmentable binder. The most common pigmented asphalt is red, which is produced by substituting ferrous oxide for the same proportion of fines (usually 0.5–6%, depending on the purity of the ferrous oxide), which has a size <0.063 mm. Appropriately coloured aggregates can be used to ensure that the overall appearance of the material is maintained when the aggregate is exposed after trafficking. The main draw-backs to colouring mixtures using conventional bitumens are that

- the only acceptable colour that can be achieved is a fairly dark red
- the quantity of the expensive ferrous oxide required to achieve an acceptable red is fairly high, which significantly increases the cost of the mixture.

There are many applications for coloured asphalts, as illustrated in Figures 8.11–8.13, including architectural applications and use in historical monuments, parks and squares, sports venues and stadiums, boardwalks and promenades (bicycle and pedestrian lanes), office buildings, car parks, bus lanes, speed regulation, crossing paths in roundabouts, school exits, tunnels, multifunctional lanes (Genardini, 1994; Gustafssen, 1988; Lohan, 1999; O'Connor, 1999).

8.3.2.1 Shell Mexphalte C®

To enable asphalt to be pigmented in colours other than red, Shell developed a range of synthetic clear binders (Mexphalte C) for various applications that contain no asphaltenes (Le Coroller and Herment, 1989; Schellekens and Korenstra, 1987). These can be pigmented as required, to produce a broad spectrum of coloured asphalts enabling pavements to be matched to the local environment. Mexphalte C products possess rheological and

Figure 8.11 Tunnel de Lorentweiler, Luxembourg

Figure 8.12 Bus lanes in Tours, France

Figure 8.13 Parking area in Orange, France

mechanical properties similar to penetration grade bitumen, and the temp-erature at which the synthetic binder is mixed with aggregates is usually lower than the corresponding penetration grades. Coloured asphalts manu-factured using Mexphalte C products require around 0.5–1% of pigment to achieve a satisfactory colour, whereas only red can be achieved with penetration grade bitumen using up to 6% ferrous oxide in the mixture.

Inorganic pigments can be mixed alone or in combination with a synthetic binder, to give a range of colours, examples being

- ferrous oxide for red, yellow, brown and black
- titanium dioxide for white
- chromium oxide for green
- cobalt oxide or cobalt aluminate for blue.

A range of Mexphalte C products is available for multiple applications.

- Mexphalte C 70/100 is a clear synthetic binder. A clear cationic synthetic emulsion modified with polymers can also be formulated and used as a slurry seal and for surface dressing applications (light traffic only).
- Synthetic binder modified with polymers: Mexphalte C LT requires lower asphalt mixing and paving temperatures, Mexphalte C P2 is used for heavy traffic and Mexphalte C 35/50 is used for special applications, but note that it has to be handled at high temperatures.
- Mexphalte C P3J is a mastic joint solution that combines the durability of a PMB and a clear synthetic binder.

8.3.3 Modified binders for industrial applications

For industrial applications, Shell has developed a wide range of high performance PMB products used in Europe and Asia, under the brand names of Shell Tixophalte®, Shell Caritop® and Shell Flintkote®.

8.3.3.1 Shell Tixophalte®

The first development of a cold applicable bituminous adhesive with improved wet or tack properties under the brand name of Tixophalte was undertaken in Shell's R&D laboratory in The Netherlands. Tixophalte was first trialled in 1981, and this product has since been reformulated several times to comply with health and safety regulations, with new variants being developed for specific applications – revisions include the bitumen, the polymer, additives, the mineral filler, and use of a non-chlorinated and non-toxic solvent (Deygout and Seive, 2004).

The current formulation of Tixophalte is a ready to use bituminous mastic that can be applied as a filler, joint sealer or adhesive for any job, from large scale industrial applications to domestic situations. Its waterproofing properties mean it can be used underwater. It is easy to apply, and offers long lasting protection and waterproofing. It is capable of withstanding movement, weathering, chemicals and other environmental factors.

The main uses for this material are

- in building applications (sealing leaks and making emergency repairs even on wet surfaces, gluing isolation panels onto metal or concrete surfaces, fixing roofing felt overlaps and many other roofing uses, plumbing and waterproofing tasks, sealing joints/connections in chimneys, guttering, skylights and water evacuations)
- in infrastructure applications (sealing joints and drainage systems in concrete, drainage channels and concrete bridge decks, sealing induction loops, sealing between tram rails and asphalt)
- in hydraulic applications (for waterproofing cracks and joints in waterways, water canals, reservoirs, dams and bridges).

8.3.3.2 Shell Caritop®
Caritop is a range of polymer modified binders designed and manufactured by Shell specifically for use in the industrial market. Examples of its use include

- the manufacture of carpet tiles (lower mixing temperature during the manufacturing process and an increased filler content or a wider use of filler types, and improved mechanical performance of the finished tile)
- the production of roofing felts (improved resistance to high and low temperatures, improved cold bend and adhesion properties, and reduced risk of damage during installation)
- the production of mastic flooring (improved high temperature stability and low temperature flexibility, increased tolerance to building movement, better durability and increased fatigue resistance).

8.3.3.3 Shell Flintkote®
From its conception over 50 years ago as a cost effective and reliable waterproof coating, the Flintkote product range has expanded to create an integrated portfolio of products (Shell Flintkote Colourflex for exposed waterproofing and decoration) that are designed to work together, offering a seamless barrier to keep water out. Their wide range of uses includes waterproofing wet areas (e.g. bathrooms, kitchens and toilets), waterproofing metal roofs, waterproofing blockwork and concrete structures, waterproofing mastic flooring, waterproofing flat concrete roofs and mastic asphalt roofs, the treatment of water tanks and the protection of steelwork.

8.3.4 *Shell Bitufresh®: a special bitumen additive for odour neutralisation*

Bitumen and asphalt operations are facing increasing public concern about odour emissions, especially for production facilities located in urban areas. Bitumen and asphalt producers are seeking appropriate solutions in order to resolve disputes, maintain their image and ensure the long term viability of their production sites. These complaints can necessitate significant investment to treat expelled gases, and may even threaten the continued operation of the asphalt plant. Although treatments of expelled gases are often effective in reducing concentrations of volatile organic compounds, these systems require modification of the facilities, making their use rather costly. It is for this reason that Shell undertook intensive research in early 2006 in collaboration with a partner that specialised in the treatment of odours. This led to the joint development of the Shell Bitufresh additive that can be introduced into hot bitumen with ease and is designed to minimise bitumen odour conditions effectively for both workers during asphalt mixing and laying operations, and also to reduce odour nuisance for local residents in areas where bitumen is being used (SIA, 2008). The mechanism employed by this product is fundamentally different from other bitumen additives that work by masking the smell. Bitufresh has proven to be effective at low dosage levels because of its well balanced mixture of neutralising components (around 40–60 ppm in bitumen, as depicted in Figure 8.14). This has been supported by both laboratory and field trials undertaken with external partners. After treatment, the characteristics of the bitumen remain unchanged and bitumen odour neutralisation continues for a period of at least 2 weeks. A patent application for this technology was filed in 2007.

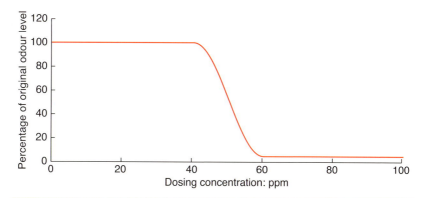

Figure 8.14 Effectiveness of Bitufresh in reducing bitumen fume odour. (From an external evaluation performed by Science Industry Australia (SIA))

8.4 Properties of PMBs and related test methods

PMBs used in road applications are normally tested using traditional bitumen tests, investigating, for example, penetration, softening point, Fraass breaking point and viscosity, and properties after ageing in a rolling thin-film oven and a pressure ageing vessel. In addition, there are often specific tests, such as for storage stability, elastic recovery, force ductility, and more complex rheological characterisation using the dynamic shear rheometer and the bending beam rheometer. The methods used for such tests may follow EN (for Europe) or ASTM (for USA, Asia) standards, depending on the country in which the PMBs are produced and sold.

Phase separation or incompatibility between the polymer and bitumen components can be demonstrated by a simple hot storage test in which a sample of the PMB is placed in a cylindrical tube and stored vertically at elevated temperature, usually between 160 and 180°C, in an oven, typically for 2 or 3 days or longer, depending on the test method or specification. At the end of the storage period, the top and bottom of the sample are separated and tested. Incompatibility is usually assessed by the difference in softening point between the top and bottom samples – if the difference is less than a range of 2–5°C, the binder is usually considered to be storage stable, depending on local specifications. The storage stability test is usually carried out as per ASTM D7173 (ASTM, 2011) or EN 13399 (BSI, 2011).

The main factors influencing storage stability and binder properties are

- the amount and molecular weight of the asphaltenes and the aromaticity of the maltene phase
- the amount of polymer present and its molecular weight and structure
- the storage conditions (temperature, time, mixing systems).

Shell Bitumen has developed a number of compatible and stable SBE-modified binders, obtained from sulfur cross-linking technology or reactive chemistry, for a variety of surfacing applications, including deformation resistant hot rolled asphalt, stone mastic asphalt and thin surface course systems. The philosophy of producing a compatible system is based on the practicalities and user friendliness of the bitumen to facilitate the medium and long term storage that is often required. The effect of compatibility on storage stability can be seen in Figure 8.15, which shows the numerical difference in the softening point of the top and bottom samples of two bitumens containing 7% SBS after 1, 3, 5 and 7 days' storage at 140°C. The results clearly show that the compatible system is extremely stable whereas the incompatible system has separated dramatically after 7 days' storage. The implications of this for storage are obvious. In practice, it is possible to use PMBs that are not storage stable. However, they must be handled with great care, and

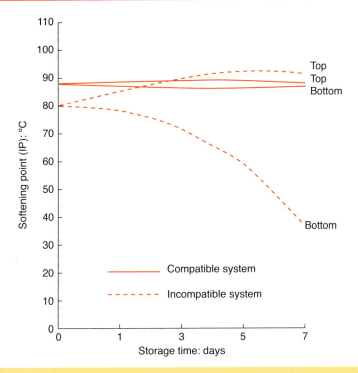

Figure 8.15 Effect of the compatibility of the polymer in the bituminous matrix on the storage stability test result

stored in tanks with stirrers or extended circulation to prevent separation of the polymer.

The elastic recovery test is a useful method for assessing whether a polymeric material that has been added to a bitumen provides a significant elastomeric characteristic. In this test, the PMB specimen is pulled to a specified distance at a specified speed and at a specified temperature (e.g. 25 or 10°C), then cut into two parts as seen in Figure 8.16, and the percentage of recoverable strain from the elongated specimen is measured after a fixed time period. The test methods used are ASTM D6084 (ASTM, 2013) and EN 13398 (BSI, 2010a).

The property of cohesion allows PMBs to be differentiated from unmodified paving grade bitumens by several test methods: force ductility, the tensile test and the Vialit pendulum reflect different characteristics of PMBs, and are not considered to be equivalent. It is intended that the most appropriate cohesion test method should be used in each case, at the discretion of the supplier. Ductility is that property of bitumen that permits it to undergo substantial deformation or elongation. The test method describes the procedure for determining the ductility of an asphalt measured by the distance by which it will

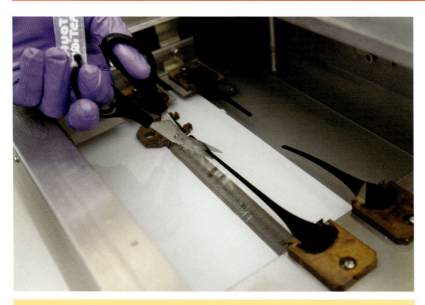

Figure 8.16 Elastic recovery test

elongate before breaking, as seen in Figure 8.17. Unless otherwise speci-fied, the test can be undertaken at ambient (25°C) or lower temperatures (5°C) and at a speed of 5 cm/min. At other temperatures, the speed should be specified. The testing can be carried out as per ASTM D113 (ASTM,

Figure 8.17 Ductility test

2007), EN 13589 (BSI, 2008), EN 13703 (BSI, 2003) or a method speci-fied in a particular country. Meeting the ductility specification is one of the major criteria for PMBs in Asia and China, and force ductility is mostly applied in Europe as per the PMB product specification EN 14023 (BSI, 2010b) using various temperatures and traction speeds. To quantify this effect, the force required to stretch the asphalt sample is recorded during the measurement phase. The area below the resulting force–distance function can be calculated, and represents the cohesive energy.

8.5 Manufacture, storage and handling of PMBs

The manufacture of PMBs can be undertaken in either fixed or mobile plants. The basic scheme, however, remains the same, although the utilities avail-able may vary according to what is available locally. Mobile plants are more dependent on associated sites for utilities, which may include hot oil, steam and electricity. These plants are mainly used at large construction sites for short periods. Fixed plants, however, are independent depots or associ-ated with refineries, and are intended to provide continuous long term production; utilities are provided on a permanent basis at such plants. Several companies specialise in constructing complete PMB plants.

PMBs generally consist of a blend of a bitumen and a polymer, usually an SBE. In some cases, other ingredients are added to the base bitumen, to assist in blending and achieving particular properties. In addition to the normal safety, quality and environmental considerations involved in handling hot bitumen, the manufacture of PMBs requires

- accurate blending of materials and control of manufacture, to ensure that the products conform to target specifications
- monitoring and control of the temperature at all phases of manufacture, storage, transport and field usage, to avoid premature deterioration of polymers at high temperatures
- maintaining the homogeneity of some PMB mixtures that may segregate in storage
- avoiding contamination with other products that may alter the performance characteristics of the end product.

A PMB manufacturing processes can be described in four main steps

1 base bitumen is heated to the required temperature of 180–190°C and fed to the tank
2 fluxing, where a suitable low viscosity non-volatile component is added to the tank (this stage is optional, and may be required for correction of the properties of the final product, or improving the compatibility between the polymer and the bitumen)

3   polymer addition and dissolution into bitumen with a high shear mill
    (using polymer pellets/beads) with simple or multiple passes, and/or
    in low shear mixing (using the polymer in a powdered form) within the
    temperature range 180–190°C

4   conditioning of the PMB blend under low shear mixing for a few
    hours, thus allowing the polymer to swell into the bitumen and ensuring
    a homogeneous mixture.

Batch or continuous processes can be used for the manufacture of PMBs.
Prior to the mill, there is almost always a pre-mix tank to permit charging
of the polymer into the bitumen, after which the combination can be
passed through the mill system to the conditioning tank. Polymer pellets are
sheared in the mill and mixed into the base bitumen. This mixture is then
permitted to swell for a certain period of time, thus allowing the polymer
to disperse fully. When the PMB product is homogeneous, it is ready for
use.

The master batch process involves the production of a PMB concentrate using
a high dosage of polymer. Usually this takes longer to manufacture, but the
PMB concentrate can be diluted with conventional bitumen to produce the
end grade required with a lower polymer content.

There are several considerations relating to the storage of PMBs. It is important to remember that both the handling of the material and the storage temperature are critical if the quality of the product is to be maintained. The tanks
for storage of the finished PMB are designed to minimise deterioration of the
product in storage, including strict control of temperature, a minimal surface
area to reduce the potential for oxidation, and mixing or circulating equipment to ensure that the product remains homogeneous during the storage
period. Shell Bitumen has handling guidelines for PMBs, with different temperature requirements depending on the length of storage for the Shell
Cariphalte® PMBs. Storage recommendations for Cariphalte are usually as
follows

- for short term storage (3–4 days), the recommended temperature is
  around 165–175°C
- for long term storage (up to 2 weeks), the recommended temperature is
  below 140–160°C
- for very long term storage (more than 2 weeks), the recommended
  temperature is below 100°C.

For truck loading of the PMB in the loading gantry, all supply lines throughout
the PMB plant, including the loading equipment, are designed and procedures established so as to avoid any contamination during a change in
product or cleaning of the supply lines. The use of oils such as kerosene,

diesel or gas oil for flushing lines must be avoided. Where lines need to be flushed, it is usually done with hot bitumen or a finished product.

8.6 Performance of PMBs for road application: Shell Cariphalte® example

PMBs provide improved balance between low and high temperature properties, such as high temperature stiffness (Figure 8.18) and low temperature flexibility. For instance, a pavement needs to be resistant to deformation at high temperatures during summer and to cracking during winter when the temperature of the pavement can fall to negative values. So, it needs both elastic and viscous components, and each needs to be dominant in the different temperature domains. In addition, increasingly aggressive traffic conditions require higher performance binders, economic considerations such as the desire for thinner pavements are the impetus for the production of premium binders, and quality issues are behind the desire for improved cohesion and adhesion properties in all pavements.

Shell PMB products are sold to customers under the brand name Cariphalte, providing solutions for different types of roads and meeting different customer needs and applications. A few examples of the uses of Cariphalte PMBs are given below:

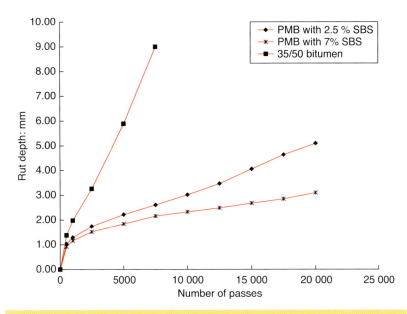

Figure 8.18 Anti-rutting performance results from the Hamburg wheel tracking test performed under warm water at 50°C (stone mastic asphalt 0/11 S granular curve using 35/50 bitumen and 2.5% and 7% SBS modified bitumens)

- in surface courses such as thin surface course systems and those made with stone mastic asphalt
- in drainage (porous) asphalt applications
- for an increased level of fuel resistance (Cariphalte Fuelsafe, specially designed to make asphalt surfaces last longer when exposed to the potential damage caused by spilt and leaking fuel)
- for motor racing circuits and test tracks (Cariphalte Racetrack®)
- for asphalts designed to possess high resistance to deformation and high flexibility, especially in binder courses
- for high performance road applications in combination with reclaimed asphalt pavement (Cariphalte RC)
- as the absorbing layer in pavements subject to reflection crack stresses (Cariphalte SAMI).

References

Akiba M and Hashim AS (1997) Vulcanization and crosslinking in elastomers. *Progress in Polymer Science* **22(3)**: 475–521.

AS (Standards Australia) (2006) AG:PT/T231. Deformation resistance of asphalt mixtures by the wheel tracking test. AS, Sydney, Australia.

AS (2008) AS 2008-1997. Residual bitumen for pavements. AS, Sydney, Australia.

ASTM (American Society for Testing and Materials) (2007) D113-07. Standard test method for ductility of bituminous materials. ASTM, West Conshohocken, PA, USA.

ASTM (2011) D7173-11. Standard practice for determining the separation tendency of polymer from polymer modified asphalt. ASTM, West Conshohocken, PA, USA.

ASTM (2013) D6084-13. Standard test method for elastic recovery of bituminous materials by ductilometer. ASTM, West Conshohocken, PA, USA.

Baumgardner GL, Masson JF, Hardee JR, Menapace AM and Williams AG (2005) Polyphosphoric acid modified asphalt: proposed mechanisms. *Proceedings of the Association of Asphalt Paving Technologists*. AAPT, Lino Lakes, MN, USA, pp. 283–305.

Bonemazzi F, Braga V, Corrieri R, Giavarini C and Sartori F (1996) Correlation between the properties of polymers and polymer modified bitumens. *Eurobitume and Eurasphalt Conference*, Strasbourg, Germany.

Brown SF, Rowlett RD and Boucher JL (1990) Asphalt modification, highway research: sharing the benefits. *Proceedings of the Conference on the United States Strategic Highway Research Program*. Thomas Telford, London, UK, pp. 181–203.

BSI (British Standards Institution) (2003) BS EN 13703:2003. Bitumen and bituminous binders. Determination of deformation energy. BSI, London, UK.

BSI (2008) BS EN 13589:2008. Bitumen and bituminous binders. Determination of tensile properties of modified bitumen by the force ductility method. BSI, London, UK.

BSI (2010a) BS EN 13398:2010. Bitumen and bituminous binders. Determination of the elastic recovery of modified bitumen. BSI, London, UK.

BSI (2010b) BS EN 14023:2010. Bitumen and bituminous binders. Specification framework for polymer modified bitumen. BSI, London, UK.

BSI (2011) BS EN 13399:2011. Bitumen and bituminous binders. Determination of storage stability of modified bitumen. BSI, London, UK.

BSI (2014) BS EN 13924-2:2014. Bitumen and bituminous binders. Specification framework for special paving grade bitumen. Multigrade paving grade bitumen. BSI, London, UK.

Bulatovic VO, Rek V and Markovic KJ (2013) Influence of polymer types on bitumen engineering properties. *Materials Research and Innovations* **17(3)**: 189–194.

Bull AL and Vonk WC (1984) Thermoplastic rubber/bitumen blends for roof and road. *Shell Chemicals Technical Manual TR 8.15*. Shell International Petroleum Company, London, UK.

Chaverot P, Godivier C, Leibler L, Iliopoulos I and Leach AK (2012) *Graft Polymer and Thermoreversibly Cross-linked Bitumen Composition Comprising Said Graft Polymer*. US Patent 0059094.

Collins JH, Bouldin MG, Gelles R and Berker A (1991) Improved performance of paving asphalt by polymer modification. *Proceedings of the Association of Asphalt Paving Technologists* **60**: 43–79.

Cong P, Chen S and Chen H (2011) Preparation and properties of bitumen modified with the maleic anhydride grafted styrene–butadiene–styrene triblock copolymer. *Polymer Engineering and Science* **51(7)**: 1273–1279.

Crews LT and Kalinowski ML (1954) *Graft Polymer-fortified Bitumen Additives*. US Patent 2812339.

Deygout F and Seive A (2004) *Bitumen Composition, a Process of Preparing a Bitumen Composition, and the Use of a Bitumen Composition as an Adhesive*. Patent WO 2004033547 A3.

Dupont (2014) Technical studies and expert papers confirm performance of Elvaloy RET. See http://www2.dupont.com/Asphalt_Modifier/en_US/Technical_References/elvaloy-ret-references.html (accessed 18/9/2014).

Edwards Y (2005) *Influence of Waxes on Bitumen and Asphalt Concrete Mixture Performance*. PhD thesis, KTH – Royal Institute of Technology, Stockholm, Sweden. See http://www.diva-portal.org/smash/get/diva2:14419/fulltext01.pdf (accessed 18/9/2014).

Fawcett AH and McNally T (2000) Blends of bitumen with various polyolefins. *Polymer Journal* **41(14)**: 5315–5326.

Gawel I (2000) Sulphur modified asphalts. In *Asphaltenes and Asphalts* (Yen TF and Chilingarian GV (eds)), vol. 2. Elsevier, Amsterdam, The Netherlands, pp. 515–535.

Genardini C (1994) Mexphalte C – for road surfacing in tunnels. *Shell Bitumen Review* **67**: 20–21.

Gillen S (2007) *Preliminary Summary Report on Ground Tire Rubber (GTR) Asphalt Pavement Demonstration Project*. Applied Research Associates, Albuquerque, NM, USA.

Gonzalez O, Munoz ME, Santamaria A, *et al.* (2004) Rheology and stability of bitumen/EVA blends. *European Polymer Journal* **40(10)**: 2365–2372.

Gustafssen P (1988) Shell Mexphalte C, Liseburg amusement park in Gothenburg, Sweden opened for the 1987 season in Shell colours. *Shell Bitumen Review* **63**: 12–14.

Isacsson U and Lu X (1999) Characterization of bitumens modified with SEBS, EVA and EBA polymers. *Journal of Materials Science* **34(15)**: 3737–3745.

Jianga Y and Zhang Y (2011) Interaction and molecular weight distribution behavior in polymer blends of styrene–butadiene–rubber and bitumen. *Advances in Building Materials* **168–170**: 973–980.

Jiqing Z, Björn B and Niki K (2014) Polymer modification of bitumen: advances and challenges. *European Polymer Journal* **54**: 18–38.

Kanabar N (2010) *Comparison of Ethylene Terpolymer, Styrene Butadiene, and Polyphosphoric Acid Type Modifiers for Ashphalt Cement.* MA thesis, Department of Chemistry, Queen's University, Kingston, ON, Canada.

Keyf S, Ismail O and Çorbacioglu OB (2007) The modification of bitumen with synthetic reactive ethylene terpolymer and ethylene terpolymer. *Petroleum Science and Technology* **25**: 561–568.

Koole R, Valkering K and Lançon D (1991) Development of a multigrade bitumen to alleviate permanent deformation. *Australian Pacific Rim AAPA Asphalt Conference*, Sydney, Australia.

Krejsa M, Koenig L and Nutting R (1937) Mechanism of rubber vulcanization with sulfur. *Industrial and Engineering Chemistry* **29(10)**: 1135–1144.

Le Coroller A and Herment R (1989) Shell Mexphalte C and Colas Colclair, a range of products for all road surfacing processes. *Shell Bitumen Review* **64**: 6–10.

Lewandowski H (1994) Polymer modification of paving asphalt binders. *Rubber Chemistry and Technology* **67(3)**: 447–480.

Lohan G (1999) Mexphalte C – adding colour to Ireland's history. *Shell Bitumen Review* **69**: 11.

Masson JF (2008) Brief review of the chemistry of polyphosphoric acid (PPA) and bitumen. *Energy and Fuels* **22(4)**: 2637–2640.

Nielsen TB and Charles M (2005) Elastomer swelling and Hansen solubility parameters. *Hansen Polymer Testing* **24**: 1054–1061.

O'Connor G (1999) Mexphalte C – bus lane differentiation in Sydney. *Shell Bitumen Review* **69**: 12–13.

Oliver J (1999) *The Use of Recycled Crumb Rubber.* Australian Road Research Board, Vermont, Australia, APRG Technical Note 10.

Oliver J and Khoo KY (2012) The effect of styrene butadiene styrene (SBS) morphology on field performance and test results: an initial study. *25th ARRB Conference.* Perth, Australia, APT 197 12 Austroads technical report.

Orange G, Martin JV, Farcas F, Such C and Marcant B (2004) Chemical modification of bitumen through polyphosphoric acid: properties – microstructure relationships. *Proceedings of the 3rd Eurasphalt and Eurobitume Congress*, Vienna, Austria.

Perez-Lepe FJ, Martinez-Boza C, Gallegos O, Gonzales ME and Munoz A (2003) Influence of the processing conditions on the rheological behavior of polymer modified bitumen. *Fuel* **82**: 1339–1348.

Pierard N, Brichant PP, Denolf K *et al.* (2013) *Recommandations Pratiques Pour le Choix des Matériaux, la conception et la Mise en Oeuvre – Détermination Objective de Leur Couleur.* Centre de recherches routières, Woluwe-Saint-Lambert, Belgium, Annexe au Bulletin CRR 97, vol. 17 (in French).

Prowell BD (2005) Field Performance of Warm Mix Asphalt at the NCAT Test Track. See http://www.pavetrack.com/documents/Pdf/E9-E10-2005.pdf (accessed 2014).

Redelius PG (2000) Solubility parameters and bitumen. *Fuel* **79**: 27–35.

Schellekens JCA and Korenstra J (1987) Shell Mexphalte C brings colour to asphalt pavements. *Shell Bitumen Review* **62**: 22–25.

SIA (Science Industry Australia) (2008) *Sustainable Infrastructure Australia – 3 December 2008: Chemical Additive to Reducing Bitumen Odour. Lab Trial.* SIA, Hawthorn, Australia, internal report SHE010.

Suleiman N (2011) *Evaluation of the Rut Resistance Performance of Warm Mix Asphalts in North Dakota.* North Dakota Department of Transportation, Bismarck, ND, USA, Report UND 2011-01 SPR-R033-004.

Taylor R and Airey G (2008) Polymer modified bitumens part one: background and history. *Asphalt Professional* **34**: 13–18.

TRB (Transportation Research Board) (2009) *Polysphosphoric Acid Modification of Asphalt Binders: A Workshop.* TRB, Minneapolis, MN, USA, Circular E-C160.

Vonk WC and Gooswilligen GV (1989) Improvement of paving grade bitumens with SBS polymers. *Proceedings of the 4th Eurobitume Symposium*, Madrid, Spain, pp. 298–303.

Weng G, Zhang Y and Zhang YX (2001) Vulcanization characteristics of asphalt/ SBS blends in the presence of sulfur. *Journal of Applied Polymer Science* **82(4)**: 989–996.

Weng G, Zhang Y, Zhang Y, Sun K and Fan Y (2002) Improved properties of SBS-modified asphalt with dynamic vulcanization. *Polymer Engineering and Science* **42(5)**: 1070–1081.

Whiteoak D (1990) *The Shell Bitumen Handbook*, 4th edn. Shell Bitumen UK, Chertsey, UK.

WRAP (Waste and Resources Action Programme) (2008) *A Review of the Use of Crumb Rubber Modified Asphalt Worldwide. WRAP Project TYR032. Rubberised Asphalt Testing to UK Standards.* WRAP, Banbury, UK.

Yetkin Y (2007) Polymer modified asphalt binders. *Construction and Building Materials* **21**: 66–72.

Yousefi AA (2003) Polyethylene dispersions in bitumen: the effects of the polymer structural parameters. *Journal of Applied Polymer Science* **90(12)**: 3183–3190.

Zhang F and Hu C (2013) The research for SBS and SBR compound modified asphalts with polyphosphoric acid and sulfur. *Construction and Building Materials* **43**: 461–468.

Zhang Q, Wand T, Fan W, Ying Y and Wu Y (2014) Evaluation of the properties of bitumen modified by SBS copolymers with different styrene–butadiene structure. *Journal of Applied Polymer Science* **131(12)**: 10.1002/app.40398.

Chapter 9

Bitumen emulsions

Mario Jair
Technical Manager, Shell Cia Argentina de Petroleo SA. Argentina

Whatever the end use, application conditions usually require bitumen to behave as a mobile liquid. In principle, there are three ways to make a highly viscous bitumen into a low viscosity liquid

- heat it
- dissolve it in solvents
- emulsify it.

Bitumen emulsions provide a convenient and environmentally friendly option in which the bitumen is liquefied in small particles by dispersing it in water.

Currently, bitumen emulsions are used for road maintenance and repairs (e.g. surface dressing and slurry seals) and for parts of structural pavements (tack coats, bond coats, prime coat, cold mixes, cold recycling and gravel emulsion).

A phase in chemistry is defined as 'a mechanically separate, homogeneous part of a heterogeneous system' (Infoplease, 2014). Bitumen is not mixed into water, it is held in suspension, so bitumen and water are part of a two phase system. They work together in the emulsion (the heterogeneous system) in a special way that maintains the homogeneous nature of each ingredient.

An emulsion is defined by the International Union of Pure and Applied Chemistry as

> A fluid colloidal system in which liquid droplets are dispersed in a liquid. An emulsion is denoted by the symbol O/W if the continuous phase is an aqueous solution and W/O if the continuous phase is an organic liquid (an 'oil'). More complicated emulsions such as O/W/O are also possible'.

The Shell Bitumen Handbook, Sixth Edition
ISBN 978-0-7277-5837-8
Shell International Petroleum Company Ltd: All rights reserved
http://dx.doi.org/10.1680/tsbh.58378.185

In the particular context of bituminous binders, an emulsion is further defined in EN 12597 (BSI, 2014) as 'an emulsion in which the dispersed phase is bituminous' (i.e. O/W) and which is 'thermodynamically meta-stable'.

The inherent thermodynamic instability of all emulsions, which is caused by a relatively high interfacial tension between the two immiscible oil and water phases, leads to a requirement for energy input to form an increased interfacial area between them (i.e. to form dispersed-phase droplets). This energy requirement can be reduced by the addition of surface-active agents (surfactants) to the system that adsorbs at the phase interface and lower interfacial tension. Surfactants, steric stabilisers (some polymers) or certain clay minerals can also kinetically stabilise an emulsion against re-coalescence. In bitumen emulsions, the surfactants and any other additives required are usually solubilised prior to the manufacture of the emulsion in the water phase. In general, surfactants that are relatively more soluble in water (hydrophilic) rather than oil (lipophilic) tend to promote the formation of O/W rather than W/O emulsions. In the case of the former, the water phase may then also be referred to synonymously as the aqueous or continuous phase of the emulsion.

By far the most common surfactants used in bitumen emulsion production are cationic in nature such as alkyl diamines, amidoimidazolines and quaternary ammonium chlorides, although anionic, non-ionic and zwitterionic surfactants and clay minerals are used for certain specialised applications. The typical bitumen droplet diameter size range of bitumen emulsions is of the order of tenths of a micrometre to a few tens of micrometres, and the droplet size distribution is more or less poly-dispersed, depending on the specific formulation and manufacturing conditions used.

The most recent worldwide update of bitumen emulsion usage was carried out for the 2010 World Emulsion Congress through an enquiry in more than 100 countries. This showed that the total production in 2009 was roughly 8 million tonnes, a similar volume to that used in the previous 5 years (Le Bouteiller, 2012). This represents between 10% and 12% of the total global road bitumen demand. Consumption by type of application around the world is shown in Figure 9.1.

9.1 A brief history of the development of bitumen emulsions

In 1906, the first patent covering the application of dispersions of bitumen in water for road building was taken out (Van Westrum, 1906). Initially, efforts were made to form emulsions by purely mechanical means. However, it rapidly became apparent that mechanical action alone was insufficient, and, since these pioneering days, emulsifiers have been used in the process (Albert and Berend, 1916).

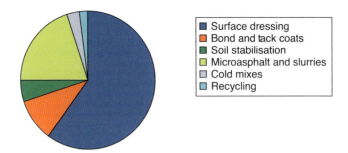

Figure 9.1 Consumption of emulsions by application

At first, the naturally occurring organic acids in the bitumen were utilised by adding sodium hydroxide or potassium hydroxide to the aqueous phase. The subsequent reaction formed an anionic soap that stabilised the dispersion (Bradshaw, 1960).

A great variety of acidic chemicals has been used to promote the stability of anionic bitumen emulsions. These include residues from fatty acid distillation, rosin acids, hydroxystearic acid and lignin sulfonates, all of which may be blended with the bitumen prior to emulsification or dissolved in the alkaline aqueous phase.

Since the early 1950s, cationic emulsifiers have become increasingly popular because of their affinity for many solid surfaces. This is an important property in road construction because the good adhesion of bitumen to different types of mineral aggregate is essential. The most widely used cationic emulsifiers are amines, amidoamines and imidazolines (Schwitzer, 1972).

In the mid-1980s, following the development of polymer modified binders, modified bitumen emulsions based on different manufacturing processes were produced. The residues of these bitumen emulsions had superior properties compared with traditional emulsions, and exhibited improved performance in specific applications (discussed further in section 9.6).

9.2 The components of emulsions

Bitumen emulsions are heterogeneous, two phase fluid systems consisting of two immiscible liquids, bitumen and water, stabilised by an emulsifier. The bitumen is dispersed throughout the continuous aqueous phase in the form of discrete particles, typically 1.0–10 μm in diameter, that are held in suspension by electrostatic charges imparted to the bitumen particles by an emulsifier (REA, 2013a). Components that are typically used to manufacture bitumen emulsions are shown in Table 9.1.

Table 9.1 Emulsion components and their functions

Component	Function
Bitumen	Conventional (commonly 180/200 pen grade; heavier grades such as 15/25 for special tack coats or 50/100 for microsurfacing or slurry seals are used) Modified (discussed in detail in section 9.6.3 of this chapter)
Solvent	Solvents may be included in the bitumen to improve emulsification to reduce settlement, the curing rate at low temperatures or to provide the right binder viscosity after curing. In addition, solvents can be used to produce primer emulsions, which, in some cases, are formulated with low bitumen content
Water	Water used to manufacture emulsions may come from various sources: municipal systems, wells, etc. Whatever the source of the water, it must contain a minimum amount of mineral and organic impurities
Adhesion promoters	Water resistance is an important property of cold asphalt mixtures and seals. The cured film from some anionic emulsions and occasionally also cationic emulsions may not have sufficient adhesion to aggregates, in which case adhesion promoters based on surface-active amine compounds are added
Calcium chlorides and sodium chlorides	Bitumen contains a small amount of salt, which can lead to an osmotic swelling of the droplets in an emulsion, as water is drawn into the droplet. This results in an increase in emulsion viscosity. Calcium chloride or sodium chloride (anionic emulsions) is included (0.1–0.2%) to reduce the osmosis of water into the bitumen and to minimise changes in the viscosity

Bitumen emulsions can be divided into four classes

- cationic emulsions
- anionic emulsions
- non-ionic emulsions
- clay-stabilised emulsions.

The first two are, by far, the most widely used.

The terms 'anionic' and 'cationic' stem from the electrical charges on the bitumen globules. This identification system originates from one of the fundamental laws of electricity – like charges repel, opposite charges attract. If an electrical potential is applied between two electrodes immersed in an emulsion containing negatively charged particles of bitumen, they will migrate to the anode. In that case, the emulsion is described as 'anionic'. Conversely, in a system containing positively charged particles of bitumen, they will move to the cathode, and the emulsion is described as 'cationic'. The bitumen particles in a non-ionic emulsion are neutral and, therefore, will not migrate to either pole. These types of emulsion are rarely used on highways.

Clay-stabilised emulsions are used for industrial applications rather than for road applications. In these materials, the emulsifiers are fine powders, often natural or processed clays and bentonites, with a particle size much less than that of the bitumen particles in the emulsion. Although the bitumen particles may carry a weak electrical charge, the prime mechanism that inhibits their agglomeration is the mechanical protection of the surface of the bitumen by the powder together with the thixotropic structure of the emulsion, which hinders movement of the bitumen particles.

9.2.1 Functions of emulsifiers

Emulsifiers perform several functions within a bitumen emulsion. They

- make emulsification easier by reducing the interfacial tension between the bitumen and water
- determine whether the emulsion formed is the water-in-oil or oil-in-water type
- stabilise the emulsion by preventing the coalescence of droplets
- dictate the performance characteristics of the emulsion such as the setting rate and adhesion.

The emulsifier is the single most important constituent of any bitumen-in-water emulsion. In order to be effective, the emulsifier must be water soluble and possess the correct balance between hydrophilic (having an affinity for water) and lipophilic (having an affinity with lipids (organic compounds that are insoluble in water but soluble in organic solvents or fats)) properties. Emulsifiers can be used singly or in combination to provide special properties.

In the emulsion, the ionic portion of the emulsifier is located at the surface of the bitumen droplet while the hydrocarbon chain orientates itself on the surface of the bitumen, and is firmly bound to it. This is illustrated in Figure 9.2.

Emulsions may also contain unbound emulsifier that can influence the final properties of the emulsion, particularly their breaking and adhesion performance. The ionic portion of the emulsifier imparts a charge to the droplets themselves, and counter-ions such as sodium or chloride diffuse into the water phase.

Emulsifying agents are large organic molecules that have two distinct parts to them. These parts are called the 'head' and the 'tail'. The head portion consists of a group of atoms that chemically have positive and negative charge areas. These two charged areas give rise to the head becoming polar (as in the poles of a magnet). Because of this polarity, and the nature of some of the atoms in this polar head, the head is soluble in water. The tail consists of a long-chain organic group that is not soluble in water but is soluble in other organic substances such as oils (bitumen). Thus, an emulsifying agent is a molecule with both water-soluble and oil-soluble portions. This unique characteristic gives the chemical its emulsifying ability.

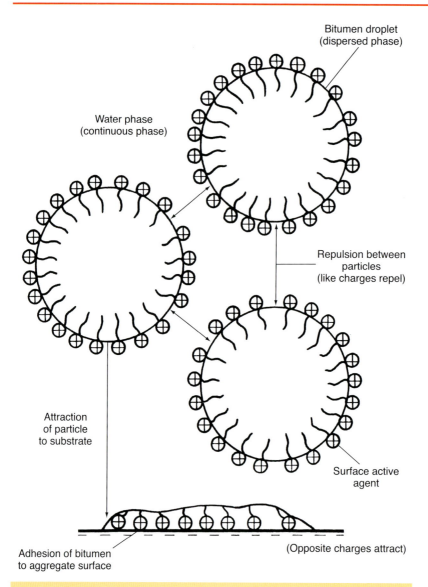

Figure 9.2 Schematic diagram of charges on bitumen droplets

In an anionic emulsion there are many billions of bitumen droplets with the emulsifying agent at the water–bitumen interface. The tail portion of the emulsifying agent aligns itself in the bitumen while the positive portion of the head floats around in the water, leaving the rest of the head negatively charged and at the surface of the droplet. This imparts a negative charge to all the droplets. An example can be represented by

R–COONa$^+$

The cationic emulsifying agent functions similarly to the anionic emulsifying agent: the negative portion of the head floats around in the water leaving a positively charged head. This imparts a positive charge to all the droplets. Because positives repel each other, all the droplets repel each other and remain as distinct bitumen drops in suspension (Bitumina, 2014). An example can be represented by

$$R-NH_3^+ Cl^-$$

In non-ionic emulsifiers, the hydrophilic head group is covalent (indicating a chemical bond that involves the sharing of electron pairs between atoms), polar and dissolves without ionisation (the process by which an atom or a molecule acquires a negative or positive charge by gaining or losing electrons). Any charge on the bitumen emulsion droplets is derived from ionic species in the bitumen itself. An example can be represented by

$$R-COO(CH_2CH_2O)_x H$$

In simple emulsifiers with the above chemical structures, R represents the hydrophobic (i.e. repelled by water) portion of the emulsifier, and is usually a long-chain hydrocarbon consisting of 8–22 carbon atoms derived from natural fats and oils such as tallow or from petroleum such as alkylbenzenes. The hydrophilic head group can variously contain amines, sulfonates, carboxylates, ether and alcohol groups. Emulsifiers with poly-functional head groups containing more than one of these types are widely used.

Complex wood-derived emulsifiers include Vinsol Resin, tannins and lignosulfonates, which contain polycyclic hydrophobic portions and several hydrophilic centres. Proteinaceous materials such as blood and casein have also been used in bitumen emulsifiers. In general, the structure of these molecules is complex, as can be seen by the structure for lignosulfonate illustrated in Figure 9.3.

Cationic emulsions constitute the largest volume of emulsions manufactured worldwide. They are produced from the following types or mixtures: monoamines, diamines, quaternary ammonium compounds, alkoxylated amines and amidoamines, and can be represented structurally as shown in Table 9.2.

Most of these are supplied in neutral basic form, and need to be reacted with an acid to become water soluble and cationic in nature. Therefore, cationic emulsions are generally acidic with a pH < 7. Hydrochloric acid is normally used, which reacts with the nitrogen atom to form an ammonium ion. This reaction can be represented as follows

$$R-NH_2 + HCl \longrightarrow R-NH_3^+ Cl^-$$
amine   acid   alkylammonium chloride

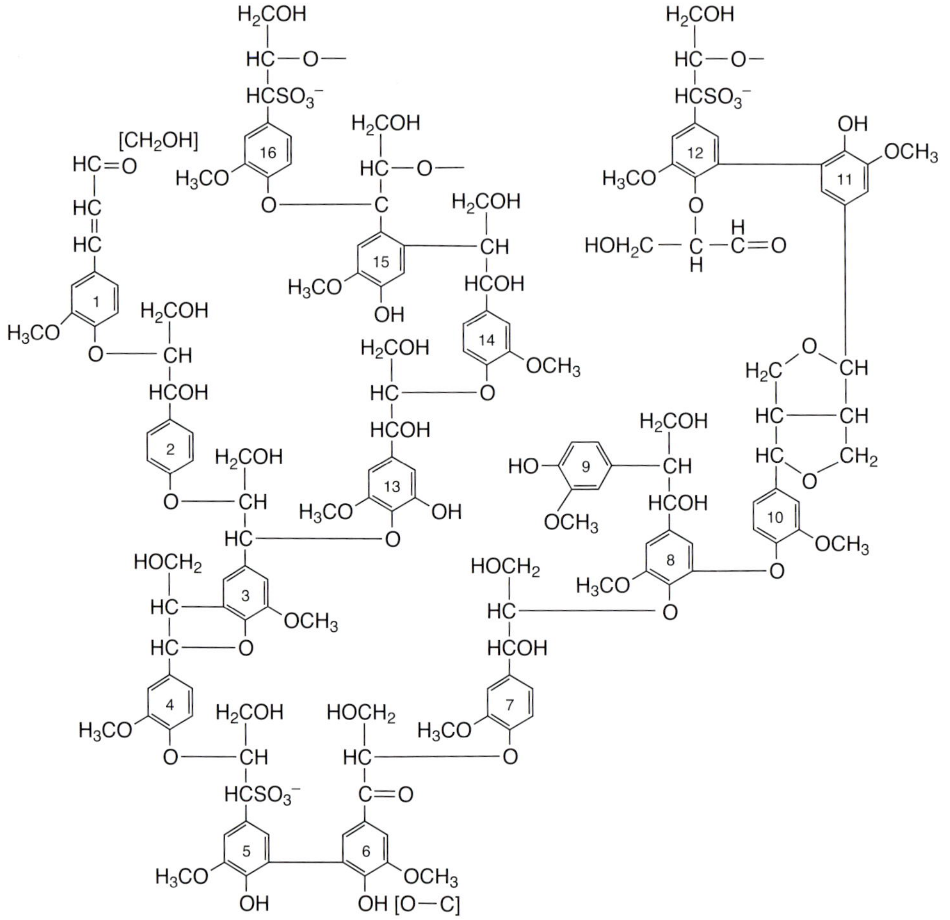

Figure 9.3 The structure of lignosulfonate

Table 9.2 Chemical components of cationic emulsions ($R=C_{8-22}$)

Example of chemical structure	Chemical type
$R-NH_2$	Monoamine
$R-NH_2CH_2CH_2CH_2NH_3^{2+}2Cl^-$	Diamine
$RCONHCH_2CH_2CH_2NH(CH_3)_2^+Cl^-$	Amidoamine
$R-\underset{\underset{CH_3}{\vert}}{\overset{\overset{CH_3}{\vert}}{N}}-CH_3Cl^-$	Quaternary ammonium compound
$R-N\begin{subarray}{l}(CH_2CH_2O)_xH \\ (CH_2CH_2O)_yH\end{subarray}$	Alkoxylated amine

Quaternary ammonium compounds do not need to be treated with acids as they are already salts and water soluble, but the water-phase pH can be adjusted with acid if required, to modify the performance of the emulsion.

Anionic emulsifiers constitute the second largest volume of emulsion produced worldwide and are usually stabilised with fatty acids or sulfonate emulsifiers.

Fatty acids are insoluble in water, and are made soluble by reacting with an alkali, normally sodium or potassium hydroxide, so that anionic emulsions are alkaline with a pH > 7. (Cationic emulsions usually have a pH in the range 2–3, with anionic emulsions having a pH in the range 10–11. A liquid having a pH value of 7 is neutral, being neither acidic nor alkaline.)

Sulfonates are usually supplied as water-soluble sodium salts. Further neutralisation is not required, but an excess of sodium hydroxide is used in order to keep the pH of the emulsion higher than 7 and also to neutralise the natural acids contained in the bitumen.

Non-ionic emulsifiers are not produced in significant quantities, and are normally only used to modify both anionic and cationic emulsions. Typical non-ionic emulsifiers include nonylphenolethoxylates and ethoxylated fatty acids.

9.3 The manufacture of bitumen emulsions

Most bitumen emulsions are manufactured by a continuous process using a colloid mill (a machine used to reduce the particle size of a solid in suspension in a liquid or to reduce the droplet size of a liquid suspended in another liquid). This equipment consists of a high speed rotor that revolves at 1000–6000 revs/min in a stator. The clearance between the rotor and the stator is typically 0.25–0.50 mm and is usually adjustable.

Hot bitumen and emulsifier solutions are fed separately but simultaneously into the colloid mill, the temperatures of the two components being critical to the process. The viscosity of the bitumen entering the colloid mill should not exceed 0.2 Pa·s (2 poise). Bitumen temperatures in the range 100–140°C are used in order to achieve this viscosity with the penetration grade bitumens that are normally used in emulsions. To avoid boiling the water, the temperature of the water phase is adjusted so that the temperature of the resultant emulsion is less than 90°C. As the bitumen and emulsifier solutions enter the colloid mill, they are subjected to intense shearing forces that cause the bitumen to break into small globules. The individual globules become coated with the emulsifier, which gives the surface of the droplets an electrical charge. The resulting electrostatic forces prevent the globules from coalescing.

When the bitumen is not a soft penetration grade bitumen or when a polymer modified binder is used, the process is more difficult. Higher temperatures are needed to allow the bitumen to be pumped to and dispersed in the mill, and dispersion of the bitumen requires more power input to the mill, which further increases the product temperature. Pressurised mills are used with bitumens having a high viscosity at normal emulsification temperatures and to allow higher throughput with normal bitumens. Emulsions with temperatures up to 130°C are produced under high pressure, and the emulsion output must be cooled below 100°C before being discharged into normal storage tanks.

As an alternative to a colloid mill, a static mixer may be used. This contains no moving parts. The high shear necessary to produce an emulsion is generated by pumping the input materials at high speed through a series of baffles designed to produce highly turbulent flow. The benefits of having no moving parts and no shaft seals are obvious; additional benefits claimed are closer control of the bitumen particle size in the emulsion produced, with consequent closer control of critical emulsion properties (discussed in section 9.5).

A batch process can be used for the production of small volumes of emulsion. The type of mixer that is used is chosen to suit the consistency of the end product: it may be a high speed propeller for low viscosity road emulsions or a slow Z-blade mixer for paste-like industrial emulsions.

Schematic diagrams of continuous and batch emulsion manufacturing facilities are shown in Figure 9.4.

9.4 Storage and handling of bitumen emulsions

The production rate of emulsion plants is generally higher than the quantity necessary to meet the current demand. Storage facilities make it possible to have longer production runs, thus improving plant productivity. Modern emulsions can be stored for up to several months without major changes in their physical properties. It is advisable to use small diameter vertical storage tanks with a minimum horizontal cross section with a dip tube filling pipe that extends to a point at or near the bottom of the storage tank. Vertical tanks are superior to horizontal tanks because they minimise the surface area exposed to air. Horizontal tanks are acceptable but not recommended because they expose the emulsion to air at its surface, which builds a crust on the product. The crust will not liquefy nor will it pump or spray. Accordingly, if a storage facility has to use horizontal tanks, keeping them full to limit exposure to air is recommended.

Emulsions of different ionic types should never be mixed, and tanks should be thoroughly cleaned before refilling with a different ionic type. Provision

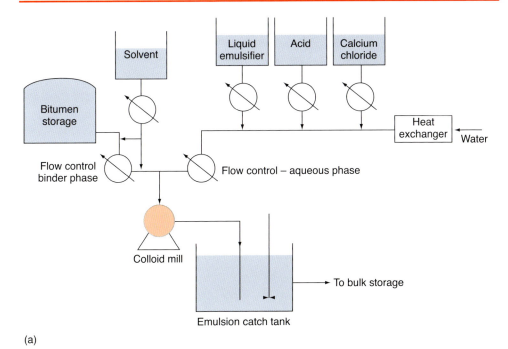

(a)

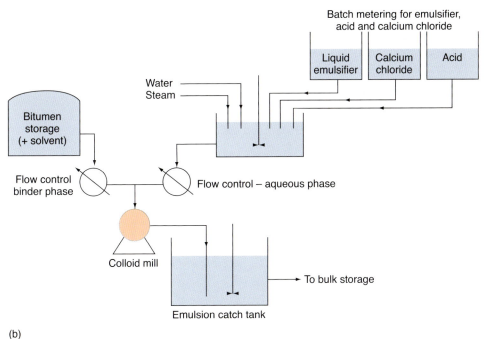

(b)

Figure 9.4 Schematic diagrams of (a) a continuous emulsion plant and (b) a batch emulsion plant

should also be made to ensure proper agitation of stored emulsion to prevent settling, decantation or creaming. Mixers may be used in the tank. They may be simple propeller mixers that revolve slowly or larger anchor sweep mixers. It is important not to mix the emulsion too frequently. It should require mixing once a week or for a short time before being transported for use. Bitumen emulsions are sensitive to frost, which can cause irreversible breaking, so appropriate measures should be put in place to prevent stored emulsion from freezing (Baumgardner, 2006).

9.5 Properties of bitumen emulsions

Surface dressing applications are, by far, the largest use for bitumen emulsions. Hence, the emulsion properties discussed below relate particularly to surface dressing requirements. Notwithstanding, the principles have general application. The most important properties of bitumen emulsions are

- stability
- viscosity (or, more accurately, rheology)
- breaking
- adhesivity.

There are conflicting requirements for the properties of bitumen emulsions. The ideal emulsion would be stable under storage, transport and application conditions but would break rapidly very soon after application, leaving a binder having the properties of the original bitumen adhering strongly to the road and the aggregates. It would have a low viscosity for ease of handling and application and would flow to minimise irregular spraying but would not flow due to road irregularities, cambers or gradients.

It is generally assumed that the bitumen produced when an emulsion breaks is the same as the bitumen that was used to produce the emulsion, but there are exceptions. The emulsifier used may modify the recovered bitumen, particularly in relation to rheological properties that are surface dependent such as adhesion. Clay-stabilised emulsions (such as Shell Flintkote®) can be formulated to produce bitumen films that have non-flow properties even at very high temperatures. These are useful for roofing and insulating applications.

9.5.1 Emulsion stability

9.5.1.1 Settlement

Emulsions, particularly those having a low bitumen content and low viscosity, are prone to settlement.

At ambient temperatures, the grades of bitumen normally used in emulsions have a density that is slightly higher than that of the aqueous phase of the emulsion. Consequently, the bitumen particles tend to fall through the aqueous phase, resulting in a bitumen-rich lower layer and a bitumen-deficient upper layer. The velocity of the downward movement of the particles can be estimated using Stokes' law (Stokes, 1851)

$$v = \frac{2}{9} g r^2 \frac{\varsigma_1 - \varsigma_2}{\eta}$$

where g is the gravitational force, r is the particle radius, ς_1 is the specific gravity of the dispersed phase, ς_2 is the specific gravity of the aqueous phase and η is the viscosity of the aqueous phase.

However, Stokes' law applies to particles that are free to move and have no inter-particular forces, conditions that are frequently not met in a bitumen emulsion.

Settlement can be reduced by equalising the densities of the two phases. One way of achieving this is to add calcium chloride to the aqueous phase. However, because the coefficients of thermal expansion of bitumen and the aqueous phase are not the same, their densities can be made equal only at one specific temperature. As large particles settle more rapidly than do small ones, settlement can be abated by reducing either the mean particle size or the range of particle sizes that are present. Increasing the viscosity of the aqueous phase will also reduce the rate of settlement. Indeed, if the aqueous phase behaviour can be made non-Newtonian by introducing a yield value, settlement can be eliminated completely. In addition to gravity, there are repulsive forces between the bitumen droplets caused by the layers of emulsifier on the droplets that impede or accelerate settlement.

Coalescence follows settlement in two stages. First, bitumen droplets agglomerate into clumps: this reversible phenomenon is called flocculation. Second, the resultant flocks fuse together to form larger globules: this irreversible process is called coalescence. This can be spontaneous or it can be induced by mechanical action.

9.5.1.2 Stability during pumping, heating and transportation
Two bitumen particles in an emulsion will coalesce if they come into contact. Contact is prevented by electric charge repulsion and the mechanical protection offered by the emulsifier. Any effect that overcomes these forces will induce flocculation and coalescence. Flow of the emulsion, caused by pumping, heating (convection currents) or transportation is one such effect. Some emulsifiers have a tendency to foam, which is itself a potential cause of coalescence because bitumen particles in the thin film of a bubble are subjected to the forces of surface tension.

9.5.2 Emulsion viscosity
As surface dressing emulsions are almost always applied by spray, their viscosity under spraying conditions is of prime importance.

The viscosity of emulsions is normally measured as the time of effluence from a flow cup with a standard orifice at different temperatures.

Saybolt Furol cups at 25 and 50°C are used according to the American Society for Testing and Materials (ASTM) standards in the USA (ASTM D244:2009 (ASTM, 2009)), and standard tar viscometers at 40 and 50°C are used in Europe (EN 12846-1:2011 (BSI, 2011)). Dynamic viscosity is used in Europe (EN 13302:2010 (BSI, 2010a)) to characterise the behaviour of the emulsion after spraying (e.g. in surface dressing or tack coat applications) or during mixing with aggregates (e.g. in coating applications).

Emulsions having a high concentration of the dispersed phase (bitumen) rarely have Newtonian viscosity characteristics (e.g. the apparent viscosity changes with the shear rate at which the viscosity is measured). In addition, the rate of change of viscosity with temperature is not the same for different emulsions. When comparing two different emulsions, it is possible for one to have a lower viscosity and better spray distribution at the spraying temperature of 85°C while also showing higher viscosity and less run off from the road at a road temperature of 30°C compared with the other emulsion.

Single point viscosity measurements can, therefore, be misleading, although more data are difficult to obtain.

In principle, there are four methods of increasing the viscosity of an emulsion

- by increasing the concentration of the dispersed phase (bitumen)
- by increasing the viscosity of the dispersed phase
- by increasing the viscosity of the continuous phase (the aqueous phase)
- by reducing the particle size distribution range.

The opposite changes will decrease the viscosity of an emulsion. As stated, emulsion viscosity is almost independent of the viscosity of the dispersed phase (bitumen). It is possible to produce emulsions based on hard bitumen (<10 pen) that are readily pourable at 10°C.

9.5.2.1 Increasing the bitumen content
At low bitumen contents, the effect is small. At high bitumen contents, a small increase in concentration can induce a dramatic change in the viscosity that may be uncontrollable.

9.5.2.2 Modification of the aqueous phase
The viscosity of a bitumen emulsion is highly dependent on the aqueous phase composition. It has been shown that, in the case of conventional cationic road emulsions, the viscosity can be increased by decreasing the acid content or increasing the emulsifier content. Additives intended specifically as viscosity modifiers can also be used.

Latex addition in the aqueous phase for modified bitumen emulsion production is another example (discussed in section 9.6.3).

9.5.2.3 Increasing the flow rate though the mill

By increasing the flow rate through the mill, the particle size distribution of the emulsion will be changed. At bitumen contents of less than 65%, the viscosity of the emulsion is virtually independent of the flow rate. However, at bitumen contents greater than 65% when the globules of bitumen are packed relatively close together, inducing a change in the particle size distribution by changing the flow rate has a marked effect on the viscosity, as shown in Figure 9.5.

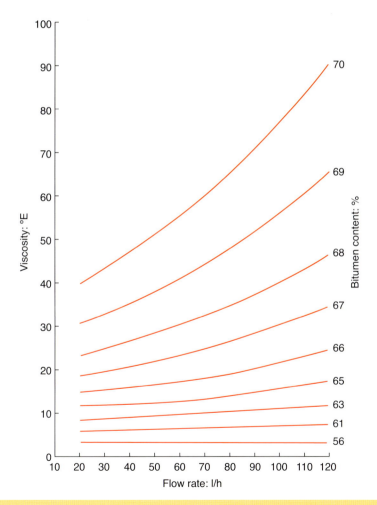

Figure 9.5 Emulsion viscosity (in degrees Engler) as a function of the flow rate for different bitumen contents

9.5.2.4 Decreasing the viscosity of the bitumen in the mill
If the viscosity of the bitumen entering the colloid mill is reduced, the particle size of the emulsion will be reduced, which, in turn, tends to increase the viscosity of the emulsion.

9.5.3 Breaking of emulsions

It is important to know when an emulsion has 'broken'. 'Breaking' is the loss of water from the emulsion. In the process of breaking, an emulsion changes from a liquid to a continuous film of bitumen. Determining whether an emulsion has broken is very easy: the colour turns from brown to black. Some specifications prohibit an emulsion to be covered by the succeeding layer unless the emulsion has broken. A truck reversing into a paver over an emulsion tack coat, or bond coat that has not broken will lift the emulsion on its tyres, thus reducing or negating the effect of the tack coat or bond coat.

Once in contact with a solid surface such as a layer of asphalt or aggregate, the emulsion is able to coalesce (set), and the water either runs off or evaporates. By careful control of the emulsion chemistry, its rate of coalescence, or setting rate, can be adjusted to be slow, medium or rapid setting in order to have optimum versatility.

According to the European standard EN 13808:2013 (BSI, 2013), there are different methods for determining the breaking behaviour. It is compulsory to declare a performance class according to one of these methods (for more details, see section 9.7.2).

European standards can be used for medium setting and fast setting emulsions, in which the breaking value is a dimensionless number corresponding to the quantity of reference filler, in grams, needed to coagulate 100 g of bitumen emulsion (EN 13075-1:2012 (BSI, 2012)). For slow setting and over stabilised emulsions, a European standard specifies a method for the determination of the mixing stability of the emulsion with cement (EN 12848:2009 (BSI, 2009)).

There are, in principle, six parameters that can be used to change the breaking properties of emulsions

- the bitumen content
- the aqueous phase composition (type and content of emulsifier, pH value)
- the particle size distribution
- environmental conditions
- aggregates
- the use of breaking agents.

How each of these parameters changes the breaking properties is set out as follows.

9.5.3.1 Bitumen content

At high bitumen contents, the bitumen particles are more likely to come into contact with each other, resulting in an increase in the breaking rate.

9.5.3.2 Aqueous phase composition

The breaking rate of a bitumen emulsion has been shown to be increased by reducing the acid content, increasing the emulsifier content or by decreasing the ratio between the acid and emulsifier contents.

9.5.3.3 Particle size distribution

The smaller the size of the bitumen particles, the finer will be the dispersion, resulting in a slower breaking rate of the emulsion.

9.5.3.4 Environmental conditions

The evaporation of water is influenced by the incident wind velocity, humidity and temperature in that order. Temperature and humidity are related: as the air temperature falls, the relative humidity increases. Working at night with emulsions can therefore be difficult at low ambient temperatures and the relative humidity reaching 100%, causing the loss of water to cease entirely.

At higher ambient temperatures, the bitumen particles in the emulsion are more mobile, and the bitumen is softer. In such circumstances, particles are more likely to come into contact and, therefore, more likely to coalesce.

9.5.3.5 Aggregates

As stated above, the spraying conditions for fast setting bitumen grades initiate the emulsion breaking process. Accordingly, it is imperative that aggregates are applied very soon after the emulsion has been applied to the surface of the road. This is necessary to ensure that the emulsion is still capable of wetting the aggregates. When the aggregates are applied, breaking is accelerated by the absorption of emulsifier onto the aggregate and the evaporation of water. The former can be completely inhibited by the use of coated aggregates. Conversely, dust can cause rapid breaking of the emulsion onto the dust with no adhesion of bitumen to the aggregates. Within these extremes, the geometry of the aggregates (i.e. their size and shape) has a considerable influence on the breaking rate of the emulsion.

The emulsion contains emulsifier molecules in both the water and on the surface of the droplets. Some of the emulsifier ions form micelles (particles of colloidal dimensions) and, in a stable emulsion, equilibrium exists as shown in Figure 9.6. If some of the emulsifier ions are removed from the solution, the balance is restored by ions from the micelles and the surface of the droplets. This occurs when an emulsion comes into contact with a

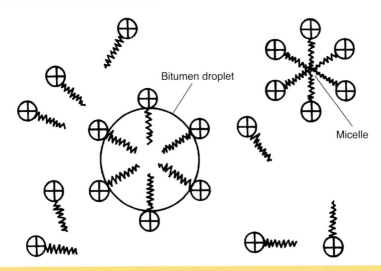

Figure 9.6 Emulsifier ions forming micelles in a stable solution

mineral aggregate. The negatively charged aggregate surface rapidly absorbs some of the ions from the solution, weakening the charge on the surface of the droplets. This initiates the breaking process, as shown in Figure 9.7. A point is reached where the charge on the surface of the droplets is so depleted that rapid coalescence takes place. The aggregate is now covered in hydrocarbon chains, and, as a result, the liberated bitumen adheres strongly to its surface.

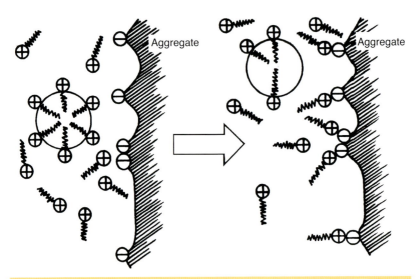

Figure 9.7 Schematic diagram of an emulsion breaking

9.5.3.6 The use of breaking agents

The use of breaking agents can accelerate the breaking of an emulsion. For surface dressing emulsions, it is possible to spray a chemical breaking agent either simultaneously with the emulsion or just after the emulsion has been applied to the road. Care is required in the use of breaking agents. Applying too little will have no effect, but applying too much may cause the emulsion to break and adversely affect its adhesivity. Poor distribution of a breaking agent can have similar effects.

9.5.4 Emulsion adhesivity

It is very important in all applications where bitumen is used as an adhesive between solid surfaces that the bitumen 'wets' the surface to create the maximum effective contact area. With dry substrates, the 'critical surface tension of wetting' of the aggregate must be high enough to ensure that the bitumen spreads easily over the surface. The resultant adhesion generally exceeds the cohesion of the bitumen. However, when the surface of an aggregate is covered with water, the wetting of the aggregate becomes a three phase phenomenon that can only occur if the balance of the interfacial energies favours wetting by the bitumen. Cationic emulsifiers are particularly efficient at reducing the free surface energy of a polar aggregate, forming a thermodynamically stable condition of minimum surface energy resulting from the emulsifier being attracted to the aggregate surface (Heukelom and Wijga, 1973; HRB, 1968).

Most cationic emulsifiers are also anti-stripping agents. Consequently, initial bonding is assured. However, the quality of the bond between the bitumen and the aggregate depends on a number of factors

- the type and amount of emulsifier
- the bitumen grade and constitution
- the pH of the emulsifier solution
- the particle size distribution of the emulsion
- the nature of the aggregate.

9.6 Modification of bitumen emulsion properties
9.6.1 Particle size distribution

The particle size distribution of bitumen emulsions influences many of the emulsion properties that are critical to achieving success in application and service. Automatic equipment is available that can measure both the mean particle size and the particle size distribution of emulsions, but bitumen emulsions can present difficulties because they usually contain a few very large particles and may contain some that are very small.

The strong influence of the particle size distribution on the properties of bitumen emulsions is due to the surface area of a spherical particle being proportional to the square of its diameter, and its mass is a function of the cube of its diameter. Many performance properties of an emulsion are influenced by the amount of 'free' emulsifier in the aqueous phase (i.e. the amount of emulsifier that has not been absorbed onto the bitumen particles). The amount of emulsifier that is absorbed onto the bitumen particles depends on the total surface area of those particles. Even a small proportion of bitumen present as sub-micrometre particles can create a large surface area.

The distribution of the emulsion droplet size is dependent on the interfacial tension between the bitumen and the aqueous phase (the lower the interfacial tension, the easier the bitumen disperses) and on the energy used in dispersing the bitumen. For a given mechanical energy input, harder bitumens will produce coarser emulsions, and high penetration or cut-back bitumens will produce finer emulsions. It is possible to influence the particle size and distribution by modifying the materials and process used to make an emulsion.

9.6.1.1 The addition of acid to the bitumen
The addition of naphthenic acids to a non-acidic bitumen is important for the production of anionic emulsions. The acids react with the alkaline aqueous phase to form soaps that are surface active and that stabilise the dispersion. The addition of naphthenic acids causes a decrease in the mean particle size of the emulsion without changing its size distribution.

9.6.1.2 Manufacturing conditions
Manufacturing conditions have a substantial influence on the resulting particle size distribution of the emulsion.

- *Temperature.* Increasing the temperature of either the aqueous phase or the bitumen normally decreases the mean particle size of the emulsion.
- *Bitumen content.* Increasing the bitumen content increases the mean particle size and tends to reduce the range of particle sizes.
- *Composition of the aqueous phase.* For cationic emulsions manufactured using hydrochloric acid and an amine emulsifier, the particle size can be decreased by increasing either the acid or the emulsifier content: if the ratio of acid to amine is kept constant, the particle size can also be reduced by increasing the amine/acid content; the size distribution does not appear to be related to the concentration of these two components.
- *Operating conditions of the colloid mill.* The gap and rotational speed of the colloid mill strongly influence the particle size and distribution of the emulsion: a small gap will result in a small particle size with a relatively narrow range of sizes; high rotational speed will produce a small particle size.

■ *Decreasing the viscosity of the bitumen*. If the viscosity of the bitumen entering the colloid mill is lowered, the particle size of the emulsion will be reduced, which will tend to increase the viscosity of the emulsion.

9.6.2 *Effects of bitumen properties*

9.6.2.1 Influence of the ionic content of bitumen in aqueous suspension

Bitumens usually contain a small amount of ionic material, principally sodium chloride (an ion is an atom or molecule, or a group of either, that has lost or gained one or more electrons and, as a result, possesses an electrical charge). Typical concentrations are less than 0.1%, which is in the low hundreds parts per million (ppm). This small amount of ionic material in conjunction with the aqueous medium in the emulsion can exert considerable influence on the mixture's viscosity. At concentrations up to 20 ppm, the ionic material has little or no effect. As the concentration increases, the viscosity of emulsions produced from the bitumen rises (for a given bitumen content) to a maximum. When the concentration exceeds approximately 300 ppm, the emulsion viscosity suddenly falls. These characteristics can be attributed to osmosis (the natural movement of water from low ionic concentrations to a higher concentration on the other side of a semi-permeable membrane to achieve an ionic concentration balance; this process can generate physical pressure). If a particle of bitumen has an ionic content it may be subject to osmosis. If this occurs, water will enter the particle, and the particle will expand under the osmotic pressure. If the emulsion has a substantial proportion of its particles susceptible to osmosis, then there will be a substantial increase in the viscosity. If the difference in ionic concentration between the particle and the surrounding emulsion medium is high enough, the particle will continue to expand to a point where it ruptures. If this point of failure is reached, it can trigger coagulation between other particles, which is undesirable. When creating an emulsion, its osmotic characteristics can be controlled. If the bitumen used to manufacture the emulsion has a low ionic content, this can be modified by adding sodium chloride to the bitumen before the process is started. If the ionic content is high, running the risk of osmotic failure, ionic material (e.g. calcium chloride) can be added to the aqueous component of the emulsion. This control of ionic concentration between the emulsion medium and the content of an osmotic particle will manage the particle size and, therefore, control the influence this parameter has on the emulsion viscosity.

9.6.2.2 Bitumen density

High bitumen density can cause rapid settlement in emulsions, leading to coagulation during static storage. The problem can be alleviated by adding a high boiling point solvent (e.g. kerosene) to bitumen that is a grade harder than would normally be used, or by increasing the specific gravity of the aqueous phase (e.g. by adding calcium chloride or by using an emulsifier that imparts a yield value to the aqueous phase).

9.6.2.3 Acid value

The presence of natural naphthenic acids in the bitumen is beneficial to the production of most anionic emulsions. The acids react with the excess of alkali in the aqueous phase during emulsification, acting as an efficient natural emulsifier. However, some industrial emulsions cannot be made with acidic bitumens: the naphthenic acids take precedence at the particle surface over the emulsifiers in the aqueous phase, and give an emulsion that is either unstable or has insufficient stability.

9.6.3 Polymer modified bitumen emulsions

9.6.3.1 Definition

Modified emulsions are those that produce a residue of modified binder. A polymer can be mixed in bitumen to obtain a polymer modified bitumen that can then be emulsified (called single phase modified emulsions). Ethylene–vinyl acetate and styrene–butadiene–styrene are most commonly used in paving grade emulsions.

Polymers can be also be added during the aqueous phase as latex (dual-phase modified emulsions). Latex is a water-based polymer, and comes in anionic, non-ionic and cationic forms. It is important that the latex type should be compatible with the emulsion. Styrene–butadiene rubber, poly-chlorophene and natural rubber latex are commonly used in road emulsions (Figure 9.8).

9.6.3.2 Properties of residual binder

The properties of the binder after the breaking of a bi-phasic modified emulsion depend on the type of bitumen, the type of modifier, their compatibility and the relative modifier/bitumen concentrations. The action caused by the modifier results in improvements in the original binder similar to those obtained with elastomeric modified bitumens, as follows

- increased cohesion at medium and high service temperatures
- improved performance at low temperatures
- increased plasticity interval (= ring and ball temperature – Fraass point)
- better ageing behaviour
- improved rheological behaviour.

In the case of single phase modified emulsions, the performance of the residual binder should be very similar to that of the modified bitumen used for its production, although the colloid mill mechanical action as well as the emulsifier physico-chemical action can modify the performance slightly.

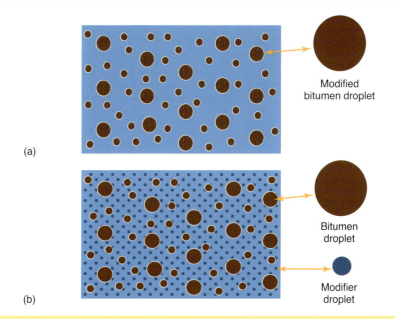

(a)

(b)

Modified
bitumen droplet

Bitumen
droplet

Modifier
droplet

Figure 9.8 (a) Single phase and dual phase (b) modified emulsions: differences

With dual phase modified emulsions, there is a concern as to how to recover material that is truly representative of what is applied in the field. The different methods of obtaining the residual binder of emulsions in the laboratory (distillation, oven stabilisation, vacuum, etc.) do not satisfactorily reproduce the breaking mechanism of these emulsions, which varies with the type of application in either surface dressing mixtures or microsurfacings. For this reason, it is preferable to check the performance of the different types of dual phase modified emulsions by an indirect method by means of tests carried out on the final product.

9.6.3.3 Manufacture
Modified emulsions are manufactured following the same basic method as conventional emulsions except for the addition of the modifying agent. As single phase modified emulsions are manufactured from previously modified bitumen, its high viscosity requires that the emulsion manufacturing temperature is higher than in the conventional emulsion manufacturing process. Accordingly, the final thermal balance leads to an emulsion temperature at the outlet of the mill in excess of 100°C. In these conditions, it is necessary to modify the conventional manufacturing equipment to make it capable of working under pressures ranging from 1.5 to 2 bar and to use a cooling system at the outlet of the mill (generally a heat exchanger). As mentioned above, in the dual phase modified emulsion case, the modifier is typically

added as a latex dispersion. It can be incorporated either in the aqueous phase or the bitumen or even added subsequently to the emulsion. Dispersion in the aqueous phase is the most commonly used method. It allows for accurate regulation of the emulsification parameters, especially pH, as well as achieving good homogeneity in the dispersion (World Road Association, 1999).

9.6.3.4 Method of use
Emulsions containing polymer modified bitumen are generally used in exactly the same way as their unmodified equivalents. It must not be assumed that the use of polymer modified products provides a safety margin for poor contracting practices. On the contrary, the modified products usually require greater care in their use and are less tolerant of unsuitable site and weather conditions. Their benefits are in the performance levels that can be obtained under high traffic stress conditions.

9.6.4 Manufacturing variables and emulsion properties
Previous sections have provided an overview of the factors that influence the properties of emulsions, and a number of alternative approaches are available to the emulsion manufacturer to adjust emulsion properties. However, it is virtually impossible to adjust one property of the emulsion without influencing others. This interdependence is illustrated in Figure 9.9.

9.7 Classification and specification of bitumen emulsions
9.7.1 US specifications
Paving grade emulsions are classified according to the sign of the charge on the droplets in anionic and cationic emulsions while the test methods are defined in ASTM specification D244 (ASTM, 2009). In accordance with an ASTM publication (ASTM, 2002), cationic emulsions are classified by an alphanumeric designation: the first part indicates the breaking rate (RS or rapid setting, QS or quick setting, MS or medium setting, and SS or slow setting) followed by numbers and text indicating the emulsion viscosity and residue properties. For example, CRS-2 would be a reactive (rapid setting) cationic emulsion of high viscosity. CSS-1h would be a non-reactive (slow setting) cationic emulsion with a hard bitumen residue.

9.7.2 European specifications
The European standard EN 13808 (BSI, 2013) specifies the requirement for the performance characteristics of conventional and modified cationic emulsions. They are defined as an expression in letters and numbers that describes the different characteristics of emulsions: their polarity, binder content, binder type, type and amount of flux (if any) and breaking value. The European standard provides a framework of specifications and classes

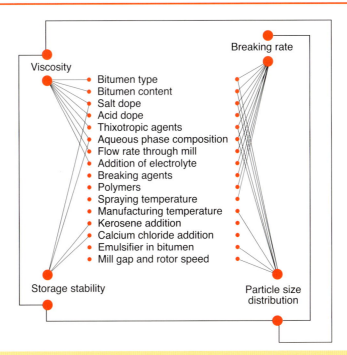

Viscosity
- Bitumen type
- Bitumen content
- Salt dope
- Acid dope
- Thixotropic agents
- Aqueous phase composition
- Flow rate through mill
- Addition of electrolyte
- Breaking agents
- Polymers
- Spraying temperature
- Manufacturing temperature
- Kerosene addition
- Calcium chloride addition
- Emulsifier in bitumen
- Mill gap and rotor speed

Breaking rate

Storage stability

Particle size distribution

Figure 9.9 Interrelationship between manufacturing variables and properties for bitumen emulsions

for the properties of cationic bituminous emulsions and is based on the use of four tables

- Table 1 defines the denomination of the abbreviated terms
- Table 2 describes 12 classes according to properties of the emulsion (breaking behaviour, viscosity and storage stability among other properties)
- Tables 3 and 4 define the properties and performance classes applicable to the residual binders obtained after distillation, recovery, stabilisation and ageing procedures.

Some examples of abbreviated terms according to the European specification are shown in Table 9.3.

9.8 Uses of bitumen emulsions
9.8.1 Road uses
The vast majority of bitumen emulsions are used in surface dressing, slurry seals or microsurfacing applications, and these are discussed in detail in Chapter 21.

However, their versatility makes them suitable for a wide variety of applications, which are briefly described below.

Table 9.3 Examples of cationic emulsion designation according to EN 13808	
Type according to EN 13808	Description
C 69 B 2	Cationic, nominal binder 69%, produced from bitumen, Class 2 breaking value
C 65 BP 3	Cationic, nominal binder 65%, produced from bitumen, containing polymers, Class 3 breaking value
C 69 BF 3 70/100	Cationic, nominal binder 69%, produced from bitumen, containing more than 3% (m/m) flux, Class 3 breaking value and 70/100 pen grade bitumen

9.8.1.1 Tack coats

Tack coating is the application of a conventional bitumen emulsion to facilitate adhesion between layers in an asphalt pavement. The function of the tack coat is to minimise the effect of residual surface dust on the existing surface and to provide an adhesive surface for the overlay.

There is considerable variation in the type of emulsion used for tack coats worldwide. In many countries, unmodified slow setting anionic or cationic emulsions are used, which may be diluted with water, but Europe and South America use unmodified rapid setting cationic emulsions. It is necessary for the tack coat to wet out any dust on the surface of the lower layer, and this favours emulsions of small particle size and some solvent content.

9.8.1.2 Bond coats

A bond coat is a proprietary polymer modified bitumen emulsion used to promote adhesion between layers in the construction of a paved area, or to bond a new surface course to an existing road surface when carrying out road maintenance or construction.

Modern road surface courses such as stone mastic asphalts, thin surface course systems and porous asphalts may be laid thinner than is the case with traditional surface courses. Such layers place greater reliance on the inter-surface bonding to reduce the risk of slippage between the two courses. Some modern surface courses are more open in texture and consequently more porous. In such cases, the bond coat has an important function in helping to waterproof and inhibit the ingress of water to the structural layers (REA, 2013b)

New developments include bond coats based on very hard binders that cure rapidly and avoid sticking to the tyres of traffic or construction equipment.

The best equipment for applying bond coats is an integral paver. This machine is equipped with an emulsion spray system that places the emulsion on the pavement surface immediately in front of the hot asphalt.

Specifications may require that, after application, the emulsion must be allowed to break before the asphalt is laid, unless it is applied by an integral paver. As pointed out in section 9.5.3, verification of an emulsion having broken is simple: it turns from brown to black in colour.

More information on bond coats can be found in section 5.5 of the UK's asphalt transport, laying and compaction standard (BSI, 2010b).

9.8.1.3 Fog seals
Fog seals are a method of adding diluted emulsions to an existing pavement surface to improve sealing or waterproofing. This may reduce or prevent further stone loss by holding the aggregate in place, or simply improve the surface appearance. However, inappropriate use can result in slick pavements and tracking of excess material.

The Asphalt Emulsion Manufacturers Association defines a fog seal as 'a light spray application of dilute asphalt emulsion used primarily to seal an existing asphalt surface to reduce ravelling and enrich dry and weathered surfaces'. Others refer to fog seals as enrichment treatments, because they add fresh asphalt to an aged surface and extend the life of the surface of the pavement. Fog seals are also useful in surface dressing (what is described in the USA as a 'chip seal') applications to hold chips in place in fresh seal coats (Caltrans, 2003).

9.8.1.4 Prime coats
Emulsion prime coats are applied to unbound subbases in order to seal the surface before the application of the asphalt layers. The primer seal prevents the ingress of water into the layer and the loss of fines from wind or water erosion, and ideally allows construction vehicles to drive over the surface without particles on the surface being picked up by tyres.

A few millimetres of penetration are readily achievable if the compacted material is not too dense, but may be very difficult in practice with fine graded and highly compacted bases. Penetration can be achieved using very slow setting cationic or anionic emulsions containing a low bitumen and solvent content, but, in some cases, deep penetration can be very difficult.

However, current thinking suggests that deep penetration may be unnecessary, as dense and highly compacted bases are already very robust and merely need to be sealed from the intrusion of water. This can be achieved using a very thin primer application with minimal penetration.

9.8.1.5 Soil stabilisation
Cationic slow setting emulsions can be used for the stabilisation of uncrushed naturally occurring gravels and sandy soils. Generally, soils with a sand

equivalence value of more than 25 (a measure of the clay content) can be treated with some degree of success for use as a base material for hot overlay or for minor roads where a seal coat may be sufficient. Materials of even lower sand equivalence can, in some cases, be treated successfully using a combination of emulsion and a hydraulic binder such as lime or cement.

9.8.1.6 Cold in-place recycling

Surface courses or even the full depth of the pavement can be recycled in place either by a specially built mobile plant or by simple equipment.

Cold recycling uses bitumen emulsions either alone or in combination with cement or lime. Typically, a cationic slow setting emulsion is used. Structural materials can be produced from emulsion and crushed aggregates or reclaimed asphalt pavement, and such mixtures can meet acceptable performance criteria comparison favourably with hot mix asphalts. Depending on the aggregate grading, medium or slow setting emulsions can be used. Cold mixtures that combine bitumen emulsion with cement can give much improved bearing capacities (AkzoNobel, 2014). Rejuvenation agents may be included, depending on either the penetration or viscosity grade of the bitumen in the existing pavement.

9.8.1.7 Cold plant mixtures

Growing concern about energy conservation and the desire for environmentally friendly processes have generated an interest in the production of cold asphalts.

Although the idea of wide-scale production of cold materials is relatively recent in countries such as the UK, elsewhere the manufacture and utilisation of 'emulsified asphalts' has been commonplace for a number of years (Bradshaw, 1974).

France has been using cold materials since the 1960s for strengthening and reprofiling lightly trafficked roads, and collaborative work between contractors and road authorities led to the development of a material called 'grave émulsion'. Although traditionally a continuously graded 20 mm material, modern grave émulsion typically comprises a 14 mm grading with a bitumen content of 4–4.5%. Grave émulsion can be stockpiled, laid using conventional paving equipment or by a blade grader, and compacted at ambient temperatures. Although the material was originally used for minor maintenance works, more recently it has been applied in structural layers of moderately trafficked roads. A national standard for grave émulsion was introduced in France in 1993 (AFNOR, 1993).

Similar developments have taken place in the USA, where environmental issues and the remoteness of some sites from asphalt plants provided the impetus for using cold mixtures. The Asphalt Institute published *A Basic Emulsion Manual* in 1986 (AI, 1986) and a third edition of its *Asphalt Cold Mix Manual* in 1989 (AI, 1989), and included cold mixtures called 'emulsified asphalt materials' in its *Thickness Design Manual* in 1991 (AI, 1991).

Cold mixture technology presents a new set of challenges to engineers who have traditionally specified hot asphalts. Whereas hot mixtures rely on the visco-elastic properties of bitumen, emulsion mixtures introduce a new series of conditions that must be met in order that such materials can be successfully produced and laid. The surface chemistry of the aggregate begins to have an important role, and emulsions must be tailored to the mineralogy of different rock types.

The classic concept of how cold mixtures work is that the emulsion breaks, either during mixing or compaction, coating the aggregate, after which there is an increase in strength over time. The strength development is a result of the expulsion of water from the aggregate matrix and the coalescence and subsequent cohesion of the bitumen particles. However, the characteristics of initial workability or being able to stockpile the material and the subsequent development of mechanical strength in situ form conflicting requirements. By tailoring the emulsion to produce a mixture that will remain workable for days or weeks, the development of cohesion in the compacted mixture and, hence, the strength gain of the matrix will also be retarded.

If the emulsion is tailored to produce a rapid break, then the mixture will quickly exhibit developing stiffness and, hence, will only have a brief workability window. Accordingly, there exists the potential for innovative technology (such as 'half-warm mixtures', discussed in section 9.9) to address such problems and develop cold mixtures that perform in a manner similar to that of hot mixed products.

Finally, cold porous asphalt mixtures based on modified emulsions can be used for hot porous asphalt repairs or maintenance.

9.8.2 *Miscellaneous uses of bitumen emulsions*
Details of miscellaneous applications for emulsions have been published by the Road Emulsion Association (REA, 2013c). Bitumen emulsions are also used in other civil engineering works and in horticultural and agricultural applications. Some examples are given below.

9.8.2.1 Slip layers and concrete curing
Bitumen emulsions are used to create a membrane between layers of concrete, the objective being to retain the strength of the upper layer by preventing water seepage into the lower layers. This avoids rigid adhesion between

layers of different ages and strengths and helps to produce a stronger upper layer by preventing water absorption into the lower layers. Bitumen emulsion is also sprayed onto the top surface of freshly laid concrete to prevent the evaporation of water.

9.8.2.2 Protective coats
Bitumen emulsions are used for protecting buried concrete, pipelines and ironwork. To enhance the adhesive and cohesive characteristics of the cured binder film, a polymer modified emulsion is normally used.

9.9 Trends and new developments in the use of bitumen emulsions
Although bitumen emulsion based techniques are known to be useful for all classes of roads, their use may also be possible in the following applications

- asphalts for heavily trafficked roads (based on the use of polymer modified emulsions)
- warm emulsion based mixtures, to improve both their maturation time and mechanical properties
- half-warm technology, in which aggregates are heated up to 100°C, producing mixtures with similar properties to those of hot asphalts
- high performance surface dressing (e.g. glass fibre reinforced chip seals, scrub seals with modified emulsified binder and special microsurfacings) on heavily trafficked roads.

9.10 Further information on bitumen emulsions
For further information on the subject of bitumen emulsions, see *Bitumen Emulsions – General Information and Applications* (SFÉRB, 1991), which is available in both French and English.

References
AFNOR (Association Française de Normalisation) (1993) *Assises de chaussées, Graves-émulsion.* AFNOR, Paris, UK, NF P 98-121 (in French).

AI (Asphalt Institute) (1986) *A Basic Emulsion Manual.* AI, Lexington, KY, USA, MS-19.

AI (1989) *Asphalt Cold Mix Manual*, 3rd edn. AI, Lexington, KY, USA, MS-14.

AI (1991) *Thickness Design: Asphalt Pavements for Highways & Street.* AI, Lexington, KY, USA, MS-1.

AkzoNobel (2014) Bitumen Emulsion, Technical Bulletin. See https://www.akzonobel.com/surface/brands_products/ (accessed 13/02/2014).

Albert K and Berend L (1916) Chem. Fabrik. Austrian Patent 72451.

ASTM (American Society for Testing and Materials) (2002) D2397-02. Standard specifications for cationic emulsified asphalt. ASTM, West Conshohocken, PA, USA.

ASTM (2009) D244-09. Tests methods and Practices for Emulsified Asphalts. ASTM, West Conshohocken, PA, USA.

Baumgardner G (2006) *Asphalt Emulsion Technology: Asphalt Emulsion Manufacturing Today and Tomorrow*. Transportation Research Board, Washington, DC, USA, Transportation Research Circular E-C102.

Bitumina (2014) Bitumen Emulsions. See http://www.bitumina.co.uk/bitumen-emulsions.html (accessed 08/06/2014).

Bradshaw LC (1960) *Paint Technology* **24**: 19–23.

Bradshaw LC (1974) Bitumen emulsion in road mixes. *Shell Bitumen Review* **45**: 8–11.

BSI (British Standards Institution) (2009) BS EN 12848:2009. Bitumen and bituminous binders. Determination of mixing stability with cement of bituminous emulsions. BSI, London, UK.

BSI (2010a) BS EN 13302:2010. Bitumen and bituminous binders. Determination of dynamic viscosity of bituminous binder using a rotating spindle apparatus. BSI, London, UK.

BSI (2010b) BS 594987:2010. Asphalt for roads and other paved areas. Specification for transport, laying, compaction and type testing protocols. BSI, London, UK.

BSI (2011) BS EN 12846-1:2011. Bitumen and bituminous binders. Determination of efflux time by the efflux viscometer. Bituminous emulsions. BSI, London, UK.

BSI (2012) BS EN 13075-1:2012. Bitumen and bituminous binders. Determination of breaking behavior. Determination of breaking value of cationic bituminous emulsions, mineral filler method. BSI, London, UK.

BSI (2013) BS EN 13808:2013. Bitumen and bitumen binders. Framework for specifying cationic bituminous emulsions. BSI, London, UK.

BSI (2014) BS EN 12597:2014. Bitumen and bituminous binders. Terminology. BSI, London, UK.

Caltrans (California Transport Division of Maintenance) (2003) *Caltrans Maintenance Technical Advisory Guide*. Caltrans, Sacramento, CA, USA, ch. 6. See http://www.dot.ca.gov/hq/maint/mtag/ch6_fog_seals.pdf (accessed 15/05/2014).

Heukelom W and Wijga PWO (1973) *Bitumen Testing*. Koninklijke/Shell Laboratorium, Amsterdam, The Netherlands.

HRB (Highways Research Board) (1968) *Effect of Water on Bitumen–Aggregate Mixtures*. HRB, Washington, DC, USA, Special Report 98.

Infoplease (2014) Phase. See http://dictionary.infoplease.com/phase (accessed 18/07/2014).

Le Bouteiller E (2012) Emulsion in the world 2012. *International Symposium of Asphalt Emulsion Technology*, Arlington, VA, USA.

REA (Road Emulsion Association) (2013a) *Bitumen Road Emulsions*. REA, Storrington, UK, Technical Data Sheet 1. See http://www.rea.org.uk (accessed 11/04/2014).

REA (2013b) *Bond Coating*. REA, Storrington, UK, Technical Data Sheet 5. See http://www.rea.org.uk (accessed 11/04/2014).

REA (2013c) *Miscellaneous Uses of Bitumen Emulsions*. REA, Storrington, UK, Technical Data Sheet 12. See http://www.rea.org.uk (accessed 17/03/2014).

Schwitzer MK (1972) *Chemistry and Industry* **21**: 822–883.

SFÉRB (Syndicat des Fabricants d'Émulsions Routières de Bitumes) (1991) *Bitumen Emulsions – General Information and Applications*. SFERB, Paris, France.

Stokes GG (1851) On the effect of internal friction of fluids on the motion of pendulums. *Trans Cambridge Phil Soc.* Vol 9, pp. 8–106.

Van Westrum S (1906) German Patent 173.639.

World Road Association (Technical Committee Flexible Roads (C8)) (1999) *Use of Modified Bituminous Binders, Special Bitumens and Bitumens with Additives in Road Pavements*. World Road Association–PIARC, Paris, France.

Chapter 10

Aggregates in asphalts

Robert Allen
Head of Infrastructure Research, Aggregate Industries (UK) Ltd

10.1 Introduction

Aggregate is the largest constituent in asphalts, typically 92–96% by mass; the type of aggregate, its mineralogy, and physical and chemical properties will have a significant impact on asphalt performance.

Suitable aggregates, and their properties, for use in conforming to European asphalt mixtures can be specified by using EN 13043 (BSI, 2002). This standard defines aggregate as a 'granular material used in construction', and separates this into one of three types – natural, manufactured or recycled – defined as follows

- Natural aggregate: 'aggregate from mineral sources that has been subjected to nothing more than mechanical processing' (e.g. crushed rock, sands and gravel, often referred to as primary aggregate).
- Manufactured aggregate: 'aggregate of mineral origin resulting from an industrial process involving thermal or other modification' (e.g. blast furnace slag).
- Recycled aggregate: 'aggregate resulting from the processing of inorganic or mineral material previously used in construction' (e.g. reclaimed asphalt).

Further categorisation of aggregates is given by the description for particle size

- coarse aggregate: substantially retained on a 2 mm test sieve
- fine aggregate: substantially passing a 2 mm test sieve
- all-in aggregate: a combination of coarse and fine aggregates
- filler aggregate: substantially passing a 0.063 mm test sieve.

In the UK, the *Specification for Highway Works* (*SHW*), Volume 1, Clause 901.3 (Highways Agency *et al.*, 2008), states that natural, recycled unbound and manufactured (artificial) aggregates for use in asphalts shall be

The Shell Bitumen Handbook, Sixth Edition
ISBN 978-0-7277-5837-8
Shell International Petroleum Company Ltd: All rights reserved
http://dx.doi.org/10.1680/tsbh.58378.217

217

clean, hard and durable and shall comply with BS EN 13043 [BSI, 2002]. Where recycled coarse aggregate or recycled concrete aggregate is used in bituminous mixtures, it shall have been tested in accordance with *SHW* Clause 710 and the content of other materials (Class X) including wood, plastic and metal shall not exceed 1% by mass.

The majority of asphalts are produced with natural crushed rock aggregate, although the trend is for ever greater use of recycled aggregate as a replacement, usually for economic or environmental reasons.

In the UK, the demand for construction aggregate from primary sources in 2011 was 145.9 million tonnes (Mt), comprising about 91 Mt of crushed rock and 55 Mt of sand and gravel (Bide *et al.*, 2013). Of this, about 19.3 Mt of crushed rock (one-fifth of all the crushed rock produced) and 1.9 Mt of sand and gravel were coated with bitumen to produce asphalt. Figure 10.1 shows the relative proportions of crushed rock used for all construction activities in 2011, with asphalt referred to as 'roadstone coated'.

10.2 Origin and type

The Earth is thought to have formed about 4.6 billion years ago, when the solar system was young. At that time the surface was very hot, but eventually, after cooling, the Earth's crust formed; today the Earth's crust is typically 7–70 km thick.

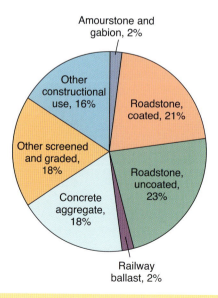

Figure 10.1 UK production (90.9 Mt) of crushed rock by end use 2011 (Bide *et al.*, 2013)

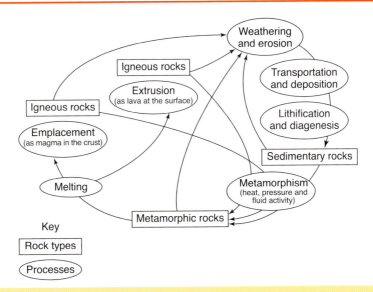

The Earth's crust is composed of 'plates' that, due to forces within the Earth, move relative to each other, resulting in the formation of mountains, ocean trenches and ridges, in a process known as 'plate tectonics'. In addition to plate tectonics, processes of erosion, weathering, and chemical and biological action have been operating over the history of the Earth, resulting in the rocks encountered today.

The formation of rocks by these processes is known as the rock cycle, and it is summarised in Figure 10.2. Overall, these processes form igneous, metamorphic and sedimentary rocks.

10.2.1 Rock types
There are three main types of rock: igneous, sedimentary and metamorphic.

10.2.1.1 Igneous rock
Igneous rock results from the solidification of the molten magma at or beneath the Earth's surface. The magma is either extruded onto the surface (volcanic, extrusive) through lines of weakness in the crust as lava, or it may solidify at depth to form large igneous bodies underground (plutonic). Those formed beneath but close to the Earth's surface, usually as smaller deposits, are known as hypabyssal.

Lavas extruded at the surface cool quickly and are characterised by their small crystalline structure (e.g. basalt, rhyolite), while intrusive (hypabyssal and plutonic) lavas cool much more slowly, and are characterised by larger crystal structures (e.g. granite, gabbro).

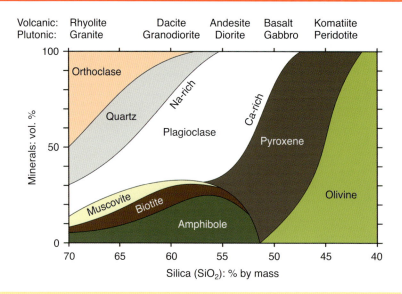

Figure 10.3 Rock type according to mineralogy

Igneous rocks can be defined by their mineralogical composition (Figure 10.3) and crystal size. Rhyolite and granite, for example, have the same composition (acidic) but very different crystal sizes, and therefore have different names, as is the case also for basalt and gabbro (basic).

The silica (SiO_2) content of igneous rocks is sometimes used to categorise rocks further as being acidic, intermediate, basic or ultrabasic on the basis of the following percentages

- >65% by mass – acidic
- 52–65% by mass – intermediate
- 45–52% by mass – basic
- <45% by mass – ultrabasic.

(Note: the terms acidic and basic used here are geology expressions and do not indicate a pH level.)

Sometimes, the slowly cooling igneous mass deep down is moved by Earth movements higher into the crust where it cools more rapidly (i.e. two or more stages of crystallisation), resulting in a porphyritic texture – large crystals surrounded by smaller ones. Porphyritic rocks are frequently used as aggregates.

Major sources of igneous rock for the production of asphalt include

- basalt
- dolerite

- granite
- andesite.

10.2.1.2 Sedimentary rock
Sediments evolve from

- the weathering and transport of particles (in air or water) and resultant sedimentation
- the precipitation of minerals from water
- the accumulation of biogenic matter (fossils).

The resultant sediments are then lithified or stuck together, in the lengthy process of diagenesis, in which the vast accumulations of sediments are transformed into rock. This can often be seen as a series of 'beds' or layers, which are characteristics of sedimentary rocks.

Sediments can now be seen in rivers, shallow seas and beaches. Glacial sand and gravel deposits can be found around many parts of the UK. These are associated with old river terraces formed from glaciers melting and depositing sediments during and at the ends of the ice ages. These sediments are unconsolidated (not stuck together), and can be easily extracted as sands and gravels.

Once sediments have been lithified and consolidated they become rocks (e.g. sands become sandstone, and sands and gravels become conglomerate).

For example, sandstone is formed from the cementation of accumulated sand that has been transported long distances and predominantly deposited in shallow seas, beaches and rivers. In the UK, the term 'gritstone' is used to describe a coarse sandstone in which the grain size is typically greater than 0.5 mm. Conglomerate is a rock formed from the cementation of sand and gravel that occurred in rivers and on beaches. Limestone is complicated, as it may be formed from a combination of fossil fragments and precipitated calcium carbonate, but always in marine conditions. Limestone can be very pure, consisting of a very high percentage of calcium carbonate, but, after formation, some of the calcium ions may be replaced by magnesium ions, which may be dissolved in the groundwater, thus transforming the limestone into a dolomite.

Major sources of sedimentary rock for the production of asphalt include

- limestone
- dolomite
- sandstone (gritstone).

10.2.1.3 Metamorphic rock

Metamorphic rock results from existing rock (sedimentary, igneous or metamorphic) being subjected to increased heat and pressure. The appearance and properties of these rocks are dependent on the varying degrees of heat, pressure and fluid activity to which they have been subjected. Small amounts of heat and pressure lead to minor textural changes (e.g. slate), and greater amounts result in substantial changes (e.g. gneiss). It should be noted that metamorphic rocks have not been melted, otherwise they would be igneous rocks. Rather, they are changed in the 'solid state'; that is, the overall chemical composition of the rock stays the same but the contents are rearranged, resulting in textural changes and growth of minerals that are stable at set temperatures and pressures. Metamorphic rocks often tend to be layered or fissile, which often makes them unsuitable for aggregate in asphalt.

Major sources of metamorphic rock for the production of asphalt include meta-quartzite.

10.2.2 Petrographic examination and classification of rocks

A basic procedure for the petrographic examination and classification of rocks for use as aggregates is given in EN 932-3 (BSI, 1997). Table 10.1 is derived from this standard, and lists some common rock types classified according to their origin. Table 10.2 shows a simplified petrographic report of a plutonic igneous rock using this standard.

10.2.3 Manufactured aggregates

Some asphalt aggregates are manufactured, typically as a by-product of other industrial processes (e.g. steel slag and blast furnace slag). Some of these aggregates have properties that are very beneficial for asphalts (e.g. skid resistance from calcined bauxite) and have a good history of use, while others, particularly those with little or no history of use, will require

Table 10.1 Simplified petrographic terms for aggregate from EN 932-3

Igneous		Sedimentary		Metamorphic
Plutonic	Extrusive/hypabyssal	Clastic	Chemical or biogenic	
Granite	Dolerite	Sandstone	Limestone	Amphibolites
Syenite	Diabase	Gritstone	Chalk	Gneiss
Granodiorite	Rhyolite	Conglomerate	Dolomite	Granulite
Diorite	Trachyte	Breccia	Chert	Hornfels
Gabbro	Andesite	Arkose	Flint	Marble
	Dacite	Greywacke		Quartzite (meta)
	Basalt	Quartzite (ortho)		Serpentinite
		Shale		Schist
		Siltstone		Slate

Table 10.2 Simplified petrographic description of a granite according to EN 932-3[a] (Courtesy of Aggregate Industries)

Discrete constituent		Particle shape	Surface texture	Coatings/ encrustations	Grade[b]
Major (≥10%)	Granite	Angular to subangular	Rough to moderately rough	None	I to II
Minor (2 to <10%)	–	–	–	–	–
Trace (<2%)	–	–	–	–	–

[a] The work was carried out by an accredited, competent, subcontracted laboratory. Based on UK experience, the above aggregate combination could be classified as potentially having low alkali–silica reactivity in accordance with BRE Digest 330 (BRE, 1997). However, a full high power microscopic examination of a representative portion of the aggregate sample is recommended in order to reach a more conclusive classification

[b] Grade I (fresh): unchanged from the original state
Grade II (slightly weathered): slight discoloration, slight weakening
Grade III (moderately weathered): considerably weakened, penetrative discoloration, large pieces cannot be broken by hand
Grade IV (highly weathered): large pieces can be broken by hand, does not readily disaggregate (slake) when a dry sample is immersed in water
Grade V (completely weathered): considerably weakened, slakes, original texture apparent
Grade IV (residual soil): soil derived by in situ weathering but retaining none of the original texture of the fabric

greater testing and need careful consideration in asphalt design and its application.

10.2.4 Recycled aggregates

Aggregate can be formed using recycled materials from other sources: for example, the use of bottom ash from municipal incinerators, foundry sand, recycled rail ballast, construction and demolition waste and milled asphalt from roads (the main source of recycled aggregate used in the production of asphalts).

The properties of these aggregates need to be carefully considered for inclusion in asphalt, not only with regard to their physico-chemical properties but also their suitability for use and their impact on the natural environment. The use of reclaimed asphalt (RA, but sometimes referred to as RAP from 'reclaimed asphalt pavement or planings') as aggregate poses the least potential risk because when it was originally manufactured it was almost certainly identified as suitable for incorporation in asphalt, and its use within European-compliant mixtures is controlled by EN 13108-8:2005 (BSI, 2005a).

The move towards performance specifications in European standards for aggregate and asphalt and the production of recycled aggregates following national quality protocols will minimise concerns and allow recycled aggregates and aggregates from primary sources to compete equally. Such an example is the quality protocol *Aggregates from Inert Waste: End of Waste Criteria for the Production of Aggregates from Inert Waste*

(Environment Agency and Waste and Resources Action Programme, 2013) applicable in Wales, Northern Ireland and England, which has the stated aim 'to provide increased market confidence in the quality of products made from waste and so encourage greater recovery and recycling'.

10.3 Aggregate processing

It is a combination of the nature of the raw material and the requirements of the market that determine the type of processing required. The essential elements of aggregate processing are breaking the rock into smaller sizes or fractions, and then separating them into the different sizes (aggregate) required by the market. The useable sizes can range from a maximum dimension of about 1 m, such as large stone blocks (armourstone) for coastal defences – which do not go through the crushing process – to fractions of a millimetre, where the rock may have gone through a multi-stage crushing process to produce the correct size and shape of aggregate.

The stages from extraction through to processing determine the aggregate shapes and sizes, but can also affect aggregate quality and integrity. Establishing a stable, controlled production process reduces the variability of products and can be achieved through the application and operation of a formal quality system.

10.3.1 Extraction

The extraction process to recover the material will depend on the characteristics of the deposit. Very often, it will be covered with soils and other material (overburden) that is not suitable for use as aggregate. This has to be removed and stored for later use (e.g. in the quarry restoration process once extraction has finished).

Sand and gravel deposits (Figure 10.4) are usually the result of ancient glacial river systems, and often occur as thin strips or sheets in lowland or offshore areas. Extraction of land and marine based deposits can be over quite large areas, but are usually not more than a few metres deep. Marine deposits (approximately 8% of Britain's primary aggregate production) are mainly recovered by suction dredging, whereas the terrestrial deposits can be recovered using hydraulic excavators or draglines. The deposit can either be worked wet (recovered from under the water) or dry (using pumps to keep the extraction area dry).

Hard rock deposits (Figure 10.5) are usually won using conventional drill and blast methods using, typically, 115–150 mm diameter holes filled with bulk ammonium nitrate fuel oil explosives, although some weaker rocks can be excavated directly into dump trucks or broken using a 'ripper' attachment on an excavator or on the back of a bulldozer. The objective is to fragment

Figure 10.4 Aggregate Industries' Newbold Quarry: a sand and gravel deposit. (Courtesy of Aggregate Industries)

Figure 10.5 Aggregate Industries' Duntilland Quarry Scotland: a large igneous rock deposit. (Courtesy of Aggregate Industries)

Figure 10.6 Large excavator loading a dump truck from a blast face. (Courtesy of Aggregate Industries)

the rock to a size suitable for loading and primary crushing (Figure 10.6), and then transporting to the plant for further processing, all using methods that are both safe and minimise environmental impacts.

Once the material has been won, it is transported to the processing plant by way of a conveyor belt or dump truck. The increased usage of modular, mobile processing equipment has meant that some or all of the required processing and loading of trucks for delivery may now take place within the extraction area itself. This equipment is able to follow the extraction area as it develops within a quarry, but can also be moved from quarry to quarry fairly easily. However, many quarries, particularly the larger ones, will have static fixed plant, where the raw feed material is processed to produce the products required for the market, remote from the extraction area and conveniently located for road transport or rail access.

10.3.2 Crushing

10.3.2.1 Crushing process

Crushing is usually the first step of processing after extraction of rock, and reduces large rock, mechanically, to more useful conforming and saleable sizes, typically 75 mm and less. Crushing is usually a sequence and can

Figure 10.7 Primary crusher being fed with blast rock. (Courtesy of Aggregate Industries)

entail one, two, three or more crushing phases, known as primary, secondary and tertiary crushing.

The first stage – primary crushing – is the first reduction in stone size. This is achieved using either a compression crusher (jaw or gyratory crusher types, see Figure 10.9) or impact crusher (see Figure 10.10). The primary crusher output is fed to a secondary crusher (e.g. a cone crusher) for further reduction in stone size. This is followed by screening its output into appropriate aggregate sizes or reprocessing with further crushing cycles (tertiary) using cone crushers to achieve the desired shape and sizes. Figure 10.7 shows a primary crusher being fed with blast rock.

Some operations provide a stage either immediately prior to or after primary crushing known as 'scalping'.

Scalping
Scalping is a coarse screening to remove the 40 mm and down fraction, and is often carried out prior to primary crushing. This removes weak rock, clays and silts, to ensure a cleaner, good quality rock feed and to improve primary crusher performance. However, on large primary crushing installations it is more common to carry out the scalping process once the rock has undergone primary crushing.

227

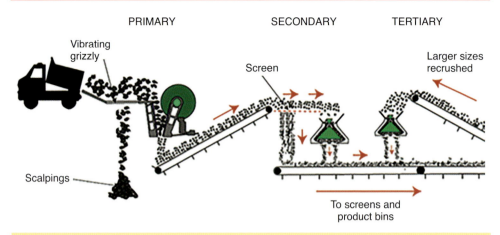

PRIMARY SECONDARY TERTIARY

Vibrating grizzly

Screen

Larger sizes recrushed

Scalpings

To screens and product bins

Figure 10.8 Simplified crushing process flow. (Courtesy of Matthew Allen)

On some sites, this scalped material is blended with a range of different sized aggregates, to produce a lower grade material, sometimes referred to as crusher runs. In very dirty rock deposits, the product derived from the scalping operation will often be diverted as a waste material.

Simplified crushing process flow
Figure 10.8 is a simplified process flow through the crushing stages before final screening and storage into the various product sizes needed by the quarry. In this process, raw material is first screened through a vibrating deck called a 'vibrating grizzly', to remove small pieces of rock along with any clay and soils followed by primary, secondary and tertiary crushings that result in the desired aggregate sizes of appropriate grade and quality, the types of crusher being dependent on the rock type.

10.3.2.2 Crusher types
Compression crushers
Compression crushers apply pressure to the rock between fixed and moving parts in a repeated squeezing action, reducing the rock in size until it is small enough to pass through a crushing chamber. There are two types of compression crusher (Figure 10.9).

■ Jaw crushers work by squeezing rock in the crushing chamber between two jaws, one of which is fixed while the other is angled and moves backwards and forwards applying the crushing force. The distance between the jaws reduces as the crushed rock travels down the chamber.
■ Gyratory crushers work by having a gyrating crushing mantle set within a bowl, and provide a continuous crushing action. Rock is crushed and reduced in size as it travels down the crushing chamber.

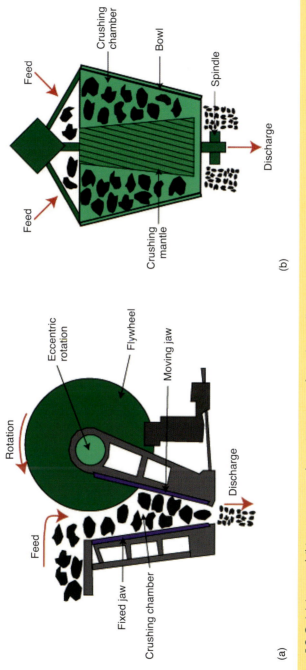

Figure 10.9 (a) Jaw and (b) gyratory crushers. (Courtesy of Mathew Allen)

Compression crushers, although relatively economical, can have a tendency to create poor particle shapes for some rock sources. In these instances, impact crushers may be used for primary crushing, but at a slightly higher cost per tonne.

Impact crusher
In impact crushers, as the name suggests, the breaking down of the rock is achieved when the high speed rotating blow bars or swinging metal hammers impact on the cascade of rock discharged into the crusher. Large rocks are either broken down by this primary action or by the secondary action of rock against rock as particles are impelled at high speed from the process. Figure 10.10 shows the simplified layout of a horizontal impact crusher.

Cone crusher
Cone crushers (Figure 10.11) are sometimes called gyratory cone crushers, and although they can be used as a primary crusher they are more often used as secondary or tertiary crushers. In operation as a secondary or tertiary crusher, pre-crushed aggregate is fed through the top of a cone crusher between two conical surfaces, called the mantle and the concave liners, into the crushing chamber.

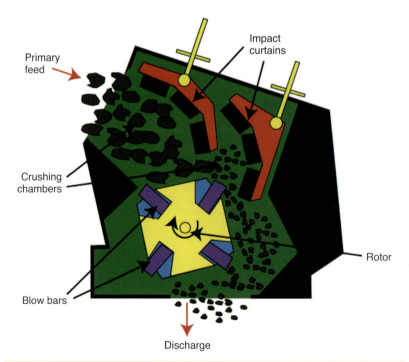

Figure 10.10 Horizontal action impact crusher. (Courtesy of Matthew Allen)

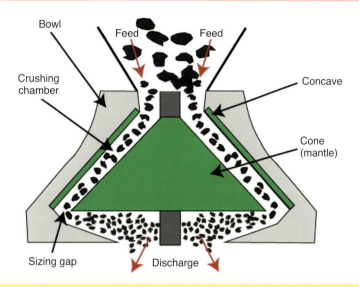

Figure 10.11 Cone crusher. (Courtesy of Matthew Allen)

The mantle rotates eccentrically within the crushing bowl, and this gyrating action causes the gap between the mantle and liners to change dimension and perform the actual crushing as the rock falls through.

10.3.3 Screening

Once scalped and crushed, the multi-sized rock is passed over a number of different screens that are typically double or triple decked, to sort the aggregate into individual aggregate sizes. These sizes will reflect the aggregate product range, and are typically 40, 28, 20, 14, 10 and 6.3 mm, with the finest fraction, less than 4 mm, being referred to as crushed rock fines.

Screen decks are modular units usually manufactured from rubber or polyurethane, but perforated metal plate and woven wire screen decks can be used to produce a high grade, low flake product for use as surface dressing or high specification asphalt surface course aggregates.

10.4 Coarse and fine aggregates

The performance of an asphalt is largely predetermined by the characteristics of its components: bitumen, aggregates and air voids. However, for this marriage to be successful, the properties of the aggregates need to be known, fully understood and adequately specified.

The properties can be considered within a group of five broad classifications, namely

■ geometrical properties (e.g. size, shape and particle packing)

231

- mechanical properties (e.g. strength and hardness)
- physical properties (e.g. particle density and water absorption)
- chemical properties (e.g. adhesion)
- durability, weathering properties (e.g. freeze–thaw resistance).

Some of these properties (i.e. those for geometric requirements, e.g. grading, size, shape and cleanliness (clay and silt content)) are partly controlled, and are a function of the aggregate processing, while others (e.g. physical, mechanical and chemical properties) are an inherent characteristic of the aggregate unaltered by quarry processing.

In Europe, the standard EN 13043 (BSI, 2002) provides a range of categories for aggregate properties and the associated test methods: this allows purchasers and designers of asphalts the means to select appropriate limiting values for the wide range of aggregates used in asphalts. It is usually the responsibility of the purchaser to determine and define appropriate EN 13043 categories for properties that are relevant to the particular end-use of an aggregate. Where a property is not relevant to both the mixture or its application, the specifier should identify this using a 'no requirement' description.

Aggregate properties referred to in EN 13043 and their applicable test methods are outlined in the following sections: note that this standard includes mechanical properties as a part of physical properties, and the following sections have been ordered in line with EN 13043.

10.4.1 Properties and testing of coarse, fine and all-in aggregate

10.4.1.1 Geometrical properties

Aggregate size

This is described using the designation d/D, in terms of lower (d) and upper (D) sieve sizes: for example, 4/10 mm single size.

Grading (particle size distribution)

Aggregate gradings are determined in accordance with EN 933-1 (BSI, 2012a) and specified as appropriate to the aggregate sizes d/D from tables within the standard. Grading categories are used and expressed as follows: G_C for coarse aggregate, G_F for fine aggregate and G_A for all-in aggregate. In addition, for coarse aggregates the numerical specification limits for the minimum percentage by mass passing the sieve size represented by D and the maximum amount passing the sieve size represented by d are stated. For fine (G_F) and all-in (G_A) aggregates, only the minimum percentage appropriate to the sieve size D is stated.

For example, when specifying aggregate size it is sufficient to quote the aggregate size and grading category (e.g. 10/20, G_C 85/35). From this example, it can be determined that the product is a coarse aggregate (G_C) and will have between 0 and 15% by mass of aggregate larger than a 20 mm sieve size (D), this fraction being known as the 'oversize'. Also, between 0 and 35% by mass of aggregate is less than the 10 mm sieve (d), this fraction being known as the 'undersize'.

Fines content

The fines content (f) specified in EN 13043 relates to the percentage by mass passing a 0.063 mm sieve for coarse and fine aggregates, and is expressed as f_x, where x is the maximum percentage passing the 0.063 mm sieve. It is determined as part of the washing and sieving test in accordance with EN 933-1.

Fines quality

Some fines such as clay minerals are harmful to asphalt and can occur in the fines fraction of some aggregate sources. The presence of clays leads to high rates of water absorption, swelling and increased moisture sensitivity of the mixture, resulting in stripping.

The methylene blue test specified in EN 933-9, Annex A (BSI, 2009a), is used to assess the quantity of potentially harmful fines in fine aggregate for the 0/0.125 mm fraction. The test is carried out by adding an aqueous solution of methylene blue dye to a fine aggregate sample held in suspension with water, and measuring the quantity of added dye at the point where adsorption of the dye has stopped. The test exploits the fact that clay minerals absorb basic dyes from aqueous solutions, and the higher the quantity of dye absorbed, the greater is the quantity of potentially harmful fines present in the fine aggregate.

Where the fines content of either fine or all-in aggregate (where $D \leq 8$ mm) is greater than 3%, the amount of harmful fines (e.g. swelling clays) needs to be considered. A measure of such presence is determined using the methylene blue test according to EN 933-9. This method describes separate tests for two size fractions

- 0/2 mm
- 0/0.125 mm (Annex A).

The results are reported with the units of g/kg, and expressed according to the appropriate category as MB_{value} for the 0/2 mm fraction, and $MB_{F\ value}$ for the 0/0.125 mm fraction. EN 13043 (BSI, 2002) specifies the requirement for the 0/0.125 mm fraction to be used for the assessment of 'fines quality' for fine and all-in aggregates.

10.4.1.2 Physical properties
Resistance to fragmentation of coarse aggregate
Aggregates must be hard and tough enough to resist fragmentation (degradation and disintegration) when being stockpiled, transported and mixed in production processes and also to withstand the stresses from their in-service applications.

The Los Angeles (LA) test to EN 1097-2 (BSI, 2010a) determines an aggregate's resistance to fragmentation by degrading a sample placed in a rotating steel drum with 11 large (400–445 g) steel balls (45–49 mm in diameter). As the drum rotates, the aggregate particles are broken down due to impacts with the steel balls and the other aggregate particles in the drum.

The result is reported as LA_{value}, calculated from the mass of aggregate that has broken down to smaller sizes expressed as a percentage of the total mass of aggregate. The lower the value, the tougher and more fragmentation resistant is the aggregate. A value of LA_{30} or less is needed to make a good roadstone aggregate.

Figure 10.13 shows the degree of fragmentation following the Los Angeles test on a 10/14 mm igneous aggregate.

Resistance to polishing of coarse aggregate for surface courses
The action of tyres on the road surface results in polishing of the exposed surface aggregate, and the state of polish is a key factor affecting the road

Before test   **After test**

Figure 10.13 Before and after a Los Angeles test on a 10/14 mm igneous aggregate. (Courtesy of Benjamin Allen)

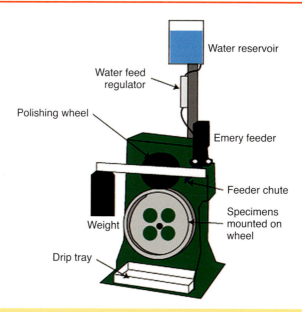

Figure 10.14 Simple diagram of polished stone value (PSV) machine. (Courtesy of Matthew Allen)

surface's skid resistance. An aggregate's resistance to polishing is therefore a most important characteristic, and aggregates with a high resistance to polishing sell at a premium relative to those with less resistance. Its resistance depends on its micro-texture characteristics, which are determined using the polished stone test in EN 1097-8 (BSI, 2009b), under standardised conditions similar to those occurring at the surface of the road.

The apparatus is illustrated in Figure 10.14: four test specimens, each consisting of between 36 and 46 (passing a 10 mm sieve but retained on a 7.2 mm flaky sieve) aggregate particles, bound in a rigid resin matrix (Figure 10.15) are clamped around the periphery of a large road wheel with which a second, smaller, solid rubber-tyred wheel has contact. Both wheels rotate, and a corn emery abrasive, similar to that used on a coarse emery paper, is fed into the interface between the two wheels, and the aggregate samples are polished for 3 h. The first cycle is followed by a second cycle of 3 h of polishing with fine emery flour instead of the corn emery. The samples are removed from the apparatus, and their degree of polishing measured using the Transport Research Laboratory portable skid resistance tester.

The results are expressed from the mean of the four test specimens as the aggregate polished stone value (PSV), in the form PSV_{value}. The higher the number, the greater the resistance to polishing, and an aggregate having a PSV of 60 or more is generally regarded as providing a good level of

Figure 10.15 PSV specimens. (Courtesy of Benjamin Allen)

polishing resistance. The in-service life of a higher PSV aggregate (i.e. the length of time from first trafficking to the time when the road surface has polished to the extent that it no longer has its anticipated degree of skid resistance) will be longer than that of a lower PSV aggregate.

Table 10.3 is an extract from the Highways Agency's Interim Advice Note (IAN) 156/12 (Department for Transport, 2012). IAN 156/12 amended HD 36/06 (Highways Agency *et al.*, 2006) for the aggregate PSV after the incorporation of the results of recent research. It shows how the Agency specifies requirements for the PSV of aggregates used in 'hot applied thin surface course' systems on the UK's trunk road network (which includes motorways) as a function of

- the traffic frequency (commercial vehicles/lane/day)
- the site (site description in terms of different degrees of risk to road users)
- the investigatory level (IL): a factor (dictated by the skid resistance policy for the road) to maintain an adequate in-service performance in relation to the skid resistance of the surface course.

Resistance to surface abrasion
For aggregates used in surface course mixtures, it is also important to ensure that the aggregate will not be worn away too quickly in service by specifying

Table 10.3 Minimum PSVs for coarse aggregates in hot applied thin surface course systems

Site category	Site description	Investigatory level (IL)	Minimum PSV required for a given IL, traffic level and type of site — Traffic at the design life: commercial vehicles/lane/day									
			0–250	251–500	501–750	751–1000	1001–2000	2001–3000	3001–4000	4001–5000	5001–6000	>6000
A1	Motorways where traffic is generally free flowing on a relatively straight line	0.30	50	50	50	50	50	50	50	53	63	63
		0.35	50	50	50	50	50	53	53	53	63	63
A2	Motorways where some braking regularly occurs	0.35	50	50	50	55	55	60	60	65	65	65
B1	Dual carriageways where traffic is generally free flowing on a relatively straight line	0.30	50	50	50	50	50	50	50	53	63	63
		0.35	50	50	50	50	50	53	53	53	63	63
		0.40	50	50	50	50	53	58	58	58	63	68+
B2	Dual carriageways where some braking regularly occurs	0.35	50	50	50	55	55	60	60	65	65	65
		0.40	55	60	60	65	65	68+	68+	68+	68+	68+
C	Single carriageways where traffic is generally free flowing on a relatively straight line	0.35	50	50	50	50	50	53	53	58	63	63
		0.40	50	53	53	58	58	63	63	63	68+	68+
		0.45	53	53	58	58	63	63	63	63	68+	68+
G1/G2	Gradients >5% longer than 50 m as per the Highways Agency's guidance HD 28 ('skid resistance')	0.45	55	60	60	65	65	68+	68+	68+	68+	68+
		0.50	60	68+	68+	HFS	HFS	HFS	HFS	HFS	HFS	HFS
		0.55	68+	HFS	HFS	HFS	HFS	HFS	HFS	HFS	HFS	HFS
K	Approaches to pedestrian crossings and other high-risk situations	0.50	65	65	65	68+	68+	68+	HFS	HFS	HFS	HFS
		0.55	68+	68+	HFS	HFS	HFS	HFS	HFS	HFS	HFS	HFS
Q	Approaches to major and minor junctions on dual carriageways and single carriageways where frequent or sudden braking occurs but in a generally straight line	0.45	60	65	65	68+	68+	68+	68+	68+	68+	HFS
		0.50	65	65	65	68+	68+	68+	HFS	HFS	HFS	HFS
		0.55	68+	68+	HFS	HFS	HFS	HFS	HFS	HFS	HFS	HFS
R	Roundabout circulation areas	0.45	50	55	60	60	65	65	68+	68+	68+	68+
		0.50	68+	68+	68+	68+	68+	68+	68+	68+	68+	68+
S1/S2	Bends (radius <500 m) on all types of road, including motorway link roads; other hazards that require combined braking and cornering	0.45	50	55	60	60	65	65	68+	68+	HFS	HFS
		0.50	68+	68+	68+	HFS	HFS	HFS	HFS	HFS	HFS	HFS
		0.55	HFS	HFS	HFS	HFS	HFS	HFS	HFS	HFS	HFS	HFS

Department for Transport (2012)

Figure 10.16 Gritstone aggregate specimens, before and after AAV testing. (Courtesy of Benjamin Allen)

a maximum aggregate abrasion value (AAV) determined in accordance with EN 1097-8:1999, Annex A (BSI, 2009c).

In this test, a sample is prepared using specimens of 6.3/14 mm aggregate particles, retained on a 10.2 mm grid sieve, bound in a rigid resin matrix. This is weighed then mounted face down against a large circular rotating steel disc. Coarse abrasive sand is fed between the disc and the faces of the resin-bound aggregates as the wheel rotates for 500 revolutions, causing the sand to abrade the aggregate specimens. At the end of the test, the sample is removed from the machine, cleaned and re-weighed. The percentage of mass lost is expressed as AAV_{value}. The lower the value, the more abrasion resistant is the aggregate. Figure 10.16 shows two gritstone AAV specimens, before and after testing for the AAV, and illustrates the abrasive effects of the test on the aggregate particles.

For a particular application, the selection of an appropriate aggregate AAV will be dependent on the amount of traffic, the design life and the mixture type. Table 10.4 illustrates this with an example from the UK's Highways Agency *Design Manual for Roads and Bridges*, HD 36/06 (Table 3.2), and shows the specification of appropriate maximum AAV levels for a number of asphalt types.

Resistance to wear of coarse aggregate

The wear of aggregates at particle points and edges due to breakdown as a result of attrition is an important aggregate property, particularly for mixtures such as porous or open graded asphalt with many point-to-point aggregate contacts.

For this aggregate property, the micro-Deval test to EN 1097-1 (BSI, 2011a) is used as a control measure, and its result calculated and reported as a micro-Deval coefficient, $M_{DE\ value}$.

Table 10.4 Maximum AAVs for aggregates or coarse aggregates in unchipped surfaces, for new surface courses

	Traffic at the design life: commercial vehicles/lane/day					
	<250	251–1000	1001–1750	1751–2500	2501–3250	>3250
Maximum AAV for aggregates for hot rolled asphalt and surface dressing, and for aggregate in slurry and microsurfacing systems	14	12	12	10	10	10
Maximum AAV for aggregate in thin surface course systems, exposed aggregate concrete surfacing and coated macadam surface course	16	16	14	14	12	12

Highways Agency *et al.* (2006)

The test method uses a rotating steel vessel containing an aggregate sample of predetermined mass, a 5 kg charge of small (9.5–10.5 mm dia.) steel balls and a volume (2.5 litres) of clean water. After 1200 revolutions, about 2 h of rotation, the now worn aggregate sample is discharged from the vessel and reweighed, and the percentage loss of mass expressed as $M_{DE\ value}$. The lower the numerical value, the more resistant to wear through attrition is the aggregate.

Figure 10.17 shows the effect and degree of 'particle rounding' that occurs as a result of this test on an igneous aggregate.

Before test   After test

Figure 10.17 Micro-Deval test on an igneous aggregate. (Courtesy of Benjamin Allen)

Resistance to abrasion from studded tyres of coarse aggregates to be used for surface courses

In countries where vehicles use studded road tyres in winter, this results in accelerated wear of the asphalt at the road surface. As a consequence, it becomes an important factor in the specification and selection of aggregates used in surface course mixtures.

Where required, the resistance to abrasion from studded tyres for the coarse aggregate fraction in asphalt is determined in accordance with EN 1097-9 (BSI, 2014b).

The test is performed on coarse aggregate between 11.2 and 16 mm that is rotated (90 ± 3 rev/min) in a drum with steel abrasive balls in the presence of water.

After a specified number of revolutions, the abrasion loss rate of the aggregate is calculated and reported as a Nordic abrasion value, AN_{value}. The lower the numerical value, the more resistant to abrasion from studded tyres is the aggregate.

Particle density and water absorption

Particle density and water absorption are determined in accordance with EN 1097-6 (BSI, 2013a), Clause 7, 8 or 9, depending on the size of the aggregate.

- Clause 7: wire basket method for aggregate particles passing the 63 mm test sieve and retained on the 31.5 mm test sieve.
- Clause 8: pyknometer method for aggregate particles passing the 31.5 mm test sieve and retained on the 4 mm test sieve.
- Clause 9: pyknometer method for aggregate particles passing the 4 mm test sieve and retained on the 0.063 mm test sieve.

The tests give a measure for particle density expressed in the units Mg/m^3 for

- the apparent particle density
- the oven-dried particle density
- the saturated and surface-dried particle density.

The aggregate's water absorption after immersion for 24 h is expressed as $WA_{24\ value}$.

Bulk density

The bulk density of an aggregate is determined in accordance with EN 1097-3 (BSI, 1998b), and is expressed in Mg/m^3.

A physical property of an aggregate not considered in EN 13043 is its thermal expansion and contraction (i.e. its thermal coefficient, α). As aggregates represent about 85% of the total volume of a typical asphalt, the thermal coefficient of asphalt is greatly influenced by that of the aggregate. The thermal coefficient of any natural aggregate depends on its mineralogy. In general, siliceous aggregates with a high quartz content exhibit a high thermal coefficient $(10.01 \times 10^{-6}$ to 13.00×10^{-6} mm/mm/°C), whereas some pure limestones that consist mainly of calcite exhibit a lower value $(5.58 \times 10^{-6}$ to 6.10×10^{-6} mm/mm/°C) (Mukhopadhyay *et al.*, 2007). The thermal coefficient of a bituminous binder is 10–20 times higher than that of the aggregate.

Thus, thermal stresses can develop in the binder film surrounding an aggregate particle as a result of differential thermal contraction. Under extreme low temperatures and/or when stiff binders are used, these thermal stresses may cause localised cracking in the asphalt. These cracks can then facilitate moisture entry into the asphalt, causing damage and a subsequent reduction in durability (El Hussein *et al.*, 1998).

10.4.1.3 Durability
Water absorption value as a screening test for freeze–thaw resistance
The water absorption value is used as an indicator of an aggregate's resistance to freezing and thawing action. If its water absorption is 2% or less, then an aggregate is considered as suitably freeze–thaw resistant without further testing.

Aggregates with water absorption values greater than 2% can be very resistant to freeze–thaw action, but need to be tested for this property more directly. Examples of this category include

- Jurassic limestone and sandstone
- blast furnace slags
- Permian limestone
- dolomite
- carboniferous sandstone.

Resistance to freezing and thawing
The vulnerability of an aggregate to damage from periods of freezing and thawing will depend on the climate, with the severity of any damage being related to the frequency and severity of these freeze–thaw cycles. This risk increases with the degree of water saturation, and is significantly increased with sea water.

EN 13043 gives guidance on the selection of an appropriate level of resistance to freezing and thawing based on climate, which can be

- Mediterranean

- Atlantic
- continental.

Also, the possible environmental conditions applying to the product in service are considered

- frost free or dry situation
- partial saturation, no salt
- saturated, no salt
- salt (sea water or road surfaces)
- airfield surfacings.

The ability of an aggregate to withstand changes in volume as a result of conditions under freeze–thaw cycling will depend on its petrographic type and the size distribution of the pores within the aggregate.

EN 13043 gives two methods for determining an aggregate's ability to resist freezing and thawing, often referred to as its soundness: EN 1367-1 (BSI, 2007) and EN 1367-2 (BSI, 2009d).

In the UK, the magnesium sulfate test is usually specified (EN 1367-2). In this test, particles of aggregate of size 10–14 mm are tested for soundness by subjecting pre-weighed samples to soaking and drying cycles in a magnesium sulfate solution, and measuring the mass of aggregate degraded to finer than that which passes a 10 mm test sieve, due to salt crystallisation and its associated cracking. This degradation is reported as a percentage, in the form MS_{value}.

Aggregates for UK conditions are considered satisfactory for general purpose use if their magnesium sulfate soundness category is MS_{25} or better (i.e. a lower value). Other categories of magnesium sulfate soundness may be more appropriate for climatic conditions outside the UK.

Vulnerable aggregates tend to be derived from highly weathered rocks and some conglomerates and breccias, and can include, for example,

- schist
- mica schist
- phyllite
- shale
- chalk.

10.4.1.4 Resistance to thermal shock

Aggregates are subjected to heating and drying as a part of the production of asphalt, and undergo an element of thermal shock as a consequence. EN 13043 provides a method, EN 1367-5 (BSI, 2011b), for determining the extent to which aggregates degrade by heating during the drying process, and knowledge of this property can be useful when assessing new and potential resources with no previous history of use in asphalt, particularly recycled and manufactured aggregates.

The test involves the preparation of two aggregate samples. One sample is tested for resistance to fragmentation using the Los Angeles test (described in section 10.4.1.2), and the other sample is tested for fragmentation after exposure to thermal shock. Thermal shock involves soaking a sample of the aggregate in water and subjecting it to thermal shock by heating it to 700 °C for 3 min.

The difference between the LA coefficient before and after thermal shock is defined as 'the loss in strength due to thermal shock'.

10.4.1.5 Affinity of coarse aggregates for bituminous binders
Good coating and adhesion of bitumen to aggregate is essential, but the loss of bond between aggregate and bitumen in service due to the presence of water, known as 'stripping', has a major detrimental effect on the integrity of an asphalt pavement. Although many factors affect the affinity of the aggregate for bitumen, it is accepted that the mineralogy and chemical composition of the aggregate are important contributory factors.

Generally, aggregates with high silica contents (acid) have a good affinity for water (hydrophilic) but not for bitumen (Figure 10.18). They are more likely to suffer from stripping after exposure to water than those with lower silica contents (basic) that have an affinity for bitumen rather than water (hydrophobic). Aggregate binder affinity for highly siliceous aggregates can be sufficiently improved with the addition of adhesion promoters.

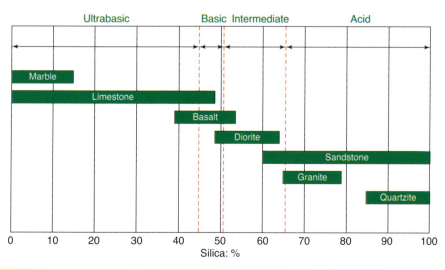

Figure 10.18 Typical range of silica content (SiO_2) for some common rocks

EN 13043 requires the affinity of coarse aggregates to bituminous binders to be determined in accordance with EN 12697-11 (BSI, 2012c). The latter standard gives three methods

A. A rolling bottle containing a loose bitumen-coated sample of aggregate, a glass rod device providing a stirring action, and water are used. After rolling for defined periods of time, the sample is removed from the roller bottle and visually examined for bitumen stripping.
B. A sample of bitumen-coated aggregate is immersed in distilled water for 48 h, and the number of particles that are no longer completely coated is assessed.
C. An aggregate sample is mixed with bitumen to give complete coverage of the aggregate. This prepared sample is first subjected to stripping in boiling water for 10 min. The proportion of exposed aggregate following this is determined by placing the sample in contact, for 5 min, with either hydrochloric acid (for calcareous aggregates) or hydrofluoric acid (for siliceous aggregates) of given concentrations. A titration method is used to determine the volume of hydrochloric acid or hydrofluoric acid consumed by the reaction of these reagents with the free surface of the aggregate. Defined calibration curves are used to estimate the proportion of the exposed aggregate surface from the volume of reagent consumed.

In the UK, the Ministry of Defence, Defence Infrastructure Organisation (DIO) Specification 13 (Ministry of Defence, 2009a) is specified in Table 3.1 using EN 12697-11 Method B, and the requirement is 'not greater than 6 particles from a 150 particle test sample'.

There are other tests or means of evaluation. In the UK, the Highways Agency's *Specification for Highway Works*, Series 900, Clause 953 (Highways Agency *et al.*, 2008), describes the saturation ageing tensile stiffness (SATS) test, which is used to assess the durability of asphalts. In this test, cylindrical specimens of the asphalt are conditioned in a pressure vessel, stiffness levels being determined before and after conditioning, to establish a percentage of retained stiffness. It is essentially a measure of the adhesion between the aggregate and the bitumen where deterioration in asphalts due to a gradual loss of adhesion results in a reduction in mixture stiffness.

10.4.1.6 'Sonnenbrand' of basalt

'Sonnenbrand' is a German word that translates into English as 'sunburn'. It is used in connection with basalt to describe a particular type of rock decay that is present in some basalt that develops over time under certain atmospheric conditions. This visual observation can take place within months of extraction or extend over many years. Affected rocks first show staining (grey/white-coloured spots) and efflorescence, and later develop hairline cracks that reduce the strength of the aggregate, in some cases resulting in total breakdown. When an aggregate is known or suspected of being afflicted with 'sonnenbrand', then the aggregate's loss of mass and resistance to fragmentation is determined using the EN 1367-3 (BSI, 2001) and EN 1097-2 (BSI, 2010a) test methods.

10.4.1.7 Chemical

The chemical requirements for coarse and fine aggregates within EN 13043 are determined in accordance with EN 932-3 (BSI, 1997). This method employs a visual examination of the aggregate to determine the constituent rock or mineral types (see Table 10.2). Where a more quantitative study is required, then an analytical approach is needed using methods such as X-ray fluorescent techniques, principally to identify an aggregate's oxide composition (see Figure 10.19).

Other chemical requirements are stated within EN 13043 essentially for slag (manufactured) aggregate and recycled aggregate. These test methods are based on EN 1744-1 (BSI, 2009e) procedures, falling under the following categories.

Coarse lightweight contaminators

The term 'coarse lightweight contaminators' refers to the presence of organic contaminators larger than 2 mm. These are not likely to occur in natural rocks, and are usually specified where manufactured or recycled aggregates are used.

Testing is carried out in accordance with Clause 14.2 of EN 1744-1, and the result expresses the amount of organic contaminators as a percentage of the coarse aggregate.

Constituents that affect the volume stability of blast furnace and steel slags

In the cases of blast furnace and steel slags, there is a specific requirement for establishing and measuring the volume stability of these materials

- air-cooled blast furnace slag aggregate shall be free from dicalcium silicate disintegration when tested in accordance with EN 1744-1
- air-cooled blast furnace slag aggregate shall be free from iron disintegration when tested in accordance with EN 1744-1
- the volume stability of steel slag aggregate shall be determined in accordance with EN 1744-1.

celtest
independent materials testing — diamond core drilling & sawing

Aggregate Industries
Bardon Hill

UKAS
TESTING
0494

Unit:	**Croft Quarry**		Date Issued :	**18th July 2013**
Address:	Bardon Aggregates			
	Marions Way			
	Leicestershire		Aggregate Type:-	Crushed granite
			Aggregate Colour:-	Grey

AGGREGATE PROPERTIES SUMMARY DATA SHEET

Test Description		Specification Reference	Type 1	STR No.	Date	Dust	STR No.	Date	10mm	STR No.	Date
Particle Density (Mg/m³)	Apparent	EN 1097-6 : 2000				2.66	326196	Jul-13			
	S.S.D					2.60					
	Oven Dry					2.58					
Water Absorption (%)						1.2					
Chemical Analysis* -	SiO2					63.87					
	TiO2					0.60					
	Al2O3					16.35					
	Fe2O3					3.80					
	MnO					0.08					
	MgO					1.81					
	CaO	X-Ray Fluorescent Techniques				4.15	326201	Jul-13			
	Na2O					4.21					
	K2O					1.79					
	P2O					0.19					
	BaO					0.04					
	SO3					0.04					
	Loss on Ignition					3.0					
Polished Stone Value		BS EN 1097-8							59	326193	Jul-13
Water Soluble Chloride Ion Content		BS EN 1744-1				<0.001	326203	Jul-13			
Water Soluble Sulfate Content SO3		BS EN 1744-1				0.02	326202	Jul-13			
Acid Soluble Sulfate Content SO3		BS EN 1744-1				0.1	326204	Jul-13			
Total Sulfur Content		BS EN 1744-1				0.3	326206	Jul-13			
Oxidisable Sulfides*		TRL				<0.01	326205	Jul-13			
Frost Heave		BS 812 : Part 124 : 1989	7.2	326221	Jul-13						
OMC/MDD (%/Mg/m3)		Optimum Dry Density	2.13	326222	Jul-13						
		Optimum Moisture Content	6.9								
Moisture Content		BS EN 1097-5	5.5	326223	Jul-13						
Horizontal Permeability		Dtp HA 41/90	5.4x10-2	326224	Jul-13						
Magnesium Sulphate Soundness Value		BS EN 1367-2				3	326199	Jul-13			
Methylene Blue (MB) Value		BS EN 933-9				1.0	326198	Jul-13			
Plastic Limit (%)		BS 1377 : Part 2 : 1990	N-P	326226	Jul-13						
Resistivity		BS 1377 : Part 3 : 1990	80	326227	Jul-13						
pH Value		BS 1377 : Part 3 : 1990				8.5	326200	Jul-13			
Calcium Carbonate Equivalent (%)		EN 196-21 : 1992				4.82	326197	Jul-13			

All tests carried out are UKAS accredited unless otherwise denoted by *
Comments:

* Full report available upon request

Figure 10.19 Summary of physical and chemical properties for a granite aggregate, part 1. (Courtesy of Aggregate Industries)

10.4.1.8 Summary of properties and testing: example of a producer's property data sheet

Figures 10.19 and 10.20 give an example of a suite of tests carried out by a UK aggregates supplier on a granite aggregate for the UK market.

This example includes not only physical and chemical properties applicable to the production of asphalts, and described in this chapter, but also includes properties relating to unbound aggregate mixtures, such as subbase for pavement foundations (described as Type 1, in Figure 10.19).

BARDON
AGGREGATES

Aggregate Industries
Bardon Hill

celtest
independent materials testing

— *diamond core drilling & sawing*

UKAS
TESTING
0494

Unit:	Croft Quarry		Date Issued :	18th July 2013
Address:	Bardon Aggregates			
	Marions Way			
	Leicestershire		Aggregate Type:-	Crushed granite
			Aggregate Colour:-	Grey

AGGREGATE PROPERTIES SUMMARY DATA SHEET

Test Description		Specification Reference	14mm	STR No.	Date							
Particle Density (Mg/m³)	Apparent	EN 1097-6 : 2000	2.67	326207								
	S.S.D		2.60									
	Oven Dry		2.57									
Water Absorption (%)			1.7									
Aggregate Crushing Value		BS 812 Pt 110	23	326194	Jul-13							
Micro Deval Coefficient		BS EN 1097-1	32	326195	Jul-13							
Aggregate Impact Value (Dry)		BS 812 Pt 112	30	326209	Jul-13							
Aggregate Impact Value (Soaked)		BS 812 Pt 112	34	326210	Jul-13							
10% Fines Value (Dry)		BS 812 Pt 111	150	326211	Jul-13							
10% Fines Value (Soaked)		BS 812 Pt 111	120	326212	Jul-13							
Aggregate Abrasion Value		BS EN 1097-8	2.0	326214	Jul-13							
Water Soluble Chloride Ion Content		BS EN 1744-1	<0.001	326217	Jul-13							
Water Soluble Sulfate Content SO3		BS EN 1744-1	<0.01	326216	Jul-13							
Acid Soluble Sulfate Content SO3		BS EN 1744-1	0.1	326218	Jul-13							
Total Sulfur Content		BS EN 1744-1	<0.1	326220	Jul-13							
Los Angeles Coefficient		BS EN 1097-2	15	326215	Jul-13							
Oxidisable Sulfides*		TRL	<0.01	326219	Jul-13							
Petrographical Examination*			*	326192	Jul-13							
Magnesium Sulphate Soundness Value		BS EN 1367-2	20	326208	Jul-13							
All tests carried out are UKAS accredited unless otherwise denoted by *												
Comments:												
* Full report available upon request												

Figure 10.20 Summary of physical and chemical properties for a granite aggregate, part 2. (Courtesy of Aggregate Industries)

10.5 Filler aggregate

The main function of filler is to act as mastic when it is combined with the binder. It provides added stiffness and stability to the mixture, as well as acting as a packing mineral to control the void structure of the mixture.

Filler aggregate is defined in EN 13043 as 'aggregate, most of which passes a 0.063 mm sieve, which can be added to construction materials to provide certain properties', and is derived according to EN 13043 as being

- added filler, 'filler aggregate of mineral origin, which has been produced separately'
- mixed filler, 'filler aggregate of mineral origin, which has been mixed with calcium hydroxide' or
- fines, 'the particle size fraction of an aggregate which passes the 0.063 mm sieve' (e.g. baghouse fines from the dust collection devices at asphalt mixing).

The requirements for filler aggregate in EN 13043 with regard to geometrical, physical and chemical properties and consistency of production apply to mixed filler, added filler and fines from the fine aggregate when this content exceeds 10% by mass of the fine aggregate.

Sometimes added filler with requirements for mineral type or mixed fillers are specified for the purpose of improving a mixture's adhesion property and durability. These are described below as 'beneficial fillers'. In the UK, guidance on the use and specification of asphalts is given in PD 6691:2010 (BSI, 2010b), and recommends that, for BS EN 13108-4 (BSI, 2006) hot rolled asphalt mixtures, any 'added filler' shall 'consist of limestone, hydrated lime or cement', and the UK's Ministry of Defence DIO Specification 40 (Ministry of Defence, 2009b), *Porous Friction Course for Airfields*, requires that

between 1.5% and 2.0% by mass of the combined aggregate/filler aggregate ... shall be CL 90-S lime to BS EN 459-1 which shall be added to the mixture as part of the fraction passing the 0.063 mm sieve. If additional material of this grading is required, it shall be crushed limestone.

EN 459-1 (BSI, 2010c) is the European standard for building lime in which lime is defined as

calcium oxide and/or hydroxide, and calcium-magnesium oxide and/or hydroxide by the thermal decomposition (calcination) of naturally occurring calcium carbonate (for example limestone, chalk, shells) or naturally occurring calcium magnesium carbonate (for example dolomitic limestone, dolomite).

Included within the scope of EN 459-1 is the use of lime for civil engineering applications, including asphalts. This standard gives definitions for the different types of lime, and their classification, and chemical and physical properties.

Using EN 459-1, the CL 90-S lime specified in the DIO Specification 40 has the following derivation

- CL: a calcium lime, mainly either calcium oxide and/or calcium hydroxide that has no hydraulic or pozzolanic activity
- 90: a characteristic content of calcium oxide and magnesium oxide equal to or greater than 90%
- S: the product is hydrated lime ($Ca(OH)_2$) and not quicklime (CaO), which is designated with the letter Q instead of an S, and that it is in powder form rather than being either a putty (S PL), slurry or milk of lime (S ML).

Other chemical and physical properties are required by EN 459-1 appropriate to this classification.

Recognised beneficial fillers for use in asphalts include

- hydrated lime
- cement
- limestone (with high calcium carbonate content)
- mixed filler (with high calcium hydroxide content).

While limestone filler is the most commonly used because of its lower cost and wide availability, both hydrated lime and cement have greater beneficial effect.

When beneficial filler is used instead of, or as a supplement to, reclaimed (baghouse) filler, the binding properties of the mastic are improved, resulting in better adhesion between the aggregate and the mastic. In the UK, the durability of adhesion for base and binder course mixtures that have been designed according to the requirements of the Highways Agency's *Specification for Highway Works*, Series 900, Clause 929 (Highways Agency *et al.*, 2008), are required to be tested and shown to achieve a SATS durability index above 80%. Mixtures that include 2% of CL 90-S hydrated lime filler are deemed to satisfy the durability requirements of this specification without the need for SATS testing.

In general, aggregates with high carbonate content such as limestone are easier to coat with binder than aggregates with high silica content. This is due to the fact that siliceous aggregates contain high concentrations of hydroxyl groups with greater affinity for carboxylic acid and water. The carboxylic acid components that are present in the binder are adsorbed by the surface of these aggregates, generating a binder–aggregate bond. However, these bonds with carboxylic acids are also prone to displacement in the presence of water (Petersen and Pancher, 1998). Such displacement is reduced when hydrated lime is used in the mixture and also, but to a lesser degree, when limestone filler is used.

10.5.1 Properties and testing
Within EN 13043 the requirements for testing filler properties is treated separately from other aggregates. The properties required through specification should be limited to those relating to the particular filler type, end use application or source of the coarse and fine aggregate.

10.5.1.1 Geometrical requirements
Grading
Gradings are carried out in accordance with EN 933-10 (BSI, 2009f), which describes the reference method used for filler grading, for type test purposes and, in cases of dispute, both for natural or manufactured filler using air jet sieving.

At least 70% filler must pass the 0.063 mm sieve, and 100% must pass the 2 mm sieve size.

Harmful fines
It is usually considered that added fillers from beneficial sources are not likely to contain harmful fines, but when harmful fines in fillers are considered likely, their presence is assessed using the methylene blue test in accordance with EN 933-9, Annex A (BSI, 2009a) (and reported as MB_{Fvalue}: a description of this method is given in section 10.4.1.1 under the heading 'Fines quality' for fine and all-in aggregate).

10.5.1.2 Physical requirements
Water content
The water content of added filler is determined in accordance with EN 1097-5 (BSI, 2008b), which limits it to a maximum of 1% by mass.

Particle density
Particle density is determined in accordance with EN 1097-7 (BSI, 2008c), and the results declared.

Stiffening properties
The addition of filler to form mastic with binder will stiffen an asphalt and affect a mixture's workability and deformation resistance. The degree of stiffening is a function of many properties, with those relating to filler within

EN 13043 including the voids in dry compacted filler (Rigden) and the 'delta ring and ball'.

Voids of dry compacted filler (Rigden)

The void content of dry compacted filler is known as the 'fractional voids', and is often referred to as the 'Rigden voids' after P. J. Rigden, who introduced the concept and a method for measuring them in 1947 (Rigden, 1947).

In EN 13043, the determination of the fractional voids is based on Rigden's original method, and is specified in accordance with EN 1097-4 (BSI, 2008d). The result is reported within an EN 13043 class as $V_{xx/yy}$, where *xx* and *yy* define the overall range for individual results. The method involves the compaction of a dry filler test portion (10 g), using a standard compactive effort, into a small metal mould, to establish a reference bulk density. Using this bulk density and the particle density of the filler, the volume and percentage of fractional voids can be calculated.

After Rigden, it is considered that the bitumen required to fill the voids in the dry compacted sample is the 'fixed' bitumen (forming the mastic phase with the filler), and the bitumen in excess of this is the 'free' bitumen – the fraction that coats the coarse aggregate and lubricates the mixture. Higher Rigden void contents lead to a greater stiffening effect, and knowledge of the amount of voids can be important in both asphalt mixture design and quality control.

French guidance on the specification for Rigden void content is given in *The Use of Standards for Hot Mixtures* (Sétra, 2008), and for the characterisation of fillers used in asphalt this recommendation is $V_{28/45}$. When specified to EN 13043, all results are expected to be within this specification range, with a further requirement that at least 90% of the last 20 values be within 4 units of each other.

'Delta ring and ball' of filler aggregate for asphalts

The 'delta ring and ball' ($\Delta_{R\&B}$) value determined in accordance with EN 13179-1 (BSI, 2013b) measures the increase in the softening point of a laboratory-prepared bitumen/filler mastic sample, respectively 62.5/37.5 by volume, relative to the softening point of the bitumen used in the test. The bitumen is 70/100 grade to EN 12591 (BSI, 2009g). The resulting increase in the softening point is reported as the $\Delta_{R\&B}$ value (to the nearest 0.5°C), and results are declared within a relevant specified category, which is appropriate to the particular application.

In the UK, the stiffening property of added fillers used in EME2 base and binder course mixtures are recommended – according to PD 6691 (BSI,

2010b) – to be category $\Delta_{R\&B}$ 8/16, with the note that 'Fillers with delta ring and ball values above 16 but not greater than 20 may be used where there is a history of satisfactory use in asphalt.'

10.5.1.3 Chemical requirements

When required, the following chemical requirements of filler in EN 13043 are tested.

- Water solubility, determined in accordance with EN 1744-1, with results declared as a percentage of solubility by mass, in the form WS_{value} for categorisation purposes.
- Water susceptibility, determined in accordance with EN 1744-4 (BSI, 2005b).
- Calcium carbonate content of limestone filler aggregate, determined in accordance with EN 196-21 (replaced by EN 196-2:2013 (BSI, 2013c)), with results declared as the calcium carbonate content, percentage by mass, in the form CC_{value} for categorisation purposes.
- Calcium hydroxide content of mixed filler, determined in accordance with EN 459-2 (BSI, 2010d), with results declared as the calcium hydroxide content, percentage by mass, in the form Ka_{value} for categorisation purposes.

10.5.2 Requirements for consistency of filler production

EN 13043 requires at least one of the following properties to be measured as a means of establishing and measuring the consistency of filler production.

- The 'bitumen number' of the added filler. This is the amount of water, in ml, needed to be added to 100 g of filler to reach a reference consistency defined by a penetration value of 5–7 mm, in accordance with EN 13179-2 (BSI, 2000). The results are declared as BN_{value}, and for consistency purposes the values are required to be within the specified range, with the additional requirement that, on the basis of the last 20 values, 90% of the results are required to be within 6 units of each other.
- The loss on ignition of coal fly ash. This is determined in accordance with EN 1744-1, Clause 17, with the additional requirement that a producer's declared range shall not be greater than 6% by mass. Also, when aggregates contain non-volatile oxidisable constituents, then the loss on ignition is corrected in accordance with EN 196-2.
- The particle density of the added filler. This is determined in accordance with EN 1097-7, with the units of Mg/m^3, and where this value is declared as a range, then this is not to exceed $0.2\,Mg/m^3$.

- The loose bulk density in kerosene. An apparent density determined using a specified measuring flask with 10 g of filler in 25 ml of kerosene. The density is derived by measuring the height of the filler sediment in the kerosene after 6 h in accordance with EN 1097-3, Annex B (BSI, 1998c). The producer's range of values is required to be between 0.5 and 0.9 Mg/m^3.

- The Blaine test. This measures the specific surface of filler, in units of m^2/kg, as specified in EN 196-6 (BSI, 2010e), with the requirement that a producer's declared range shall be not greater than 140 m^2/kg.

10.6 Assuring aggregate quality

The preceding sections describe the characterisation and categorisation of aggregate, and how it is measured and specified. However, this is of little value if the producer is unable to supply consistently to the declared grade.

Consistency is achieved through the application of a suitable and independently verifiable quality system that underpins the appropriate level of processing and inspection needed to control production. The establishment of aggregate declared properties is achieved by the application of a statistically based sampling and testing procedure, which is not always demanded by standards or specifications. This provides a more reliable measure of an aggregate's true performance level than non-statistically based methods.

All aggregate used in the production of asphalt will have undergone some processing with subsequent assessment and testing, to deliver a final product that can be shown to meet a product specifier's demands. It is the specifier's responsibility to ensure that their specification can deliver a product 'fit for its intended purpose', and it is the aggregate producer's responsibility to deliver a consistent supply of aggregate that meets their claimed performance levels.

With effect from 1 July 2013 in all European member states, the Regulation 2011 (EC, 2011) made it mandatory that all products are produced to harmonised European standards – this includes aggregates used in asphalts to EN 13043, which are required to be CE marked.

CE marking ensures that aggregates used within construction products not only satisfy the requirements of the relevant European standards but are also safe to use. The adherence to recognised systems of factory production control (FPC) gives the assurance that both of these requirements are being fulfilled.

Annex B (Normative) of the aggregate standard EN 13043:2002 deals with the FPC requirements, and specifies a system for aggregate production that ensures they conform to the relevant requirements of the standard. The essential elements of this are

- organisation
  - responsibility and authority
  - management representative for factory production control
  - management review
- control procedures
  - document and data control
  - sub-contract services
  - knowledge of the raw material
- management of production
- inspection and test
  - general
  - equipment
  - frequency and location of inspection, sampling and tests
- records
- control of non-conforming products
- handling, storage and conditioning in production areas
- transport and packaging
  - transport
  - packaging
- training of personnel.

A new European standard is currently being developed that will specify requirements for both the initial type testing (ITT) and the FPC of aggregates. The FPC system will describe methods of controlling the sourcing and processing of aggregate, and will be combined with sampling and testing routines to provide ongoing assurances that aggregate products consistently conform to those properties that were determined in the ITT.

References

Bide T, Brown TJ and Hobbs SF (2013) *United Kingdom Minerals Handbook 2012*. British Geological Survey, Keyworth, UK, Open Report OR/13/024.

BRE (Building Research Establishment) (1997) *Alkali-silica Reaction in Concrete*. BRE, Garston, UK, BRE Digest 330.

BSI (British Standards Institution) (1997) BS EN 932-3:1997. Tests for general properties of aggregates. Part 3: Procedure and terminology for simplified petrographic description. BSI, London, UK.

BSI (1998a) BS EN 933-5:1998. Tests for geometrical properties of aggregates Determination of percentage of crushed and broken surfaces in coarse aggregate particles. BSI, London, UK.

BSI (1998b) BS EN 1097-3:1998. Tests for mechanical and physical properties of aggregates Determination of loose bulk density and voids. BSI, London, UK.

BSI (1998c) BS EN 1097-3:1998. (Annex B). Tests for mechanical and physical properties of aggregates Determination of loose bulk density and voids. BSI, London, UK.

BSI (2000) BS EN 13179-2:2000. Test for filler aggregate used in bituminous mixtures Bitumen number. BSI, London, UK.

BSI (2001) BS EN 1367-3:2001. Tests for thermal and weathering properties of aggregates boiling test for Sonnenbrand basalt. BSI, London, UK.

BSI (2002) BS EN 13043:2002. Aggregates for bituminous mixtures and surface treatments for roads, airfields and other trafficked areas. BSI, London, UK.

BSI (2005a) BS EN 13108-8:2005. Bituminous mixtures; Material specifications; Part 8: Reclaimed asphalt. BSI, London, UK.

BSI (2005b) BS EN 1744-4:2005. Tests for chemical properties of aggregates. Determination of water susceptibility of fillers for bituminous mixtures. BSI, London, UK.

BSI (2006) BS EN 13108-4:2006. Bituminous mixtures. Material specifications. Hot rolled asphalt. BSI, London, UK.

BSI (2007) BS EN 1367-1:2007. Tests for thermal and weathering properties of aggregates. Determination of resistance to freezing and thawing. BSI, London, UK.

BSI (2008a) BS EN 933-4:2008. Tests for geometrical properties of aggregates. Determination of particle shape. Shape index. BSI, London, UK.

BSI (2008b) BS EN 1097-5:2008. Tests for mechanical and physical properties of aggregates. Determination of the water content by drying in a ventilated oven. BSI, London, UK.

BSI (2008c) BS EN 1097-7:2008. Tests for mechanical and physical properties of aggregates. Determination of the particle density of filler. Pyknometer method. BSI, London, UK.

BSI (2008d) BS EN 1097-4:2008. Tests for mechanical and physical properties of aggregates. Determination of the voids of dry compacted filler. BSI, London, UK.

BSI (2009a) BS EN 933-9:2009+A1:2013. Tests for geometrical properties of aggregates. Assessment of fines. Methylene blue test. BSI, London, UK.

BSI (2009b) BS EN 1097-8:2009. Tests for mechanical and physical properties of aggregates. Determination of the polished stone value. BSI, London, UK.

BSI (2009c) BS EN 1097-8:2009 (Annex A). Tests for mechanical and physical properties of aggregates. Determination of the polished stone value. BSI, London, UK.

BSI (2009d) BS EN 1367-2:2009. Tests for thermal and weathering properties of aggregates. Magnesium sulfate test. BSI, London, UK.

BSI (2009e) BS EN 1744-1:2009+A1:2012. Tests for chemical properties of aggregates. Chemical analysis. BSI, London, UK.

BSI (2009f) BS EN 933-10:2009. Tests for geometrical properties of aggregates. Assessment of fines. Grading of filler aggregates (air jet sieving). BSI, London, UK.

BSI (2009g) BS EN 12591:2009. Bitumen and bituminous binders. Specifications for paving grade bitumens. BSI, London, UK.

BSI (2010a) BS EN 1097-2:2010. Tests for mechanical and physical properties of aggregates. Methods for the determination of resistance to fragmentation. BSI, London, UK.

BSI (2010b) PD 6691:2010. Guidance on the use of BS EN 13108. Bituminous mixtures. Material specifications. BSI, London, UK.

BSI (2010c) BS EN 459-1:2010. Building lime. Definitions, specifications and conformity criteria. BSI, London, UK.

BSI (2010d) BS EN 459-2:2010. Building lime. Test methods. BSI, London, UK.

BSI (2010e) BS EN 196-6:2010. Methods of testing cement. Determination of fineness. BSI, London, UK.

BSI (2011a) BS EN 1097-1:2011. Tests for mechanical and physical properties of aggregates. Determination of the resistance to wear (micro-Deval). BSI, London, UK.

BSI (2011b) BS EN 1367-5:2011. Tests for thermal and weathering properties of aggregates. Determination of resistance to thermal shock. BSI, London, UK.

BSI (2012a) BS EN 933-1:2012. Tests for geometrical properties of aggregates. Determination of particle size distribution – Sieving method. BSI, London, UK.

BSI (2012b) BS EN 933-3:2012. Tests for geometrical properties of aggregates. Determination of particle shape – Flakiness index. BSI, London, UK.

BSI (2012c) BS EN 12697-11:2012. Bituminous mixtures. Test methods for hot mix asphalt. Determination of the affinity between aggregate and bitumen. BSI, London, UK.

BSI (2013a) BS EN 1097-6:2013. Tests for mechanical and physical properties of aggregates. Determination of particle density and water absorption. BSI, London, UK.

BSI (2013b) BS EN 13179-1:2013. Tests for filler aggregate used in bituminous mixtures. Delta ring and ball test. BSI, London, UK.

BSI (2013c) BS EN 196-2:2013. Methods of testing cement. Chemical analysis of cement. BSI, London, UK.

BSI (2014a) BS EN 933-6:2001. Tests for geometrical properties of aggregates. Assessment of surface characteristics. Flow coefficient of aggregates. BSI, London, UK.

BSI (2014b) BS EN 1097-9:2014. Tests for mechanical and physical properties of aggregates. Determination of the resistance to wear by abrasion from studded tyres. Nordic test. BSI, London, UK.

Department for Transport (2012) *Revision of Aggregate Specification for Pavement Surfacing*. Highways Agency, London, UK, Interim Advice Note 156/12.

Dunford A (2013) *Friction and the Texture of Aggregate Particles Used in the Road Surface Course*. PhD thesis, University of Nottingham, Nottingham, UK.

EC (2011) Regulation (EU) No 305/2011 of the European Parliament and of the Council of 9 March 2011 laying down harmonised conditions for the marketing of construction products and repealing Council Directive 89/106/EEC. *Official Journal of the European Union* **L88/5**. See http://eurlex.europa.eu/LexUriServ/LexUriServ. do?uri=OJ:L:2011:088:0005:0043:EN:PDF (accessed 25/03/2014).

El Hussein HM, Kim KW and Phonia J (1998) Assessment of localized damage associated with exposure to extreme low temperatures. *Journal of Materials in Civil Engineering* **10(4)**: 269–274.

Environment Agency and Waste and Resources Action Programme (2013) *Aggregates from Inert Waste: End of Waste Criteria for the Production of Aggregates from Inert Waste*. Environment Agency, Bristol, UK. See http://cdn.environment-agency.gov.uk/LIT_8709_c60600.pdf (accessed 25/03/2014).

Highways Agency, Scottish Executive, Welsh Assembly Government and Department for Regional Development Northern Ireland (2006) *Design Manual for*

Roads and Bridges. Volume 7, *Pavement Design and Maintenance*. Section 5, *Surfacing and Surfacing Materials*. Part 1, *Surfacing Materials for New and Maintenance Construction*. The Stationery Office, London, UK, HD 36/06.

Highways Agency, Scottish Government, Welsh Assembly Government and Department for Regional Development Northern Ireland (2008) *Manual of Contract Documents for Highway Works*. Volume 1, *Specification for Highway Works*. *Road Pavements – Bituminous Bound Materials*. The Stationery Office, London, UK, Series 900.

Kutay E, Arambula E, Gibson N and Youtcheff J (2010) Three-dimensional image processing methods to identify and characterize aggregates in compacted asphalt mixtures. *International Journal of Pavement Engineering* **11(6)**: 511–528.

Ministry of Defence (2009a) *Marshall Asphalt for Airfields*. Defence Infrastructure Organisation, London, UK, DIO Specification 13.

Ministry of Defence (2009b) *Porous Friction Course for Airfields*. Defence Infrastructure Organisation, London, UK, DIO Specification 40.

Mukhopadhyay AK, Neekhra S and Zollinger DG (2007) *Preliminary Characterization of Aggregate Coefficient of Thermal Expansion and Gradation for Paving Concrete*. Texas Transportation Institute, Texas A&M University, College Station, TX, USA, Report FHWA/TX-05/0-1700-5.

Petersen JC and Plancher H (1998) Model studies and interpretative review of the competitive adsorption and water displacement of petroleum asphalt chemical functionalities on mineral aggregate surfaces. *Petroleum Science Technology* **16(1–2)**: 89–131.

Rigden PJ (1947) The use of fillers in bituminous road surfacing – a study of filler-binder system in relation to filler characteristics. *Journal of Society of Chemical Industry* **66(9)**: 299–309.

Sétra (Service D'études Techniques des Routes et Autoroutes) (2008) *The Use of Standards for Hot Mixes*. Sétra, Paris, France.

Chapter 11

Choosing asphalts for use in flexible pavement layers

Flavien Teurquetil
Senior Application Specialist,
Société des Pétroles, Shell.
France

Dr Sridhar Raju
Product Researcher,
Shell India Markets Pvt. Ltd.
India

There are three phases to the process of designing an asphalt pavement

- choosing the particular asphalt to be used in each layer of the pavement
- design of the asphalt mixture
- design of the layers in a flexible pavement.

Design of the asphalt mixture and design of the pavement layer thicknesses are described in Chapters 12 and 13, respectively. This chapter discusses the asphalts that are used for particular roles in a flexible pavement in different regions of the world.

11.1 Asphalts used in flexible pavements in Europe
11.1.1 Surface courses
Different asphalts are available for the top layer of a road constructed with a flexible pavement. The top layer is normally called the surface course, and the asphalt chosen for that role depends on the requirements for the road surface.

The surface course constitutes the top layer of the pavement, and should be able to withstand high traffic and environment-induced stresses without exhibiting unacceptable levels of cracking and rutting. It should also have an even profile for the comfort of the user (in the UK, surface regularity) and possess a suitable texture to ensure adequate skid resistance. Depending on the applicable specification in any particular region, functional characteristics such as skid resistance, noise reduction and durability are often required for surface courses. In some cases, rapid drainage of surface water is achieved by having a porous structure, while in other cases the surface course should be impermeable in order to keep water out of the pavement structure. As indicated, the surface layer is important for the pavement performance but no single material can provide all the desired characteristics

The Shell Bitumen Handbook, Sixth Edition
ISBN 978-0-7277-5837-8
Shell International Petroleum Company Ltd: All rights reserved
http://dx.doi.org/10.1680/tsbh.58378.261

(e.g. being porous and impermeable at the same time). A wide range of surface layer products can, therefore, be considered appropriate, depending on specific requirements

- asphalt concrete (AC)
- asphalt concrete for very thin layers (AC-TL)
- asphalt for ultra-thin layers (AUTL)
- stone mastic asphalt (SMA)
- hot rolled asphalt (HRA)
- porous asphalt (PA)
- double layered porous asphalt (2L-PA)
- mastic asphalt (MA)
- soft asphalt (SA).

11.1.1.1 Asphalt concrete
AC is asphalt in which the aggregate particles are continuously graded or gap graded to form an interlocking structure. Dense asphalt concretes are often used in the base layers in flexible pavements.

11.1.1.2 Asphalt concrete for very thin layers
AC-TL is asphalt for surface courses with a thickness of 20–30 mm in which the aggregate particles are generally gap graded to form stone to stone contact and to provide an open surface texture. This mixture is often used in France where it is called béton bitumineux très mince (BBTM).

11.1.1.3 Asphalt for ultra-thin layers
AUTL is a surface course that is laid on a bond coat (a proprietary tack coat made with polymer modified bitumen) at a nominal thickness of 10–20 mm with properties suitable for the intended use. The method of bonding is an essential part of the process, and the final product is a combination of the bonding system and the asphalt. AUTL is an asphalt in which the aggregate particles are generally gap graded to form stone to stone contact with an open surface texture. Several varieties of this layer are often used to provide a good, new, low noise running surface.

11.1.1.4 Stone mastic asphalt
SMA is a gap graded asphalt with bitumen as a binder, composed of a coarse crushed aggregate skeleton bound with a mastic mortar. This mixture is often used as a surface layer in cases where high stability is needed. The surface structure generates low noise levels.

11.1.1.5 Hot rolled asphalt (HRA)
HRA is a dense, gap graded asphalt in which the mortar of fine aggregate, filler and high viscosity binder are major contributors to the performance

of the laid material. Coated chippings (nominally single size aggregate particles with a high resistance to polishing, which are lightly coated with high viscosity binder) are always rolled into and form part of an HRA surface course. This durable surface layer is often used as a surface layer in the UK.

11.1.1.6 Porous asphalt
PA is an asphalt with a very high level of interconnected voids (18–25%), which allow the passage of water. It generates low noise levels and very little spray.

11.1.1.7 Double layered porous asphalt
The top layer of porous asphalt typically uses an aggregate with nominal size of 4–8 mm and is laid about 25 mm thick. The second/bottom layer is a thicker layer of porous asphalt with a larger aggregate nominal size (11–16 mm). As a result of the finer texture at the top (resulting in fewer tyre vibrations), this material is designed to produce lower noise levels than a porous asphalt laid in a single layer.

11.1.1.8 Mastic asphalt
MA is designed to be of low void content. The binder content is adjusted so that the voids are completely filled. MA is pourable and can be spread in its working temperature condition. It requires no compaction on site. This mixture is very durable and has often been used as a surface layer in certain countries.

11.1.1.9 Soft asphalt
SA is a flexible mixture of aggregate and soft bitumen grades and is used in the Nordic countries for secondary roads.

The selection of the surface course is a matter of identifying the most appropriate material during the design process. The functional requirements can conflict. For example, low noise generation may require the use of a double layered porous asphalt but that would conflict with the requirement for the surface to be very durable. The durability of surface layers can be improved by using higher quality materials. The higher costs of these components are recouped, at least in part, by the lower costs of traffic measures and user costs.

11.1.2 Bases and binder courses
Throughout Europe, in fully flexible pavements, the most dominant type of asphalt used in these layers is asphalt concrete (AC). Different variations of AC are used and further details are given in this chapter.

Table 11.1 CEN asphalt standards	
Standard	Title
BS EN 13108-1:2006	Bituminous mixtures. Material specifications. Asphalt concrete
BS EN 13108-2:2006	Bituminous mixtures. Material specifications. Asphalt concrete for very thin layers
BS EN 13108-3:2006	Bituminous mixtures. Material specifications. Soft asphalt
BS EN 13108-4:2006	Bituminous mixtures. Material specifications. Hot rolled asphalt
BS EN 13108-5:2006	Bituminous mixtures. Material specifications. Stone mastic asphalt
BS EN 13108-6:2006	Bituminous mixtures. Material specifications. Mastic asphalt
BS EN 13108-7:2006	Bituminous mixtures. Material specifications. Porous asphalt
BS EN 13108-8:2005	Bituminous mixtures. Material specifications. Reclaimed asphalt
BS EN 13108-20:2006	Bituminous mixtures. Material specifications. Type testing
BS EN 13108-21:2006	Bituminous mixtures. Material specifications. Factory production control

11.1.3 European asphalt standards

On 1 January 2008, the European Committee for Standardisation (CEN) adopted a unified approach to mixture design, manufacture and CE marking of asphalts.

The ultimate aim is to specify asphalts in terms of fundamental properties based on performance. However, given the differences in knowledge and experience with the basic specifications for this type of mixture in Europe, it is not currently possible to adopt a fundamental approach.

Eight different asphalts are covered by the various parts of BS EN 13108 (BSI, 2005, 2006a–g) and are shown along with the type testing and factory production control standards (BSI, 2006h–i) in Table 11.1.

11.2 UK pavements

Layers in UK flexible pavements are shown in Figure 11.1.

Pressures are applied at the surface of a road by the tyres of vehicles running over the pavement. The structural wear caused by lighter traffic (i.e. cars and lighter goods vehicles) is considered to be negligible according to paragraph 2.6 of the UK's traffic assessment standard (Highways Agency *et al.*, 2006a). It is only heavier commercial vehicles (i.e. those having a gross vehicle weight in excess of 3.5 tonnes) (also defined in paragraph 2.6 of the UK's traffic assessment standard that are considered to cause structural damage to a pavement). The highest pressure will be exerted by heavy goods vehicles having six or more axles, and the pressure will typically be of the order of 8.5 bar, and may well be higher for particular loads. The function of a road pavement is to dissipate such pressures down through the pavement, the foundation and the subgrade such that none of the materials in each of these layers deteriorates as a result of the effects of traffic.

The role played by each layer in a road with a flexible pavement, together with the effects that the location has on the choice of asphalt, is discussed below.

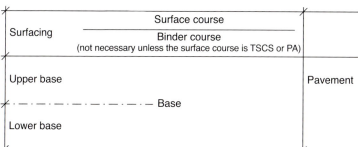

Surface course, binder course and base are asphalts

Foundation
(Capping or subbase, or subbase on capping)

Existing subsoil/fill – subgrade

Figure 11.1 UK road constructed with a fully flexible pavement: PA, porous asphalt; TSCS, thin surface course system

11.2.1 Subgrade and foundation

The subgrade will play a significant role in determining the service life of a pavement (see Table 13.2). In addition, the foundation also plays a key role in the performance of the road. However, foundations are not discussed in this book in any detail. Advice on foundations can be found in Interim Advice Note 73/06 (Highways Agency et al., 2009) and Chapter 32 of the *ICE Manual of Highway Design and Management* (Payne and Walsh, 2011).

11.2.2 Base

The primary function of the base is to act as the main structural layer in the pavement. Although UK pavement specifications allow a wide range of materials to be chosen for use as a base, including many cementitious mixtures, the material found in the vast majority of bases are asphalts. Properly designed asphalts provide the stiffness necessary to dissipate the pressures imposed by traffic without causing distress in the pavement layers, the foundation or subgrade. Mixtures formulated for bases need to have an adequate stiffness modulus and the required degree of resistance to fatigue. The formulation of these mixtures has evolved over the years, through a reduction in the size of the aggregates, and the use of harder bitumens.

The thickness of the base in a heavily trafficked pavement will typically be of the order of 200 mm or more. Although UK specifications no longer place a limit on the maximum thickness of asphalt layer that can be placed in a single lift, it would be wise to lay in two layers if the thickness of the base exceeds 120 mm. The reason for laying a thickness no greater than this is the difficulty of achieving the required void contents (the usual UK specification has an upper limit of void content) and the surface regularity specification, with the latter being the crucial parameter. The thicker a layer is, the more difficult it is to achieve the surface regularity tolerance. Thus, bases on pavements carrying heavy traffic will usually be constructed in two layers, with the first being described as the lower base and the second being described as the upper base, as depicted in Figure 11.1.

11.2.3 Binder course

The surface regularity of a finished road is a key determinant of its quality. It is a measure of the number of undulations in the surface. The fewer the undulations or irregularities, as the normal UK specification describes them in Clause 702 of the *Specification for Highway Works* (Highways Agency *et al.*, 2008), the more comfortable is the ride for those travelling on the road. UK surface regularity specifications require that the surface course is within a specified tolerance of the specified finished level. The most common tolerance is ± 6 mm of the specified finished level. However, a tolerance of +6 to 0 mm is not unusual (on racing circuits the tolerance may be even smaller). Asphalts are placed in layers using specialist items of equipment called 'pavers'. These are discussed in detail in Chapter 15. Every time a layer of asphalt is laid by a paver, the action of the paver reduces any irregularities in the laid mat. Thus, the irregularities at the top of the upper base are less than those in the lower base. Similarly, the irregularities on top of the binder course are less than those in the upper base. Finally, the irregularities in the surface course are less than those in the binder course. Tolerances apply at each layer, with the tolerance becoming tighter closer to the surface. Compliance after each layer has been placed is a specification requirement but achieving regularity limitations is most important on top of the surface course. It is for this reason that a binder course is necessary. Without a fourth layer, asphalt contractors may find it difficult to achieve compliance with specified regularity requirements.

Secondary, albeit important, benefits of having a binder course are

- the contribution made by the binder course in dissipating pressure from traffic (this property is described as the stiffness of the asphalt, i.e. a measure of its load spreading ability; this is discussed further in Chapter 13)
- preventing water reaching the base and foundation layers and the subgrade.

11.2.4 Surface course

The surface course is the visible portion of the road structure. It is this layer that suffers most from the effects of climate aggression and vehicle loads.

The formulation of the surface course asphalt will be the factor that is key in determining the characteristics of the surface layer of the pavement. These attributes include

- durability
- surface regularity
- skid resistance
- noise reduction
- resistance to rutting
- resistance to cracking
- water drainage
- appearance.

The appearance of the finished surface is rarely, if ever, given any thought. However, one of the many benefits of a thin surface course system that has been properly designed and skilfully laid is that it looks so much more attractive than does a chipped hot rolled asphalt surface course.

Investment in research and development of surface course asphalt has brought major innovations over the last 25 years. Examples of these initiatives are improved methods of choosing aggregates and the use of special bitumens, including, most importantly, polymer modified bitumens, which have expanded the range of types of surface course. Transport Scotland's work on surface courses is worthy of particular note (Transport Scotland, 2010).

11.3 Asphalts used in flexible pavements in the UK

Referring to Figure 11.1, each of the layers in the pavement has to perform specific functions as discussed below.

11.3.1 Base

The base is the main structural layer in the pavement, although the other layers will also contribute to the structural strength of the finished pavement. Many materials are available to act as a base in a modern pavement.

It is permissible to construct bases using a group of materials described as 'hydraulically bound mixtures' (HBMs). Many of these mixtures are cementitious.

However, the bulk of bases, as specified and as constructed, are asphaltic. The asphalts that are permitted for use as bases on UK trunk roads (the major arterial routes in the UK) are

- DBM125
- DBM50/HDM50
- EME2
- HRA50.

The names used for particular asphalts in the UK have arisen over many years. Current descriptions continue to be used, with dense bitumen macadam (DBM) and heavy duty macadam (HDM) (discussed below) being examples. DBM and HDM will become 'dense bituminous material' and 'heavy duty material' in 2015. The family of standards governing asphalts is the BS EN 13108 series (see Table 11.1).

Generically, DBM is an asphalt concrete, so production of this material is governed by BS EN 13108-1 (BSI, 2006a). The word 'macadam' comes from John Loudon McAdam (1756–1836) – the spelling of the surname is correct. McAdam is credited with inventing the system of placing increasingly smaller aggregate sizes and finally blinding the surface with dust. (In fact, the Romans used a similar system 2000 years ago (Mitchell, 2011)). The provisions of BS EN 13108-1 (like all the other parts of BS EN 13108) apply throughout Europe. However, the UK has a document that does not have the status of a European standard but is the core of any asphalt production specification in the UK. It is designated PD 6691 (BSI, 2010b). PD stands for 'published document' and the role of this document is explained in the second paragraph of its Introduction as: 'This Published Document gives the recommended choices for the mixtures most commonly used in the UK'.

HDM is also an asphalt concrete, so its production is governed by BS EN 13108-1 (BSI, 2006a).

Two options in bases are DBM50 and HDM50. The '50' in the description indicates that these asphalts are manufactured with 40/60 pen bitumen as per BS EN 12591 (BSI, 2009). DBM50 and HDM50 can be treated as identical materials. These materials account for the majority of bases in heavily trafficked UK roads. Table 11.2 gives typical declared gradings for a DBM50 and an HDM50. (The standards that apply in European member states set limits that apply to particular mixtures but it is for individual producers to 'declare' gradings within those limits, hence the caption for Table 11.2.)

As can be seen from Table 11.2, the mixture details for DBM50 and HDM50 are identical, and this will be the case with the DBM50 and HDM50 asphalts offered by many asphalt manufacturers. So why bother

Table 11.2 Examples of declared mixture details for DBM50 and HDM50

	AC 32 dense base 40/60	
	(DBM50)	(HDM50)
Grading (passing):		
40 mm sieve	100%	100%
31.5 mm sieve	99%	99%
20 mm sieve	86%	86%
6.3 mm sieve	52%	52%
2 mm sieve	30%	30%
0.25 mm sieve	13%	13%
0.063 mm sieve	6.0%	6.0%
Binder content	$B_{min4.0}$ (4.0%)	$B_{min4.0}$ (4.0%)
Temperature of the mixture	150–190°C	150–190°C

having two different asphalts available when they are in fact identical? The reason is that these mixtures have been available historically and customers expect them to continue to be available. Over time, one or both designations may disappear following a rationalisation exercise within PD 6691 (BSI, 2010b).

DBM125 is a dense bitumen macadam, like DBM50, but is made with a 100/150 pen bitumen.

EME2 stands for enrobé à module élevé class 2. EME2 is treated as an asphalt concrete, and thus its production falls under BS EN 13108-1 (BSI, 2006a). This asphalt contains very hard bitumen (15–20 pen is the target, which is not a range supported in the hard grade bitumen standard in BS EN 13924 (BSI, 2006j), in which the relevant classes are 10/20 and 15/25 pen). EME2 was developed following studies by the Highways Agency in France. French highway engineers have been using mixtures made with very hard bitumens since 1985 (Williams, 2007), and so have amassed a great deal of valuable experience in the behaviour of asphalts made with very hard bitumens. More information on EME2 can be found in Sanders and Nunn (2005) and Williams (2007).

HRA stands for hot rolled asphalt. Generically, HRA is a 'hot rolled asphalt' so its production is governed by BS EN 13108-4 (BSI, 2006d). The '50' in the description HRA50 indicates that these asphalts are manufactured using 40/60 pen bitumen as per BS EN 12591 (BSI, 2009). HRA has been in use in the UK for many years, the first standard for the material being published in January 1935 (BSI, 1935). Traditionally, in the UK there were two categories of asphalt, described as gap graded and continuously graded. Gradings, of course, describe the proportion of aggregate passing a particular sieve size; an example of this is shown in Table 11.3. Where a mixture included a proportion of all or most of the sieve sizes, the material was

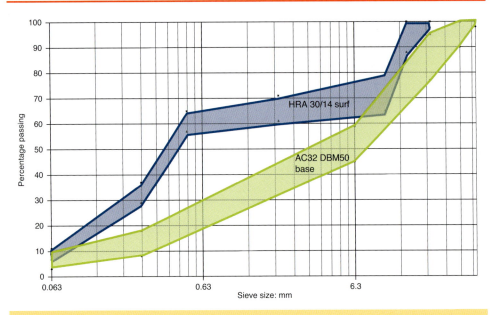

Figure 11.2 Grading envelopes for a continuously graded asphalt (AC32 DBM50 base) and a gap graded asphalt (HRA 30/14 surf)

categorised as 'continuously graded' (i.e. asphalt concretes (previously called 'bitumen macadams')). Where the grading was discontinuous, the material was categorised as 'gap graded' (i.e. hot rolled asphalts). An example of both categories is shown in Figure 11.2, which is drawn from the data presented in Table 11.3. However, the distinction based on the gaps or otherwise in gradings is not easy to perceive and this method of classification, while of interest historically, is of little practical assistance in categorising asphalts.

The designer, therefore, has, in effect, a choice of four materials to use in the

Table 11.3 Gradings for a continuously graded asphalt (AC32 DBM50 base) and a gap graded asphalt (HRA 30/14 surf)

	AC32 DBM50 base			HRA 30/14 surf		
	Target	Min.	Max.	Target	Min.	Max.
40	100	98	100			
31.5	99	90	100			
20	86	77	95	100	98	100
14				96	88	100
10				72	65	79
6.3	52	45	59			
2	30			66	61	71
0.5				61	57	65
0.25	13	8	18	33	29	37
0.063	6	3	9	9	7	11

Table 11.4 Advantages and disadvantages of available base materials

Asphalt	Advantages	Disadvantages
DBM125	Good mechanical interlock, low bitumen content, reasonable crack and rut resistance	Soft bitumen, low stiffness, relatively high cost
DBM50/HDM50	Good mechanical interlock, relatively low cost, medium stiffness, high deformation resistance	
EME2	High stiffness, impermeability	Relatively high cost
HRA50	Impermeability	Relatively high cost, low deformation resistance

base. Some designers offer all options to contractors pricing tenders and accept whichever option the winning bidder offers, which is the approach required by HD 26/06 (Highways Agency *et al.*, 2006b). Other designers may decide to be more selective by taking account of the advantages and disadvantages of each option. Table 11.4 summarises the advantages and disadvantages of each available option.

Note that the low deformation resistance of HRA50 is not relevant in a base, as deformation in thicker pavements is restricted to the surface course and binder course (Nunn, *et al.*, 1997). However, the binder course is usually made of the same generic material as the base, and an HRA50 as a binder course would carry with it significant concerns about deformation reaching a critical level (i.e. where replacement is warranted) much earlier than would be the case with the other options.

11.3.2 Binder course

As explained above, the binder course is present primarily to allow the asphalt contractor to achieve the very tight tolerances applied to the finished levels of the surface course. In addition, however, binder courses usually also have a higher quantity of bitumen in the asphalts used for their construction. This increases the durability of the layer. It also improves the waterproofing characteristic of the layer. section 4.2.1 of RN42 (Nicholls *et al.*, 2008) states that

> There is a need for the properties required for the binder course, particularly impermeability to limit any flow of water from the surface course downwards and/or from the base upwards. With surfacing materials being made thinner and more open and with base layers often being designed with relatively low binder contents, the need for an impermeable binder layer becomes vital.

The types of asphalt used for binder courses are much the same as those used for bases, with one interesting addition

- DBM125
- DBM50/HDM50

- EME2
- HRA50
- SMA.

It is important to note that DBM125 is available as a base or a binder course. The difference between these two mixtures is that the binder course variant has approximately 0.5% more bitumen than the base variant. The same is true of DBM50 and HDM50. In contrast, for any particular source, both EME2 and HRA50 mixtures used in base and binder courses are the same (i.e. they will have the same proportion of bitumen in the asphalt).

SMA stands for stone mastic asphalt, an asphalt that was developed in Germany in the 1960s. At that time in Germany, studded tyres were widely used in winter, and the Germans had a choice of two running surfaces – asphalt concrete and gussasphalt. As has been explained, asphalt concrete is continuously graded, and has a relatively low binder content. It was found to wear rapidly under studded tyres, failing by way of a mechanism called ravelling, which means the asphalt is abraded away. Attempts to counter this tendency by increasing the binder content simply resulted in the material failing in deformation in the wheel tracks. The performance of gussasphalt, however, was quite different. Being gap graded, it has a relatively high binder content but, because of this, it was expensive to produce and lay.

Thus, in the late 1960s, splitt mastic asphalt appeared, splitt being the German for 'stone'. Stone mastic asphalt was a proprietary gap graded material, with a high stone content and a high binder content. The mixture is durable and, being negatively textured, generates lower levels of noise and spray. A positively textured surface is one where some constituents of the asphalt project upwards from the surface. A negatively textured surface is one that has voids in the surface. Negatively textured surface courses generate significantly lower noise levels, and result in lower noise levels than is the case with positively textured surface courses. An example of a positively textured surface course would be a chipped hot rolled asphalt surface course. An example of a negatively textured surface course would be an asphalt concrete close graded surface course. Studded tyres were banned in Germany in 1975 but the benefits of SMAs were, by that time, apparent. SMA did not find its way into the German national specification until 1984. The use of SMA spread to other countries in Europe throughout the 1980s. Sweden, France, Belgium, Denmark and other countries all used variations of this material. In the early 1990s, formulations based on SMA gradings appeared in the UK, and some are in use today as thin surface course systems (TSCS) complying with Clause 942 of the standard UK highway specification (Highways Agency et al., 2008).

SMA also exists as binder courses and regulating courses in the UK. The very useful advantage of SMA binder courses is that they can be laid as thin as 30 mm. In the design of pavements, the total asphalt thickness is derived using procedures that are described in HD 26/06 (Highways Agency *et al.*, 2006b). Sometimes it is very helpful in the process to be able to specify a thin binder course. Most asphalt binder courses have to be at least 50 mm thick, as per Table 1A of BS 594987:2010 (BSI, 2010a). Table 1A applies to asphalt concrete binder courses; hot rolled asphalt binder courses can be as thin as 45 mm but are not recommended because of their propensity to rut relatively easily.

Generally, the binder course is made of the same generic material as the base, albeit that the binder course material may have more bitumen than that of the base. So, for example, the base would be HDM50 and the binder course would be HDM50, or the base would be EME2 and the binder course would be EME2.

11.3.3 Surface course

On heavily trafficked UK roads, two types of surface course predominate

- chipped HRA surface course
- thin surface course systems.

Chipped HRA was the only surface course on major roads for many years, until the introduction of thin surface course systems in the 1990s.

11.3.3.1 Chipped hot rolled asphalt surface course

Chipped HRA consists of a layer of HRA into which precoated chippings are rolled. The most common type of HRA surface course consists of approximately

- 30% of 14 mm single-sized aggregate
- 55% sand
- 9% very fine aggregate (called filler)
- 6.5% bitumen binder.

After the HRA surface course has been placed and before it has been rolled, precoated single-sized chippings (usually 20 mm) are spread onto the surface. The chippings are then pushed into the surface as part of the compaction process. The chippings are coated with bitumen of the same type as was used in the manufacture of the asphalt itself. After being coated, the chippings are allowed to cool, and are then transported to the site to be stored until they are added to the surface course at a specified rate of spread. The chippings are necessary to give the surface adequate skid resistance, thus allowing vehicles to brake in the shortest possible length. Good frictional characteristics are essential, and will reduce injuries and fatalities. It is most important for those involved in highways to be familiar with the factors that affect skid resistance.

HD 28/04 (Highways Agency *et al.*, 2004) is entitled *Skid Resistance*, and provides essential information on this very important topic.

A chipped HRA surface course using the above formulation is laid such that, when the chippings have been added and the material has been rolled (more meaningfully, compacted), it is 40 mm thick, in accordance with Table 1B of BS 594987:2010 (BSI, 2010a).

11.3.3.2 Thin surface course systems

Thin surface course systems are proprietary materials that can be laid as the running surface of a carriageway. They can usually be laid at thicknesses from 40 mm down to a few millimetres.

This class of materials first appeared in late 1992, when two UK companies imported French technology (the products were called Safepave, which was produced by Associated Asphalt, and *UL-M* (ultra-mince being French for 'very thin'), which was introduced by Alfred McAlpine). Other companies followed, with Tarmac, for example, undertaking trials in 1992 and 1993 of its Masterpave. A major difficulty in the usage of these asphalts on trunk roads was that the national specification did not recognise the possibility that such materials may appear.

The emergence of the proprietary Safepave and UL-M mixtures and others, the effects of harmonisation activities in relation to standards, combined with stories of this wonderful stone mastic asphalt material from Germany stimulated significant interest in a new breed of what were generically referred to in the industry as 'stone mastic asphalts'.

Indeed, the level of interest generated was such that the Road Engineering and Environmental Division of the Highways Agency commissioned a demonstration trial at the TRL. The resultant report (Project Report 65, commonly referred to as PR 65), was published in 1994 (Nunn, 1994). This report was extremely successful in that it served to highlight the availability of alternative mixtures to the industry.

However, the difficulty remained that the national specification did not recognise these mixtures, and thus their use was limited to non-trunk road locations. As a result of the pressure for usage from producers, external customers and Highways Agency and TRL staff who had realised the benefits of these materials, the Highways Agency set about considering how best to allow the use of such materials on trunk roads. The solution was to put in place a system that allowed the use of proprietary mixtures, which had been rigorously tested in every single respect relevant to the life and performance of surface courses. This started with close control over the ingredients through to the establishment of controls over the way the material is laid. This system is known as HAPAS (Highway Authority

Table 11.5 Permitted pavement surface course materials on UK trunk roads

Country	Default surface course
England	Thin surface course system
Northern Ireland	Thin surface course system Chipped hot rolled asphalt Dense macadam
Scotland	Chipped hot rolled asphalt
Wales	Thin surface course system

Notes
1. In Northern Ireland, asphalt concrete surface courses can only be used on roads that are not high speed.
2. In Scotland, surface dressing may be permissible on roads that are not high speed.
3. All countries may permit other treatments subject to a departure from standard.

Product Approval Scheme) or the BBA HAPAS (British Board of Agrément), the UK body that provides reassurance to manufacturers, users, specifiers, insurers and regulators of construction products and systems. More information on HAPAS can be found in section 23.3.3.

The *Specification for Highway Works* now accepts the use of thin surface course systems, and these materials are addressed in Clause 942 (Highways Agency *et al.*, 2008). It should be noted that this clause requires the contractor to guarantee the integrity of the surface course and the workmanship for a period of 5 years from the date of opening to traffic, but it is very important to be aware that this applies to the trunk road network only. The 5 year guarantee includes defects such as fretting, ravelling, stripping and loss of chippings.

Table 11.5 summarises the choice of surface course on trunks roads in the different countries of the UK as per Tables 2.2E, 2.2W, 2.2S and 2.2NI in HD 36/06 (Highways Agency *et al.*, 2006c).

So, on trunk roads, the default surface course is, in most cases, a thin surface course system. However, some authorities continue to specify chipped HRA. Primarily, this is because of a belief that the lifespan of HRA surface courses is significantly better than that of TSCS. A detailed study undertaken by TRL (Nicholls *et al.*, 2010) states in section 2.5.1.1 of the report that a TSCS will last for, on average, 12–13 years (for thin asphalt concretes with 0/14 and 0/10 mm gradings, generally with polymer modified bitumen). A chipped HRA surface course is expected to have a lifespan of around 18 years according to the *ICE Manual of Highway Design and Management*, which states a range of 14–24 years (Payne and Walsh, 2011: Table 2 of Chapter 50). The surface regularity on TSCSs is superior to that of chipped HRA. TSCSs are also superior in relation to traffic noise, which may well be approximately two-thirds of that on chipped HRA. TSCSs also exhibit reduced spray as compared with chipped HRA. TSCSs are superior in aesthetic terms to a chipped HRA surface course. However, they do use a higher proportion of high polished stone value (PSV) aggregate compared with chipped HRA surface courses. Thus, the only disadvantages of a TSCS compared with a chipped HRA surface course are the lower lifespan and the greater usage of premium aggregate.

11.3.3.3 Other asphalt surface courses
A number of other asphalts are available for use as surface courses

- asphalt concrete
- porous asphalt
- generic stone mastic asphalt.

Asphalt concrete surface courses
Asphalt concrete surface courses can be used on roads having low traffic levels. There are five types of asphalt concrete surface course available

- dense
- close graded
- medium graded
- fine graded
- open graded.

Generally speaking, asphalt concrete surface courses are used on roads that carry light traffic. Open graded mixtures were widely used in the UK, particularly in England, prior to the emergence of heavier commercial vehicles. Having an open texture means that they are accessible to water and are likely to degrade rapidly where there are frequent freeze–thaw weather cycles. Their durability is a function of the nature of the pavement itself and the type of traffic to which the road is subject. The use of a surface dressing may be necessary as a sealant. Close graded, dense and medium graded surface courses are less susceptible to intrusion and damage by surface water. However, they are only suitable for light traffic or applications such as footways, car parks, playgrounds etc. Fine graded surface course was previously described as fine cold asphalt, and it is only suitable for footways etc.

Porous asphalt
Porous asphalt is a high void asphalt concrete surface course. The void content can be between 18 and 25% (most asphalts have void contents between 4% and 8%). The high voidage in porous asphalt results in the material being pervious, so rainwater passes right through the layer. This is illustrated in Figure 11.3.

The rainwater passing through the porous asphalt surface course runs along the top of the impervious asphalt concrete binder course and into a drain

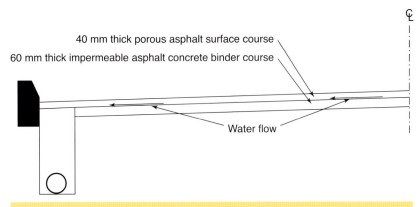

40 mm thick porous asphalt surface course

60 mm thick impermeable asphalt concrete binder course

Water flow

Figure 11.3 Principle of porous asphalt surface courses

placed below the channel to catch this water. Another feature of the material is that it has exceptionally low traffic noise characteristics, generating around half the noise of a chipped hot rolled asphalt surface course.

The disadvantage of porous asphalt is its lifespan, which is approximately 8 years. A number of trials have been carried out in the UK but the material has performed disappointingly and, as a consequence, it appears unlikely that it will enjoy widespread usage in the UK in the foreseeable future.

High stone content hot rolled asphalt
A number of specifiers favour this material, which is also known as medium temperature asphalt (MTA). The asphalt in a chipped HRA surface course often has 30% of 14 mm aggregate (sometimes represented as 30/14). High stone content asphalts are either 55% of 10 mm or 14 mm aggregate (i.e. 55/10 or 55/14). It is generally easier to lay than a traditional chipped HRA surface course. It has no precoated chippings as it is textured when laid properly.

Generic stone mastic asphalt
This material would not have the frictional characteristics favoured in the UK. Accordingly, it is not used in the UK.

11.4 Asphalts used in flexible pavements in France
11.4.1 The evolution of asphalts in France
Most roads in France are flexible pavements and are constructed using asphalt in the base, binder course and surface courses. France is considered to have a very high quality road network as a result of knowledge and experience of asphalt and pavement performance acquired over many years by both highway administrators and road contractors. Several of the paving materials developed in France, such as high modulus asphalt and thin

surface course layers, have been adopted in other countries. The most frequently used paving grade bitumens are 35/50 and 50/70 pen bitumens. In addition, since the 1980s, special harder grades of bitumen (e.g. 10/20 or 20/30 pen) have been used in specific circumstances to meet the demands of heavy traffic loading or climatic conditions. Polymer modified bitumens are used extensively in France.

In common with other European countries, during the 1950s most asphalts used in France were generic 'recipe' type mixtures, and the limitations of such an approach for coping with increasing traffic levels was beginning to manifest itself in the form of pavement damage through rutting, surface cracking and premature ageing of bitumen. This was exacerbated because many of the recipe mixtures at that time were based on relatively small nominal size aggregates (16 mm maximum size) and were manufactured using natural sand and crushed gravel. In addition, the bitumens commonly used were relatively soft, with 80/100 and 180/220 pen grades being used to construct pavements at that time.

By the mid-1960s, French roads were exhibiting unacceptable levels of rutting due to increased levels of loading from commercial vehicles. The monitoring of badly rutted roads and subsequent laboratory studies resulted in a different approach to the manufacture of asphalts, which dictated the direction taken by asphalt technology in France. Instead of using natural sands and gravel, asphalts were to be manufactured with crushed rock. In addition, harder penetration bitumen grades of 40/50 and 60/70 pen were employed. These designs formed the basis for the semi-coarse asphalt concretes (bétons bitumineux semi-grenus (BBSG)) used today.

For surface courses, the invention of polymer modified bitumen in the 1970s led to the development of mixture designs suitable for use in very thin layers designed to ensure the highest levels of grip and durability. This began with 40 mm nominal thickness surface course and, in the 1980s, to the development of very thin layer asphalt layers known as béton bitumineux trés mince (BBTM). These asphalts are now widely used on many parts of the road network, in layer thicknesses of 15–20 mm. In the early 1980s, béton bitumineux drainant (porous asphalt) was introduced to improve driving visibility and skid resistance by reducing surface water. A further adaptation of thin asphalt and porous asphalt has led to the development of small nominal size aggregate asphalts (0/6 mm), known as béton bitumineux mince, for the reduction of traffic noise.

For base and binder course materials, in the 1980s, high modulus asphalts, enrobé à module élevé (EME), were first introduced with the aim of reducing thicknesses and optimising aggregate resources. These mixtures have a nominal aggregate size of 10–20 mm, the most commonly used size being 14 mm. A hard bitumen grade of 10/20 or 20/30 pen is used with relatively high binder contents of 5.5–6.0% by mass of aggregate in the mixture, to ensure good durability and protect against fatigue damage and water sensitivity, while producing high stiffness modulus mixtures capable of withstanding high traffic loading.

BBME (bétons bitumineux à module élevé – high modulus asphalt) first appeared in the mid-1990s, using gradings with nominal sizes of 0/10 or 0/14 mm. It is designed using 20/30 pen bitumen. This asphalt is another mixture type used in binder course or surface course to overcome the phenomenon of rutting.

11.4.2 Current practice for asphalt pavements in France

All asphalts in France are required to meet European (CEN) specifications. Table 11.6 summarises the majority of asphalts used today in France.

The CEN standards allow two different approaches to asphalt mixture design. The first is an empirical approach, in which the final composition is based on the characteristics of the mixture (e.g. the nominal size of aggregate, the air void content, the particle size distribution and the target binder content). Alternatively, the fundamental approach measures a number of material properties of the mixture (e.g. complex stiffness modulus, fatigue and resistance to rutting), and includes requirements for water sensitivity of the mixture and the percentage of air voids within the mixture. The latter is the approach that is most often used in France, and a mixture design procedure has been developed over many years to assess the properties of asphalt mixtures. In this approach, the target binder content and the grading envelope are not predetermined but are designed to achieve the performance criteria. Furthermore, the measured mechanical performance of the mixture is given a performance class, ranging from 1 to 4, with 4 being the highest levels of loading and/or most severe climatic conditions. The pavement design, and hence selection of the type and thickness of binder course and surface course, will reflect the performance class and thickness of base layer proposed.

Although EME and BBME are in widespread use in France, in particular on the major road networks, the most common material used for the asphalt base layers is grave bitume. This asphalt has a grading of nominal size 0/14 or 0/20 mm, and typically contains 35/50 pen bitumen with a binder content between 3% and 5%. Variation of the bitumen content and the grading curve is permitted to achieve varying defined performance classes (from 2 to 4).

Béton bitumineux semi-grenus (BBSG) (semi-coarse asphalt concrete) remains the most widely used asphalt in France. These mixtures use a continuous grading curve with a nominal size of 0/10 or 0/14 mm, reconstituted from at least three granular fractions combined to meet the specification. The bitumen content of paving grade (typically 35/50 or 50/70 pen) is 5–5.8% by mass of

Table 11.6 Asphalts in France

Normes EN	Nom	Grading	Class	Thickness Min.	Thickness Moy.	Disc.	k	% content ppc	Level 1 PCG Test / Duriez	Level 2 Thick.	%vides	Rutting test %	Level 3 Modulus (15°C, 10 Hz) MPa	Level 4 Fatigue (10°C, 25 Hz) µdef	Test in situ Défor. règle 3m	Accr	HSv	%vides
EN 13108-1	BBSG EB	0/10	1	4	5-7		3.4	5.7	10g ≥11; 25-60g 5-10; r/R ≥0.70	10 cm	5-8	30000: <10 / ≤7.5 / ≤5	≥5500 / ≥7000 / ≥7000	≥100	2 cm	≥250	>0.4 / >0.5	4-8
			2															
			3															
		0/14	2	5	6-9		3.2	5.4	80g 4-9			30000: <10 / ≤7.5 / ≤5	≥5500 / ≥7000 / ≥7000					
			3															
	BBA C EB	0/10	1	4	6-7		3.6		Roulement ≥0.80; Liaison ≥0.70; Liaison 4-8 Roulement 3-7	10 cm	4-7	10000: ≤15 / ≤10 / ≤7.5	≥5000 / ≥7000 / ≥5000	≥130 / ≥100	2 cm	≥250	Voies de circulation ≥0.4 / Pistes ≥0.6	3-7 Joints ≤9
			2															
			3										≥130 / ≥100					
		0/14	1	5	7-9		3.5						≥5000 / ≥7000	≥100				
			2							5 cm				≥130 / ≥100				
			3															
	BBA D EB	0/10	1	4	4-5	2/6.3 ou 4/6.3	3.4		Liaison >11 Roulement >10	5 cm		10000: ≤7.5 / ≤15 / ≤10	≥7000 / ≥5000	≥130 / ≥100 / ≥130				
			2						>9	10 cm			≥5000	≥100				
			3						>10				≥7000 / ≥5000	≥130 / ≥100				
		0/14	1	5	5-7		3.2		5-9									
			2															
			3															
	BBM A EB	0/10	1	2.5	3-4		3.3	5.5	≥11; r/R ≥0.70		7-10	3000: ≤15; 30000: <10			1.5 cm	≥250	≥0.7	5-10
			2					5.3	5-9			10000: ≤15					≥0.5	
			3									3000: ≤15						
		0/14	1	3	3.5-5	2/6.3	3.2		6-11									
			2															
			3															
	BBM B EB	0/10	1	2.5	3-4		3.3	5.5	≥11; r/R ≥0.70	5 cm	8-11	3000: ≤15; 30000: <10					≥0.7	7-12
			2					5.3	7-12			10000: ≤15					≥0.5	
			3									3000: ≤15						
		0/14	1	3	3.5-5	4/6.3	3.2											
			2															
			3															
	BBM C EB	0/10	1	2.5	3-4		3.3	5.5	8-13									
			2															

Table 11.6 Continued

Normes EN	Nom	Grading	Class	Min.	Moy.	Disc.	k	% content ppc	PCG 10 g	25 g	40 g	60 g	80 g	100 g	120 g	200 g	Duriez r/R	L2 Thick.	%voids	Rut 3000	Rut 10 000	Rut 30 000	Modulus (15 °C, 10 Hz) MPa	Fatigue (10 °C, 25 Hz) μdef	Défor. règle 3 m	Accr	HSv	Vp	%voides
EN 13108-7	BBDr	0/6	1	2	3-4		3.4	4.9			20-26					≥14	≥0.80								1 cm	≥350		0.6	
							3.7	4.7			26-30					≥20												0.9	
		0/10	2	2	4-5		3.4	4.8			20-26					≥14												0.8	
							3.7	4.6			26-30					≥20												1.2	
EN 13108-1	BBS 1	0/10	1	2	4-5		3.4	5.7	≥9			4-9													2 cm	≥250			4-9
	BBS 2	0/10	2		4-6		3.7	6	≥10		4-9						≥0.80												
	BBS 3	0/14	3		8	2/6.3	3.4	5.5	≥10					4-9															
	BBS 4	0/14	4	3	10-12		3.1	5.2	≥10						4-9		≥0.75								3 cm				
EN 13108-2	BBTM	0/6	1	1.5	2-3		3.5			12-19							≥0.75	5 cm	16-22	≤20						≥300	0.7		Non significatif
			2							20-25																			
		0/10	1				3.4			10-17									9-16	≤15							0.9		
			2							18-25																			
EN 13108-1	GB EB	0/14	2	6	8-14		2.5	4.2							<11		≥0.70	10 cm	8-11		<10		≥9000	≥80		≥250			≤11
			3				2.8	4.7							<10				7-10		<10		≥9000	≥90					≤9
			4				2.9	4.9							<9				5-8				≥11000	≥100					≤8
		0/20	2	8	10-16		2.5	4.2							<11				8-11		<10		≥9000	≥80					≤11
			3				2.8	4.7							<10				7-10		<10		≥9000	≥90					≤9
			4				2.9	4.9							<9				5-8				≥11000	≥100					≤8
EN 13108-1	EME EB	0/10	1	5	6-8		2.5	4.2						<10	<10		≥0.70	10 cm	7-10			<7.5	≥11000	≥100	2 cm	≥250	≥0.4		≤10
			2				3.4	5.7						<6	<6				3-6				≥14000	≥130					≤6
		0/14	1	6	7-13		2.5	4.2											7-10				≥11000	≥100					≤10
			2				3.4	5.7											3-6				≥14000	≥130					≤6
		0/20	1	8	9-15		2.5	4.2											7-10										≤10
			2				3.4	5.7											3-6										≤6
EN 13108-1	BBME EB	0/10	1	4	5-7		3.5	5.8	≥11				4-9				≥0.80	10 cm	5-8		≤10		≥9000	≥100	2 cm	≥250	≥0.5		4-8
			2																		≤7.5		≥11000	≥100					
			3																			≤5	≥11000	≥100					
		0/14	1	5	6-9		3.3	5.5													≤7.5		≥9000	≥100					
			2																			≤5	≥11000						
			3																										

aggregate. The mixtures are traditionally laid between 50 and 80 mm thick and are designed to accommodate the requirements of each pavement layer.

Following many years of the increasing use of harder grades of bitumen in all layers of the pavement, some softer grades are now being used more commonly in France. This is, in part, due to the increasing incorporation of reclaimed asphalt pavement (RAP) in asphalts. Concurrently, there has been a rise in the use of warm asphalt mixtures in France, with lower temperature versions of the common mixtures outlined in this section being regularly produced and used.

11.5 Asphalts used in flexible pavements in Germany

In Germany, an asphalt pavement is constructed in a traditional manner (i.e. a base, a binder course and a surface course). All asphalts must comply with BS EN 13108 (BSI, 2005, 2006a–g). Implementation of the requirements of BS EN 13108 in Germany is specified in the German national standard TL Asphalt-StB (FSGV, 2013a). Layers in a flexible pavement in Germany are shown in Figure 11.4 (see Table 11.7).

The base is the lowest layer in a German asphalt pavement. It is placed on a foundation that is typically formed of an unbound material.

For the base and binder courses, asphalt concretes are normally used. The composition of the asphalt concretes in the base and binder courses differ significantly in terms of their void content, binder content and grading curve. To distinguish between base and binder course asphalts the base is called

AC T (asphalt concrete Tragschicht (base))

and the binder course is called

AC B (asphalt concrete Binderschicht (binder course)).

For the surface course, a range of asphalts is available

- AC D (asphalt concrete Deckschicht (surface course))
- SMA (stone mastic asphalt)

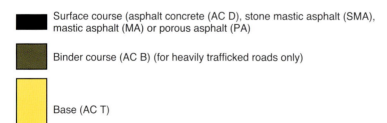

Surface course (asphalt concrete (AC D), stone mastic asphalt (SMA), mastic asphalt (MA) or porous asphalt (PA))

Binder course (AC B) (for heavily trafficked roads only)

Base (AC T)

Figure 11.4 Layers in a German flexible pavement

Table 11.7 German asphalt layers and their thicknesses

Layer in pavement	Asphalt	Layer thickness according to ZTV Asphalt-StB (FSGV, 2013b): mm
Surface course	AC 5 D L	20 to 30
	AC 8 D N, AC 8 D L	30 to 40
	AC 11 D N, AC 11 D L	35 to 45
	AC 8 D S	30 to 40
	AC 11 D S	40 to 50
	AC 16 D S	50 to 60
	SMA 5 N	20 to 30
	SMA 8 N	20 to 35
	SMA 8 S	30 to 40
	SMA 11 S	35 to 40
	MA 5 S, MA 5 N	20 to 30
	MA 8 S, MA 8 N	25 to 35
	MA 11 S, MA 11 N	35 to 40
	PA	45 to 60
Binder course	AC 16 B N	50 to 60
	AC 16 B S	50 to 90
	AC 22 B S	70 to 100
Base course	AC 22 T S, AC 22 T N, AC 22 T L	$\geq$80
	AC 32 T S, AC 32 T N, AC 32 T L	$\geq$80

- PA (porous asphalt)
- MA (mastic asphalt).

Materials used in Germany for particular layers in flexible pavements and their thickness ranges are shown in Table 11.7.

The thickness of these layers depends on climatic conditions and the applicable traffic load

- surface course – up to 40 mm
- binder course – 60–80 mm (for heavily trafficked roads only)
- base – up to 220 mm.

All asphalt layers are classified as L, N or S types, depending on the anticipated loading

- L – low loading
- N – normal loading
- S – severe loading.

11.5.1 Stone mastic asphalt

SMA was developed in Germany in the 1960s. It was originally designed to resist studded tyres. Today, SMA is used in Europe, Australia, the USA and Canada to reduce deformation in heavy duty asphalt pavements.

Typically, SMA is a gap graded mixture and normally contains cellulose fibres to prevent binder drainage (the fibres hold the bitumen in the mixture). As SMA is mainly used for heavily trafficked asphalt pavements, the use of polymer modified bitumen should be given careful consideration.

11.6 Asphalts used in flexible pavements in the USA

Over 93% of paved surfaces in the USA are asphalts (usually described in the USA and areas that have adopted US practice as hot mix asphalts (HMAs)). As in Europe, the surface transportation system is mature, which places the emphasis on the maintenance, repair, rehabilitation or reconstruction of existing pavements, rather than new pavement construction.

In the USA, there is no national specification used throughout the country for road construction. The Federal Highway Administration (FHWA) and the American Association of State Highway and Transportation Officials (AASHTO) publish specifications, test methods and guidelines for a multitude of activities, but a state department of transportation (DOT) typically administers highway projects and publishes its own specifications. Local governments and other specifiers generally follow the lead of the state DOT specification, or use it outright for their projects. For airfield pavement construction, the Federal Aviation Administration publishes a standard specification that is modified for use on local projects.

Asphalt selection includes identification of the type, classification and specific aggregate and bitumen requirements that may be appropriate for a particular project. For this discussion, mixture type refers to general grading descriptions (dense, gap or open graded), while mixture classification refers to the nominal maximum aggregate size (NMAS) of a mixture, which is defined as 'one sieve size larger than the first sieve to retain more than 10 per cent' in the AASHTO standard M 323-13 (AASHTO, 2013), which uses the standard sieve nest for characterising the grading of aggregates used in asphalt in the USA.

As there is such a wide range of conditions in the USA, no single mixture type or classification can provide requirements for component materials that apply across the entire country. Consequently, specifiers need to select the appropriate bitumen grade and any aggregate quality requirements appropriate for the particular use of a particular asphalt. Tools such as LTPPBind, Version 3.1 (FHWA, 2003) are publicly available and provide a means of selecting the proper grade of bitumen according to the anticipated range of pavement temperatures, traffic characteristics and desired reliability for projects throughout North America. When selecting the bitumen grade, consideration should also be given to the proportion and nature of any reclaimed asphalt permitted for use in the mixture design.

Aggregate quality requirements often vary depending on the location of the asphalt within the pavement structure, the asphalt type and regional considerations reflecting the availability of particular materials. Test procedures and criteria for aggregate quality are described in Chapter 10.

11.6.1 Dense graded mixtures

As in other regions, dense graded asphalts have been widely used for paving in the USA since early in the twentieth century. Mixture design systems have evolved over time in a quest to provide a means of comprehensively analysing mixture qualities that are related to pavement performance.

Superpave (superior performing asphalt pavement) was a product of the Strategic Highway Research Program, a programme of focused research into long term pavement performance studies, cement and concrete, maintenance and bitumen that took place between 1987 and 1992. It includes a system for grading and selecting the correct bitumen for a particular asphalt, and a volumetric mixture design procedure that uses a gyratory compactor for moulding laboratory specimens. The performance graded (PG) bitumen specification was adopted first, with most state DOTs including either the AASHTO M 320-10 (AASHTO, 2010a) specification or a modification thereof. The Superpave mixture design procedure came into use more gradually but is currently the most commonly used method in the USA, although the Marshall method continues to be used, especially for airfield pavements. Superpave is discussed in detail in section 12.4.5.

Although no single standard specification is used throughout the USA, recommendations for mixture type selection have been jointly published by the National Asphalt Pavement Association (NAPA) and the FHWA in a document entitled *HMA Pavement Mixture Type Selection Guide* (NAPA and FHWA, 2001). This guide references the Superpave mixture design system for dense graded asphalt, which identifies mixture classifications according to nominal maximum particle size. Cross sections through US flexible pavements are shown in Figure 11.5.

Although there are significant differences in how Superpave has been implemented from one agency to another, the use of NMAS (nominal maximum aggregate size – nominal stone size (NSS) in the UK) to classify dense graded mixtures is commonplace, and is an important consideration for mixture selection, as lift thickness recommendations are directly related to the NMAS. Table 11.8 lists typical applications and compacted lift thicknesses commonly used for NMAS designations used with dense graded mixtures. Not all of these mixture classifications are used throughout the USA.

Dense graded mixtures may also be classified according to whether or not they are used as a surface course or according to the anticipated traffic level.

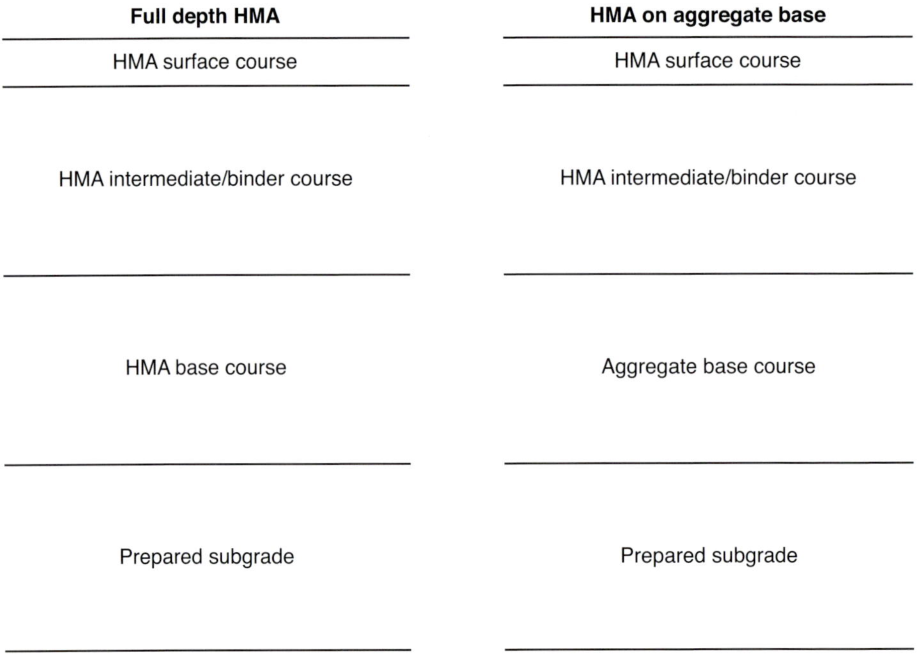

Full depth HMA	HMA on aggregate base
HMA surface course	HMA surface course
HMA intermediate/binder course	HMA intermediate/binder course
HMA base course	Aggregate base course
Prepared subgrade	Prepared subgrade

Figure 11.5 Layers in flexible pavements in the USA (NAPA and FHWA, 2001)

Surface course mixtures used where there is high speed traffic often require more stringent criteria for coarse aggregates in order to resist polishing and the loss of surface friction in wet conditions.

Performance related tests for resistance to rutting and various forms of cracking are increasingly used in state DOT specifications. Examples of these tests are listed in Table 11.9.

11.6.2 Speciality asphalts

Open graded mixture and SMA (stone matrix asphalts in the USA, stone mastic asphalts in the UK) have gained acceptance in parts of the USA for the

Table 11.8 Superpave mixture classifications

NMAS: mm	Typical application(s)	Typical range of compacted lift thickness	
		mm	in
4.75	Thin surface course	19–38	$\frac{3}{4}$–$1\frac{1}{2}$
9.5	Surface course	38–63	$1\frac{1}{2}$–$2\frac{1}{2}$
12.5	Surface course, binder course	50–100	2–4
19.0	Binder/base course	75–125	3–5
25.0	Binder/base course	100–150	4–6
37.5	Base course	125–175	5–7

Table 11.9 Examples of performance related tests used in US asphalt specifications

Description	Test method	Distress mode(s)	Requirement(s)
Hamburg wheel track testing of compacted hot mix asphalt (HMA)	AASHTO T 324 (AASHTO, 2014)	Rutting, sensitivity to moisture damage	Maximum deformation after a prescribed number of passes – the minimum number of passes to a defined inflection point. Usually performed at 50°C.
Asphalt Pavement Analyzer	AASHTO T 340 (AASHTO, 2010b)	Rutting	Maximum deformation after a prescribed number of passes at the high pavement temperature used for bitumen selection.
Indirect tensile strength	AASHTO T 322-07 (AASHTO, 2007)	Cracking	Minimum, or range of indirect tensile strengths measured at a prescribed temperature.
Texas overlay test	Tex-248-F (TxDOT, 2014)	Cracking	Minimum cycles to failure.

same reasons as described for European countries. In 1990, industry, government and research organisations organised a tour of six western European countries to determine what could be learned from their asphalt paving practices that could be readily applied to the USA. The findings from this initiative led to the introduction of SMA, and the revision and reintroduction of permeable surface layer mixtures, borrowing and adapting the requirements for each mixture to conditions in the USA. As with dense graded mixtures, individual state DOTs have established their own specifications and requirements for these mixtures suited to the local climate and available materials.

There has been a great deal of interest in developing thin asphalts, particularly for use in pavement preservation. Examples include a thin overlay mixture (TOM) classification, developed by the Texas Department of Transportation, and thin, bonded overlay mixtures that usually require placement using an integral paver. Integral (or integrated) pavers are equipped with a bar that sprays a tack coat on the surface immediately before the asphalt is laid. While there are no standard definitions or accepted nomenclature for these asphalts, there is a trend towards their increased use due to the need to preserve and improve the functional characteristics of existing pavements using an approach that is less expensive and time consuming than traditional cold milling and asphalt inlay, and more durable under high traffic conditions than surface treatments.

11.7 Asphalts used in flexible pavements in India

A flexible pavement in India consists of asphalt layers, granular layers and the subgrade layer. A typical cross section of a flexible pavement in India for a national highway pavement is shown in Figure 11.6.

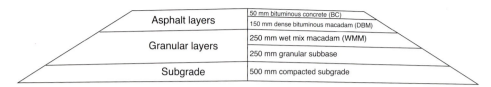

Asphalt layers	50 mm bituminous concrete (BC)
	150 mm dense bituminous macadam (DBM)
Granular layers	250 mm wet mix macadam (WMM)
	250 mm granular subbase
Subgrade	500 mm compacted subgrade

Figure 11.6 Typical cross section of a national highway pavement in India

For national highways and expressways, the thickness of the asphalt layers is in the range 100–200 mm. The asphalt course is often constructed in three lifts: an upper layer, a middle layer and a lower layer. For other carriageways, such as class 2 highways, provincial highways, and urban or municipal roads, there are typically two asphalt layers: an upper layer and a lower layer. The upper layer, commonly 40–50 mm thick, is described as the surface course. The middle layer and lower layer, commonly 50–100 mm thick, are considered as acting as a binder course.

The most commonly used asphalts in India are specified in a standard published by the India Roads Congress (IRC, 2009)

- dense graded bituminous concrete (BC) surface course
- dense graded dense bituminous macadam (DBM) binder course.

Dense bituminous macadam (DBM) is mainly used as a base/binder course and profile corrective courses. The work consists of single or multiple layers of DBM on a previously prepared base or subbase. The thickness of individual layers is in the range 50–100 mm.

Bituminous concrete for use in surface courses can be a single layer of bituminous concrete on a previously prepared bituminous bound surface. A single layer can be 25, 40 or 50 mm thick.

The asphalt mixture design uses the Marshall method as specified in the Ministry of Road Transport and Highways (MoRT&H, 2013) specifications for roads and bridges in India. The Marshall method is discussed in detail in section 12.4.3. Aggregate gradings for DBM and BC mixtures are listed in Table 11.10.

In India, unmodified bitumen is graded by viscosity, and the commonly used grades are VG 30 and VG 40. The commonly used modified bitumens are

Table 11.10 Grading limits for DBM and BC asphalts in India (MoRT&H, 2013)

| Asphalt | Percentage passing sieve size: mm, listed below: % | | | | | | | | | | | | |
	45.0	37.5	26.5	19.0	13.2	9.5	4.75	2.36	1.18	0.6	0.3	0.15	0.075
DBM-I	100	95–100	63–93	–	55–75	–	38–54	28–42	–	–	7–21	–	2–8
DBM-II	–	100	90–100	71–95	56–80	–	38–54	28–42	–	–	7–21	–	2–8
BC-I	–	–	100	90–100	59–79	52–72	35–55	28–44	20–34	15–27	10–20	5–13	2–8
BC-II	–	–	–	100	90–100	70–88	53–71	42–58	34–48	26–38	18–28	12–20	4–10

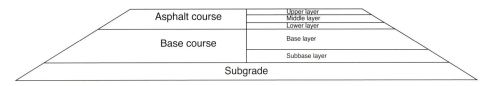

Figure 11.7 Typical cross section of a road pavement in China

polymer modified bitumens (PMBs) and crumb rubber modified bitumens (CRMBs). The mechanical requirements for asphalts using unmodified and modified binders are given in IRC:111-2009 (IRC, 2009).

IRC:111-2009 recommends that when the traffic exceeds 2000 commercial vehicles per day (CVPD) per lane and the highest daily mean temperature exceeds 40°C, VG 40 or modified bitumen is used in BC and top layers of DBM.

11.8 Asphalts used in flexible pavements in China

Roads in China commonly consist of a flexible pavement with a semi-rigid base constructed with aggregate that is stabilised with lime or concrete. A typical flexible pavement in China is shown in Figure 11.7.

For major highways, including expressways, the thickness of the asphalt course is in the range 120–180 mm. The asphalt course is often constructed in three lifts: an upper layer, a middle layer and a lower layer. For other carriageways, such as class 2 highways, provincial highways, and urban or municipal roads, there are typically two layers, an upper layer and a lower layer. The upper layer, commonly 30–50 mm, is regarded as a surface course. The middle layer, commonly 40–60 mm thick, and the lower layer, 50–100 mm thick, are considered as acting as a binder course.

The asphalts commonly used in China are AC (asphalt concrete, a dense graded asphalt), ATB (asphalt treated base, a dense graded bitumen stabilised aggregate), ATPB (asphalt treated permeable base, an open graded asphalt), AM (asphalt macadam, semi-open graded mixture), SMA (stone mastic asphalt) and OGFC (open graded friction course). In some circumstances, Superpave mixtures (see section 12.4.5) are employed. For semi-rigid flexible pavements, the asphalts for individual layers are generally selected from those shown in Table 11.11.

Table 11.11 Asphalts used in flexible pavements in China, as per JTJ F40-2004 (Research Institute of Highways, 2004)

Asphalt course	Asphalt
Upper layer	AC-13, AC-16, SMA-13, SMA-16, SUP-13
Middle layer	AC-16, AC-20, SUP-20
Lower layer	AC-25, AM-25, SUP-20, ATB-25, ATB-30

Table 11.12 Grading limits for commonly used asphalts in China, as per JTJ F40-2004 (Research Institute of Highways, 2004)

Asphalt mixture	Percentage passing sieve size: mm, listed below: %												
	31.5	26.5	19.0	16.0	13.2	9.5	4.75	2.36	1.18	0.6	0.3	0.15	0.075
AC-13				100	90–100	68–85	38–68	24–50	15–38	10–28	7–20	5–15	4–8
AC-16			100	90–100	76–92	60–80	34–62	20–48	13–36	9–26	7–18	5–14	4–8
AC-20		100	90–100	74–92	62–82	50–72	26–56	16–44	12–33	8–24	5–17	4–13	3–7
AC-25	100	90–100	70–90	60–83	51–76	40–65	24–52	14–42	10–33	7–24	5–17	4–13	3–7
SMA-13				100	90–100	50–75	20–34	15–26	14–24	12–20	10–16	9–15	8–12
SMA-16			100	90–100	65–85	45–65	20–32	15–24	14–22	12–18	10–15	9–14	8–12
ATB-25		100	90–100	60–80	48–68	42–62	32–52	20–40	15–32	10–25	8–18	5–14	3–10
ATB-30	100	90–100	70–90	53–72	44–66	39–60	31–51	20–40	15–32	10–25	8–18	5–14	3–10
AM-25	100	70–98	50–85	–	32–62	20–50	6–29	6–18	3–15	2–10	1–7	1–6	1–4

Design of individual asphalts is usually based on the Marshall method, as required by the national specification JTJ F40-2004 (Research Institute of Highways, 2004). Aggregate gradings for commonly used asphalts are listed in Table 11.12.

The most commonly used bitumens are heavy traffic bitumens AH-70 and AH-90 (penetration grade), used in the lower and middle layers, and sometimes in the upper layer. The commonly used polymer modified bitumens (PMBs) use SBS and SBR modifiers. In China, these modified binders are generally used in the upper layer for high class highways, such as expressways or class 1 or 2 highways. If the expressway has heavy duty traffic, PMB is also used in the middle layer.

References

AASHTO (American Association of State and Highway Transportation Officials) (2007) T 322-07. Standard method of test for determining the creep compliance and strength of hot mix asphalt (HMA) using the indirect tensile test device. AASHTO, Washington, DC, USA.

AASHTO (2010a) AASHTO M 320-10. Standard specification for performance-graded asphalt binder. AASHTO, Washington, DC, USA.

AASHTO (2010b) AASHTO T 340. Standard method of test for determining the rutting susceptibility of hot mix asphalt (APA) using the Asphalt Pavement Analyzer (APA). AASHTO, Washington, DC, USA.

AASHTO (2013) AASHTO M 323-13. Standard specification for Superpave volumetric mix design. AASHTO, Washington, DC, USA.

AASHTO (2014) AASHTO T 324. Standard method of test for Hamburg wheel-track testing of compacted hot mix asphalt (HMA). AASHTO, Washington, DC, USA.

BSI (British Standards Institution) (1935) BS 594:1935. Rolled asphalt. Fluxed lake asphalt and asphaltic bitumen. Hot process. BSI, London UK.

BSI (2005) BS EN 13108-8:2005. Bituminous mixtures. Material specifications. Reclaimed asphalt. BSI, London, UK.

BSI (2006a) BS EN 13108-1. Bituminous mixtures. Material specifications. Asphalt concrete. BSI, London, UK.

BSI (2006b) BS EN 13108-2:2006. Bituminous mixtures. Material specifications. Asphalt concrete for very thin layers. BSI, London, UK.

BSI (2006c) BS EN 13108-3:2006. Bituminous mixtures. Material specifications. Soft asphalt. BSI, London, UK.

BSI (2006d) BS EN 13108-4. Bituminous mixtures. Material specifications. Hot rolled asphalt. BSI, London, UK.

BSI (2006e) BS EN 13108-5:2006. Bituminous mixtures. Material specifications. Stone mastic asphalt. BSI, London, UK.

BSI (2006f) BS EN 13108-6:2006. Bituminous mixtures. Material specifications. Mastic asphalt. BSI, London, UK.

BSI (2006g) BS EN 13108-7:2006. Bituminous mixtures. Material specifications. Porous asphalt. BSI, London, UK.

BSI (2006h) BS EN 13108-20:2006. Bituminous mixtures. Material specifications. Type testing. BSI, London, UK.

BSI (2006i) BS EN 13108-21:2006. Bituminous mixtures. Material specifications. Factory production control. BSI, London, UK.

BSI (2006j) BS EN 13924. Bitumen and bituminous binders – Specifications for hard paving grade bitumens. BSI, London, UK.

BSI (2009) BS EN 12591. Bitumen and bituminous binders – Specifications for paving grade bitumens. BSI, London, UK.

BSI (2010a) BS 594987:2010. Asphalt for roads and other paved areas. Specification for transport, laying and compaction. BSI, London, UK.

BSI (2010b) PD 6691. Guidance on the use of BS EN 13108 Bituminous mixtures – Material specifications. BSI, London, UK.

FHWA (Federal Highway Administration) (2003) LTPPBind 3.0/3.1. See http://www.fhwa.dot.gov/research/tfhrc/programs/infrastructure/pavements/ltpp/dwnload.cfm (accessed 27/10/2014).

FSGV (Forschungsgesellschaft für Straben und Verkehrswesen) (2013a) TL Asphalt-StB 07/13, Asphaltbauweisen FSGV 797. Technische Lieferbedingungen für Asphaltmischgut für den Bau von Verkehrsflächenbefestigungen. FSGV, Cologne, Germany.

FSGV (2013b) ZTV Asphalt-StB 07/13, Asphaltbauweisen FSGV 799. Zusätliche Technische Vertragsbedingungen und Richtlinien für den Bau von Verkehrsflächen aus Asphalt. FSGV, Cologne, Germany.

Highways Agency, Scottish Executive, Welsh Assembly Government and Department for Regional Development Northern Ireland (2004) *Design Manual for Roads and Bridges*. Volume 7, *Pavement Design and Maintenance*. Section 3, *Pavement Maintenance Assessment*, Part 1, *Skid Resistance*. The Stationery Office, London, UK, HD 28/04.

Highways Agency, Scottish Executive, Welsh Assembly Government and

Department for Regional Development Northern Ireland (2006a) *Design Manual for Roads and Bridges*. Volume 7, *Pavement Design and Maintenance*. Section 2, *Pavement Design and Construction*, Part 1, *Traffic Assessment*. The Stationery Office, London, UK, HD 24/06, Table 2.1.

Highways Agency, Scottish Executive, Welsh Assembly Government and Department for Regional Development Northern Ireland (2006b) *Design Manual for Roads and Bridges*. Volume 7, *Pavement Design and Maintenance*. Section 2, *Pavement Design and Construction*, Part 3, *Pavement Design*. The Stationery Office, London, UK, HD 26/06, Table 2.1.

Highways Agency, Scottish Executive, Welsh Assembly Government and Department for Regional Development Northern Ireland (2006c) *Design Manual for Roads and Bridges*. Volume 7, *Pavement Design and Maintenance*. Section 5, *Surfacing and Surfacing Materials*, Part 1, *Surfacing Materials for New and Maintenance Construction*. The Stationery Office, London, UK, HD 36/06.

Highways Agency, Scottish Government, Welsh Assembly Government and The Department for Regional Development Northern Ireland (2008) *Manual of Contract Documents for Highway Works*. Volume 1, *Specification for Highway Works. Road Pavements – Bituminous Bound Materials*. The Stationery Office, London, UK, Series 900.

Highways Agency, Scottish Executive, Welsh Assembly Government and Department for Regional Development Northern Ireland (2009) *Design Guidance for Road Pavement Foundations*. The Stationery Office, London, UK, Interim Advice Note 73/06 Rev 1.

IRC (India Roads Congress) (2009) IRC:111-2009. Specifications for dense graded bituminous mixes. IRC, New Delhi, India.

Mitchell H (2011) *Unwinding the Long, Long Trail. The History of Roads in the UK*. Part 5, *Civil Engineering Surveyor*. Chartered Institution of Civil Engineering Surveyors, Sale, UK.

MoRT&H (Ministry of Road Transport and Highways) (2013) Specifications for Road & Bridge Works, 5th revision. IRC, New Delhi, India.

NAPA (National Asphalt Pavement Association) and FHWA (Federal Highway Administration) (2001) *HMA Pavement Mixture Type Selection Guide*. NAPA/FHWA, Lanham, MD/Washington, DC, USA.

Nicholls JC, McHale MJ and Griffiths RD (2008) *Best Practice Guide for Durability of Asphalt Pavements*. TRL Ltd, Crowthorne, UK, Road Note RN42.

Nicholls JC, Carswell I, Thomas C and Sexton B (2010) *Durability of Thin Asphalt Surfacing Systems*. Part 4, *Final Report After Nine Years' Monitoring*. TRL Ltd, Crowthorne, UK, Report TRL674.

Nunn ME (1994) *Evaluation of Stone Mastic Asphalt (SMA): A High Stability Wearing Course Material*. Transport Research Laboratory, Crowthorne, UK, Project Report 65.

Nunn ME, Brown A, Weston D and Nicholls JC (1997) *Design of Long-life Pavements for Heavy Traffic*. Transport Research Laboratory, Crowthorne, UK, TRL Report 250 (TRL250).

Payne IR and Walsh ID (2011) *ICE Manual of Highway Design and Management* (Walsh ID *et al.* (eds)). Institution of Civil Engineers, London, UK.

Research Institute of Highways (2004) Industrial Standards of People's Republic of China. JTJ F40-2004. Technical specification for construction of highway asphalt pavements. China Communications Press, Beijing, China.

Sanders PJ and Nunn ME (2005) *The Application of Enrobé à Module Élevé in Flexible Pavements*. TRL Ltd, Crowthorne, UK, Report 636.

Transport Scotland (2010) *Surface Course Specification & Guidance*. Transport Scotland, Glasgow, UK, TS 2010.

TxDOT (Texas Department of Transportation) (2014) Tex-248-F. Overlay test. TxDOT, Austin, TX, USA.

Williams J (2007) EME2 – Intégralement Français. *Asphalt Professional*, **25**: 23–24.

Chapter 12

Design of asphalt mixtures

Nilanjan Sarker
Senior Application Specialist (Asia), Shell India Markets Private Ltd. India

In many countries, asphalts that have traditionally been found to perform satisfactorily for acceptable periods have evolved over the years. Indeed, these empirical mixtures continue to be used the world over. However, the best approach in terms of maximising the functionality and minimising the environmental effects is to design the asphalt for a particular purpose. Both European and US engineers are moving in that direction. This chapter discusses the methods used to design asphalt mixtures themselves. It begins with a very brief look at the history of the development of asphalt mixture design methods, considers the very important concepts of volumetrics and grading, and then describes the major asphalt design methods.

US practice is to use the word 'asphalt' for bitumen. In this chapter. the word 'bitumen' has been used except where the word 'asphalt' features in a quote, the title of an organisation, a process or a reference. In the sections on volumetrics and grading, US symbols have been employed. Note that Appendix 8 contains a comparison of US and EU symbols used in matters associated with mixture volumetrics.

12.1 A very brief history of asphalt mixture design methods

In the early days of asphalt pavements, the mixture employed would either have been one that the contractor chose or recommended or it may have been a proprietary mixture. In 1905, Clifford Richardson published his book *The Modern Asphalt Pavement*, with a second edition in 1908 (Richardson, 1908). It is clear that Richardson carried out tests in order to settle on the constituents for a particular asphalt.

In acknowledgement of the debt he owed to other scientists, Sir Isaac Newton wrote in 1676, 'if I have seen further it is by standing on the shoulders of giants'. It is apt to begin here by acknowledging that today's

The Shell Bitumen Handbook, Sixth Edition
ISBN 978-0-7277-5837-8
Shell International Petroleum Company Ltd: All rights reserved
http://dx.doi.org/10.1680/tsbh.58378.295

achievements in the world of asphalt mixture design would not have been possible without the discoveries of the past.

In the mid-1920s, Charles Hubbard and Frederick Field in conjunction with the newly created Asphalt Association (which later became the *Asphalt Institute*) developed a system for designing asphalts called the Hubbard Field method of design. It was used by US state highway departments in the 1920s and 1930s, and continued in some states until the 1960s.

In 1927, Francis Hveem, a resident engineer in the California Division of Highways, began development of a method of assessing the bitumen content based on a calculation of the surface area of the constituent aggregate blend, something he achieved by 1932.

Bruce Marshall, an employee of the Mississippi Department of Highways, is probably the most well known pioneer associated with asphalt mixture design. He devised the Marshall method in the 1930s. Indeed, versions of his method remain in use today in areas throughout the world.

In 1987, the American Association of State Highway and Transportation Officials (AASHTO) member states began the 5-year research programme entitled the *Strategic Highway Research Program* (SHRP). It began because of an increasing number of premature pavement failures. In an attempt to deal with this problem, the states initiated the development of a coordinated, well funded, national research effort to derive improved specifications for bitumens and, ultimately, asphalts. It consisted of four areas of activity

- bitumen (i.e. asphalt in the USA)
- concrete and structures
- highway operations
- long term pavement performance.

One of the outcomes of the bitumen study area was the Superpave system, the final product of the bitumen and asphalt mixture research programme. Superpave is a method of designing asphalt mixtures, and is used throughout the USA and areas that adopt US asphalt technology practices.

On a worldwide basis, the most widely implemented asphalt mixture design methods are as follows

- Hubbard Field
- Hveem or California
- Marshall or Corps of Engineers
- Superpave.

In order to understand the various methods of designing asphalt mixtures, users must be familiar with the concepts of both mixture volumetrics and aggregate grading (often described as gradation). This chapter begins with sections on these important subject areas and then considers the development history, an outline and usage of a number of different asphalt design mixtures.

Much has been published about the development of asphalts and asphalt mixture design methods (Hveem, 1970; White, 1985; Goetz, 1989; Leahy and McGennis, 1999).

12.2 Mixture volumetrics

As has been explained, volumetric analysis is a key element of asphalt mixture design. Accordingly, a sound understanding of this topic is necessary in order to design high quality, functional asphalts.

Pavement engineers often consider asphalts in terms of the proportion by weight of bitumen and/or aggregate. However, in relation to asphalt mixture design, it is important to consider an asphalt in terms of its three components

- aggregate
- bitumen
- air.

The complex interplay of these three components is the key to understanding the behaviour of asphalts both in the laboratory and in service.

The composition of an asphalt can be expressed either in terms of weight or volume. However, as air has no mass, if all three components are to be considered, then constituents need to be expressed in volumetric terms. While a typical asphalt may contain 5% bitumen and 95% aggregate by weight, the scenario changes completely when composition is considered volumetrically. Figure 12.1 denotes the composition by volume of three types of asphalt (Advanced Asphalt Technologies, 2011), stone matrix asphalt would be described as stone mastic asphalt in the UK.

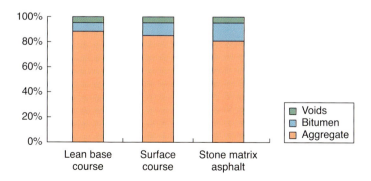

Figure 12.1 Typical composition by volume of different asphalts (Advanced Asphalt Technologies, 2011)

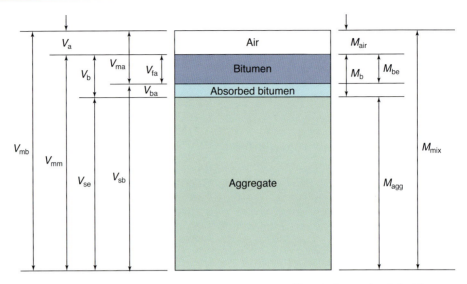

V_{ma} = volume of voids in mineral aggregate
V_{mb} = bulk volume of compact mix
V_{mm} = voidless volume of paving mix
V_{fa} = volume of voids filled with asphalt
V_a = volume of air voids
V_b = volume of asphalt binder
V_{ba} = volume of absorbed asphalt binder
V_{sb} = volume of mineral aggregate (by bulk specific gravity)
V_{se} = volume of mineral aggregate (by effective specific gravity)

M_{mix} = total mass of asphalt mixture
M_b = mass of asphalt binder
M_{be} = mass of effective asphalt binder
M_{agg} = mass of aggregate
M_{air} = mass of air = 0

Figure 12.2 Symbols for constituents in an asphalt

The simplified layout of the component diagram shown in Figure 12.2 (sometimes described as a phase diagram) helps users to visualise the volumetric and mass relationships that are used in the analysis of an asphalt.

There now follows an explanation of some of the abbreviations used in Figure 12.2.

12.2.1 Air voids (VIM/V_a)

The air void content is a measure of the total volume of air expressed as a percentage of the bulk volume of the compacted mixture. The air is distributed throughout a compacted asphalt, mostly between the coated aggregate particles. The air void content does not include pockets of air within individual aggregate particles or air contained in microscopic surface voids or capillaries on the surface of the aggregate.

Air voids define the performance of an asphalt. Air voids are a key consideration in designing the asphalt, and will have a profound effect on the life of the asphalt in service. If the air void content is too high, the resulting mixture may exhibit additional permeability to air and water, resulting in moisture damage and age hardening. When air void content is too low, the binder content is likely to be high, resulting in an asphalt that is prone to bleeding and plastic deformation.

Determination of the air voids is one of the prime objectives of volumetric analysis, but there is no direct means of ascertaining its value. Air voids are determined by comparing the specific gravity (or relative density – specific gravity and relative density are the same property here but note that 'relative density' is the preferred term now) and is the ratio of the density (mass of a unit volume) of a substance to the density of a given reference material (in this case water, specific gravity usually means relative density with respect to water) of a compacted specimen with the maximum theoretical density of the mixture used to make that specimen. For example, if the compacted density of an asphalt specimen is 95.3% of the theoretical maximum specific gravity, the air void content is $100 - 95.3 = 4.7\%$.

The air void content can be determined using the following equation

$$\text{VIM} = 100\left(\frac{G_{mm} - G_{mb}}{G_{mm}}\right) \tag{12.1}$$

where VIM is the air voids in the compacted mixture expressed as a percentage of the total volume, G_{mm} is the maximum specific gravity of the mixture and G_{mb} is the bulk specific gravity of the compacted mixture.

12.2.2 Bitumen content (P_b)
The bitumen content of an asphalt is one of its most important characteristics. Using a bitumen content lower than required may result in a dry, stiff mixture that is difficult to place and compact and will be prone to cracking and other durability problems while possessing lower resistance to fatigue cracking. Too much bitumen is likely to make the asphalt uneconomic because of the relatively high cost of bitumen and, in performance terms, it is likely to make the asphalt prone to bleeding and deformation.

P_b is expressed and specified as a percentage of the total weight of the asphalt. However, expressing bitumen content by weight has two major drawbacks (Advanced Asphalt Technologies, 2011)

- it is the bitumen content by volume and not by weight that dictates performance
- most aggregates tend to absorb some bitumen, and that bitumen does not significantly contribute to the durability of a mixture.

12.2.3 Bitumen absorption (P_{ba})
Any bitumen that is absorbed into the aggregate particles (P_{ba}) does not influence the performance characteristics of an asphalt but has the effect of

changing the specific gravity of the aggregate. Bitumen absorption is expressed as a percentage proportion by weight of the aggregate (not the asphalt!), and is calculated using the following equation

$$P_{ba} = \frac{100\left(G_{se} - G_{sb}\right)G_b}{G_{se}G_{sb}}$$

(12.2)

where P_{ba} is the absorbed bitumen expressed as the percentage proportion of the total weight of the aggregate, G_{se} is the effective specific gravity of the aggregate, G_{sb} is the bulk specific gravity of the aggregate and G_b is the specific gravity of bitumen.

12.2.4 Effective bitumen content (P_{be})

The term effective binder content (P_{be}) is used to describe the amount of bitumen in a mixture excluding that portion of the bitumen that is absorbed by the aggregate. It is the effective binder content that is used in mixture volumetrics, and it is the portion of the total bitumen content that remains as a coating on the outside of the aggregate particles. It is this bitumen content on which the service performance of an asphalt depends. It is calculated using the following equation

$$P_{be} = P_b - \frac{P_{ba}P_s}{100}$$

(12.3)

where P_{be} is the effective bitumen content expressed as the percentage proportion of the total weight of the mixture, P_b is the bitumen content expressed as the percentage proportion of the total weight of the mixture, P_{ba} is the absorbed bitumen expressed as the percentage proportion of the total weight of the aggregate and P_s is the aggregate content expressed as the percentage proportion of the total weight of the mixture.

12.2.5 Voids in the mineral aggregate (VMA)

The voids in the mineral aggregate (VMA) is the volume of void space between the aggregate particles of a compacted asphalt. VMA is numerically equal to the air void content plus the effective binder content by volume, and can be calculated using the following equation

$$VMA = 100 - \frac{G_{mb}P_s}{G_{sb}}$$

(12.4)

where VMA is the voids in the mineral aggregate, G_{mb} is the bulk specific gravity of the compacted mixture, G_{sb} is the bulk specific gravity of the aggregate and P_s is the aggregate content, expressed as the percentage proportion of the total weight of the mixture.

Establishing a single design air void content (such as the value of 4.0% used in Superpave mixtures) and then controlling VMA is the same as controlling the effective binder content. For example, a Superpave 12.5 mm mixture designed at 4.0% air voids with 14.0% minimum VMA has a minimum effective binder content of $14.0 - 4.0 = 10.0\%$ by volume.

12.2.6 Voids filled with bitumen (VFB)

Voids filled with bitumen (VFB) is the percentage of VMA filled with bitumen, the balance being air voids (VIM). The bitumen content in the expression of VFB is the effective bitumen content. VFB can be calculated using the following equation

$$VFB = 100\left(\frac{VMA - VIM}{VMA}\right) \tag{12.5}$$

where VFB is the voids filled with bitumen expressed as a percentage of the VMA, VMA is the voids in the mineral aggregate expressed as a percentage proportion of the bulk volume and VIM is the air voids in the compacted mixture expressed as a percentage proportion of the total volume.

In designing asphalts, VFB is closely related to both VMA and P_{be}. This occurs because the designed air voids is constant at about 4.0%, and, as VMA increases, both P_{be} and VFB increase. Thus, in most cases, VFB can be considered to be simply an indicator of the richness of the mixture, as can VMA or P_{be} (Advanced Asphalt Technologies, 2011).

It is not entirely clear what aspects of performance are related to VFB that are not also strongly related to other volumetric factors, especially V_{be}. Some engineers have suggested that fatigue resistance increases with increasing VFB. However, VFB and P_{be} are closely related. Recent research strongly suggests that P_{be} is a better overall indicator of resistance to fatigue cracking in asphalts. Therefore, in order to control or evaluate resistance to fatigue cracking, engineers should either use P_{be} or VMA at a constant design air voids content. There is then little need to specify VFB independently (Advanced Asphalt Technologies, 2011).

12.2.7 Bulk specific gravity (G_{sb}) of an aggregate

The bulk specific gravity is the ratio of the weight in air of a unit volume of a permeable aggregate (including both permeable (to water and bitumen) and impermeable voids within the aggregate particles) at a particular temperature to the weight in air of an equal volume of gas free distilled water at the same temperature. It can be calculated using the following equation

$$G_{sb} = \frac{\text{weight of a unit volume of permeable aggregate in air}}{\substack{\text{weight of a unit volume of permeable aggregate} \\ \text{in gas free distilled water}}} \tag{12.6}$$

As an asphalt contains a mixture of coarse aggregate, fine aggregate and filler, the bulk specific gravity of the mixture can be determined using the following equation

$$G_{sb} = \frac{P_1 + P_2 + \cdots + P_n}{P_1/G_1 + P_2/G_2 + \cdots + P_n/G_n}$$ (12.7)

where G_{sb} is the bulk specific gravity for the total aggregate; P_1, P_2, ..., P_n are individual percentages by weight of aggregates, where P_1 is the percentage by weight for aggregate fraction 1 and so on; and G_1, G_2, ..., G_n are individual bulk specific gravities of the aggregates, where G_1 is the bulk specific gravity of the aggregate in aggregate fraction 1 and so on.

12.2.8 Apparent specific gravity (G_{sa}) of an aggregate

The apparent specific gravity of an aggregate is the ratio of the weight in air of a unit volume of an impermeable aggregate at a particular temperature to the weight in gas free distilled water of a unit volume of the same impermeable aggregate at the same temperature. It can be calculated using the following equation

$$G_{sb} = \frac{\text{weight of a unit volume of impermeable aggregate in air}}{\substack{\text{weight of a unit volume of impermeable aggregate} \\ \text{in gas free distilled water}}}$$

(12.8)

12.2.9 Effective specific gravity (G_{se}) of an aggregate

The effective specific gravity of aggregate is the ratio of the weight in air of a unit volume of a permeable aggregate (excluding voids permeable to bitumen) at a fixed temperature to the weight in air of an equal volume of gas free distilled water at a stated temperature. It can be calculated using the following equation

$$G_{se} = \frac{100 - P_b}{100/G_{mm} - P_b/G_b}$$ (12.9)

where G_{se} is the effective specific gravity of the aggregate, G_{mm} is the maximum specific gravity of mixed material (no air voids), P_b is the bitumen content expressed as the percentage proportion of the total weight of the mixture at which the ASTM D2041 test (G_{mm}) (ASTM, 2011) was performed and G_b is the specific gravity of the bitumen.

It is clear from the above that, for a given aggregate, the apparent specific gravity will have the highest value followed by the effective specific gravity and then the bulk specific gravity. Mathematically, it may be represented as

$$G_{sa} > G_{se} > G_{sb}$$ (12.10)

If the results of testing do not follow this relationship, then those results should be rejected and the material resampled and/or retested.

12.2.10 Specific gravity of a bitumen (G_b)

The specific gravity of a bitumen is the ratio of the weight in air of a unit volume of bitumen at a particular temperature to the weight in air of an equal volume of gas free distilled water at the same temperature. The methodology of determining the specific gravity of bitumen is given in AASHTO T 228 (AASHTO, 2009) or ASTM D70 (ASTM, 2009).

12.2.11 Bulk specific gravity of an asphalt (G_{mb})

The bulk specific gravity of an asphalt is the ratio of the weight of a particular volume of asphalt including the permeable and impermeable voids in the specimen to the weight of the same volume of gas free distilled water. The standard procedure for determining the bulk specific gravity of a compacted asphalt involves weighing the specimen in air and in water. If the water absorption of the aggregates is less than 2.0%, saturated surface dry specimens are used (the procedure is outlined in AASHTO T 166) (AASHTO, 2013a). For specimens having high water absorption values (>2%), paraffin coated specimens should be used in the specific gravity determination (the procedure is given in AASHTO T 275) (AASHTO, 2007a).

G_{mb} can be calculated using the following equation

$$G_{mb} = \frac{A}{B - C} \tag{12.11}$$

where G_{mb} is the bulk specific gravity of the compacted specimen, A is the mass of the dry specimen in air (g), B is the mass of the saturated surface dry specimen in air (g) and C is the mass of the specimen in water (g).

The set up for determining G_{mb} is shown in Figure 12.3.

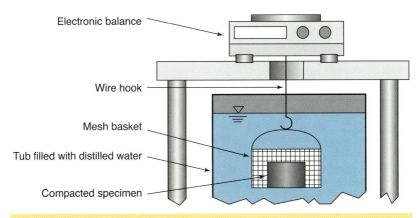

Electronic balance

Wire hook

Mesh basket

Tub filled with distilled water

Compacted specimen

Figure 12.3 Arrangement for determining the bulk specific gravity (G_{mb})

12.2.12 Maximum specific gravity of the loose mixed material (G_{mm})

The theoretical maximum specific gravity of an asphalt (G_{mm}) is the specific gravity of the mixture at zero air voids content. It is one of the most important tests to be performed in the design of an asphalt mixture, and high levels of repeatability are required for the mixture designer to draw accurate conclusions and make appropirate adjustments to the mixture design. Like bulk specific gravity, the theoretical maximum specific gravity does not affect the performance of an asphalt. However, it is essential in determining volumetric factors that are good indicators of performance, such as the air void content and VMA (Advanced Asphalt Technologies, 2011). The maximum specific gravity is determined by measuring the specific gravity of the loose asphalt having removed all of the air trapped in the mixture by subjecting it to a partial vacuum (vacuum saturation). The complete procedure is outlined in ASTM D2041 (ASTM, 2011). Following this procedure, the maximum specific gravity of a loose asphalt specimen is calculated using the following formula

$$G_{mm} = \frac{A}{A + D - E} \qquad (12.12)$$

where G_{mm} is the theoretical maximum specific gravity of the loose mixture, A is the mass of the oven dry specimen in air (g), D is the mass of container filled with water at 25°C to the calibration mark (g) and E is the mass of the container with the specimen filled with water at 25°C to the calibration mark (g).

G_{mm} can also be calculated using the following equation

$$G_{mm} = \frac{100}{P_s/G_{se} + P_b/G_b} \qquad (12.13)$$

where G_{mm} is the maximum specific gravity of the mixture at no air voids, P_s is the aggregate content expressed as the percentage proportion of the total weight of the mix, P_b is the bitumen content expressed as the percentage proportion of the total weight of the mix, G_{se} is the effective specific gravity of aggregate and G_b is the specific gravity of bitumen.

12.3 Grading

12.3.1 Historical development of packing theory

Aggregates typically constitute some 92–96% of an asphalt by mass. Clearly, the nature and grading of the aggregate in a particular asphalt will have a profound effect on the performance properties of that asphalt. Kandhal and Parker (1997) suggested that aspects of the aggregate can be related to particular asphalt defects. Table 12.1 is based on Kandhal and Parker's findings.

Aggregate properties are considered in detail in Chapter 10 except for aggregate grading, which constitutes the subject area of this section.

Table 12.1 Aggregate properties related to asphalt performance

Aggregate property/test	Performance measure
Grading and size	Permanent deformation and fatigue cracking
Uncompacted void content of coarse aggregate	Permanent deformation and fatigue cracking
Flat or elongated particles in coarse aggregate	Permanent deformation and fatigue cracking
Uncompacted void content of fine aggregate	Permanent deformation
Methylene blue test of fine aggregate (passing the 0.075 mm sieve)	Permanent deformation resulting from loading and environmental effects (i.e. asphalt stripping)
Micro-Deval (magnesium sulfate soundness)	Ravelling, potholes

Kandhal and Parker (1997)

Aggregate grading (often described as gradation) is the distribution of particle sizes in a batch of aggregates. It is one of the most important mixture design properties. A change in the size distribution of the aggregates can result in a different load distribution over the surface of the aggregate particles. The best grading for an asphalt concrete is that which gives the densest particle packing, thus increasing stability by having more inter-particle contacts and reducing the air voids. (This is obviously not the case for asphalt concretes that are designed to be permeable, e.g. porous asphalt or open graded friction course.) In dense mixtures, however, there must be sufficient air voids to permit the bitumen to be incorporated and assure durability without filling all the space to avoid bleeding and/or rutting (Roberts et al., 1996) (a condition sometimes described as voids over-filled with binder).

Arranging aggregates in an asphalt mixture is akin to packing matter into a confined space. Packing of spheres into a confined space and measuring its density has intrigued scientists and mathematicians since Isaac Newton (1643–1727). The German mathematician Johannes Kepler (1571–1630) worked on sphere packing in three dimensional Euclidian space, and proposed an arrangement known as close packing (either face centred cubic or hexagonal close packing, both of which have average densities of $\pi/(3\sqrt{2})$ approximately 0.740480489 . . .). This is known as the Kepler conjecture, and states that no other arrangement of spheres has a higher density.

Fuller and Thompson (1907) studied different combinations of stones and sand with the aim of producing the densest concrete. The main objective of their study was to improve the quality of the mixture and decrease its cost by using the optimum quantity of cement. As a guide to obtaining the best concrete with constant cement content, the authors concluded that the stones should be evenly graded from fine to coarse, as an excessive amount of fine or middle sizes is very harmful to strength. They also concluded that the diameter of the largest grain of sand should not exceed one tenth of the diameter of the largest stone, and that the coarser the stone used, the coarser the sand must be – an approach that provided a more dense and watertight concrete mixture.

Following this work, Talbot and Richart (1923) and Weymouth (1938) evaluated the size distribution of spheres that maximised density. They determined that when plotting on a semi-log basis (i.e. a linear scale on one axis and a log scale on the other axis) with the percentage passing a particular sieve (plotted on the linear y axis) versus the particle size (plotted on the logarithmic x axis) the maximum packing was obtained when a straight line on the graph had a slope of 0.5. This can be represented by the following equation

$$P = 100\left(\frac{d}{D}\right)^n$$ (12.14)

where P is the percentage of material by weight passing a given sieve with an opening of size d, D is the maximum particle size in a given aggregate blend and n is the an exponent that affects the coarseness or fineness of the grading given by the slope.

Nijboer (1948) investigated the effect of particle size, taking shape into account, for asphalts and, as determined previously, he found that the densest packing was produced by a straight line in a semi-log plot of the percentage by weight passing a sieve size versus the size of the sieve. Empirically, it was found that for asphalts and aggregates as rough, shaped material this line had a slope of 0.45. Goode and Lufsey (1962) published a validation of Nijboer's work, and described the numerical procedure for drawing a semi-power chart for gradings. This is the chart commonly referred as the 0.45 power chart (Meininger, 1992). The 0.45 power chart has become a cornerstone, as its use has been widely accepted worldwide including in the Superpave design method discussed later in this chapter.

12.3.2 Aggregate grading determination methodologies using packing theory

Although the development of semi-log graphs of the percentage of material passing each sieve versus sieve size is still used for determining aggregate gradings, an alternative method was proposed by Robert Bailey (formerly of the Illinois Department of Transportation) using particle packing concepts. This approach is discussed below.

It is a fact that granular particles such as aggregates cannot be placed to achieve 100% packing. Thus, there will always be voids between the aggregate particles. Vavrik et al. (2001) summarised the factors on which the degree of packing of aggregates depend

- *Type and amount of compactive energy.* Several types of compactive force can be used, including static pressure, impact (e.g. Marshall hammer) or shearing (e.g. gyratory shear compactor or California kneading compactor). Higher density can be achieved by increasing the compactive effort: for example, higher static pressure, more blows of the hammer, or more tamps or gyrations.

- *Shape of the particles.* Irregular elongated particles tend to resist packing in a dense configuration. Regular, blocky particles tend to arrange in dense configurations.
- *Surface texture of the particles.* Particles with smooth textures will more easily re-orient into denser configurations. Particles with rough, textured surfaces will resist sliding against one another, leading to lower density configurations.
- *Size distribution (gradation) of the particles.* Single sized particles will not pack as densely as a mixture of particle sizes.

12.3.2.1 Bailey's method

In the early 1980s, Robert Bailey developed a method of grading analysis that takes into account the packing characteristics of individual aggregates, providing a quantified criterion that can be used to control mixture properties such as workability, segregation and compactibility (Vavrik *et al.*, 2001). In a mixture design with a given compactive effort, three aggregate properties control the packing characteristics

- grading
- surface texture
- shape.

Changing the grading of a mixture will influence the amount of void space in the aggregate skeleton.

The Bailey method marks a significant departure by defining coarse and fine aggregates as follows

- *coarse aggregate* – large aggregate particles that when placed in a unit volume create voids
- *fine aggregate* – aggregate particles that can fill the voids created by the coarse aggregate in the mixture.

The Bailey method determines the coarse fraction as those particles that create voids, while the fine fraction is those particles that fit into the voids created by the coarse aggregate. It is this premise on which the method is based. It is also clear from the above that more than a single aggregate size is needed to define the descriptions coarse or fine. Figure 12.4 shows how the Bailey method makes use of a break point to define coarse and fine aggregates. The break point that determines the sieve size separating the coarse and fine aggregates is designated as the primary control sieve

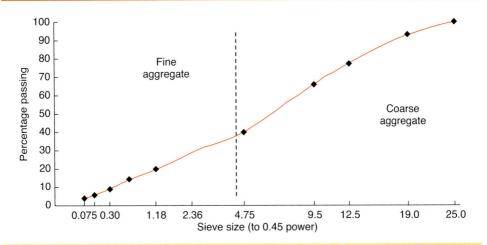

Figure 12.4 Example of break between coarse and fine aggregate for 19.0 mm NMPS mixture

(PCS), which is determined using the following equation

$$PCS = NMPS \times 0.22 \tag{12.15}$$

where PCS is the primary control sieve for the overall blend and NMPS is the nominal maximum particle size for the overall blend being one sieve size larger than the first sieve to retain more than 10% of the aggregate.

In addition, the term NMAS (nominal maximum aggregate size) is defined (AASHTO, 2014a) as the sieve size that is one size larger than the largest sieve that retains more than 10% aggregate. The significance of the NMAS in the selection and performance of a layer is examined later in this chapter.

Vavrik *et al.* (2001) explain that the value of 0.22 in the above equation was determined empirically. Various particle shape combinations and the resultant void sizes were studied. The analysis revealed the spacing as shown in Figure 12.5 and Table 12.2.

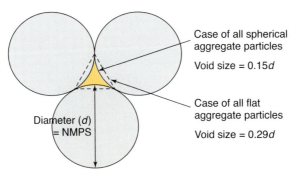

Figure 12.5 Determination of void size on the basis of particle size combinations

Table 12.2 Void sizes for various scenarios

Details of particles	Void size
All round	0.15d
2 Round/1 flat	0.20d
1 Round/2 flat	0.24d
All flat	0.29d

The void sizes vary from 0.15d to 0.29d, where d is the NMPS of the aggregate. Four cases were studied, considering all three particles to be round and replacing each round particle with a flat particle until all three particles are flat. Aschenbrener (2002) states that these figures were a result of a two dimensional analysis that gave particle diameter ratios of 0.155 (all round) to 0.289 (all flat). The void sizes are listed in Table 12.2.

A value of 0.22 was chosen, because it is the average of all four scenarios above, and the primary control sieves for Superpave gradings (the Superpave mixture design method is discussed below) are as shown in Table 12.3.

Table 12.3 clearly demonstrates that the coarse aggregate void size is a function of particle size and shape, and therefore the average size of the coarse aggregate voids in, say, a 9.5 mm NMAS mixture is smaller than the voids in a 37.5 mm NMAS mixture. Because of this change in average void size, a different size of material will be required to fit into the voids created. The primary control sieve determines not only which aggregates are considered as coarse and fine individually but also what portion of each individual aggregate is considered as coarse and fine for use in a given blend. The sieve found to be closest to the calculated value serves as the primary control sieve for a given blend. The amount of material above and below this specific sieve directly determines the volumes of coarse aggregate and fine aggregate (Vavrik et al., 2001).

Table 12.3 Primary control sieve for each corresponding nominal maximum particle size in the Superpave method

Mixture NMPS: mm	NMPS × 0.22	PCS: mm
37.5	8.250	9.5
25	5.500	4.75
19	4.180	4.75
12.5	2.750	2.36
9.5	2.090	2.36
4.75	1.045	1.18

12.3.3 *Impact of aggregate grading on mixture volumetrics*

This section considers how a compliant grading can result in the alteration of the mixture volumetrics of an asphalt (Advanced Asphalt Technologies, 2011). As an example, a grading is classified as coarse or fine depending on whether it passes below or above the maximum density line, respectively. In reality, both the gradings are dense, but one is coarser than the other in comparison with the maximum density gradation.

A concept based on packing theory and maximum density grading is the continuous maximum density (CMD) grading, which can be calculated using the following equation

$$P_{CMD}(d_2) = P(d_1)\left(\frac{d_1}{d_2}\right)^{0.45} \tag{12.16}$$

where $P_{CMD}(d_2)$ is the percentage passing the continuous maximum density grading for sieve size d_2, d_1 is one sieve size larger than d_2 and $P(d_1)$ is the percentage passing sieve d_1.

The usefulness of the CMD grading is that it allows a comparatively detailed analysis of how closely a grading follows the maximum density grading calculated using Equation 12.16. Figure 12.6(a) shows a 9.5 mm grading compared with the standard maximum density grading as calculated using Equation 12.14, while Figure 12.6(b) shows the deviation using Equation 12.16. While both graphs show deviation, the graph in Figure 12.6(b) is much clearer in the way the deviation is shown. As an example, the aggregate grading actually follows the maximum density grading below the 1.18 mm sieve (as is evident from Figure 12.6(b)) while the upper graph (Figure 12.6(a)) shows a different picture.

This is further exemplified in Figure 12.7, where several gradings are considered and corresponding CMD plots are developed.

Figures 12.6 and 12.7, demonstrate that while there is a significant difference in their gradings, the fine aggregate fractions follow the maximum density grading very closely. The importance of the CMD plot is exemplified by the following explanation. Consider the dense/coarse grading plot in Figure 12.7 – the chart in Figure 12.7(a) indicates that the fine aggregate portion deviates significantly from the maximum density grading, and changing this portion may help to bring the VMA within limits, whereas the chart in Figure 12.7(b) shows that this portion already follows the maximum density grading and that the portion from 2.36 to 9.5 mm needs to be changed. This concept may be extended to the discussion on the modification of existing asphalt designs in which the changes will mostly consist of altering the air voids or the binder content, which are, primarily, modifications in VMA. The CMD plots can be used as a tool by the mixture designer to alter

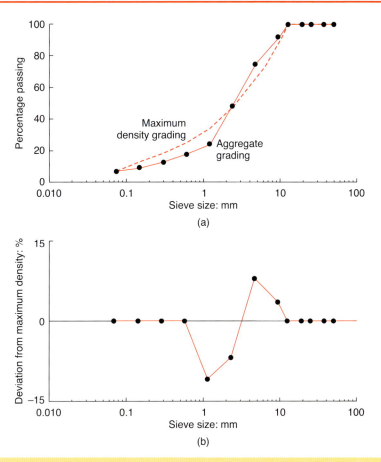

Figure 12.6 (a) A 9.5 mm aggregate grading compared with the maximum density grading; (b) deviation from the continuous maximum density grading for the same 9.5 mm grading (Advanced Asphalt Technologies, 2011)

that portion of the grading that will have an impact on the VMA, as shown in Figure 12.8.

12.4 Mixture design methods

This section briefly considers a number of mixture design methods. The evolution, the method itself and an indication of usage is given for each of the following asphalt mixture design methods

- Hubbard Field
- Hveem or California
- Marshall (US Army Corps of Engineers)
- Smith or Asphalt Institute
- Superpave.

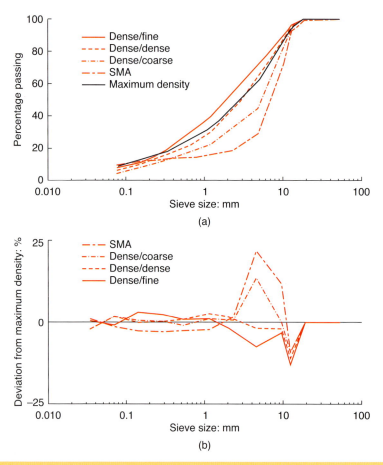

Figure 12.7 (a) Four different 12.5 mm gradings; (b) percentage deviation from continuous maximum density grading for these same blends (Advanced Asphalt Technologies, 2011)

While American design methods are highly regarded worldwide, it should be noted that other methods exist that take a different approach, most notably the French mixture design approach whose 'methodology is based on component characteristics, water-sensitivity testing, void content assessments using gyratory compaction, resistance to permanent deformation, stiffness and fatigue resistance' (Laboratoire Central des Ponts et Chaussées, 2007). It has not been possible to include all of the various methods within this chapter.

12.4.1 Hubbard Field method

By the early 1920s, the principles of soil compaction and the effect of water on density had been quantified and the Proctor method of evaluating soil compaction properties had been developed. Prevost Hubbard and Fredrick Field of the Asphalt Association (later to become the Asphalt Institute) developed a

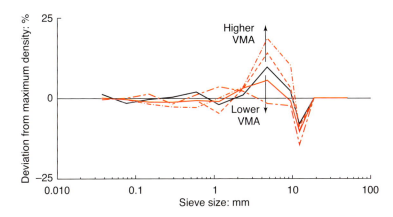

Figure 12.8 Effect of changes in aggregate grading as shown on a CMD plot (Advanced Asphalt Technologies, 2011)

method of mixture design in which asphalt samples were compacted using a Proctor hammer (Hubbard and Field, 1935) and by direct compression. The Hubbard Field mixture design method was developed for sheet asphalt and sand base mixtures with at least 65% of the aggregate passing the 2 mm sieve and 100% of the aggregate passing the 5 mm sieve. Sheet asphalts were pre-mixed asphalts made with bitumen and clean, angular sand dosed with mineral filler. Sheet asphalts were used as surface courses in the late 19th and early 20th century and consisted of a pre-mixed layer of graded aggregate, mineral filler and binder. They were laid in thicknesses varying from 40 to 55 mm and provided a smooth riding surface. Hubbard and Field demonstrated remarkable engineering wisdom in developing the traffic simulation test for the selection of the asphalt mixture. As the traffic at that time was dominated by wagons with steel wheels, Hubbard and Field used a punching shear test to simulate the punching effect of the steel wheels on the sheet asphalt mixtures. The punching test was conducted at 60°C, which was considered to be the highest pavement service temperature. The maximum load required to force a 50 mm diameter by 25 mm high compacted sample through a 43.75 mm diameter orifice under conditions of this test was reported as the stability value. It is interesting to note that some of the systems that were adopted in the Hubbard Field method have been followed in subsequent mixture design methodologies. Some of those methodologies are as follows

1. Measure the bulk density of the specimens.
2. Develop the following plots
   a. bulk density versus bitumen content
   b. stability versus bitumen content
   c. percentage voids in the total mixture versus bitumen content
   d. percentage voids in the aggregate mass versus bitumen content.

3. These plots were then used to select the design bitumen content by comparing them against design criteria.
4. The criteria, which included limits on stability values and percentage voids in the total mixture, were established by comparing the test values with the pavement performance.
5. The bitumen content would be selected based on the mid-point of the voids range: for example, if the voids requirement is from 3 to 5%, the bitumen content corresponding to a void content of 4% would be chosen. If the stability criteria are met in that mixture, it was adopted as the bitumen content.
6. The plot of bitumen content versus voids in the aggregate mass was used to adjust the grading when this was necessary to meet the voids criteria (Goetz, 1989).

Although the shortcomings of the punching shear type of test became apparent when applied to coarser mixtures, the principles of a design procedure incorporating stability and durability criteria in a systematic, quantitative way became well established. It has been reported that the Hubbard Field method was still in use by several states in the USA in the 1970s (Prowell and Ray Brown, 2006).

12.4.2 Hveem or California method

Francis Hveem is credited with the development of the Hveem or California method of asphalt mixture design. Hveem joined the California Division of Highways in 1917, and developed the stabilometer test during the late 1920s and early 1930s. Around 1927, Hveem was given the task of developing a method to determine the appropriate amount of 'oil' (bitumen) to add to aggregates in order to produce a 'hard and smooth' road surface that would not deform under traffic (Vallerga and Lovering, 1985). While working with oil–aggregate mixtures, Hveem observed that coarser gradings tended to require less oil than finer gradings, and made the connection that the surface area of the aggregate varied with the grading. He went on to identify a method for estimating the surface area of aggregate developed by a Canadian engineer, Captain L. N. Edwards, for concrete mixtures (Hveem, 1942, 1970).

The usage of 'oil–aggregate mixtures' to study the determination of design bitumen content using surface area concepts was first developed in the western USA, with the highway departments of California and Nebraska developing formulae based on the surface area concept (Goetz and Wood, 1960; Vallerga, 1953). The California Division of Highways (later the California Department of Transportation) carried forward work on the surface area concept, and used the Hveem method for the development of its highway network.

Hveem realised that, in addition to the surface area, the optimum bitumen content, or at least the point where the optimum bitumen content was exceeded and the stability decreased, was affected by the surface texture of the aggregate. Accordingly, a surface factor was used by Hveem in combination with the calculated surface area to determine the optimum bitumen content. Hveem observed that a smaller bitumen film thickness was required for smaller particles than for larger particles. Hveem continued to work on the surface area concept of aggregates in his quest for a rational mixture design method, and developed some protocols to measure the surface area factors accurately for fine and coarse aggregates. Hveem used the centrifuge kerosene equivalent (CKE) test to determine the surface area of the fine aggregates, taking into account the surface area, aggregate absorption and adjustment for surface texture. For coarse aggregates, the surface area is determined by the amount of SAE 10 oil that is retained on the aggregate. Hveem stated that the CKE method indicated the optimum bitumen content in 95% of cases (Hveem, 1970; War Department, 1948).

Hveem also evaluated the stability of asphalts. He hypothesised that, depending on the roughness and angularity of the aggregate, the film thickness at which the particles would become overly lubricated by the bitumen and therefore unstable would vary (War Department, 1948). Hveem developed his first stabilometer in 1930. The Hveem stabilometer evolved into a hydraulic device into which a compacted sample of asphalt was placed. The sample was loaded vertically on its flat surface, and the radial force transmitted to the surrounding hydraulic cell was measured. The stability value was calculated using the following equation

$$S = \frac{22.2}{P_h D_2 / (P_v - P_h) + 0.222} \tag{12.17}$$

where P_v is the vertical pressure (= 400 psi), P_h is the horizontal pressure at a vertical pressure of 400 psi and D_2 is the displacement of the sample (= number of turns of the handle).

The assumptions made by Hveem in developing the test (Stanton and Hveem, 1934) were that

- asphalt is a plastic solid, and a measure of its resistance to plastic deformation should be a measure of its stability
- the degree of distortion of an asphalt concrete pavement 'is the accumulated results of a large number of quick shoves or pushes, each lasting a small fraction of time'.

The optimum bitumen content using the Hveem method was determined as follows. First, the bitumen contents for which moderate to heavy bleeding was observed on the surface of the compacted sample were eliminated. Next, any bitumen contents that failed the minimum stability value were eliminated. Finally, the highest bitumen content that had at least 4% air voids was selected as the optimum. A swell test was also conducted to determine the resistance of the specimens to water induced damage (Vallerga and Lovering, 1985).

Vallerga and Lovering (1985) quote Hveem's own summary of his mixture design philosophy in 1937 as follows

> For the best stability, a harsh, crushed stone with some gradation, mixed with only sufficient asphalt to permit high compaction with the means available. For greatest resistance to abrasion, raveling, aging and deterioration, and imperviousness to water, a high asphalt content, broadly speaking, the richer the better. For impermeability, a uniformly graded mixture with a sufficient quantity of fine sand (fine sand is more important than filler dust). For non-skid surfaces, a large quantity of the maximum sized aggregate within the size limits used. For workability and freedom from segregation, a uniformly graded aggregate. To reduce the above factors to as simple a consideration as possible, it seems to be the best rule to use a dense, uniformly graded mixture without an excess of dust and to add as much oil or asphalt as the mixture will tolerate without losing stability.

In this quotation, the word 'asphalt' is that used in the USA for bitumen.

Much more detail of the development of the Hveem mixture design method can be found in documents published by Caltrans (1983) and Vallerga and Lovering (1985). For a step by step explanation of the Hveem mixture design method, a chapter in Manual Series No. 2 by the Asphalt Institute (1997) provides that information.

The Hveem mixture design method was used by some authorities prior to the implementation of the Superpave method. A study by Kandhal and Koehler carried out in 1984 in the USA indicated that, at that time, ten US states were using the Hveem mixture design method (Kandhal and Koehler, 1985). The performance of the Hveem method has been summarised by Vallerga and Lovering (1985)

> It can be concluded that the Hveem method has served the state of California well over the past 50 years in the design and construction of a paved State highway system second to none.

12.4.3 Marshall or Corps of Engineers method
The basic concepts of the Marshall mixture design method were developed by Bruce Marshall around 1939 while he was an employee of the Mississippi State Highway Department. Marshall used many of the principles found in the Hubbard Field method. Instead of using a Proctor hammer, which covered

only part of the specimen face, the hammer size was increased to match the diameter of the specimen. The weight and drop of the hammer were increased to maintain a similar amount of compactive energy. Air voids, and later the VMA, were calculated for the compacted specimens. Marshall changed the orientation of his mechanical property test by turning the specimen on its side, placing it between two nearly semi-circular loading heads and loading it until it failed. The peak load sustained by the specimen was known as the Marshall stability. White (1985) has published an outstanding account of the development of the Marshall mixture design method.

During World War II, the Corps of Engineers was charged with selecting a method of asphalt mixture design to cope with the increasing tyre pressures found on military aircraft as aircraft weights began increasing during that period. The first research efforts addressing mixture designs for airfields within the US War Department were conducted at the US Army Tulsa District (War Department, 1943). Four mixture design methods were evaluated for their suitability, and the following conclusions were drawn

> The Hubbard field test is considered the most satisfactory method for general utility. The Texas punching method is similar but apparently over emphasizes gradation and is not reproductable with different personnel. The Hveem stability test should be good for research but is not self-sufficient for routine testing since it measures chiefly internal friction. Experience has indicated that it is a good method of determining the most satisfactory bitumen content for a given aggregate. The Skidmore test is also not self-sufficient and it is believed to measure chiefly cohesion and interlock.

A continuation of the above work was assigned to the Waterways Experiment Station (WES) in September 1943. The first step was to develop 'a simple apparatus suitable for use with the present California bearing ratio (CBR) equipment to design and control asphalt paving mixtures' (War Department, 1944). The Marshall method at the time of the WES study (1943) had been used in southern US states by several highway agencies for about 4 years (War Department, 1944).

In the first phase of the study begun in 1943 (White, 1985), comparisons were made between the Hubbard Field and Marshall mixture design methods using a wide range of asphalts. From this study it was concluded that the Marshall stability test gave results that were comparable to the Hubbard Field stability test. Furthermore, the Hubbard Field test was not readily adaptable to the field CBR equipment, and the Marshall apparatus was also more portable. Therefore, the Marshall method was selected for additional study to evaluate some additional objectives. Some of the reasons given for recommending the adoption of the Marshall procedure for the design and field control of asphalt paving mixtures (War Department, 1944) were

a. The Marshall stability test stresses and fails the entire specimen. Definite shear planes divide the specimen into four separate pieces. The stability value is a measure of the resistance of the specimen to the development of internal shear planes. This resistance is considered directly to the degree to which masses of particles are bonded together mechanically or with bituminous material. The Hubbard Field method produces a shear around the perimeter at the base of the specimen. Therefore stability is measured by a small fraction of the specimen. Several rock fragments may be contained within the specimen and would reflect increased stability only if they are in close proximity to the lower edge. Erratic stability values between identical specimens are possible as the laws of chance place more or less rock particles within the shear section. A number of check tests are therefore required for accurate details.

b. Tests can be made more rapidly with less effort by the Marshall stability method.

c. The equipment of the Marshall stability method weighs less and is more compact.

d. The density secured by the Marshall method appears to duplicate that obtained in the field more closely than does the Hubbard Field method. Therefore the amount of asphalt required in a mix is more accurately determined by the Marshall test.

Marshall design stability and flow criteria have evolved over the years and an example of the criteria used in the late 1970s is shown in Table 12.4.

For the present day user, the Marshall criteria have evolved significantly to provide criteria based on the traffic that the pavement will carry, and are presented in Table 12.5.

A brief description of the Marshall mixture design method (the test procedure can be found in Manual Series No. 2 by the Asphalt Institute (1997)) is as follows

1. *Aggregate selection.* The aggregate selection consists of conducting the grading and other tests such as the aggregate specific gravity and so on to determine the suitability of an aggregate for asphalt mixture design.

Table 12.4 Typical Marshall design criteria (Asphalt Institute, 1979)

Mix criteria	Light traffic ($<10^4$ ESALs)		Medium traffic (10^4–10^6 ESALs)		Heavy traffic ($>10^6$ ESALs)	
	Min.	Max.	Min.	Max.	Min.	Max.
Compaction (number of blows on each end of the sample)	35		50		75	
Stability (minimum)	224 N (500 lbs)		336 N (750 lbs)		6672 N (1500 lbs)	
Flow (0.25 mm (0.01 inch))	8	20	8	18	8	16
Percentage air voids		5	3	5	3	5

Table 12.5 Current criteria for the Marshall mixture design method

Marshall method mix criteria[a]	Light traffic surface and base[b]		Medium traffic surface and base		Heavy traffic surface and base	
	Minimum	Maximum	Minimum	Maximum	Minimum	Maximum
Compaction: No. of blows to each end of the specimen[c]		35		50		75
Stability, N (lb)	3336 (750)		5338 (1200)	–	8006 (1800)	–
Flow: 0.25 mm (0.01 in)[d]	8	18	8	16	8	14
Air voids: %[e]	3	5	3	5	3	5
VMA: %[f]	See Table 5.3 in Asphalt Institute (1997)					
VFB: %	70	80	65	78	65	75

Asphalt Institute (1997).
[a] All criteria, not just the stability value alone, must be considered in designing an asphalt paving mix. Hot mix asphalt bases that do not meet these criteria when tested at 60°C (140°F) are satisfactory if they meet the criteria when tested at 38°C (100°F) and are placed 100 mm (4 in) or more below the surface. This recommendation applies only to regions having a range of climatic conditions similar to those prevailing throughout most of the USA. A different lower test temperature may be considered in regions having more extreme climatic conditions.
[b] Traffic classifications:
Light   Traffic conditions resulting in a design equivalent axle load (EAL) $<10^4$
Medium   Traffic conditions resulting in a design EAL between 10^4 and 10^5
Heavy   Traffic conditions resulting in a design EAL $>10^6$
[c] Laboratory compaction efforts should closely approach the maximum density obtained in the pavement under traffic.
[d] The flow value refers to the point where the load begins to decrease.
[e] The portion of the asphalt concrete lost by absorption into the aggregate particles must be allowed for when calculating the air voids.
[f] The voids in the mineral aggregate are to be calculated on the basis of the ASTM bulk specific gravity for the aggregate.

Aggregates are chosen based on compliance with local specifications. The traditional Marshall mixture design requires aggregates to have a maximum size of 25 mm. For aggregate sizes larger than 25 mm and up to 38 mm, the modified Marshall method should be followed. This is outlined in section 5.16 of MS-2 (Asphalt Institute, 1997).

2. *Binder selection.* The binder is selected based on the local specifications, which may require the binder to be selected based on penetration, absolute viscosity or the Superpave performance grading.

3. *Preparing laboratory specimens.* The Marshall method requires that cylindrical samples of approximately 64 mm in height and 102 mm in diameter are prepared. The mixture designer may start the sample preparation with about 1.2 kg of aggregates per mould. If the height is more than the permissible 64 mm, the following equation can be used to adjust the height of the specimen

$$\text{adjusted mass of aggregate} = 63.5 \times \frac{\text{mass of aggregate used}}{\text{specimen height (mm) obtained}}$$

(12.18)

In order to determine the optimum binder content, specimens at six different binder contents around the expected binder content are prepared. For example, if the optimum binder content is expected to be around 4.5%, specimens are prepared with binder contents of 3.0%, 4.0%, 4.5%, 5.0%, 5.5% and 6.0%. Three samples are prepared for each binder content chosen. As a starting point, the binder content can be estimated from the following equation (Asphalt Institute, 1997)

$$DBC = 0.035a + 0.04b + Kc + F \qquad (12.19)$$

where DBC is the approximate design bitumen content, as the percentage by total weight of the mixture; a is the percentage by weight of the mineral aggregate retained on a 2.36 mm sieve; b is the percentage by weight of the mineral aggregate passing the 2.36 mm sieve and retained on a 0.075 mm sieve; c is the percentage by weight of the mineral aggregate passing the 0.075 mm sieve; $K = 0.15$ for 11–15% passing the 0.075 mm sieve, 0.18 for 6–10% passing the 0.075 mm sieve and 0.20 for 5% or less passing the 0.075 mm sieve; and $F = 0$–2% (this factor is based on the absorption of bitumen and, in the absence of other data, a value of 0.7 is suggested).

Once the quantity of graded aggregate is fixed, the batched aggregates are heated to constant weight and the binder is heated to the mixing viscosity. MS-2 requires that temperatures at which the bitumen attains kinematic viscosities of 170 ± 20 cSt kinematic and 280 ± 30 cSt kinematic shall be established as the mixing and compaction temperatures, respectively. The Marshall method does not require any ageing/curing of the asphalt prior to compaction. The Marshall specimens are prepared using the standard Marshall compactor that consists of a flat faced hammer that is 98.3 mm in diameter, weighs 4.5 kg and falls freely by a distance of 457 mm onto the asphalt placed within the specimen mould. The Marshall compactor is required to impart a fixed number of blows per face on the specimen, usually 75 blows (for details refer to Table 12.5). After compaction, the specimens are extruded from the mould and then allowed to cool to room temperature (preferably overnight).

4. *Mechanical testing of the asphalt.* The Marshall method requires that the specimens are tested for stability and flow. Testing for stability and flow utilises the *Marshall stabilometer,* which applies a load at a constant rate of vertical strain rate of 51 mm per minute through cylindrical testing heads, as shown in Figure 12.9. The maximum load taken by the specimen at 60°C is noted as the Marshall stability value, and the amount of deformation at the maximum load is designated as

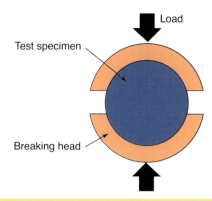

Load

Test specimen

Breaking head

Figure 12.9 Schematic diagram of load application in the Marshall stability test

the *flow*. The stability and flow are measured using calibrated proving rings, and the values are reported in newtons and millimetres, respectively. Prior to placing the extruded Marshall specimen in the Marshall stabilometer breaking head, the specimens are placed in a water bath at $60 \pm 1°C$ for 30–40 min prior to the test, and the stability and flow test should be completed within 30 s of extracting the specimen from the water bath.

5. *Density and voids analysis*. The Marshall asphalt specimens and loose asphalt specimens are then tested to determine the following parameters
   a. The average bulk density of the Marshall asphalt specimens (G_{mb}) (AASHTO, 2013a) and the average specific gravity of the binder (G_b) (ASTM, 2009).
   b. The theoretical maximum specific gravity (G_{mm}) as per ASTM D2041 (ASTM, 2011) at two binder contents close to the optimum binder content. The G_{mm} value will be used to determine the effective specific gravity of the aggregate (G_{se}) (Equation 12.9).
   c. Using the three aggregate specific gravities, the specific gravity of the binder, the G_{mm} value and the three key volumetric parameters (i.e. air voids (V_a) (Equation 12.1), voids filled with bitumen (VFB) (Equation 12.5) and voids in the mineral aggregate (VMA) (Equation 12.4)).
   The formulae for calculating the above can be found earlier in this chapter in section 12.2 on mixture volumetrics – corresponding equation numbers and references have been appended.
6. *Data interpretation*. The test data are interpreted as follows
   a. The stability values are standardised based on the specimen height.
   b. The following plots are developed
      i. V_a versus binder content
      ii. VFB versus binder content
      iii. VMA versus binder content

iv. stability versus binder content

v. flow versus binder content

vi. G_{mb} of the mix versus the binder content.

Typical results from the above tests are given in Figure 12.10, which shows typical trends of the plots as noted above. These graphs are used to determine the optimum binder content, by plotting them as

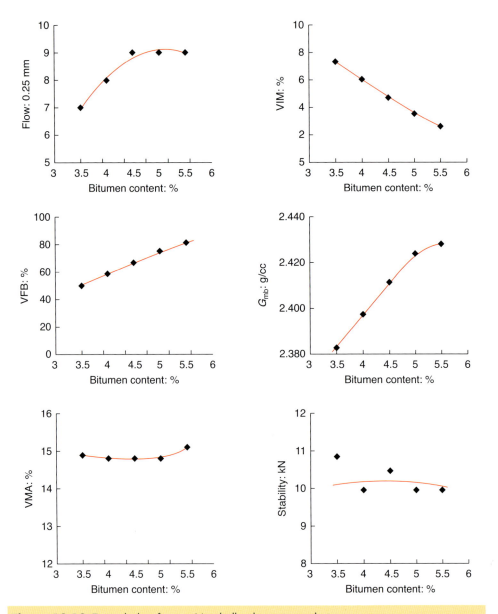

Figure 12.10 Typical plots from a Marshall volumetric analysis

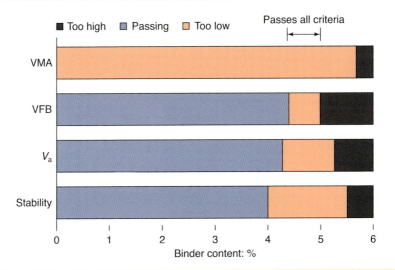

Figure 12.11 Examples of the acceptable range of values

shown in Figure 12.11, which helps the mixture designer to observe the influence of binder content on the various mixture volumetric parameters at a glance and to determine the binder content suitable for the project.

7. *Selection of the optimum binder content.* The test data and plots are further analysed, and the initial binder content is selected as a binder content that satisfies all strength and volumetric requirements. The Asphalt Institute (1997) recommends a starting point as the median of the percentage of air voids, say the binder content at 4% air voids. Specimens are prepared at this binder content and evaluated for the strength and volumetric requirements, and further adjusted.

8. *Selection of the final mixture design.* The final mixture should be that which is the most economical and which meets all the criteria. However, the mixture designer must use his expertise to ensure that all mixture properties are balanced. Low binder contents are financially attractive. However, the designer must ensure that unduly low binder contents are avoided, as this can have an adverse effect on the properties of the asphalt. The Asphalt Institute (1997) suggests that a narrow range of binder content satisfying all binder requirements should be selected. This is shown in Figure 12.11. However, it must be borne in mind that the requirements of most projects are unique, and the mixture designer must make the necessary adjustments to ensure that these requirements are satisfied prior to deciding on the value of the optimum binder content. The plot shown in Figure 12.11 enables the mixture designer to view all the responses of the change in

binder content versus all volumetric parameters. This unique plot helps the mixture designer not only to validate the design but also to decide on the values of tolerances that should apply to a particular project.

The Marshall method has enjoyed widespread usage the world over for the design of dense graded asphalts. Indeed, a number of authorities continue to use it today in conjunction with their local specifications.

12.4.4 Smith or Asphalt Institute method

The Hubbard Field, Hveem and Marshall mixture design methods are all empirical, with criteria developed from the correlation of laboratory and field studies. V. R. Smith and the Asphalt Institute applied the principles of the tri-axial compression test to the testing of asphalts (Goetz and Wood, 1960). Smith used a closed system triaxial test wherein the vertical load was applied in static increments and the lateral transmitted pressure was measured when equilibrium had essentially been reached. Specimens of rational size (at least twice their diameter) were formed by spading and double plunger compaction or by the use of the kneading compactor. The test was conducted at room temperature, as the effects of temperature and the rate of loading are virtually non-existent under the conditions of the test. The data obtained were used to generate a plot of vertical applied load versus lateral transmitted pressure, from which the parameters angle of internal friction (Φ) and cohesion (c) were calculated. From theoretical considerations, Smith developed the relationship between the angle of internal friction and the cohesion for varying applied stresses, and presented this in the form of supporting power curves.

By testing many mixtures in the laboratory whose performance in the field was known, an evaluation chart was devised that delineated regions of satisfactory and unsatisfactory mixtures in a plot of the angle of internal friction versus the cohesion. Boundary conditions were set on the basis of traffic severity and the minimum angle of internal friction required for satisfactory performance while applying voids criteria.

Goetz (1989) opines on the test

> while a triaxial test is a rational one that generates the measurement of parameters of a more fundamental nature than those measured previously and does use theoretical concepts in applying these measurements to mixture design, in the final analysis, it depends upon correlation between field and laboratory results for application. The measured parameters can be determined in a rational way for known stress conditions in the laboratory, but the pavement is not loaded in the rational way in service and stress conditions in the pavement are unknown … In spite of such shortcomings from a design point of view, a design based on fundamental

measurements was considered progressive in anticipation of the day when stress conditions in the pavement could be more accurately defined and rational strength test data applied.

Schaub and Goetz (1961) were of the view that

> the triaxial shear strength test has been recognized as the most fundamental in that it uses field-simulated confined loading where stress conditions are known and statically determinant methods of analysis can be applied.

The complete description of the Smith mixture design method can be found in two publications by Smith (1949, 1951), and a complete theoretical consideration of the triaxial state of stress can be found in a paper by McLeod (1950). A comparison of the Marshall and Smith methods can be found in a paper by Goetz (1951).

The Smith or Asphalt Institute method was used more for the purposes of research rather than in the field.

12.4.5 Superpave

12.4.5.1 Background to Superpave

The Strategic Highway Research Program (SHRP) Asphalt Research Program was developed in the USA in the early 1980s because of an increasing number of premature asphalt pavement failures. In recognition of this problem, US states initiated the development of a coordinated, well funded, national research effort to develop improved specifications for bitumens and, ultimately, asphalts. The importance of and need for specification development in the SHRP Asphalt Research Program originated in Transportation Research Board Special Report 202, *America's Highways: Accelerating the Search for Innovation* (TRB, 1986a), popularly known as the blue book, which presented the conclusions and recommendations of the Strategic Transportation Research Study (STRS). The authors of the blue book stated the objective of the bitumen research programme as follows

> To improve pavement performance through a research program that will provide increased understanding of the chemical and physical properties of asphalt cements and asphalt concretes. The research results would be used to develop specifications, tests ... needed to achieve and control the pavement performance desired.

This emphasis was reinforced and further defined in the May 1986 Transportation Research Board report *Strategic Highway Research Program Research Plans* (TRB, 1986b), popularly known as the brown book. This report stated a specific constraint or guideline for the bitumen programme: 'the final product will be performance-based specifications for asphalt, with or without modification, and the development of an asphalt–aggregate mixture analysis system (AAMAS)' (where 'asphalt' means 'bitumen').

The brown book further described the programme's Project 1-4, 'Preparation of Performance-based Specifications for Asphalt and Asphalt–aggregate Systems'. This project consisted of two tasks: to develop the bitumen specification and to develop the AAMAS.

Finally, the SHRP executive committee in 1987 approved *A Contracting Plan for SHRP Asphalt Research* (SHRP, 1987). This blueprint became the strategic plan for the SHRP Asphalt Research Program, and took precedence over earlier research plans when issues of proper technical direction arose. The contracting plan combined the many tasks identified in the 1986 brown book into a coordinated, manageable structure of eight main contracts.

The responsibility for the development of the Superpave performance based binder specification, performance based bitumen–aggregate mixture specification and mixture design system, and for the technical direction and coordination of the entire programme, was assigned to the A-001 contractor, The University of Texas at Austin, in October 1987.

Between 1987 and 1993, the Asphalt Research Program was established and carried out through the award of the eight major research contracts and an additional 15 supporting studies. Throughout this period, the goals of the programme remained substantially unchanged from those originally articulated in the blue book, although details were changed as the programme evolved.

Four significant changes to the research and contracting plans evolved through an ongoing dialogue in the highway community that participated in the development of the programme:

- The term asphalt was broadened to asphalt binder in recognition of the fact that the specification encompasses modified as well as unmodified binders.
- The original concept of an SHRP specification for an asphalt–aggregate mix analysis system (AAMAS) evolved to a performance based specification for asphalt–aggregate mixtures supported by a distinct mixture design system. (The term 'AAMAS' was introduced by NCHRP Project 9-6 (Von Quintus *et al.*, 1991). The SHRP adopted the term Superpave in order to distinguish the two research efforts and to better describe the specific goal of the SHRP research.)
- The term Superpave was chosen in 1991 to signify the integrated structure of performance based specifications, test methods, equipment and protocols, and a mixture design system. The Superpave system is the principal product of the SHRP Asphalt Research Program.

Table 12.6 The eight major contracts of the SHRP Asphalt Research Program

SHRP contract number	Contract title (contractor)
A-001	SHRP Asphalt Research Program: Technical Direction, Specification and Superpave® Development (The University of Texas at Austin)
A-002A	Binder Characterization and Evaluation (Western Research Institute)
A-002B	Novel Approaches for Investigation of Asphalt Binders (University of Southern California)
A-002C	Nuclear Magnetic Resonance Investigation of Asphalt (Montana State University)
A-003A	Performance-related Testing and Measuring of Asphalt–aggregate Interactions and Mixtures (University of California at Berkeley)
A-003B	Fundamental Properties of Asphalt–aggregate Interaction Including Adhesion and Adsorption (Auburn University)
A-004	Asphalt Modification Practices and Modifiers (Southwestern Laboratories)
A-005	Performance Models and Validation of Test Results (Texas Transportation Institute)

Kennedy et al. (1994)

- Also in 1991, the Superpave software was developed to provide a unifying framework for the entire system.

The eight major contracts of the SHRP Asphalt Research Program are listed in Table 12.6.

The major products of the SHRP Asphalt Research Program were

- a performance based specification for asphalt binders and the supporting test methods and equipment
- a performance based mixture design system with supporting test methods and equipment
- a modifier evaluation protocol
- the Superpave specification, design and support software.

These products are an integrated, coordinated system, and are tied together by the Superpave software. The Superpave system provides a means for the rational design of asphalt paving mixtures, and provides a framework and mechanism to tie mixture and structural design together.

Superpave was developed to replace the diverse specifications and mixture design methods used by the 50 US states and other transportation agencies. Superpave provides a single, performance based system that can account for the distinct traffic and environmental conditions found throughout the USA, Canada and other parts of the world. It was developed to address and minimise permanent deformation, fatigue cracking and low temperature

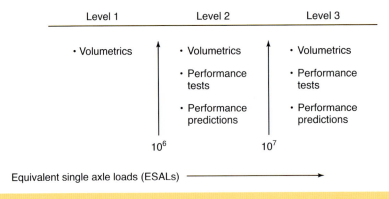

Figure 12.12 Hierarchical structure of Superpave

cracking, and it considers how the effects of ageing and moisture damage contribute to the development of these three distress modes.

12.4.5.2 Overview of Superpave

This section is based on the National Research Council publication SHRP-A-410, *Superior Performing Asphalt Pavements (Superpave): The Product of the SHRP Asphalt Research Program*, which was published in 1994 (Kennedy *et al.*, 1994).

Superpave mixture design encompasses the performance based tests and performance prediction models developed in the SHRP Asphalt Research Program using mixture volumetrics. Superpave has been developed for ready application to all classes of roadway, from rural or urban residential streets to roads carrying substantial numbers of heavy commercial vehicles. The Superpave hierarchical approach is illustrated conceptually in Figure 12.12. It matches the appropriate level of mixture design effort and technology to the pavement being designed.

The Superpave mixture design is an example of a hierarchical mixture design method that utilises defined subsystems to define the requirements with which the mixture design must comply. The three levels of design are defined based on traffic and the use to which the pavement is put. While the suggested boundary values are 1 million and 10 million equivalent single axle loads (ESALs), the actual traffic levels are established by the user. All three design levels include a volumetric mix design phase, while level 2 and level 3 designs require performance based tests to allow the mixture design to be optimised for resistance to deformation, fatigue cracking and low temperature cracking – the three primary distress modes that SHRP binders and mixtures are designed to counter. (Although the criteria for level 2 and level 3 in Figure 12.12 appear to be identical, the detailed requirements that have to be met for level 3 are much more demanding than those for level 2.)

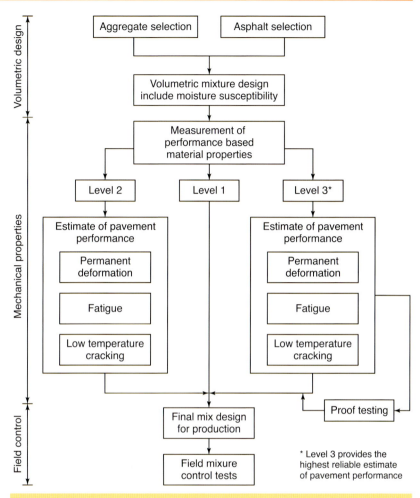

Figure 12.13 Flowchart for Superpave mixture design (Kennedy *et al.*, 1994)

A general flow chart of the Superpave mixture design system is shown in Figure 12.13. For level 1, the procedure involves only volumetric design, which evaluates aggregates and asphalt binders in order to select a grading and binder content that satisfy specified criteria for air voids, voids in the mineral aggregate and voids filled with asphalt. For levels 2 and 3, performance based tests are conducted, and estimates of distress with time are made. This allows the mixture design to be optimised with regard to one or more of the three distress modes: deformation, low temperature cracking and fatigue cracking. Subsequent to completion of the mixture design, modifications may be required because of changes to the material properties once the mixture is produced in a plant. These modifications may or may not require retesting.

Table 12.7 Comparison of level 2 and level 3 mixture designs

	Permanent deformation/fatigue cracking	Low temperature cracking
Test types	Level 3 considers more states of stress, and requires two additional test methods	No difference between level 2 and level 3
Test temperatures	Level 3 considers a range of temperatures from 4 to 40°C Level 2 uses one effective temperature for fatigue cracking and one for permanent deformation	Level 3 considers three temperatures Level 2 considers tensile strength at one temperature only
Performance prediction	Level 3 breaks the year into seasons Level 2 considers the entire year as a single season	No difference between level 2 and level 3

Kennedy *et al.* (1994)

During construction, control tests are undertaken at the production plant to verify that the mixture being manufactured accords with the laboratory mixture design. Level 3 also includes an optional proof testing scheme that allows the mixture to be subjected to tests that simulate the actual traffic and environmental conditions to confirm that the mixture actually performs as required.

In level 2 mixture design, a smaller number of tests are performed at a smaller number of temperatures than for level 3 mixture design. Performance based tests for rutting are undertaken at a single effective temperature for deformation. Similarly, tests to predict fatigue cracking are performed at a single effective temperature for fatigue cracking. Effective temperatures for deformation and fatigue cracking are defined as the single test temperature at which the amount of deformation or fatigue cracking that occurs is equivalent to that obtained by considering each season separately throughout a year. These two temperatures will be different from each other, and can be calculated using equations contained in Chapter 4 of SHRP-A-408 (Cominsky *et al.*, 1994). Low temperature tensile strength is measured at a single temperature in level 2 design.

Level 3 mixture design simulates the entire year by breaking it into representative seasons. Performance based tests for deformation and fatigue cracking are performed at a range of temperatures. A greater number of tests are conducted to evaluate more rigorously mixture response across a greater range of stresses. Deformation and fatigue cracking are predicted using mixture properties in each of the representative seasons. A summary comparison of level 2 and level 3 is shown in Table 12.7.

12.4.5.3 Laboratory compaction
Superpave marked the departure from impact compaction systems to a gyratory mechanism for the preparation of laboratory specimens. The entire mixture design system, including field control, is based on the use of the

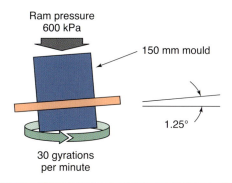

Ram pressure
600 kPa

150 mm mould

1.25°

30 gyrations
per minute

Figure 12.14 The Superpave gyratory compactor

Superpave gyratory compactor (SGC). The adoption of the gyratory mechanism for the laboratory compaction of asphalt specimens followed consideration of the topic including use by the French of this method of compaction. The SHRP committee defined the following as being the key advantages of the SGC (Kennedy *et al.*, 1994)

- It is relatively inexpensive, portable and capable of quickly moulding specimens with minimal variation between individual samples.
- The performance properties of the compacted specimen simulate the performance properties of cores from pavements constructed with the same bitumen–aggregate combination.
- It allows the compactibility of the mixture to be evaluated including an estimate of the final air voids content under traffic (the probability of the mixture becoming plastic under traffic) and a measure of the grading of the aggregate in the mixture.

The SGC (Figure 12.14) has the following characteristics

- an angle of gyration of $1.25 \pm 0.02°$
- a rate of 30 gyrations per minute
- a vertical pressure during gyration of 600 kPa
- the capability of producing specimens that are 150 mm in diameter having a height of 150 mm.

The trend line that is obtained by plotting the change in relative density with the increasing number of gyrations, known as the densification curve, also defines the three different values of compactive effort specified in the Superpave mixture design procedure

- N_{ini}, the initial compactive effort
- N_{design}, the design compactive effort
- N_{max}, the maximum compactive effort.

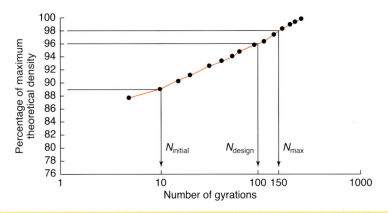

Figure 12.15 Typical densification curves using the SGC

where N is the number of gyrations to which the asphalt specimen is subjected (Figure 12.15).

The compactive efforts, N_{ini} and N_{max}, are used to evaluate the compactability of the mixture, while N_{design} is used to select the bitumen content. Densities that correspond to each value of compactive effort are designated

- C_{ini}
- C_{design}
- C_{max}.

These densities are expressed as a percentage of the maximum theoretical specific gravity (AASHTO T 209 (AASHTO, 2011a)).

In the original Superpave manual (Kennedy *et al.*, 1994) the values of N_{ini}, N_{design} and N_{max} were defined as indicated below.

The number of gyrations for N_{ini} varies from 7 to 10, and is given by the following equation

$$\log N_{ini} = 0.45 \log N_{design} \tag{12.20}$$

The N_{design} values are selected from a matrix that defines N_{design} values corresponding to traffic (in millions of standard axles (msa)) and the average design air temperature (in °C). It will be recalled that Superpave applies to all classes of road, from residential streets to heavily trafficked expressways, thus, the traffic values varied from 0.3 to >100 msa, with the temperature varying in the following ranges: <39°C and 39–45°C with increments of 2°C.

The number of gyrations for N_{max} varies from 104 to 287, and is given by the following equation

$$\log N_{max} = 1.10 \log N_{design} \tag{12.21}$$

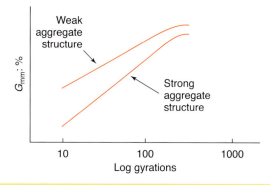

Figure 12.16 Influence of the aggregate structure on the densification curve

The importance of aggregate structure (at the same binder content) is shown in Figure 12.16, and is indicated by the slope of the densification curve. Mixtures exhibiting relatively steep slopes and low C_{ini} values are indicative of mixtures that have developed a good aggregate structure or internal resistance to densification. To ensure adequate structure, it is suggested that the specifications require that

$C_{ini} \leq 89\%$

where the number of gyrations N_{max} varies from about 7 to 10 according to Equation 12.21.

A maximum density requirement at N_{max} ensures that the mixture will not compact excessively under the anticipated traffic, become plastic and result in excessive deformation. Thus, the specification requires that

$C_{max} \leq 98\%$ or air voids $\geq 2\%$

As N_{max} represents a compactive effort that would be equivalent to traffic that greatly exceeds the design traffic (ESALs), excessive compaction under traffic will not occur. Thus, the air voids will not drop below 2%, and the mixture will not become susceptible to plastic deformation.

12.4.5.4 Selection of aggregate
Consensus properties
In order to finalise the aggregate properties, the SHRP committee adopted the Delphi method for developing consensus. It was the view of the SHRP pavement researchers that certain aggregate characteristics were critical and needed to be achieved in all cases to arrive at asphalts that performed well. These characteristics were described as consensus properties because there was wide agreement in their use and specified values. Those properties are

- coarse aggregate angularity (CAA)
- fine aggregate angularity (FAA)
- flat, elongated particles
- clay content.

Tests for CAA and FAA were made mandatory in the Superpave method.

Coarse aggregate angularity
This property ensures a high degree of aggregate internal friction and rutting resistance. It is defined as the percentage by weight of aggregates larger than 4.75 mm with one or more fractured faces. The test procedure for measuring the CAA is described in ASTM D5821 (ASTM, 2013). The procedure involves manually counting particles to determine the number of fractured faces for a defined size of sample. A fractured face is defined as any fractured surface that occupies more than 25% of the area of the outline of the aggregate particle visible in that orientation.

Fine aggregate angularity
This property ensures a high degree of fine aggregate internal friction and rutting resistance. It is defined as the percentage of air voids present in loosely compacted aggregate smaller than 2.36 mm. Higher void contents mean more fractured faces. The test procedure used to measure this property is described in AASHTO T 304 (AASHTO, 2011b). Both the CAA and FAA limits are defined by the location of the layer in the pavement and the traffic intensity to which it will be subjected.

12.4.5.5 Grading
Superpave uses the 0.45 power chart to define a permissible grading. Superpave uses a standard set of ASTM sieves. The definitions below are related to the size of the aggregate

- maximum size: one sieve size larger than the nominal maximum size (see Figure 12.17)
- nominal maximum aggregate size (NMAS): one sieve size larger than the first sieve to retain more than 10% (see Table 12.6).

Superpave specifies that two additional features are added to the 0.45 power chart: control points and a restricted zone (Figure 12.17). Control points function as master ranges through which gradings must pass. They are placed on the nominal maximum size, an intermediate size (2.36 mm) and the dust size (0.075 mm).

The restricted zone resides along the maximum density grading between the intermediate size (either 4.75 or 2.36 mm, depending on the maximum size) and the 0.3 mm size. It forms a band through which gradings should not

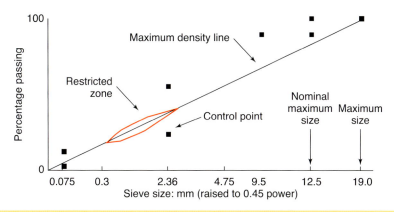

Figure 12.17 Maximum density line for an NMAS 12.5 mm grading indicating the restricted zone and the control points

pass. Gradings that pass through the restricted zone have often been called humped gradings because of the characteristic hump in the grading curve that passes through the restricted zone. In most cases, a humped grading indicates a mixture that possesses too much fine sand in relation to total sand. Such a grading practically always results in tender mixture behaviour, which manifests itself as a mixture that is difficult to compact during construction and offers reduced resistance to deformation during its service life. Gradings that violate the restricted zone may possess weak aggregate skeletons that depend too much on binder stiffness to achieve mixture shear strength. These mixtures are also very sensitive to bitumen content and can easily become plastic. Superpave recommends, but does not require, mixtures to be graded below the restricted zone. However, in 2002 the Transportation Research Board (TRB) issued its Technical Circular E-C043, which recommended the discontinuation of the restricted zone (TRB, 2002).

The term used to describe the cumulative distribution of aggregate particle sizes is the design aggregate structure. A design aggregate structure (Figure 12.18) that lies between the control points and avoids the restricted zone meets the Superpave grading requirements. Superpave defines five mixture types by their nominal maximum aggregate size (Table 12.8).

12.4.5.6 Summary of Superpave mixture design

This section describes, in broad terms, the Superpave mixture design procedure. For a step by step procedure, reference should be made to AASHTO M 323 (AASHTO, 2013b).

The Superpave asphalt mixture design procedure consists of the following four key steps

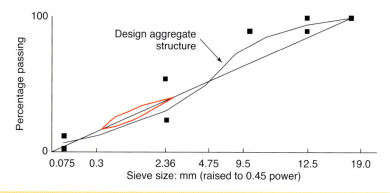

Figure 12.18 Design aggregate structure

- select the materials
- select the design aggregate structure
- select the design binder content
- evaluate the moisture resistance.

Step 1: select the materials

1. Binder requirements
   a. The binder shall be a performance-graded (PG) binder, meeting the requirements of AASHTO M 320 (AASHTO, 2010).
2. Aggregate requirements
   a. Superpave requires that the combined aggregate shall have an NMAS of 4.75–19.0 mm for asphalt surface courses, and not larger than 37.5 mm for asphalt subsurface courses. The NMAS definitions in section 12.3.2.1 of this chapter shall apply.
   b. The control points for the various grading requirements are defined in Table 3 of AASHTO M 323 (AASHTO, 2013b), reproduced here as Table 12.9.
   c. The aggregates are then checked for compliance with the consensus properties as per Table 12.10. Some consensus properties are defined by the location of the layer within the pavement (e.g. the base).

Table 12.8 NMAS values for Superpave mixtures

Superpave mixture designation	Nominal maximum size: mm	Maximum size: mm
37.5 mm	37.5	50
25 mm	25	37.5
19 mm	19	25
12.5 mm	12.5	19
9.5 mm	9.5	12.5

Table 12.9 Aggregate grading control points per AASHTO M 323 (© 2013, the American Association of State Highway and Transportation Officials, Washington, DC. Used by permission)

Sieve size: mm	Nominal maximum aggregate size: mm									
	37.5 mm		25.0 mm		19.0 mm		12.5 mm		9.5 mm	
	Min.	Max.	Min.	Max.	Min.	Max.	Min.	Max.	Min.	Max.
50	100	–	–	–	–	–	–	–	–	–
37.5	90	100	100	–	–	–	–	–	–	–
25.0	–	90	90	100	100	–	–	–	–	–
19.0	–	–	–	90	90	100	100	–	–	–
12.5	–	–	–	–	–	90	90	100	100	–
9.5	–	–	–	–	–	–	–	90	90	100
4.75	–	–	–	–	–	–	–	–	–	90
2.36	15	41	19	45	23	49	28	58	32	67
0.075	0	6	1	7	2	8	2	10	2	10

AASHTO (2013b)

Step 2: select the design aggregate structure

1. The grading is classified (Table 12.11) by using the primary control sieve (PCS) control point as defined in Table 12.9. The combined aggregate grading shall be classified as coarse graded when it passes below the PCS control point, and fine graded for all other combinations.
2. Superpave requires that a minimum of three trial blend gradings are prepared. Each of the blends is plotted on the 0.45 power chart to check whether they conform to the AASHTO M 323 requirements.
3. The compliant trial blends are tested against the requirements for aggregates in section 6 of AASHTO M 323.
4. Replicate asphalt specimens are prepared using the initial trial binder content and the chosen trial aggregate blend grading. The asphalt

Table 12.10 Superpave aggregate consensus property requirements (From AASHTO M 323. © 2013, the American Association of State Highway and Transportation Officials, Washington, DC. Used by permission)

Design traffic: 10^6 ESALs	Coarse aggregate angularity, minimum: % with one fractured face/% with two fractured faces		Uncompacted void content of fine aggregate, minimum: %		Sand equivalent, minimum: %	Flat and elongated, maximum: %
	Thickness ≤100 mm	Thickness >100 mm	Thickness ≤100 mm	Thickness >100 mm		
<0.3	55/–	–/–	–	–	40	–
0.3 to <3	75/–	50/–	40	40	40	
3 to <10	85/80	60/–	45	40	45	
10 to <30	95/90	80/75	45	40	45	10
≥30	100/100	100/100	45	45	50	

Table 12.11 Grading classification defining the PCS concept per AASHTO M 323 (© 2013, the American Association of State Highway and Transportation Officials, Washington, DC. Used by permission)

Nominal maximum aggregate size: mm	Primary control sieve: mm	PCS control point: % passing
37.5	9.5	47
25.0	4.75	40
19.0	4.75	47
12.5	2.36	39
9.5	2.36	47

An aggregate gradation that passes below the PCS control point is classified as 'coarse graded'. A gradation that passes above the PCS control is classified as 'fine graded'.

specimens are prepared using the SGC following the guidelines per AASHTO T 312 (AASHTO, 2014b). The number of gyrations is given in Table 12.12.

5. The mixture volumetrics are then determined for the compacted specimens.

Step 3: select the design binder content
The replicate specimens are prepared at four binder contents using the number of gyrations from Table 12.12

- the estimated design binder content, P_b (design)
- 0.5% below P_b (design)

Table 12.12 Number of gyrations for varying levels of traffic (From AASHTO M R35. © 2014, the American Association of State Highway and Transportation Officials, Washington, DC. Used by permission)

Design ESALs × 10⁶	Compaction parameters			Typical roadway application
	$N_{initial}$	N_{design}	N_{max}	
<0.3	6	50	75	Applications include roadways with very light traffic volumes such as local roads, country roads and city streets where truck traffic is prohibited or at a very minimal level. Traffic on these roadways would be considered local in nature, not regional, intrastate or interstate. Special purpose roadways serving recreational sites or areas may also be applicable to this level
0.3 to <3	7	75	115	Applications include many collector roads or access streets. Medium trafficked city streets and the majority of country roadways may be applicable to this level
3 to <30	8	100	160	Applications include many two-lane, multilane, divided and partially or completely controlled access roadways. Among these are medium to highly trafficked city streets, many state routes, US highways and some rural Interstates
≥30	9	125	205	Applications include the vast majority of the US interstate system, both rural and urban in nature. Special applications such as truck weighing stations or truck climbing lanes on two-lane roadways may also be applicable to this level

- 0.5% above P_b (design)
- 1.0% above P_b (design).

The objective is to determine the binder content that produces a target air void content (V_a) of 4% at N_{design}. For this, the following steps are followed

1. The mixture volumetric parameters (i.e. V_a, VMA and VFB at N_{design}) are calculated using the standard equations as given in AASHTO R 35 (AASHTO, 2014a).
2. For the duplicate specimens, the average V_a, VMA, VFB and relative density at N_{design} against binder content are plotted (Figure 12.19).
3. The binder content to the nearest 0.1% at which the target V_a is equal to 4.0% is determined by graphical or mathematical interpolation.
4. A check is carried out using interpolation where necessary to ensure that volumetric parameters are satisfied at the binder content determined in the previous step.
5. A densification curve is prepared to check its slope and so on, followed by a determination of the corrected specimen relative densities at $N_{initial}$ (percentage of $G_{mm\ initial}$) to confirm that its value satisfies the design requirements in AASHTO M 323 (AASHTO, 2013b) (Table 12.13).

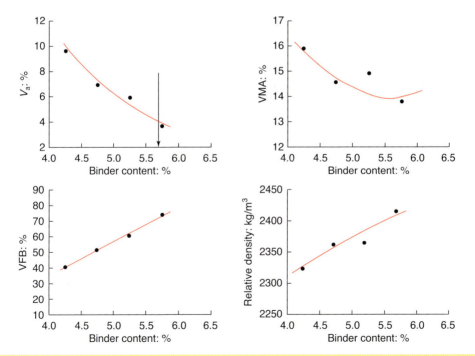

Figure 12.19 Results of plotting average V_a, VMA, VFB and relative density at N_{design} against binder content

Table 12.13 Superpave mixture design requirements (© 2013, the American Association of State Highway and Transportation Officials, Washington, DC. Used by permission)

Design ESALs × 10⁶ [a]	Required relative density: % of the theoretical maximum specific gravity			VMA, minimum: % Nominal maximum aggregate size: mm						VFB range: % [b]	Dust-to-binder ratio range [c]
	$N_{initial}$	N_{design}	N_{max}	37.5	25.0	19.0	12.5	9.5	4.75		
<0.3	≤91.5	96.0	≤98.0	11.0	12.0	13.0	14.0	15.0	16.0	70–80 [d]	0.6–1.2
0.3 to <3	≤90.5	96.0	≤98.0	11.0	12.0	13.0	14.0	15.0	16.0	65–78	0.6–1.2
3 to <10	≤89.0	96.0	≤98.0	11.0	12.0	13.0	14.0	15.0	16.0	65–75 [e]	0.6–1.2
10 to <30	≤89.0	96.0	≤98.0	11.0	12.0	13.0	14.0	15.0	16.0	65–75 [e]	0.6–1.2
≥30	≤89.0	96.0	≤98.0	11.0	12.0	13.0	14.0	15.0	16.0	65–75 [e]	0.6–1.2

[a] Design ESALs are the anticipated project traffic level expected on the design lane over a 20-year period. Regardless of the actual design life of the roadway, determine the design ESALs for 20 years.
[b] For 37.5 mm nominal maximum size mixtures, the specified lower limit of the VFB range shall be 64% for all design traffic levels.
[c] For 4.75 mm nominal maximum size mixtures, the dust-to-binder ratio shall be 0.9–2.0.
[d] For 25.0 mm nominal maximum size mixtures, the specified lower limit of the VFB range shall be 67% for design traffic levels <0.3 million ESALs.
[e] For design traffic levels >3 million ESALs, the specified VFB range for 9.5 mm nominal maximum size mixtures shall be 73–76% and for 4.75 mm nominal maximum size mixtures shall be 75–78%.

6. Eight duplicate specimens are prepared having the design aggregate structure at the design binder content, to confirm that the percentage of $G_{mm\ max}$ satisfies the design requirements in AASHTO M 323.

Step 4: evaluate the moisture resistance
The moisture resistance of the asphalt is evaluated in accordance with AASHTO T 283 (AASHTO, 2007b).

The mixtures are then adjusted after which a report is issued.

12.4.5.7 Use of Superpave
The Superpave system of mixture design uses gradings that pass below the maximum density grading. Research at the National Center for Asphalt Technology and, especially, results of the WesTrack study have shown that some coarse graded Superpave mixtures can exhibit very poor rut resistance (Brown *et al.*, 1998; Kandhal *et al.*, 1998). At the same time, durability problems have been observed in a significant number of pavements constructed using Superpave surface course mixtures. A study in Florida has documented relatively high permeability of Superpave surface course mixtures (Choubane *et al.*, 1998). NCHRP Project 1-42 has been initiated to evaluate the increasing occurrence of top down cracking in asphalt pavements since the implementation of Superpave.

The TRB published a report in 2005 on the status of implementation of Superpave, entitled *Superpave Performance by Design* (TRB, 2005). The report notes that the first ever Superpave pavement was constructed on

8 July 1992, when the Mathy Construction Company of Onalaska, Wisconsin, and the Wisconsin Department of Transportation placed the first 150 m of asphalt conforming to the then prototype Superpave bitumen and mixture specifications. The report summarises that, in 2005, the Superpave binder specifications were adopted by 52 US state transportation organisations, with the mixture design method being adopted by 36 state transportation organisations.

12.4.5.8 Asphalt mixture design quotations

The following are quotes from famous asphalt designers on various aspects of asphalt mixture design and performance.

Leahy and McGennis (1999) provide a rare quote of Marshall's own mixture design philosophy

> The ultimate result in the improvement of aggregate gradation is the reduction of the VMA. VMA should be reduced to the lowest practical degree. This reduction results in a superior pavement structure as well as to reduce the quantity of asphalt required in the mixture. No limits can be established for VMA, for universal application, because of the versatile application of bituminous materials to many types and gradations of aggregates.

White (1985) concludes his paper on the history of Marshall mixture design by stating that

> the two variables that stand out in the design and performance of pavement mixtures are asphalt content and density. In the field it is the highest satisfactory asphalt content at a density achieved under traffic that is significant. In the laboratory an important feature is selecting a compaction procedure that represents traffic induced density and then selection of responsive properties that can be averaged to give an asphalt content that will produce satisfactory performance even if further densification takes place under traffic … producing a mix density comparable to traffic induced densities appears to be the key to Marshall or any other mix design procedure.

Cominsky et al. (1994) note that

> compaction is considered the single most important factor affecting the performance of asphalt pavements.

Hughes (1989) states that

> It is important that the density of laboratory-compacted specimens approximate that obtained in the field in terms of (a) the structure of the mix and (b) the quantity, size, and distribution of the air voids.

Monismith *et al.* (1989) opine that

> Proper selection of the mix components and their relative proportions, that is, asphalt or 'binder' content, requires a knowledge of the significant properties and performance characteristics of asphalt paving mixtures and how they are influenced by the mix components.

References

AASHTO (American Association of State Highway and Transportation Officials) (2007a) T 275-07. Bulk specific gravity (G_{mb}) of compacted hot mix asphalt (HMA) using paraffin-coated specimens. AASHTO, Washington, PA, USA.

AASHTO (2007b) T 283-07. Standard method of test for resistance of compacted asphalt mixtures to moisture-induced damage. AASHTO, Washington, PA, USA.

AASHTO (2009) T 228-09. Standard method of test for specific gravity of semi-solid asphalt materials. AASHTO, Washington, PA, USA.

AASHTO (2010) M 320-10. Standard specification for performance-graded asphalt binder. AASHTO, Washington, PA, USA.

AASHTO (2011a) T 209-11. Standard method of test for theoretical maximum specific gravity and density of hot-mix asphalt (HMA). AASHTO, Washington, PA, USA.

AASHTO (2011b) T 304-11. Standard method of test for uncompacted void content of fine aggregate. AASHTO, Washington, PA, USA.

AASHTO (2013a) T 166-13. Standard method of test for bulk specific gravity of compacted hot-mix asphalt using saturated surface-dry specimens. AASHTO, Washington, PA, USA.

AASHTO (2013b) M 323-13. Standard specification for superpave volumetric mix design. AASHTO, Washington, PA, USA.

AASHTO (2014a) R 35-14. Standard practice for Superpave volumetric design for hot-mix asphalt (HMA). AASHTO, Washington, PA, USA.

AASHTO (2014b) T 312-14. Standard method of test for preparing and determining the density of hot-mix asphalt (HMA) specimens by means of the Superpave gyratory compactor. AASHTO, Washington, PA, USA.

Advanced Asphalt Technologies (2011) *A Manual for Design of Hot Mix Asphalt with Commentary*. Advanced Asphalt Technologies, Sterling, VA, USA, National Cooperative Highway Research Report 673.

Asphalt Institute (1979) *Mix Design Methods for Asphalt Concrete and Other Hot-Mix Types*, 4th edn. Asphalt Institute, Lexington, KY, USA, Manual Series No. 2 (MS-2).

Asphalt Institute (1997) *Mix Design Methods for Asphalt*, 6th edn. Asphalt Institute, Lexington, KY, USA, Manual Series No. 2 (MS-2).

Aschenbrener TB (2002) *Bailey Method for Gradation Selection in Hot-mix Asphalt Mixture Design*. Transportation Research Board, Washington, DC, USA, Circular E-C044.

ASTM (American Society for Testing and Materials) (2009) D70-09e1. Standard test method for density of semi-solid bituminous materials (pycnometer method). ASTM, West Conshohocken, PA, USA.

ASTM (2011) D2041-11. Standard test method for theoretical maximum specific gravity and density of bituminous paving mixtures. ASTM, West Conshohocken, PA, USA.

ASTM (2013) D5821-13. Standard test method for determining the percentage of fractured particles in coarse aggregate. ASTM, West Conshohocken, PA, USA.

Brown R, Michael L, Dukatz E *et al.* (1998) *Performance of Coarse-graded Mixes at WesTrack – Premature Rutting.* Federal Highway Administration, Washington, DC, USA, Final Report, FHWA-RD-99-134.

Caltrans (California Department of Transportation) (1983) *Highway Recollections of F. N. Hveem, Caltrans Committee on Preservation of Historical Heritage.* Caltrans, Sacramento, CA, USA.

Choubane B, Page G and Musselman J (1998) Investigation of water permeability of coarse-graded superpave pavements. *Journal of the Association of Asphalt Paving Technologists* **67**: 254.

Cominsky R, Leahy RB and Harrigan ET (1994) *Level One Mix Design: Materials Selection, Compaction, and Conditioning.* National Research Council, Washington, DC, USA, SHRP-A-408.

Fuller WB and Thompson SE (1907) The laws of proportioning concrete. *Transactions American Society of Civil Engineers* **59(2)**: 67–143.

Goetz WH (1951) Comparison of triaxial and Marshall test results. *Journal of the Association of Asphalt Paving Technologists* **20**: 200–245.

Goetz WH (1989) The evolution of asphalt concrete mix design. In *Asphalt Concrete Mix Design: Development of More Rational Approaches* (Gartner Jr W (ed.)). American Society for Testing and Materials, West Conshohocken, PA, USA, pp. 5–14.

Goetz WH and Wood LE (1960) *Highway Engineering Handbook*, 1st edn. (Woods KB (ed.)). McGraw-Hill, New York, USA, pp. 18-55–18-92.

Goode JF and Lufsey LA (1962) A new graphical chart for evaluating aggregate gradations. *Journal of the Association of Asphalt Paving Technologists* **31**: 176–207.

Hubbard P and Field FC (1935) *Stability and Related Tests for Asphalt Paving Mixtures.* The Asphalt Institute, Lexington, KT, USA, Research Series Number 1.

Hughes CS (1989) *Compaction of Asphalt Pavement.* Transportation Research Board, National Research Council, Washington, DC, USA, NCHRP Synthesis No. 152.

Hveem FN (1942) Use of the centrifuge kerosene equivalent as applied to determine the required oil content for dense graded bituminous mixes. *Journal of the Association of Asphalt Paving Technologists* **13**: 9–41.

Hveem FN (1970) Asphalt pavements from the ancient East to the modern West. *5th Annual Nevada Street and Highway Conference, Reno, NV, USA.*

Kandhal PS and Koehler WS (1985) Marshall mix design method: current practices. *Journal of the Association of Asphalt Paving Technologists* **54**: 284–303.

Kandhal PS and Parker F (1997) *Aggregate Tests Related to Asphalt Concrete Performance in Pavements.* Transportation Research Board, National Research Council, Washington, DC, USA, NCHRP 4-19 Final Report.

Kandhal PS, Foo KY and Mallick RB (1998) *A Critical Review of VMA Requirements in Superpave*, NCAT Report No. 98-1. NCAT, Auburn, AL, USA.

Kennedy TW, Huber GA, Harrigan ET *et al.* (1994) *Superior Performing Asphalt Pavements (SUPERPAVE): The Product of the SHRP Asphalt Research Program.* National Research Council, Washington, DC, USA, SHRP-A-410.

Laboratoire Central des Ponts et Chaussées (2007) *LPC Bituminous Mixtures Design Guide.* LCPC, Paris, France.

Leahy RB and McGennis RB (1999) Asphalt mixes: materials, design and character-ization. *Journal of the Association of Asphalt Paving Technologists* **68A**: 70–127.

McLeod NW (1950) A rational approach to the design of bituminous paving mixtures. *Journal of the Association of Asphalt Paving Technologists* **19**: 82–187.

Meininger RC (1992) *Effects of Aggregates and Mineral Fillers on Asphalt Mixture Performance*. American Society for Testing and Materials, Philadelphia, PA, USA, STP 1147.

Monismith CL, Finn FN and Vallerga BA (1989) A comprehensive asphalt concrete mixture design system. In *Asphalt Concrete Mix Design: Development of More Rational Approaches* (Gartner Jr W (ed.)). American Society for Testing and Materials, Philadelphia, PA, USA, pp. 39–71, STP 1041.

Nijboer LW (1948) *Plasticity as a Factor in the Design of Dense Bituminous Road Carpets*. Elsevier, Amsterdam, The Netherlands.

Prowell BD and Ray Brown E (2006) *Appendixes to NCHRP Report 573: Super-pave Mix Design: Verifying Gyration Levels in the N$_{design}$ Table*. Transportation Research Board, National Research Council, Washington, DC, USA, NCHR Web Only Document 96. See http://onlinepubs.trb.org/onlinepubs/nchrp/nchrp_w96.pdf (accessed 29/10/2014).

Richardson C (1908) *The Modern Asphalt Pavement*. Wiley, New York, USA.

Roberts FL, Kandhal PS, Brown ER, Lee DY and Kennedy TW (1996) *Hot Mix Asphalt Materials, Mixture Design, and Construction*. National Asphalt Pavement Association Education Foundation. Lanham, MD, USA.

Schaub JH and Goetz WH (1961) Strength and volume change characteristics of bituminous mixtures. *Proceedings of the Highway Research Board* **40**: 371–405.

SHRP (Strategic Highway Research Program) (1987) *A Contracting Plan for SHRP Asphalt Research*. National Research Council, Washington, DC, USA.

Smith VR (1949) Triaxial stability method for flexible pavement design. *Journal of the Association of Asphalt Paving Technologists* **18**: 63–94.

Smith VR (1951) *Application of the Triaxial Test to Bituminous Mixtures*. American Society for Testing and Materials, West Conshohocken, PA, USA, STP 106.

Stanton TE and Hveem FN (1934) Role of the laboratory in the preliminary investi-gation and control of materials for low cost bituminous pavements. *Proceedings of the Highway Research Board, Washington, DC, USA*, Part II.

Talbot AN and Richart FE (1923) *The Strength of Concrete and its Relation to the Cement, Aggregate and Water*. University of Illinois Engineering Experiment Station, Urbana-Champaign, IL, USA, Bulletin 137.

TRB (Transportation Research Board) (1986a) *America's Highways: Accelerating the Search for Innovation*. TRB, Washington, DC, USA, Special Report 202.

TRB (1986b) *Strategic Highway Research Program Research Plans*. TRB, Washington, DC, USA.

TRB (2002) *Significance of Restricted Zone in Superpave Aggregate Gradation Specification*. TRB, Washington, DC, USA, Technical Circular E-C043.

TRB (2005) *Superior Performing Asphalt Pavement: Superpave Performance by Design*. TRB, Washington, DC, USA, Final Report of the TRB Superpave Committee.

Vallerga BA (1953) *Notes on the Design Preparation and Performance of Asphaltic Pavements*. Institute of Transportation and Traffic Engineering, University of California, Berkeley, CA, USA.

Vallerga BA and Lovering WR (1985) Evolution of the Hveem stabilometer method of designing asphalt paving mixtures. *Journal of the Association of Asphalt Paving Technologists* **54**: 243–265.

Vavrik WR, Pine WJ, Huber G, Carpenter SH and Bailey R (2001) The Bailey method of gradation evaluation: the influence of aggregate gradation and packing characteristics on voids in the mineral aggregate. *Journal of the Association of Asphalt Paving Technologists* **70**: 132–175.

Von Quintus HL, Scherocman JA, Hughes CS and Kennedy TW (1991) *Asphalt–aggregate Mixture Analysis System: AAMAS*. Transportation Research Board, National Research Council, Washington, DC, USA, NCHRP Report 338.

War Department (1943) *Comparative Laboratory Tests on Rock Asphalts and Hot Mix Asphaltic Concrete Surfacing Materials*. US Engineers Office, Tulsa District, Tulsa, OK, USA.

War Department (1944) *A Stability Method and Apparatus for the Design and Control of Asphalt Paving Mixtures in the Field. Investigation of Stability of Asphalt Paving Mixtures*. US Waterways Experiment Station, Vicksburg, MS, USA, Interim Report No. 1.

War Department (1948) *Investigation of the Design and Control of Asphalt Paving Mixtures*. US Waterways Experiment Station, Vicksburg, MS, USA, Technical Memorandum 3-254, vol. 1.

Weymouth CAG (1938) A study of fine aggregate in freshly mixed mortars and concretes. *Proceedings of ASTM International, West Conshohocken, PA, USA*.

White TD (1985) Marshall procedures for design and quality control of asphalt mixtures. *Journal of the Association of Asphalt Paving Technologists* **54**: 265–284.

Chapter 13

Design of flexible pavements

David Strickland
Technical Business Development Manager, Shell International Petroleum Co. Ltd.
UK

Road pavements are often placed into three main categories

- flexible pavements
- rigid pavements
- composite pavements.

Flexible pavements are constructed with several layers of asphalt. The description arises because, when a heavy commercial vehicle traverses the pavement, it flexes vertically by an imperceptible amount. In the UK, the maximum gross vehicle weight of a commercial vehicle is 44 tonnes, with a maximum axle load of approximately 8 tonnes. However, these values vary widely throughout the world. Rigid pavements are constructed entirely with a cementitious material. Composite pavements consist of a combination of asphalt and cementitious layers.

Flexible pavements are, by far, the most common pavement type in the UK. There is a wide range of asphalts available, and appropriate materials can be chosen based on a range of factors, including the nature of the traffic to be carried and the availability of materials locally.

Although this chapter discusses design practices for flexible pavements as adopted in the UK, with a short explanation of how pavement design is undertaken in the USA, the principles have universal application.

13.1 Introduction
In the UK, there are three generic types of pavement

- flexible
- rigid (continuous)
- rigid (jointed).

The Shell Bitumen Handbook, Sixth Edition
ISBN 978-0-7277-5837-8
Shell International Petroleum Company Ltd: All rights reserved
http://dx.doi.org/10.1680/tsbh.58378.347

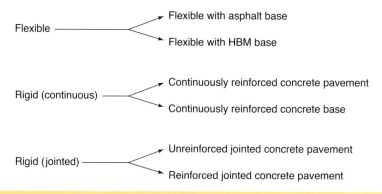

Figure 13.1 Pavement options

There are two types within each of these categories, and these are illustrated in Figure 13.1. As can be seen, there are two types of flexible pavement. 'Flexible with HBM base' is constructed entirely in asphalt, except for the lower part of the base (called the 'lower base'), which is a hydraulically bound mixture (HBM). The upper base, the binder course and surface course are all asphalts. A wide range of materials is acceptable as HBMs with many being cementitious mixtures. Notwithstanding, a cross section through a road having the most common type of pavement is that shown in Figure 13.2.

The functions of each of the layers shown in Figure 13.2 are as follows.

The foundation is the platform on which the more expensive layers are placed. On a unitary basis, the cost of each layer is usually higher than its predecessor. The foundation carries the load bearing layers of the pavement.

The base is the main structural layer in the pavement, and its main function is to distribute the stresses generated within it efficiently to the foundation. It must be able to sustain the stresses and strains generated within itself without excessive or rapid deterioration of any kind.

The binder course is necessary to allow asphalt contractors to achieve the very high standards of surface regularity required on modern road pavements. (Surface regularity is a measure of rideability. In other words, an assessment of the smoothness of the journey experienced by persons travelling in a vehicle across a particular pavement. It is a very important feature of a finished pavement.) Asphalts are laid by specialist machines called pavers, and one of the results of each pass of a paver is the reduction of any irregularities in the upper surface of the material being covered. The binder course may well also provide a waterproof layer to protect the base from the ingress of water. Water and pavements simply do not mix. Pavements must be designed such that water is prevented from entering the pavement layers

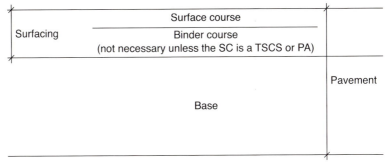

Surface course, binder course and base are asphalts

| Surfacing | Surface course |
| | Binder course
(not necessary unless the SC is a TSCS or PA) |

Base

Pavement

Foundation
(capping or subbase or subbase on capping)

Existing subsoil/fill – subgrade

Figure 13.2 A road constructed with a 'flexible with asphalt base' pavement: SC, surface course; TSCS, thin surface course system; PA, porous asphalt

or indeed the foundation (unless it is a pavement specifically designed to allow water to pass through without being trapped, such as is the case with a porous asphalt surface course and specialist SuDS (sustainable drainage systems) or reservoir pavements). Many thin surface course systems (TSCS) (defined in Clause 942 of *Specification for Highway Works* (Highways Agency *et al.*, 2008)), often incorrectly described as 'stone mastic asphalts', are permeable, and water simply flows through them. If the water reaches the base, it may strip the bitumen from the aggregate in the base asphalt and lead to premature failure of the weakened asphalt. The current UK pavement design standard HD 26 allows, per Note 8 to Figure 2.1 (Highways Agency *et al.*, 2006a), omission of the binder course if the surface course is a chipped hot rolled asphalt surface course. Many engineers wisely insist on the inclusion of a binder course to give asphalt contractors the opportunity to achieve the very tight surface regularity tolerance. Although binder courses are often more expensive than their base versions because they have some 0.6% more bitumen than the base material (e.g. an HDM50 binder course

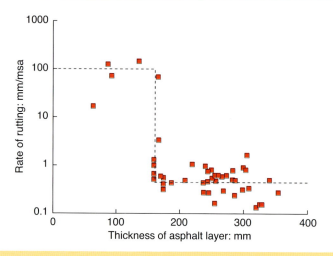

Figure 13.3 Rate of rutting versus thickness

has around 0.6% more than an HDM50 base), many engineers feel that a binder course is essential in an asphalt pavement. (Indeed, most binder course materials (DBM125, HDM50 and DBM50) have an additional 0.6% or so more bitumen than the base equivalent, the exceptions being EME2 and HRA50, which have the same bitumen content regardless of whether they are used in a base or a binder course.)

The surface course has to provide a smooth skid resistant running plane having good surface regularity and adequate resistance to the onset of defects.

The most common structural defects (a structural defect is one that requires replacement of the surface course, binder course and base to correct the defect) are deformation and cracking. Thicker pavements, according to Transport Research Laboratory (TRL) Report 250, are those in which the depth of asphalt is at least 180 mm – see Figure 13.3, which is based on Figure C2 in the TRL report (Nunn *et al.*, 1997). This is a very important figure. It explains the basis of long-life pavements (i.e. pavements that have an indeterminate life). Examination of Figure 13.3 indicates that if the thickness is above about 180 mm, then the effect of loading in producing deformation is reduced by a factor of about 200, given that the *y* axis is logarithmic.

How long is indeterminate? Who knows? It is expected to be perhaps 80, 100 or 120 years, maybe longer. It is simply not known. However, what TRL Report 250 established and states, in paragraph 1 of section 13 (Nunn *et al.*, 1997), was that

A well constructed, flexible pavement that is built above a defined threshold strength will have a very long structural service life provided that distress, in the form of cracks and ruts appearing at the surface, is detected and remedied before it begins to affect the structural integrity of the road.

Note that prompt action to deal with cracks and ruts (the two most common reasons for structural failure) is vital if long life is to be achieved.

From Figure 13.3 it can be seen that the minimum asphalt thickness for long life is 180 mm. Different authorities in the UK, and indeed all over the world, specify different thicknesses for a road to be considered a long-life pavement ('perpetual pavement' in the USA).

13.2 Background to the design of pavements

The aim of pavement design is to produce a structure that will distribute traffic loads efficiently while minimising the whole life cost of the pavement. The term 'whole life' when applied to a pavement refers to all the costs incurred in connection with a pavement and throughout its useful life (i.e. the period during which it will be 'in service'). Thus, it includes the works costs (construction, maintenance and residual value) and user costs (traffic delays, accidents at roadworks, skidding accidents, fuel consumption/tyre wear and residual allowance).

Designing a pavement is essentially a structural evaluation process, and is needed to ensure that traffic loads are distributed such that the stresses and strains developed at all levels in the pavement and the subgrade are within the capabilities of the materials at those levels. It involves the selection of materials for the different layers and the calculation of the required thicknesses. The load-carrying capacity of a pavement is a function of both the thickness of the material and its stiffness. Consequently, the mechanical properties of the materials that constitute each of the layers in a pavement are important in designing the structure. As moisture may affect the subgrade and the subbase (and also the base if it is unbound), and temperature affects the bitumen bound layers, it is essential that the design process takes account of the prevailing climatic conditions. As has been suggested in other chapters, pavements and water simply do not mix. The drainage is a vital element of the system, and must be designed to ensure that the pavement and the foundation remain dry.

Figure 13.4 illustrates the two classic modes of failure caused by trafficking of a pavement. It assumes two failure modes

- fatigue cracking at the underside of the base
- cumulative deformation.

It is the modes illustrated in Figure 13.4 that pavement engineers have traditionally sought to hold to acceptable limits within the design life. The concept of a design life was regarded as being particularly important for pavements,

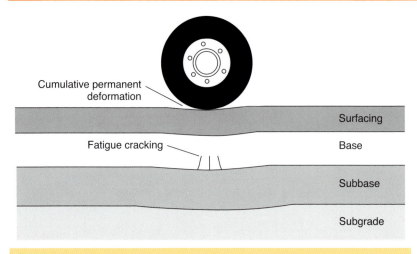

Figure 13.4 Traditional pavement failure modes

because it is known that they do not fail suddenly but gradually deteriorate over a period of time. However, TRL Report 250 (Nunn *et al.*, 1997) demonstrated that these failure modes did not apply to thicker pavements. It was explained above that thicker pavements are those of at least 180 mm in thickness. TRL Report 250 states that

> No evidence of conventional roadbase [now 'base'] failure or structural deformation was found in well constructed pavements. The observed rutting and cracking were found to originate in the surface layers and, as long as it is treated in a timely manner, it is unlikely to lead to structural deterioration.

The findings of TRL Report 250 were incorporated in the UK's pavement design standard HD 26/01 (Highways Agency *et al.*, 2001) and its current successor, HD 26/06 (Highways Agency *et al.*, 2006a). The latter states in paragraph 2.15 that

> Monitoring the performance of all types of flexible pavements that are heavily trafficked has indicated that deterioration, in the form of cracking or deformation, is far more likely to be found in the surfacing, rather than deeper in the structure. Generally for 'long life' it is not necessary to increase the pavement thickness beyond that required for 80 msa (millions of standard axles), provided that surface deterioration is treated before it begins to affect the structural integrity of the road.

In this statement, 'surfacing' means the combination of the surface course and the binder course. The number of commercial vehicles carried by a pavement is usually measured in terms of millions of standard axles (msa). In the UK, a standard axle is one imposing, exerting or applying a force of 80 kN

(Highways Agency *et al.*, 2006b). Other countries have different values for a standard axle.

13.3 The importance of stiffness

Stiffness can be defined as a measure of the load spreading ability of a material, and applies whether the material is granular, asphaltic or cementitious. It is a fundamental and important parameter that must be fully appreciated by the pavement designer. It can be readily understood by considering the situations depicted in Figure 13.5.

In Figure 13.5(a), material 1 has a lower stiffness value than material 2. If both are laid at the same thickness, then the pressure resulting from a particular load will be higher at the underside of the layer of material 1 compared with the pressure resulting from the same load at the underside of material 2. This, in turn, subjects the layer underneath to a higher value of stress. If this value exceeds that which can be tolerated by the underlying layer, then it will fail. In order to get the same pressure on the underside, it will be necessary to lay a greater thickness of material 2, which has a lower stiffness. Compare that situation with that shown in Figure 13.5(b), where a thicker layer of the material having a lower stiffness is required to produce an acceptable value of the resultant pressure at the underside.

This explains why different pavement materials require different thicknesses for a particular design. An example of such a design is shown in Figure 13.6.

Figure 13.6 illustrates what happens when contract tenders for pavements are issued. Usually, the client will offer tendering contractors the opportunity to price any of a number of options. There may be four or more options, and not all of them will necessarily be flexible pavements. All the options are deemed to be structurally equivalent. Figure 13.6 shows two options. One is based on the binder course and base being EME2 while the other is based on the binder course and base being HDM50. The materials used in these example pavements are as follows.

- TSCS is a thin surface course system, which is a proprietary surface course conforming to Clause 942 of the UK *Specification for Highway Works* (Highways Agency *et al.*, 2008).
- EME2 is a relatively high bitumen content (usually about 6% by weight of the total mixture) asphalt based on a French specification. EME is an abbreviation of 'enrobé à module élevé', which translates into English as 'high modulus base'. It uses an unmodified bitumen with a penetration of 15–20.
- HDM50 is 'heavy duty macadam' asphalt concrete, and is probably the most common UK binder course and base asphalt. It has a mid-

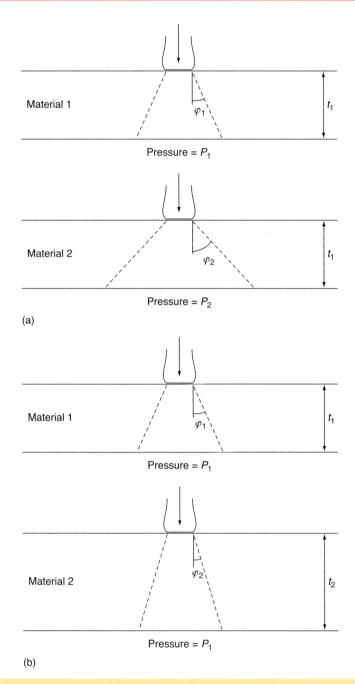

Figure 13.5 Stiffness of pavement layers: (a) equal thicknesses, different material stiffnesses, different resultant pressures; (b) different thicknesses, different material stiffnesses, equal resultant pressures. (Courtesy of Dr Robert N Hunter)

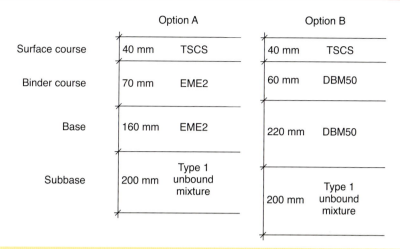

Figure 13.6 Comparison of two fully flexible pavements, one constructed using EME2 and the other constructed with DBM50

range bitumen content of typically 4% in HDM50 base and 4.5% in HDM50 binder courses, and is usually made with crushed rock. It uses an unmodified bitumen with a penetration of 40–60, and following modern convention should be called HDM40/60, but tradition remains, and it is still called HDM50.

■ Type 1 unbound mixture is a graded crushed rock (it can also be crushed slag, crushed concrete, recycled aggregates or well burnt non-plastic shale) previously known as type 1 subbase (subbase was previously called 'sub-base') complying with Clause 803 of the *Specification for Highway Works* (Highways Agency et al., 2008).

The base is 60 mm thicker in the HDM50 variant compared with its value in the EME2 pavement. EME2 has a higher stiffness than HDM50. A pressure is applied over a relatively small area by heavy commercial vehicles running over the surface courses, and Clause 2.3.2 of TRL Report 615 (Nunn, 2004) states that the

> UK design method uses a multi-layer, linear elastic response model to calculate the critical stresses or strains induced under a single standard wheel load (40 kN) that is represented by a circular patch (0.151 m radius) with a uniform vertical stress.

The additional thickness in the HDM50 pavement is necessary because this pressure reduces as it travels down through the pavement. Because EME2 has a higher stiffness than HDM50, the pressure reduces more quickly as it passes through the EME2 pavement. Accordingly, a thinner pavement is

Table 13.1 Presumptive values for elastic characterisation of asphalt mixtures

Asphalt mixture type	Binder type	Volume of binder: %	Asphalt modulus at heavy vehicle operating speed: MPa			
			10 km/h	30 km/h	50 km/h	80 km/h
OG10	A5S	N/A	800	800	800	800
OG14	A5S	N/A	800	800	800	800
SM14	A5S	13.0	1000[a] (600)	1000[a] (900)	1100	1300
DG10	C320	11.5	1000[a] (900)	1300	1600	1900
DG10	A5S	11.5	1000[a] (600)	1000[a] (800)	1000	1200

From State of Queensland (2013), Table Q6.5
[a] Indicated values have been limited to a value of 1000 MPa

needed with an asphalt having a higher stiffness to give structural equivalence.

It will be recalled from other chapters that the value of stiffness is affected by temperature and the period of loading. UK design methods assume a standard frequency in service (i.e. a standard vehicle speed). Thus, UK roads carrying slow-moving traffic will be under-designed because the stiffness decreases as the speed decreases. The publication *The London Bus Lane and Bus Stop Construction Guidance* states in section 2.2 that 'where traffic is slow moving and/or channelised this traffic load has to be increased by a factor of 3' (London Technical Advisors Group, 2013). Australian specifications take account of vehicle speed, and Table 13.1 is an extract from Table Q6.5 of *Pavement Design Supplement* issued by the State of Queensland (2013).

13.4 The elements of a flexible pavement

As was shown in Figure 13.2, a road consists of three elements

- the subgrade
- the foundation
- the pavement.

13.4.1 The subgrade

The subgrade is the material below the foundation, and will either be existing soil or imported fill. In areas of cut, it will be the existing soil unless the existing soil is very weak and has to be improved by some means: see Clauses 5.16–5.21 in Interim Advice Note (IAN) 73 (Highways Agency *et al.*, 2009) for advice on the options available to do this. In areas of fill, the subgrade will be imported fill made of material that has characteristics suitable for its intended role, laid and compacted in layers. It may be very tempting to think that the characteristics of the subgrade will have little effect on the useful lifespan of a pavement. However, nothing could be further from

Table 13.2 Comparison of rates of rutting	
Base material	Life for a 10 mm rut: msa
HRA or DBM where the subgrade CBR <5%	17
HRA or DBM where the subgrade CBR >5%	28

DBM, dense-bitumen macadam; HRA, hot rolled asphalt
From Nunn *et al.* (1997), Table A1

the truth, as is demonstrated by Table 13.2. The key role of the subgrade material can be demonstrated by considering Table 13.2, which has been taken from Table A1 in TRL Report 250 (Nunn *et al.*, 1997).

It has long been the case that maintenance engineers will consider that a carriageway begins to warrant consideration for replacement or renewal when deformation ('deformation' and 'rutting' are interchangeable terms) reaches 10 mm (what was sometimes described as a 'critical condition'). This was established following a study published in 1972 (see Table 1 in TRL Report LR 375) (Lister, 1972).

Table 13.2 relates to pavements where the base is either hot rolled asphalt or asphalt concrete (DBM; dense bitumen macadam), which will have been the vast majority of pavements at the time when the investigation associated with TRL Report 250 was undertaken. What these data suggest is that if the California bearing ratio (CBR; a measure of the strength of a soil, a test devised by US engineers and used all over the world for roads and airfield construction) is less than 5%, then the pavement will carry some 17 msa before a 10 mm depth of rut is formed (BSI, 1990a, 1990b). However, if the CBR exceeds 5%, then the pavement will carry some 28 msa before a 10 mm depth of rut is formed. Thus, the increase in the CBR results in the pavement carrying an additional 65% or so of traffic before a 10 mm rut is generated. This is stark proof that the subgrade will have a significant effect on the service life of the pavement.

This book does not address subgrades to any extent. Suffice to say that great care needs to be taken in constructing the subgrade and/or preparing it to receive the foundation.

13.4.2 The foundation

The foundation will also play a very important role in determining the performance of the pavement, including the length of its service life.

The foundation has to perform in two different sets of circumstances.

■ During construction it has to protect the subgrade and withstand the relatively high stresses generated within it by construction traffic (i.e. excavators, lorries, pavers, etc.). Although the number of stress repetitions will be lower than when it is in service and the line travelled by plant is not as channelised as that taken by traffic when the pavement is in service, the intensity of individual loads will be much higher because site traffic travels directly on the foundation. In addition, it may well be exposed to adverse weather conditions without the protection of the overlying pavement layers.

■ In service, traffic, in terms of the line travelled, is much more disciplined. The load intensity will be reduced because of the effect of the intervening pavement. However, the number of stress repetitions will be very much higher than was the case during construction.

Like the subgrade, constructing the foundation requires great care. The material chosen must be appropriate for the role it has to fulfil. In addition, it has to be placed in strict conformance with the specification if it is to perform adequately (layer thickness, compaction passes, etc.).

As can be seen from Figure 13.2, the foundation can now be formed of a subbase on capping, entirely capping or entirely subbase.

Materials used in UK foundations range from low grade fill to a number of hydraulically bound mixtures that may contain cement. However, the material most commonly used in the UK is type 1 unbound mixture (formerly type 1 subbase), a graded crushed rock.

This book does not consider foundations in any detail, but their importance cannot be overstated. Advice on foundations can be found in IAN 73 (Highways Agency *et al.*, 2009) and Chapter 32 of the *ICE Manual of Highway Design and Management* (Payne, 2011).

13.4.3 The pavement
Fully flexible pavements invariably consist of the base, on top of which is the binder course, and above which is the surface course. Each of these is discussed below.

13.4.3.1 Bases
The base is the main structural element in the pavement, although the binder course and surface course contribute to the strength of the pavement. The function of the base is to distribute the imposed loading so that the underlying materials are not overstressed. It must resist deformation and cracking caused by fatigue through repeated loading. It must also be capable of withstanding stresses induced by temperature gradients through the structure. The base in a flexible pavement is usually a dense asphalt.

In the UK, the options for base asphalts are

■ EME2
■ DBM50 or HDM50

- HRA50
- DBM125.

In pavements carrying heavy traffic, the base thickness will typically be 200 mm or more. Layers of this thickness would normally be laid in two layers, with the first layer being described as the 'lower base' and the second layer being described as the 'upper base'.

13.4.3.2 Binder courses

The combination of the binder course and the surface course is described as the surfacing (see Figure 2.1 in HD 23 (Highways Agency *et al.*, 1999)). The purpose of the binder course is to allow surfacing contractors to achieve the very high standards of level control (often described in contracts as the 'surface regularity') required by current highway maintenance and construction contracts. It may also act as an impermeable layer protecting the base from the ingress of water.

Pavement layers are usually placed by a specialised piece of equipment called a 'paver'. The nature of these machines is such that, as each layer is placed, they tend to reduce any irregularities on the top of the layer being laid. A major factor in determining the quality of the ride experienced by persons travelling across a pavement is its 'surface regularity'. The smoother the ride, the better the surface regularity. It is for this reason that surface regularity requirements are often very tight: for example, in the UK the surface course more often than not must be within ± 6 mm of the designed finished level at the top of the pavement (Highways Agency *et al.*, 2009, Table 7.1). Indeed, some contracts require compliance to $+6/-0$ mm.

The thickness of the binder course is often around 60 mm. The thickness adopted should be the minimum because, unless it is EME2 or HRA50, the binder course variant will have around 0.6% more bitumen than the equivalent base material.

In the UK, the options for binder course asphalts are

- EME2
- DBM50 or HDM50
- HRA50
- DBM125
- SMA.

The difference between the options available for bases and binder courses is that SMA (stone mastic asphalt – a material originally developed in Germany) can be used. The reason for this is that SMA can be laid as thin as 30 mm, while the other options cannot be laid much thinner than 50 mm. (The minimum thickness of an asphalt layer is governed by a rule of thumb that it should be at least 2.5–3 times the nominal stone size, with the nominal stone size being the maximum aggregate size in the mixture – minimum layer thicknesses are often specified by reference to Tables 1a and 1b of BS 594987 (BSI, 2010).)

13.4.3.3 Surface courses

The layer on which traffic runs is termed the surface course. It is this layer, and this layer only, that is visible to the road user. The surface course has to meet a formidable list of requirements. It must

- resist deformation by traffic
- be durable, resisting the effects of weather, abrasion by traffic and fatigue
- provide an acceptable level of skid resistance
- not result in unacceptable noise levels being generated
- provide a surface of acceptable riding quality
- except in the case of porous asphalt, preferably be impermeable to the passage of water through the layer, thus protecting the lower layers of the pavement
- contribute to the strength of the pavement structure
- not cause unacceptable levels of spray to be produced.

Surface courses on heavily trafficked UK roads (i.e. those carrying 80 msa or more in their design lives) are usually either a TSCS, as described above, or a chipped hot rolled asphalt surface course (complying with Clause 911 or 943 of the UK's *Specification for Highway Works* (Highways Agency *et al.*, 2008). (TSCSs should be described as such. In the UK, they are often described as SMAs, but that is incorrect because they are not stone mastic asphalts.)

The pavement design standard mostly commonly used in the UK (HD 26) makes a binder course optional where the surface course is a chipped hot rolled asphalt surface course. However, this standard states that a binder course is always required when the surface course is a TSCS, many of which are permeable. The reason for this approach is to ensure that pavement bases are protected from the ingress of water through the upper layers of the pavement. Notwithstanding, in order to give surfacing contractors the best chance of achieving surface regularity compliance, many engineers advocate the use of a binder course in all asphalt pavements. This is a wise approach. However, it is a more expensive option, and the client's approval for their inclusion should be canvassed when necessary.

13.5 Stages involved in pavement design

The stages involved in designing a UK pavement are shown in Figure 13.7, and are based on the use of HD 24 (Highways Agency *et al.*, 2006b),

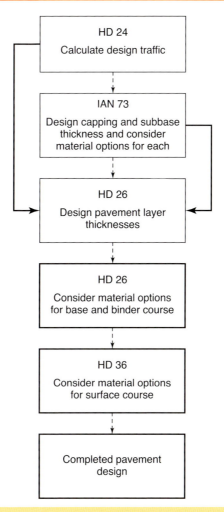

Figure 13.7 Pavement design procedure. (Courtesy of Dr Robert N Hunter)

IAN 73 (Highways Agency *et al.*, 2009), HD 26 (Highways Agency *et al.*, 2006a) and HD 36 (Highways Agency *et al.*, 2006c), which are all standards published by the UK government. However, the approach depicted in Figure 13.7 could well be applied in any country by application of the appropriate applicable standard. The steps in carrying out this process are discussed below.

13.5.1 Design traffic

The first stage in most structural designs is to estimate the loading that the structure is expected to carry, and that approach is no different in the case of the design of pavements. The load carried by a UK pavement is described as the 'design traffic'.

The design traffic is the total loading that the structure (i.e. the pavement) is designed to carry throughout its service life. In the UK, calculation of the design traffic is usually undertaken on the basis of the method outlined in HD 24 (Highways Agency *et al.*, 2006b). The service life of the pavement is 40 years on the basis that this period has generally proved to be the most economic, particularly where traffic flow is high, all as per paragraph 2.14 of HD 24.

Loads are applied to the pavement through contact between vehicle tyres and the surface course. Obviously, the degree and nature of the traffic loading is one of the major factors affecting the design and performance of pavements. Both the magnitude and the number of loadings contribute to the overall damage to the pavement. The tyre pressure primarily affects the stresses (and strains) developed at the surface and within the upper layers of the pavement. (In road pavement design, a typical commercial vehicle tyre pressure would be between 5 and 9 bar (0.5 and 0.9 MPa), whereas aircraft tyres inflict pressures up to 30 bar (3.0 MPa). These values have a significant influence on the design of the materials used in road and airfield pavements.)

Although a great deal of publicity has been given to the physical size and gross weight of commercial vehicles, it is the axle load that is critical in pavement design and performance. Heavier vehicles are usually carried on a larger number of axles, thereby maintaining or reducing the axle load.

Most countries limit the axle loadings that can be carried on their carriageways. However, these limits are difficult to enforce, and there is the possibility that some lorries carry a higher payload than their design or the applicable regulatory framework permits.

Some countries base their pavement design on the total number of vehicles of any size carried by the pavement. However, that approach is incorrect. The loading exerted by cars and lighter goods vehicles does not contribute to any structural deterioration in a pavement. Only commercial vehicles, which in the UK for pavement design purposes are defined as vehicles having a gross vehicle weight in excess of 3.5 tonnes, cause any structural deterioration.

Normal traffic on conventional roads is mixed in composition and the magnitudes of axle loads, and it is therefore necessary for design purposes to simplify the real situation by converting actual axle loads to an 'equivalent' loading system. This idea originated with the American Association of State Highway Officials (AASHO) road test in the USA, which was undertaken in 1958 (Highway Research Board, 1962; Liddle, 1962). The mixed axle load spectrum is converted to an equivalent number of 'standard axles', an example being the value of 80 kN adopted in the UK (Road Research

Laboratory, 1970). This is based on a concept of equivalent damage to the pavement. Thus, on an axle of a heavy lorry having a single tyre at each end, the very common super single tyre, the load on each tyre will be 40 kN (i.e. approximately 4 tonnes). The loading exerted (and thus the damage) for each vehicle can then be computed in terms of a number of equivalent standard axles by using the 'fourth power law' given below

wear/axle $\propto$ axle load^4

It has been shown (Currer and Connor, 1979) that on weak pavements the exponential power can be as high as 6 if trafficked by heavy axle loads at or above the current legal limit. However, if the UK is used as an example again, the value of 4 is accepted in the national traffic assessment standard HD 24 (Highways Agency *et al.*, 2006b), and has been so for many years. This approach has proved to lead to the construction of highways that have successfully withstood the enormous growth in traffic in the UK.

Thus, using the above relationship, under normal conditions it takes the application of 16 40 kN axle loads to cause the same damage as a single application of an 80 kN axle load. Furthermore, one application of a 160 kN axle load induces 16 times the damage of a standard 80 kN axle load.

13.5.2 Foundation
This is an extremely important element of pavement design, and must be undertaken by experienced, qualified engineers. However, it is not discussed in any detail in this book (see section 13.4.2).

13.5.3 Pavement design
Once the design traffic has been ascertained, the pavement thickness can be designed. In the UK this is done by reference to Figure 2.1 of HD 26 (Highways Agency *et al.*, 2006a). Flexible pavement design in HD 26 is based on TRL Report 615 (Nunn, 2004).

Figure 13.8 is an excerpt from Figure 2.1 of HD 26. The procedure for designing the pavement is as follows

1 on the upper x axis, select the calculated design traffic
2 drop vertically to the point that cuts the applicable Foundation Class
3 turn right until the EME2 line is intercepted
4 drop vertically down to the lower x axis
5 read off the applicable 'total asphalt thickness' for EME2
6 repeat step 3 above, but turn right until the DBM50/HDM50 line is intercepted
7 drop vertically down to the lower x axis
8 read off the applicable 'total asphalt thickness' for DBM50/HDM50

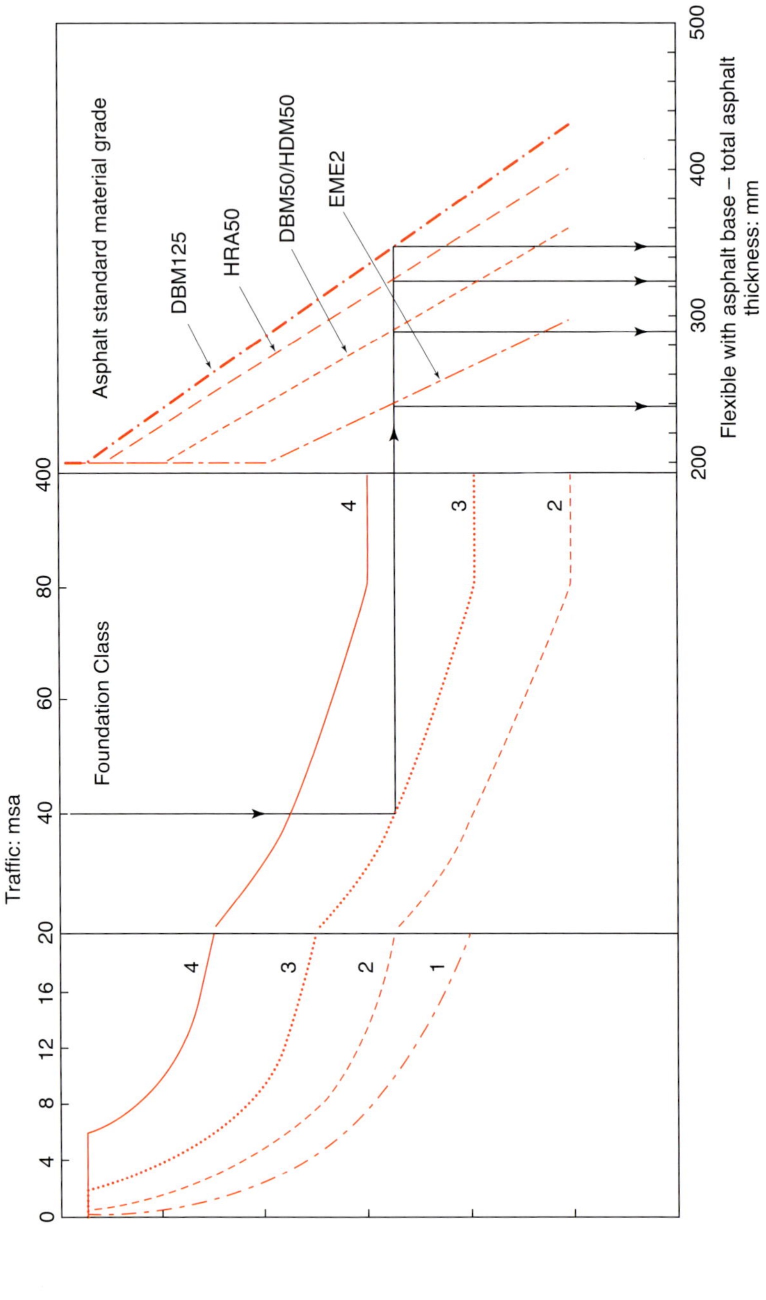

Figure 13.8 Design thickness for flexible pavements

9 repeat step 6 above, but turn right until the HRA50 line is intercepted
10 drop vertically down to the lower x axis
11 read off the applicable 'total asphalt thickness' for HRA50
12 repeat step 9 above, but turn right until the DBM125 line is intercepted
13 drop vertically down to the lower x axis
14 read off the applicable 'total asphalt thickness' for DBM125
15 repeat for any other applicable foundation classes.

An example is a road with a design traffic of 40 msa on a class 3 foundation

1 go to the upper x axis at the point where the 'traffic' (it should actually be labelled 'design traffic') is 40 msa
2 drop down to where the Foundation Class 2 line is encountered
3 turn right to the point where the EME2 line is intercepted
4 drop down to the lower x axis, and read off the 'total asphalt thickness' which is around 240 mm
5 return to the point where the EME2 line was crossed, and move horizontally right to the point where the DBM50/HDM50 line is intercepted
6 drop down to the lower x axis, and read off the 'total asphalt thickness', which is just below 300 mm and would be rounded up to 300 mm
7 repeat the process for HRA50 and DBM125.

The reading for EME2 is 240 mm which is the 'total asphalt thickness'. The EME2 pavement option will then be 40 mm of surface course, 70 mm of AC 14 EME2 binder course (this minimum thickness value of EME2 is stated in Table 1A of BS 594987 (BSI, 2010)) and 130 mm of AC 20 EME2 base, making a 'total asphalt thickness' of 240 mm. Table 13.3 shows the full set of results for the entire range of asphalt options.

Figure 2.1 in HD 26 (Highways Agency *et al.*, 2006a) has, attached to it, a series of notes relating to the materials featuring in Figure 2.1. These notes must always be considered as part of the design process.

Inclusion of the above thicknesses in a tender allows the pavement alternatives to be put out to tender as part of a submission for highway maintenance

Table 13.3 Example of pavement design results

Asphalt	EME2	DBM50/HDM50	HRA50	DBM125
Total asphalt thickness: mm	240	300	320	350
Surface course thickness: mm	40	40	40	40
Binder course thickness: mm	70	60	60	60
Base thickness: mm	130	200	220	250

or highway construction or as an element of contract for other work types.

It is worth considering some aspects of Figure 2.1 and the notes attached thereto

- the thickness is to be rounded up to the next 10 mm
- where the asphalt design thickness is 300 mm or less, the material is to be laid with no negative tolerance
- the figure assumes that the binder course and base will be constructed of the same asphalt (e.g. HDM50)
- once the design traffic reaches a value of 80 msa, then the same design applies up to a design traffic of 400 msa
- for low traffic values up to 5 msa, the thickness for different asphalts is the same
- the minimum design traffic is 1 msa.

Finally, it is worth noting the following about the asphalt options available.

- Although permitted, it may not be wise to use a DBM125 asphalt, as the penetration, being somewhere between 100 and 150, is significantly softer than the 40/60 bitumen used in DBM50/HDM50 and HRA50 and the 15/20 penetration bitumen in EME2.
- In roads carrying heavy traffic, it may be wise to be very wary of specifying HRA50 as a binder course because it is prone to deforming far more rapidly than the other options.

In summary, the UK pavement design method is a mechanistic–empirical method, and, therefore, that fact alone means that the method can only ever be an approximation to the thicknesses necessary to carry the applied loading. This notwithstanding, it has served the UK well, having provided roads that have carried very high traffic levels for many years.

13.6 Pavement designs in the USA

The most widely used empirical pavement design methods for roadway pavements were developed from data collected in the AASHO road test, which took place over a 25 month period starting in October 1958 near Ottawa, Illinois. Among the objectives of the road test were the determination of relationships between axle load applications and the performance of different types and thicknesses of pavement layer materials; and the organisation and storage of data that could be further analysed were essential for the development of the *AASHTO Interim Guide for Design of Pavement Structures* (AASHTO, 1972) and the revisions and versions that followed. The most recent version was published in 1993, and AASHTO has developed software for using the design procedure (AASHTO, 1993).

The basic pavement performance models developed from road test data are used in the 1993 guide. To design a flexible pavement structure using the AASHTO method, the designer must characterise

- the pavement performance (serviceability change over time)
- the traffic (characterised as equivalent single axle loads)
- the roadbed soil (subgrade) resilient modulus, M_R
- the pavement layer materials (layer coefficients, a_i)
- the drainage
- reliability.

The AASHTO flexible pavement design procedure ultimately results in a structural number for the pavement that provides a pavement design adequate for the assumed conditions. The structural number for a specific pavement cross-section is determined using the following equation

$$SN = a_1 D_1 + a_2 D_2 m_2 + a_3 D_3 m_3$$

where a_i is the layer coefficient for layer 1, 2 or 3 (dimensionless); D_i is the thickness for layer 1, 2 or 3 (inches); and m_2 and m_3 are the drainage coefficients for the base and subbase layers, respectively (dimensionless).

Commonly used values for layer coefficients are shown in Table 13.4.

Many combinations of different pavement layer materials and thicknesses can result in the same structural number for a pavement section that would probably perform very differently. Consequently, pavement designers using the 1993 AASHTO guide must apply engineering judgement to develop appropriate pavement cross sections to consider for a particular project. One way to evaluate and compare different pavement sections is to use an analytical or mechanistic–empirical procedure.

The AASHTO *Mechanistic–Empirical Pavement Design Guide* (MEPDG) was developed to overcome many of the shortcomings of the empirical method (TRB, 2014). This procedure requires the use of the computer program AASHTOWare Pavement ME to analyse a pavement structure and predict its performance from the input variables.

Table 13.4 Commonly used values for layer coefficients

Layer material	Layer coefficient range	Typical value assumed
Asphalt surface courses, less than 4 in. (100 mm) thick	0.40–0.55	0.44
Asphalt base	0.20–0.34	0.30
Unbound aggregate base	0.12–0.18	0.14
Unbound aggregate subbase	0.08–0.12	0.11
Portland cement treated base	0.10–0.28	0.20

Inputs to this design method include the pavement layer thickness, the layer material properties (e.g. stiffness, strength), project climate/environmental data and traffic axle load spectra. Accumulated damage is predicted, with the output from the analysis including predicted levels of roughness (as the international roughness index, IRI), rutting, load-associated cracking and low temperature cracking.

The MEPDG has been implemented for use by several state departments of transportation, and is currently being evaluated for implementation by most highway agencies in the USA and Canada. The evaluation requires calibration of the various performance models for conditions specific to the state or province.

There are other pavement design methods used in the USA, such as the MnPave (Mechanistic–Empirical Flexible Pavement Design) program or the California Department of Transportation empirical method among others.

13.7 Empirical and analytical pavement design

There are two approaches to pavement design: empirical and analytical. Most design methods in current practice around the world are empirical, being based on experience accumulated in practice and from specially constructed test sections. There is an assumption within this approach that the performance of the pavement will be the same irrespective of the properties of the pavement or the constituent materials. This approach to pavement design is somewhat removed from engineering principles, and cannot cope with circumstances beyond those included in the trial section on which the method is based. In contrast, the analytical approach uses theoretical analysis of the mechanical properties of materials, and is capable, in principle, of dealing with any design situation. Efficient pavement design can only be achieved through the use of accurate analytical methods. Hopefully, in the foreseeable future, pavements will be designed analytically based on an accurate assessment of their actual physical properties.

13.8 Analytical pavement design

The philosophy of an analytical approach to pavement design is that the structure should be treated in the same way as other civil engineering structures. The basic procedure is as follows

1. assume a form for the structure, usually involving rational simplification of the actual structure to facilitate analysis
2. specify the loading
3. estimate the size of the components
4. carry out a structural analysis to determine the stresses, strains and deflections at critical points in the structure

5 compare these values with the maximum allowable values to assess whether the design is satisfactory

6 adjust the materials or geometry repeating steps 3, 4 and 5 until a satisfactory design is achieved

7 consider the economic feasibility of the result.

13.9 The contribution of Shell Bitumen to the development of pavement design

Shell Bitumen has a long association with the subject of pavement design that stretches back many decades.

The nomographs for asphalt characteristics (stiffness modulus and fatigue properties) are based on those originally developed by Shell in the 1950s. The bitumen stiffness nomograph was developed by Van der Poel in 1954 (Van der Poel, 1954), and is shown in Figure 13.9. This has proved to be an excellent tool for predicting bitumen stiffness over a wide range of loading conditions. The stiffness of the bitumen is calculated at a specific temperature and loading time (or frequency of loading) from the penetration and softening point of the bitumen. It has been found that there is good agreement between the measured and predicted stiffnesses of asphalts containing conventional penetration grade bitumens. Accordingly, the nomograph is still considered accurate enough for engineering purposes.

The Van der Poel nomograph was later superseded by PONOS, a computer program that was able to calculate the bitumen properties based on the nomographs, which enabled a wider range of bitumen characterisation to be used as input. PONOS has been adapted for use in the BANDS (bitumen and asphalt nomographs developed by Shell) software package.

BANDS was devised to predict the bitumen properties as described above, using the Van der Poel nomograph and PONOS. However, the improvement with BANDS is that it has been developed to be used on modern computer systems such as Windows. Thus, BANDS is far easier to use than its predecessor.

Over time, there have been ongoing efforts to develop methods for the prediction of the mechanical properties of asphalts from the properties and the relative volumes of the constituent materials, hence the interest in the Van der Poel nomograph.

In the late 1950s, there was a great deal of interest in analytically-based pavement design procedures. This interest grew, and, in 1963, Shell published a set of design charts that allowed the calculation of stress and strain in a structure based on a multi-layer linear elastic analysis. (A multi-layered elastic analysis calculates stresses, strains and deflections at any point in a pavement structure resulting from the application of a surface load.

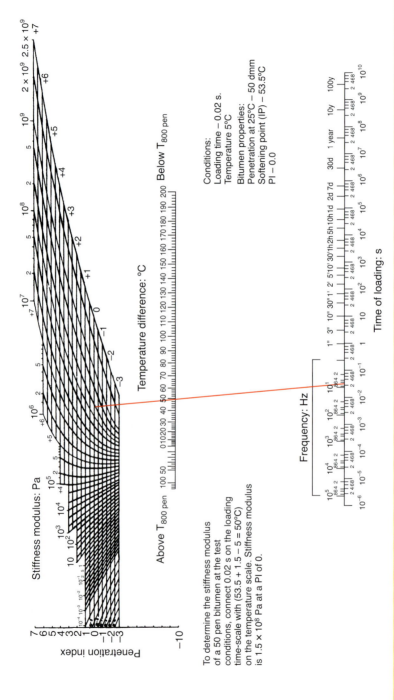

Figure 13.9 Van der Poel nomograph for determining the stiffness modulus of bitumens

Layered elastic models assume that each pavement structural layer is homogeneous and linearly elastic throughout the layer. A multi-layered elastic analysis examines a structure made up of a number of structural layers.) In 1978, this system was extended to incorporate all relevant major design parameters that were derived from empirical design methods and the AASHO road test and laboratory data, and published as the *Shell Pavement Design Manual* (*SPDM*) (Shell International Petroleum Company, 1978, 1985). A photograph of the first delivery of these design charts is shown in Figure 13.10. *SPDM* allowed for the effects on the pavement of temperature, traffic density, and the physical properties of the bitumen and aggregates while standardising the asphalts with respect to stiffness and fatigue properties. *SPDM* was presented in the form of graphs, charts and tables, and, to keep the number of graphs, charts and tables manageable, the number of variable parameters used was limited, with the pavement engineer being expected to interpolate using these data.

In the early 1970s, Shell developed BISAR (bitumen stress analysis in roads) mainframe computer program. This program was designed to calculate and analyse the stresses and strains within the layers of a pavement. BISAR, whose forerunner was the 1963 design charts, was used in drawing up the *SPDM* design charts that were issued in 1978.

Figure 13.10 First delivery of Shell's pavement design charts in 1978

In 1992, the first release of SPDM-PC (Shell pavement design method on a personal computer) was issued. This software followed the same design method as the 1978 manual. However, the computer program allowed the use of a wide variety of data without the need for cumbersome interpolations.

Since then, with the huge increase in computer power available to virtually every individual within an organisation and at home, the limitations of the original Shell analytical software and the manual procedures introduced by Shell over the last 40 years have now been removed and replaced with user-friendly computer packages that can be integrated with other computer-based applications.

As already mentioned, one of the benefits of the analytical approach is the ability to design for individual sets of conditions. Therefore, it is very important that the characteristics and properties of the materials that are used in the design are appropriate for the loading and climatic conditions.

13.10 Example pavement designs using SPDM-PC

Figure 13.11 shows a pavement designed for 20 msa (Strickland, 2000) using three different methods.

■ The first pavement was designed in accordance with the current UK design method (DMRB) using this assumed value of design traffic

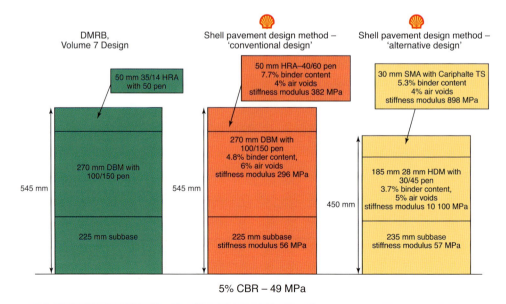

Figure 13.11 Example pavement designs

followed by the procedures set out in IAN 73 (Highways Agency et al., 2009) and HD 26 (Highways Agency et al., 2006a).

■ The second pavement was designed with the SPDM-PC computer software with 'conventional materials'.

■ The third pavement was designed with SPDM-PC using 'alternative materials'. SPDM-PC requires criteria for subgrade strain and fatigue failure to be input as part of the design process, and the values of these parameters were taken from RRL Report 1132 (usually described as 'LR 1132') (Powell et al., 1984). The subgrade strain and fatigue failure criteria given in LR 1132 provide details for the conditions that prevail in the UK. However, this can easily be changed within SPDM-PC for the criteria that apply in different countries. Thus, by adopting the applicable criteria, SPDM-PC can be used in any country.

The subgrade is taken to have a CBR of 5%, a value that is typical for a silty clay. The CBR has been converted to an equivalent stiffness modulus, E, using the relationship specified in paragraph 5.11 of IAN 73 (Highways Agency et al., 2009). The relationship is

$$E = 17.6(CBR)^{0.64} \text{ MPa}$$

This relationship was used for all three designs, which calculated the stiffness modulus of the subgrade as 49 MPa.

The designs of foundations for the three pavements are shown in Table 13.5.

For the three pavement constructions, the designed asphalt thicknesses are shown in Table 13.6.

The above example demonstrates that pavements designed in accordance with the current UK pavement design method (DMRB) and pavements designed with SPDM-PC are very similar. However, as SPDM-PC is an

Table 13.5 Foundation design

Current UK pavement design (DMRB)	SPDM-PC 'conventional design'	SPDM-PC 'alternative design'
By following IAN 73, two subbase designs were possible: ■ 150 mm subbase on 250 mm capping ■ 225 mm subbase, no capping For this design the second option was selected	225 mm of subbase was selected to compare with the current UK pavement design method The stiffness modulus of the subbase was calculated to be 56 MPa with an 85% confidence level	The subbase was increased to 235 mm to reduce the cost of the pavement by decreasing the amount of asphalt needed The stiffness modulus of the subbase was calculated to be 57 MPa with an 85% confidence level

Table 13.6 Asphalt pavement design		
Current UK pavement design (DMRB)	SPDM-PC 'conventional design'	SPDM-PC 'alternative design'
50 mm of chipped hot rolled asphalt (HRA) surface course with 40/60 pen bitumen	50 mm of the following chipped HRA surface course: ■ 40/60 pen bitumen ■ 7.7% binder content (17% volume) ■ 4% air voids Stiffness modulus of 382 MPa	30 mm of SMA with the following properties: ■ Cariphalte TS bitumen ■ 6.3% binder content (12.4% volume) ■ 4% air voids Stiffness modulus of 898 MPa
Per HD 26/06, 270 mm of dense base macadam (DBM125) made with 100/150 pen bitumen. This material is an asphalt concrete	270 mm of the following 20 mm DBM was calculated as being required: ■ 100/150 pen bitumen ■ 4.8% binder content (11.5% volume) ■ 6% air voids Stiffness modulus of 2964 MPa	185 mm of HDM, an asphalt concrete, with the following properties: ■ 30/45 pen bitumen ■ 3.7% binder content (9.0% volume) ■ 5% air voids Stiffness modulus of 10 100 MPa

analytical procedure, it has the advantage of being able to take into account variations in material properties. The example also shows that, with the SPDM-PC computer package, thinner pavements constructed using alternative materials can be designed to carry the same number of standard axles (i.e. the same design traffic). This can significantly reduce the cost of the pavement or, alternatively, provide a pavement that, having the same thickness, will have a longer service life.

13.11 A glimpse of the future for pavement design?

It can be seen from the section above that true analytical design offers a much more accurate method of designing pavements. This fact coupled with the flexibility that such a method offers means it has many advantages over the current empirical methods in use around the world. Figure 13.12 shows a section through what is probably the most common pavement construction in the UK.

Figure 13.12 is based on a flexible pavement (i.e. the surface course, binder course and base are all asphalts). The load is provided by a heavy goods vehicle equipped with super single tyres pressurised to 8.5 bar. The values of pressure down through the pavement and foundation layers were calculated by BISAR. Figure 13.12 shows how the pressure is reduced as it passes through each of the layers. As can be seen, 2.5% of the applied stress is lost as the pressure passes through the surface course, and 17.4% as it passes through the binder course. As would be expected, the majority of the pressure is lost as the loading passes through the base. Finally, 2.8%,

Material in layer	Elastic modulus: MPa	Poisson's ratio	Thickness: mm			Pressure	Pressure drop in layer	% Drop in layer
						8.500 bar		
HRA SC	3100	0.35	40	Surface course		8.285 bar	0.215 bar	2.5%
HDM50	4700	0.35	60	Binder course			1.479 bar	17.4%
						6.806 bar		
HDM50	4700	0.35	200	Base			6.309 bar	74.2%
						0.497 bar		
Type 1 unbound mixture	150	0.35	200	Subbase			0.239 bar	2.8%
						0.258 bar		
Capping	75	0.35	600	Capping			0.162 bar	1.9%
						0.096 bar		
Soil	40	0.35	N/A	Soil			N/A	1.1%

Figure 13.12 Pressure–stress dissipation down through a typical UK flexible pavement. (Courtesy of Dr Robert N Hunter)

1.9% and 1.1% are lost through the subbase, capping and subgrade respectively. Note that 94.1% of the pressure is lost by the time the pressure reaches the foundation.

Thus, this powerful program enables engineers to begin to design pavements in the way they should be designed as set out in section 13.10 above.

Other analytical pavement design software packages are available, including Amadeus, CIRCLY, ALYSEE and CAPA 3D.

References
AASHTO (American Association of State Highway and Transportation Officials) (1972) *AASHTO Interim Guide for Design of Pavement Structures*. AASHTO, Washington, DC, USA.
AASHTO (1993) *AASHTO Guide for Design of Pavement Structures*. AASHTO, Washington, DC, USA.

BSI (British Standards Institution) (1990a) BS 1377-4:1990. Methods of test for soils for civil engineering purposes. Compaction-related tests. BSI, London, UK.

BSI (1990b) BS 1377-9:1990. Methods of test for soils for civil engineering purposes. In-situ tests. BSI, London, UK.

BSI (2010) BS 594987:2010. Asphalt for roads and other paved areas. Specification for transport, laying, compaction and type testing protocols. BSI, London, UK.

Currer EWH and Connor MGD (1979) *Commercial Traffic: Its Estimated Damaging Effect, 1945–2005*. Transport Research Laboratory, Crowthorne, UK, Report 910.

Highways Agency, Scottish Office Development Department, Welsh Office and Department of the Environment for Northern Ireland (1999) *Design Manual for Roads and Bridges*. Volume 7, *Pavement Design and Maintenance*. Section 1, *Preamble*. Part 1, *General Information*. The Stationery Office, London, UK, HD 23/99.

Highways Agency, Scottish Office Development Department, Welsh Office and Department for Regional Development Northern Ireland (2001) *Design Manual for Roads and Bridges*. Volume 7, *Pavement Design and Maintenance*. Section 2, *Pavement Design and Construction*. Part 3, *Pavement Design*. The Stationery Office, London, UK, HD 26/01.

Highways Agency, Scottish Executive, Welsh Assembly Government and Department for Regional Development Northern Ireland (2006a) *Design Manual for Roads and Bridges*. Volume 7, *Pavement Design and Maintenance*. Section 2, *Pavement Design and Construction*. Part 3, *Pavement Design*. The Stationery Office, London, UK, HD 26/06.

Highways Agency, Scottish Executive, Welsh Assembly Government and Department for Regional Development Northern Ireland (2006b) *Design Manual for Roads and Bridges*. Volume 7, *Pavement Design and Maintenance*. Section 2, *Pavement Design and Construction*. Part 1, *Traffic Assessment*. The Stationery Office, London, UK, HD 24/06.

Highways Agency, Transport Scotland, Welsh Assembly Government and Department for Regional Development Northern Ireland (2006c) *Design Manual for Roads and Bridges*. Volume 7, *Pavement Design and Maintenance*. Section 5, *Surfacing and Surfacing Materials*. Part 1, *Surfacing Materials for New and Maintenance Construction*. The Stationery Office, London, UK, HD 36/06.

Highways Agency, Scottish Government, Welsh Assembly Government and The Department for Regional Development Northern Ireland (2008) *Manual of Contract Documents for Highway Works*. Volume 1, *Specification for Highway Works. Road Pavements – Bituminous Bound Materials*. The Stationery Office, London, UK, Series 900.

Highways Agency, Scottish Executive, Welsh Assembly Government and The Department for Regional Development Northern Ireland (2009) *Design Guidance for Road Pavement Foundations*. The Stationery Office, London, UK, Interim Advice Note 73/06.

Highway Research Board (1962) *The AASHO Road Test. Report 5, Pavement Research*. Highway Research Board, Washington, DC, USA, Special Report 61E.

Liddle WJ (1962) Application of AASHO road test results to the design of flexible pavement structures. *Proceedings of the 1st International Conference on the Structural Design of Asphalt Pavements*, Ann Arbor, MI, USA, pp. 42–51.

Lister NW (1972) *Deflection Criteria for Flexible Pavements*. Transport and Road Research Laboratory, Crowthorne, UK, Report 375.

London Technical Advisors Group (2013) *London Bus Lane and Bus Stop Construction Guidance*. London Technical Advisors Group, London, UK.

Nunn M (2004) *Development of a More Versatile Approach to Flexible and Flexible Composite Pavement Design*. Transport Research Laboratory, Crowthorne, UK, Report 615.

Nunn ME, Brown A, Weston D and Nicholls JC (1997) *Design of Long-life Flexible Pavements for Heavy Traffic*. Transport Research Laboratory, Crowthorne, UK, Report 250.

Payne IR (2011) Site investigation and foundation design. In *ICE Manual of Highway Design and Management* (Walsh ID (ed.)). Institution of Civil Engineers, London, UK, pp. 253–272.

Powell WD, Potter JF, Mayhew HC and Nunn ME (1984) *The Structural Design of Bituminous Roads*. Transport and Road Research Laboratory, Crowthorne, UK, Report 1132.

Road Research Laboratory (1970) *A Guide to the Structural Design of Pavements for New Roads*, 3rd edn. Road Research Laboratory, Crowthorne, UK, Road Note 29.

Shell International Petroleum Company (1978) *Shell Pavement Design Manual*. Shell International Petroleum Company, London, UK.

Shell International Petroleum Company (1985) *Addendum to the Shell Pavement Design Manual*. Shell International Petroleum Company, London, UK.

State of Queensland (2013) *Pavement Design Supplement. Supplement to 'Part 2: Pavement Structural Design' of the Austroads Guide to Pavement Technology*. State of Queensland (Department of Transport and Main Roads), Queensland, Australia.

Strickland D (2000) *Shell Pavement Design Software for Windows*. Shell International Petroleum Company, London, UK.

TRB (Transportation Research Board) (2014) *NCHRP Synthesis 457. Implementation of the AASHTO Mechanistic–Empirical Pavement Design Guide and Software: A Synthesis of Highway Practice*. TRB, Washington, DC, USA.

Van der Poel C (1954) A general system describing the visco-elastic properties of bitumen and its relation to routine test data. *Journal of Applied Chemistry* **4**: 221–236.

Chapter 14

Asphalt production plants

John Moore
Consultant (UK)

14.1 Introduction

The principles of making asphalt have not changed since the middle of the nineteenth century (Jones, 1986). Different size aggregates are dried, heated and then mixed with bitumen to make a saleable product. The aggregates require drying and heating to enable the bitumen to adhere to the surface of the aggregate.

However, although the principle has not changed, the methods of manufacture, and therefore the production plants, have changed significantly in their design. Capacities have increased from just a few tonnes per hour up to 800 tonne/h of asphalt. Worldwide environmental regulations have resulted in particulate emissions from current asphalt production plants being undetectable with the naked eye, whereas early asphalt plants had a bad reputation due to clouds of dust surrounding the works. The main recent developments have been in the controls: early plants were controlled manually but modern asphalt plants are completely computer controlled. The quality of the asphalt produced has also improved beyond recognition from the early days, when 'hot, black and sticky' were the criteria. Now, the quality of a mixed asphalt can be guaranteed, with an accurate bitumen content and a consistent asphalt product.

In the first production of asphalt, coal was used as the fuel to dry and heat the aggregates and the bitumen. Later, oil was used, and now there are numerous fuels available as alternatives to oil, including natural gas, liquid propane gas, butane and coal. The choice of fuel is influenced by the fluctuating cost of oil and the availability of the alternatives.

Each of the different types of asphalt plant is available in fixed, free standing and mobile configurations. The fixed plants are bolted to reinforced concrete foundations. The free standing plants are mounted on a flat, levelled surface that can withstand a predetermined pressure, and no concrete foundations

The Shell Bitumen Handbook, Sixth Edition
ISBN 978-0-7277-5837-8
Shell International Petroleum Company Ltd: All rights reserved
http://dx.doi.org/10.1680/tsbh.58378.379

are required. Mobile plants are also mounted on a flat, levelled surface, and the tyres are usually positioned on steel base plates; again concrete foundations are not required. In some parts of the world the term 'portable' is used instead of 'mobile', as the plants are not mobile during the production of asphalt. The choice of using a fixed, free standing or a mobile asphalt plant will depend mainly on the size and location of the contract.

14.2 Types of mixing plants

There are two basic categories of asphalt plant: batch and continuous. The terminology can be confusing, as both categories are continuous in operation. The categorisation refers to the method of mixing the dried and heated aggregates with the bitumen. In a batch plant the ingredients are mixed in discrete batches, whereas in a continuous plant the ingredients are mixed in an uninterrupted continuous process.

All of the batch and continuous asphalt plants operating in Europe are designed to meet the European safety standard BS EN 536:1999 (BSI, 1999).

The selection of which type of asphalt plant to use is influenced by many factors.

- *Location*. Is the plant to be located in a quarry, where the aggregate source is consistent and can be controlled, or is the plant to be located on a depot site, where the aggregate supply could change?
- *Market requirements*. How many different mixtures are to be produced? What type of mixtures and how many different mixtures will be required in a day's production?
- *Plant capacity*. What is the estimated output required from the plant, and what are the operating conditions, including the moisture content of the feed materials and the final mixture temperatures?
- *Competition*. What other types of asphalt plant are in the surrounding area? Can the plant offer a niche product mixture?
- *Budget*. How much funding is available for the purchase of the plant?
- *Operational costs*. Does the plant selected manufacture an economical product based on the operational costs, including maintenance, fuel, labour etc.?
- *Site restrictions*. Does the type of plant meet all the site restrictions, such as height limits, space available, noise parameters and emission regulations?
- *Customer preference*. Is there a preference in the country or area in which the plant is to be located for a specific type of asphalt plant? Different customers and different continents have different requirements and different preferences.

Asphalt batch plants can be divided into two further categories: conventional asphalt batch plants and batch heaters. Batch heater plants are found only in the UK. Continuous asphalt plants consist of the drum mix, the counter flow drum mix, the double drum and the continuous mix plant. Each of the different types of asphalt plant is described in detail below, highlighting the influences on selection, the benefits of each type of plant, and where appropriate, the limitations.

14.2.1 Asphalt batch plants

14.2.1.1 Conventional asphalt batch plant

The conventional asphalt batch plant is the most common type of asphalt plant in Europe, Central Africa, China, the Middle East and Australasia. Although the process is continuous in operation, each batch of mixed material is made separately and the production then repeated. The batch size relates to the quantity of material being mixed in the paddle mixer. Capacities of from 50 up to 450 tonne/h are achievable, depending on the size of the equipment, the moisture content of the feed aggregates and the final mixture temperature. The rotary dryers range from 1 up to 3.1 m in diameter, and paddle mixer batch sizes range from 500 kg up to 5 tonnes and even larger. A conventional asphalt batch plant is shown in Figure 14.1.

Figure 14.1 A 320 tonne/h conventional asphalt batch plant. A TBA-4000 with a 4 tonne mixer and a 2.8 m dryer. (Courtesy of Benninghoven UK Ltd.)

The mixture recipe being produced provides the percentages and the tolerance limits of each aggregate size, the filler content and the bitumen content in the final mixture.

There are two methods of feeding the aggregates into an asphalt batch plant. The first, and most common, is to store the different aggregate sizes in separate stockpiles, or loading bays, and then transfer the aggregates using a loading shovel into the plant feed hoppers. The feed hoppers are usually fitted with safety grids to prevent people accidentally falling into the hoppers. In Europe, the loading bays are generally covered to help reduce unwanted water reaching the aggregates, especially the fines, and therefore improve plant efficiency. The feed hoppers are also covered on three sides in Europe to help prevent the aggregates and fines becoming airborne when transferred from the loading shovel to the hopper, this covering also adds protection from the elements.

The second method of feeding the aggregates and fines into the asphalt plant is direct from the crushing plant storage bins. The asphalt plant must obviously be located in a quarry for this method to be adopted. This method is the ideal solution, as the crushed stones and fines have the minimum possible moisture content, usually less than 2%, and the efficiency of the drying and heating process is therefore considerably increased.

A flow diagram of a conventional asphalt batch plant is shown in Figure 14.2. Each different aggregate size is fed into a dedicated feed hopper. The number of feed hoppers typically ranges from eight to 20, or more, depending on the different mixture recipes to be produced. Variable speed belt feeders are predominantly used to control the flow of the feed materials from the hoppers, or the crushed stone bins, onto the collecting conveyor before they progress to the dryer. The individual belt feeder speed is set in the controls to match the mixture recipe percentage of the particular aggregate being handled. The aggregates and fines are fed through the plant in the proportions required in the final mixture, thus ensuring economic use of the feed materials.

The dryer consists of a rotating steel cylinder inclined at 3–5°, with a burner mounted at the discharge end. The inside of the dryer cylinder is fitted with lifters, which aid the passage of the feed materials along the dryer, while also exposing the aggregates to the heat from the burner flame. The lifters at the feed end create a curtain of the aggregates, and the lifters at the discharge end form an annulus of the aggregates inside the dryer cylinder around the burner flame, which enables the flame to develop fully. The rotating dryer incorporates a two stage drying and heating process: first the aggregates are dried to remove the water in the feed materials, and then the aggregates are heated to the required temperature, usually between 120°C

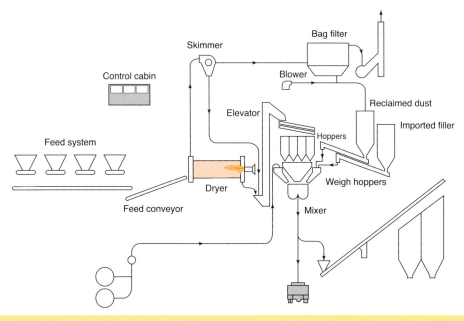

Figure 14.2 Flow diagram of a conventional asphalt batch plant

and 200°C, depending on the mixture recipe. The dryer cylinders can be insulated to prevent heat loss, with the added advantage of also reducing noise levels.

In order to support the burner flame, to remove the products of combustion and to remove the water vapour from the feed aggregates, an exhaust system is connected to the dryer. The exhaust is extracted from the feed end of the dryer. As the feed aggregates are passing along the dryer in the opposite direction to the exhaust air, this is termed a 'counter flow air system'. The counter flow air system enables the hot exhaust gases to preheat the cold incoming feed materials. The exhaust system creates a flow of air inside the dryer, which consequently also picks up dust particles from the feed materials. These dust particles must be removed before releasing the exhaust air to the atmosphere.

The exhaust system generally consists of a two stage process. The first stage removes the coarse particles of dust, usually plus 75 micron, and returns these to join the heated product from the dryer discharge. In the UK the first stage or primary dust collector is usually a skimmer, which has the added benefit of being able to alter the coarse/fine dust split. Other primary collectors include knock out boxes, drop out boxes and cyclones. The second stage is the collection of fine dust using a bag filter. The bag filter consists of a large number of filter bags that collect the fine dust. The cleaned air then

383

passes through the exhaust fan and is discharged to the atmosphere through a high level stack. Particulate emission levels of less than 20 mg/m^3 are easily achieved using a bag filter, and these dust emissions are not visible with the naked eye. The fine dust is removed from the filter bags by reverse air cleaning, a blower or pneumatic jets, and is transferred to a reclaimed filler silo for later addition to the mixture, or for disposal to waste, again depending on the final mixture recipe.

The heated aggregates, together with the coarse dust, are elevated from the dryer discharge by a hot stone elevator to a vibrating screen. The vibrating screen separates the feed materials into the sizes specified in the mixture recipe, and the sized aggregates are discharged into individual hot stone bin compartments. There are usually six individual compartments in the hot stone bins, although in some plants there are four or five compartments. There is a trend towards separating the dust and sand in the hot stone bins by way of a diverter door under the screen fines deck, resulting in seven hot stone compartments. A separate compartment is also usually incorporated in the hot stone bins for the screen bypass material – material that is a blended product and does not require screening. The hot stone bins incorporate integral overflow chutes that enable the plant to continue production if the individual compartments are full. Continuous level indicators are fitted to each of the hot stone bin compartments to provide information on the trend in the filling and discharging criteria of the sized aggregates. The operator can then adjust the feed rates accordingly, to control the overflow of aggregates and unwanted wastage of the hot materials. The overflow is collected in a separate storage hopper together with the screen rejects, or discharged to the ground. The hot stone bins can be insulated to prevent heat loss; again this also reduces the noise levels.

The aggregates are discharged separately from each hot stone bin compartment into the aggregate weigh hopper in accordance with the percentages specified in the mixture recipe. The aggregate weigh hopper is mounted on load cells, and positioned under the hot stone bins and directly above the paddle mixer. At the same time as the aggregate is being weighed off, the bitumen and filler are also weighed in dedicated weigh hoppers. In some US plant designs, the filler is weighed together with the aggregates in the aggregate weigh hopper.

The bitumen is stored in horizontal or vertical heated tanks, with each grade of bitumen being stored in a different tank. The heating of the tanks can be indirect, by using a hot oil heater with transfer oil, or electric heating elements inside the tanks. The tanks are insulated to reduce heat losses, and the piping through which the bitumen passes is also heated and insulated. The bitumen is circulated in a ring main at tank level; alternatively, each grade of bitumen can be circulated up to the bitumen weigh hopper on the plant. A three way valve is used to discharge the circulating bitumen into the bitumen weigh hopper. The bitumen weigh hopper is mounted, or suspended on load cells, and the quantity of bitumen is delivered according to the amount specified in the final mixture. Two stage weighing of the bitumen can sometimes be used to provide a bitumen top-up relative to the actual amount of aggregates and filler being held in the aggregate and filler weigh hoppers. The weighed bitumen is either gravity fed to the paddle mixer, usually through a spray bar, or pumped under pressure through a spay bar across the mixer.

Reclaimed filler from the bag filter and imported filler to the plant are stored in separate filler silos. These silos discharge by means of screw conveyors to the load-cell-mounted filler weigh hopper, the amount discharged depending on the quantity specified in the mixture recipe. In Europe, if the reclaimed filler is to be discharged to waste, it is handled through a dust conditioner, where water is mixed with the reclaimed dust to prevent airborne particles.

The aggregate, bitumen and filler are weighed simultaneously, and then discharged into the paddle mixer in a pre-set order. The paddle mixer consists of two rotating shafts with arms and tips attached. The ingredients are mixed until a homogeneous mixture is produced, which is usually achieved in 30–40 s. The mixing and weighing cycles overlap, so that while one batch is being mixed the next batch is being weighed. Most asphalt batch plants are rated using a 45 s mixing cycle, which is the time from when the mixer discharge door opens to the next time it opens. The 45 s cycle results in 80 batches of mixture produced in one hour. A 3 tonne paddle mixer can therefore produce 240 tonne/h. The mixing cycle time is variable, and can be adjusted within the control system. However, the mixing time should be sufficient to enable a thorough coating of the bitumen, but not too long in order to avoid oxidation of the bitumen in the mixture. Recovered bitumen contents of $\pm 0.1\%$ of the target bitumen value can be achieved in an asphalt batch plant. The mixer discharges the finished product either directly to trucks, or to a mixed material storage system.

The mixed material storage can be positioned directly underneath the paddle mixer, using flap doors to direct the mixed material into different storage hoppers, or the paddle mixer can discharge into a horizontal skip that traverses and discharges into individual storage hoppers. This design can be found in many European asphalt batch plant sites where space is limited. Alternatively, the mixed material storage hoppers can be positioned adjacent to the mixing tower, where the paddle mixer discharges into an inclined skip that then feeds the storage hoppers. The mixed material silos are insulated and heated with electric elements or hot oil, to help prevent heat loss. European asphalt plants usually store mixed materials for a maximum of

12–24 h, the heating and insulation ensuring minimum heat loss. Extended storage times of up to 4 days are possible in the mixed material silos without loss of mixture quality if the top and bottom doors are sealed to prevent oxidation of the mixture. This extended storage is a requirement in the USA, but not all mixtures can be stored for extended periods: polymer modified and open graded mixtures are excluded from plant manufacturers' long term storage guarantee (for literature on storage silos, see Astec Inc., 2014). Mixed material storage silos can also incorporate automatic load out systems into the trucks. The silos are mounted on load cells, or a weighbridge is positioned underneath the silos, and the operator can pre-select the amount to be discharged into a truck.

In the majority of developed countries, the asphalt batch plant tower is enclosed, or sheeted. The sheeting contains any airborne dust inside the mixing tower, to comply with environmental regulations, and also results in a more aesthetically acceptable design. However, if employed in countries that have very high ambient temperatures, sheeted mixing towers must be vented, as the sheeting also contains the heat generated in the process, which can lead to premature bearing failures.

A variant of the asphalt batch plant is the multiple hot stone bin plant. The concept is to store two different types of hot material for weighing; for example, six hard stone aggregates and six soft stone aggregates. Instead of the standard six or seven hot stone bin compartments located underneath the vibrating screen, a central division plate enables the storage of 12, 13 or more hot aggregates. The vibrating screen has a large diverter chute underneath, which directs the aggregates to the hard or soft stone compartments. An alternative is to use two vibrating screens instead of one, and the hot stone elevator then incorporates a two way diverter chute. The asphalt batch plant would be fed first with soft stone to fill the relevant hot stone compartments, and then with hard stone. Usually, the hot stone bin capacity is larger in a multiple hot stone bin plant, and the increased surge capacity results in a more economical drying process. This type of plant is ideal for complete flexibility of production, allowing quick recipe changes without having to empty the hot stone bins if a hard stone mixture is required directly after a soft stone mixture. In Europe, this type of plant is used instead of the batch heater type, which is found only in the UK. However, multiple hot stone bin plants are gradually replacing the batch heater type in the UK. A multiple hot stone bin plant is shown in Figure 14.3.

Up to 25% reclaimed asphalt pavement (RAP) can be added to the batch plant directly into the paddle mixer, the percentage depending on the moisture content of the RAP and the mixture temperature. A flow diagram for a cold RAP feed is shown in Figure 14.4. The RAP must be in manageable

Figure 14.3 A multiple hot stone bin plant with 360 tonne/h capacity. A Universal S plant with a single vibrating screen and a 14-compartment hot stone bin, located in Rubigen, Switzerland. (Courtesy of Ammann)

pieces of less than 30 mm size, and may need to be processed through a breaker to reduce the size of the sections being handled. The RAP is fed into a steep sided feed hopper, through a belt feeder to a conveyor that is fitted with a belt weigher. The conveyor discharges directly into the paddle mixer. The conveyor operates on a start–stop basis, depending on the percentage of RAP required. The weighed virgin aggregates are superheated and mixed with the RAP in the paddle mixer until the specified mixture temperature has been reached and a homogeneous mixture is produced. The paddle mixer requires an evacuation system to remove the water vapour produced from the heated RAP. This evacuation system is connected to the dryer exhaust system.

A small percentage of RAP (5–10%) can be added directly to the aggregate dryer by way of a recycling collar or a slinger conveyor positioned at the discharge end. The heated RAP is elevated to the vibrating screen, separated into the hot stone bins, weighed off in the aggregate weigh hopper and then mixed in the paddle mixer with the bitumen and filler. It is common to find a dual RAP feed to an asphalt batch plant, utilising the cold feed method direct to the paddle mixer and the RAP feed to the aggregate dryer. Percentages as high as 35% RAP can be added to the batch plant using the dual feed method.

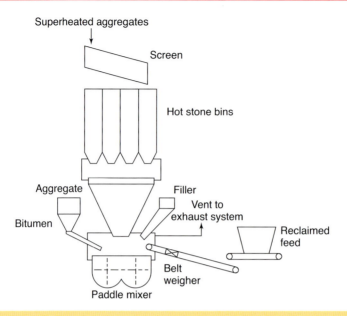

Figure 14.4 Flow diagram for a cold RAP feed (Hunter, 2000)

Higher percentages of RAP, up to 60%, can be added to a batch plant using a RAP dryer. Figure 14.5 shows a flow diagram for a typical hot RAP feed. The RAP dryer consists of a parallel flow drum, with the burner mounted at the feed end. The flow in the dryer is termed 'parallel flow', as the RAP and the exhaust air travel in the same direction through the drum. The RAP, which must be less than 30 mm in size, is fed into a steep sided feed hopper by means of a belt feeder onto a conveyor, or a vertical elevator, and discharged into the RAP dryer. The dryer heats the RAP to between 80°C and 120°C to remove the moisture, and it is then stored in a surge hopper before being discharged into a dedicated RAP weigh hopper. The RAP is then discharged directly into the paddle mixer, where it is mixed with the virgin aggregates, bitumen and filler. The RAP dryer has a large drop-out box mounted at the discharge end to allow the fine particles to be returned to the RAP. A scavenger fan removes the exhaust air, complete with the fumes, from the RAP dryer, and feeds the air and fumes into the virgin dryer burner flame for incineration. A high level RAP dryer is shown in Figure 14.6.

The conventional asphalt batch plant is the asphalt plant of choice in Europe, Central Africa, China, the Middle East and Australasia, and therefore the majority of development for this type of plant has emanated from Europe. As the plant incorporates a vibrating screen, one of its major advantages is that high quality mixed material can still be produced even if the sizing

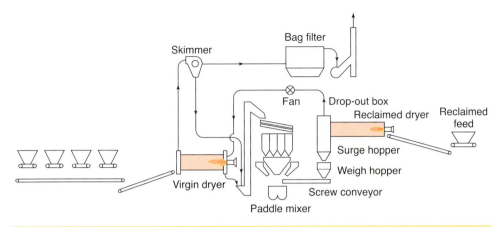

Figure 14.5 Flow diagram for a hot RAP feed (Hunter, 2000)

of the feed aggregates cannot be guaranteed. Although the capacity suffers, this makes the plant universally accepted in any working environment. There are many benefits of an asphalt batch plant

■ the ability to manufacture all European (BS EN 13108:2006 (BSI, 2006a)) and US individual state Department of Transportation's

Figure 14.6 A high percentage RAP dryer on a 320 tonne/h TBA-4000-U asphalt plant, with a 4 tonne mixer, a 2.8 m dryer and a 2.4 m high level RAP dryer. (Courtesy of Benninghoven UK Ltd.)

(DOT) mixture specifications, and therefore worldwide mixture specifications

- the emissions to the atmosphere are within European and US emission regulations, and therefore worldwide emission regulations
- inconsistent feed aggregate sizes can be accommodated while maintaining mixture quality, although plant capacity is reduced
- small and large tonnages are possible
- mixed material storage is not essential
- a high percentage of RAP can be added to the mixture.

Asphalt batch plants have few shortcomings, and these are only relative to the other types of production plant available

- capacity is restricted to the mixer size and the mixing cycle
- heat is wasted from the overflow and rejected materials
- high production costs relative to a drum mix plant.

14.2.1.2 Batch heater plant

Batch heater plants are only found in the UK. Each batch of material is produced individually throughout the process. Capacities range from 50 tonne/h up to about 200 tonne/h, with paddle mixer capacities typically ranging from 1 tonne up to 5 tonnes. The mixing cycle times can be up to 2 min, depending on the feed moisture content and the mixture temperature. A 3 tonne paddle mixer would, therefore, perform 30 cycles/h and achieve 90 tonne/h. Figure 14.7 shows a batch heater plant during construction, and completed and ready for production.

Pre-graded aggregates must be used in a batch heater plant as no further screening will take place during the process. The batch heater plant is fed from feed hoppers, the number of which is determined by the number of different aggregate sizes and different types of aggregate to be used. The number of feed hoppers can range from eight up to 20, or more, and each is fitted with a safety grid and a weather cover. Alternatively, the batch heater plant can be fed directly from the crushing plant storage bins. The latter is the ideal feed arrangement for this type of plant, as the moisture content in the feed aggregates is low, usually less than 2% for crushed stone. Batch heater plants have limited drying and heating capabilities compared with conventional batch plants.

The pre-graded aggregates are fed into the plant in a single batch using time-controlled belt feeders. Each feeder is calibrated for a specific aggregate type and size, the time set in the control system for each feeder depending directly on the final mixture recipe. Variable speed belt feeders can be used, but this is considered to be uneconomical and unnecessary. The batch is

Figure 14.7 A batch heater plant with 4 tonne paddle mixer shown during construction and fully sheeted ready for production. (Courtesy of Hanson UK)

layered onto the collecting conveyor and fed to a check weigh hopper. A flow diagram of a batch heater is shown in Figure 14.8.

The check weigh hopper is mounted on load cells and is used to measure the wet weight of the batch. An allowance is made in the control system for

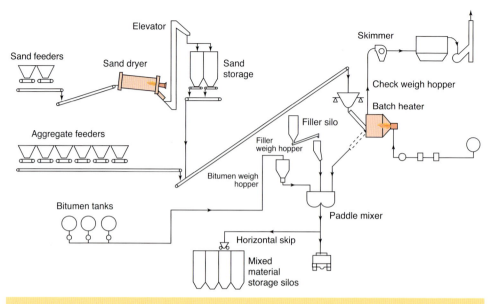

Figure 14.8 Flow diagram for a batch heater (Hunter, 2000)

the moisture content in the feed material to determine the dry weight, the bitumen and filler contents are calculated from the dry weight of the batch. The weighed batch is then discharged into the batch heater drum, which is a large diameter rotating cylinder, short in length, with a burner mounted at the opposite end. Inside the rotating drum are lifters that expose the aggregates to the heat from the burner flame. The aggregates are dried and heated inside the drum. The retention time inside the drum determines the discharge temperature.

An exhaust system is attached to the batch heater drum to remove the water vapour and to support the combustion process. The coarse dust is collected by a skimmer and fed back into the paddle mixer in discrete batches. The fine dust is collected in a bag filter. The different types of bag filter have been described in section 14.2.1.1 on conventional asphalt batch plants.

At the same time as the aggregates are being dried and heated, the bitumen and filler are weighed off in separate weigh hoppers using the calculated dry batch weight. The bitumen and filler systems are similar to those in conventional asphalt batch plants.

The heated aggregates are discharged from the rotating batch heater from the same end as the material feed, directly into the paddle mixer. In a pre-set order, the bitumen and filler are also discharged into the paddle mixer until a homogeneous mixture is produced. The mixed material can discharge from the paddle mixer either directly into trucks or to a mixed material system. The mixed material systems on a batch heater plant can be high level or low level, and these systems feature the same characteristics as described for conventional asphalt batch plants.

Some later batch heater plants incorporate an aggregate weigh hopper after the batch heater drum, and a surge hopper replaces the wet batch weigh hopper. The bitumen and filler contents are then calculated from the true dry weight of the batch, reducing the possibility of errors.

If the fines or sands being used in a batch heater plant have a high moisture content, greater than 2%, then a separate dryer is used. The wet fines are dried in a counter flow dryer and stored in silos. The dried fines are then considered as an additional ingredient, and join the batch being made at the conveyor stage.

Between 10% and 15% RAP can be added to a batch heater plant using a weigh conveyor and fed directly into the paddle mixer. The process is described in more detail in section 14.2.1.1 on conventional batch plants. The percentage of RAP used is less than in a conventional batch plant because of the difficulty of superheating the aggregates in the batch heater drum.

The batch heater plant is ideal for complete production flexibility, where small batches of many different material specifications are required in a day's work. The batch heater is unique to the UK and initially emerged because of the many different material specifications that existed prior to the advent of the European material specifications. The multiple hot stone bin plant has been developed as the European alternative to the batch heater. The benefits of the batch heater are

- an ability to manufacture all European (BS EN 13108:2006 (BSI, 2006a)) mixture specifications
- the emissions to atmosphere are within European emission regulations
- small batches can be economically produced
- quick recipe change possible
- there is no material wastage
- small and medium tonnages are possible
- mixed material storage is not essential
- a low percentage of RAP can be added to the process.

The limitations of the batch heater plant are relative only to the conventional asphalt batch plant, as the batch heater does not compete with the continuous plants

- pre-graded single-sized feed materials are essential, otherwise the mixture quality cannot be guaranteed
- a maximum 2% moisture content of the feed materials is necessary, or a pre-dryer is required for the fines
- the capital cost is high, especially if a pre-dryer is required.

14.2.2 Continuous asphalt plants

14.2.2.1 Drum mix plant

Drum mix plants were used in the USA as early as the 1920s, although they did not come into common use until the 1970s. A US patent was applied for as early as 1929 by Donald McKnight Hepburn, and was granted in 1931 for a process that mixed the aggregates and bitumen inside a rotating drum (US Patent, 1931). The early drum mix plants were used for the production of 'warm mixtures', and further developments enabled drum mix plants to be used for the majority of mixture recipes.

Drum mix plants are ideal for high capacity production ranging up to 800 tonne/h. The equipment on a drum mix plant is minimal by comparison to a batch type asphalt plant, there is no grading screen, no hot stone bins, no weigh hoppers and no paddle mixer. The maintenance and running costs on a drum mix plant are therefore less than for a batch plant. However, it is essential on all continuous plants that pre-graded, single-sized feed aggregates are used, otherwise the final mix will not meet the necessary specification requirements and rejection of the finished product will result.

Variable speed volumetric belt feeders are used to control the flow of aggregates to the plant. Each feeder is used for a specific aggregate size and type of aggregate. The number of feed hoppers and feeders can be from five up to 20 or more. A flow diagram for a drum mix is shown in Figure 14.9. As for batch plants, in Europe the feed hoppers are fitted with safety grids and weather covers. The feeders are calibrated and set during production to achieve the percentage of the individual ingredients in the final mixture. As the aggregate feed to a drum mix plant is so important to ensuring a good mixture quality, the feeders are fitted with 'no flow' indicators to confirm to the operator that the feed is being discharged. A 'no flow' would result in the plant being stopped. Wet sands and fines can be difficult to control on a continuous plant, and on these occasions weigh feeders can be used.

A scalping screen is usually fitted between the feed section and the drum mix feed conveyor, which removes any unwanted oversized material or possible debris from the feed. This is especially useful if the aggregates have been stored as ground stockpiles prior to being fed to the plant.

The drum mix feed conveyor is fitted with a belt weigher that accurately determines the wet weight of the feed aggregates. The operator manually inputs

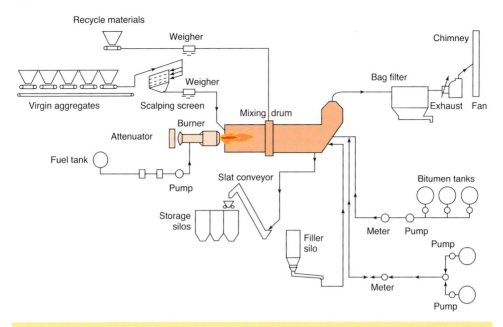

Figure 14.9 Flow diagram for a drum mix (Hunter, 2000)

the overall moisture percentage of the feed material into the control system, and the controls then calculate the dry weight of the feed aggregates being delivered to the plant. This dry weight is used to calculate the bitumen and filler contents that are to be added later to the mixture.

The aggregates are fed into a rotating steel cylinder with a burner mounted at the feed end. The drum is angled at between 3° and 5°, and is fitted internally with a series of lifters to expose the aggregates to the heat from the burner flame, to aid the aggregate flow through the drum and finally to mix the ingredients inside the drum. There are two distinct zones within the drum: the drying and heating zone; and the mixing zone, where the hot aggregates are mixed with the bitumen and filler. The two zones can be physically separated with a flame/target plate, or the design of the heating lifters separates the two zones by creating a dense material curtain out of the cascading aggregates. The length of the rotating cylinder is longer on a drum mix plant than on a batch plant dryer to cater for the mixing zone.

The exhaust system is connected to the discharge end of the rotating cylinder, and the opposite end to the burner. The exhaust system removes the water vapour, the products of combustion and the collected dust from the process. This is a parallel flow drum mix, as the aggregates and the exhaust air flow in the same direction through the drum. As the exhaust air passes through the mixing chamber on a drum mix plant the percentage of dust collected is considerably less than in a batch type plant, as a large quantity of the airborne dust particles are absorbed by the bitumen in the mixing zone. Dust collection usually consists of a large drop-out box at the discharge end of the drum, the coarse dust particles dropping back into the process. The fine dust is collected in a bag filter, the different types of bag filter are as described in section 14.2.1.1 on conventional batch plants.

The proportion of bitumen and filler to be added to the mixture is calculated from the dry weight of the aggregates passing over the dryer weigh conveyor. The bitumen flow rate is controlled by a variable speed pump, or variable orifice control valves. The bitumen is added into the drum mixing chamber from the discharge end. It is important not to expose the bitumen to the naked burner flame, hence the reason for fitting the burner at the feed end of the drum. Degradation of the bitumen can occur if direct contact is made with the burner flame, and blue smoke results.

The imported filler and/or the reclaimed collected fine dust can be either blown or screwed back into the drum mixing chamber from the discharge end. These fillers can be accurately fed back into the drum using separate 'loss-in-weight' systems. The filler is conditioned to ensure a constant density, and is continuously reverse weighed against a known calibration rate. The bitumen and filler are added at the beginning of the mixing zone inside the drum, ensuring that the fillers mix with the bitumen and do not become airborne.

The mixed material discharges from the end of the drum mix, usually to a mixed material system. Small capacity drum mix plants can discharge into a surge hopper and then to trucks but, being continuous plants, the majority discharge into mixed material systems as described for conventional asphalt batch plants. The US manufacturers' preference is to use slat conveyors instead of a skip to transport the mixed material; alternatively, vertical elevators can be used if site restrictions apply. The slat conveyor is ideal for handling high capacities of mixed material.

Between 15% and 20% of RAP can be added in a drum mix plant by means of a collar mounted around the rotating cylinder at the start of the mixing zone. The RAP must be less than 30 mm in size. It is fed into a steep sided feed hopper through a variable speed feeder, and onto a RAP weigh conveyor. The signal from the drum mix feed conveyor controls the amount of RAP added to the process. The virgin aggregates are superheated in the drying and heating section, and the RAP is mixed with the virgin aggregates, the filler and the bitumen in the mixing section. The percentage of RAP is limited because of its tendency to produce fumes or blue smoke when the RAP is mixed with the superheated virgin aggregates.

Drum mix plants have been the plant of choice in the USA, India, Malaysia, New Zealand, Norway and France, where long production runs of one material specification are required. In addition, drum mix plants have been, and still are, used for smaller capacities where funds are limited and where mixture specifications are not of paramount importance. As stated above, the success of drum mix plants, and all continuous plants, relies totally on using pre-graded single-sized aggregates as feed materials. If this cannot be guaranteed, the finished mixed material product is likely to be out of specification and rejected. The need to produce high temperature mixtures and to add high percentages of RAP has led to the development of counter flow drum mix, double drum and continuous mix plants, which are all variants of the drum mix plant. The benefits of drum mix plants are

- they are economical for long production runs of one material specification
- high capacities are readily achieved
- they have relatively low maintenance costs
- the amount of dust removed from the process is minimal
- they are easily adapted to a mobile design
- a low percentage of reclaimed material can be added to the mixture.

Drum mix plants are still in use currently in many different countries, and are performing a useful role. However, the production of high temperature mixtures and the requirement to add high percentages of RAP can result in degradation of the bitumen on a drum mix plant, and blue smoke results. These emissions adversely affect the performance of the bag filter, production rates suffer considerably over a period of time, and the fumes, or blue smoke, emitted into the atmosphere are not acceptable. The limitations of drum mix plants are

- pre-graded, single-sized feed materials are essential, otherwise mixture quality cannot be guaranteed
- they are unable to produce high temperature mixtures
- they are unable to produce a high percentage of RAP
- there is wastage of heated material at the beginning and end of production
- there is a possibility of blue smoke emissions
- small quantities are uneconomical
- a mixed material system is essential on higher capacity plants.

14.2.2.2 Counter flow drum mix plants
Counter flow drum mix plants are a direct development of the drum mix plant, and overcome some of the shortcomings so that high temperature mixtures can be produced and high percentages of RAP can be added to the mixture.

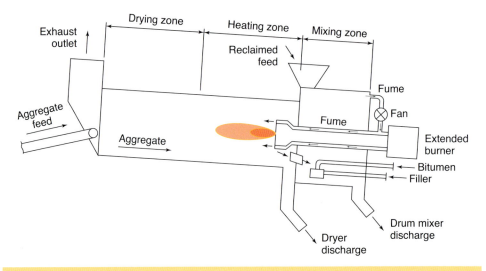

Figure 14.10 Flow diagram of a counter flow drum mix cylinder. (Hunter, 2000)

The counter flow drum mix consists of three basic categories: the counter flow drum mix, the double drum and the continuous mixture plants. All the ancillary equipment, such as the feed section, the weigh conveyor, the dust collector, the bitumen supply, the filler feed and the mixed material system, is the same as in a drum mix plant. The only difference is the drum mix rotating cylinder. The internal structure of a counter flow drum mix cylinder is shown in Figure 14.10.

The counter flow drum mix consists of a long steel rotating cylinder with an extended burner mounted at the discharge end. The burner extends inside the rotating cylinder so that the flame develops towards the centre of the drum. The rotating cylinder consists of two distinct and separate zones: the drying and heating zone, and the mixing zone, which is mounted behind the burner flame. The drying and heating section is based on the same principle as in a conventional batch type dryer, the exhaust air travels in the opposite direction to the aggregate flow, hence the term 'counter flow drum mix'. The heating and drying is therefore more efficient than in a standard drum mix, as the feed aggregates are preheated by the hot exhaust air. The bitumen and fillers are added in the mixing zone and discharged when a homogeneous mix is produced. The mixing time depends on the length of the mixing chamber, the design of the mixing lifter pattern and the angle of the drum, which is usually between 3° and 5°. Typical mixing retention times, which is the time the constituents are mixed, can range from 45 s up to 60 s or more. The mixing zone incorporates a scavenger extraction system for any fumes or blue smoke that may be created, and feeds this into the extended burner. The fumes are then incinerated in the burner flame.

Figure 14.11 Counter flow double barrel drum mix asphalt plant. (Courtesy of Astec Inc.)

Up to 60% RAP that has been reduced to less than 30 mm in size can be added to the mixing zone when called for in the mixture recipe. The reclaimed material is fed into the rotating cylinder through a collar wrapped around the drum. The aggregates are superheated to enable heat to be transferred to the RAP, first removing any moisture present in the RAP, and then heating the combined mixture. The water vapour, together with any fumes from the RAP, is extracted by the scavenger fan and fed into the extended burner for incineration.

Another counter flow drum mix solution is the double drum (Astec Inc., 2014). Figure 14.11 shows a double barrel drum mix plant. The double drum consists of a rotating dryer with the burner mounted at the discharge end. After passing through the dryer, the dried and heated aggregates are fed into an outer stationary shell of the drum. The rotating dryer has mixing paddles attached to the outside of the drum. The bitumen, filler and RAP are added to the outer stationary shell. Up to 60% RAP can be added to the double drum. Water vapour, together with any fumes created by the RAP addition, are drawn into the burner and incinerated. The internal structure of the double drum is illustrated in Figure 14.12.

The counter flow drum mix and the double drum are both US design solutions for the new breed of drum mix plant. They can both also be used as conventional dryers, with the dried and heated aggregates being discharged

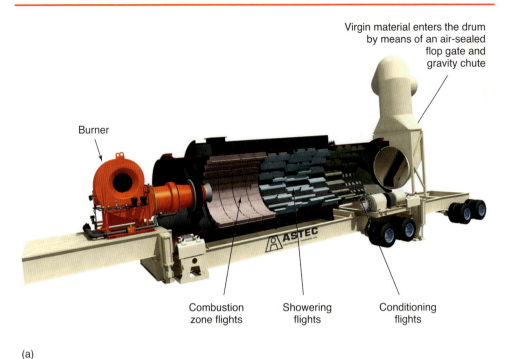

Virgin material enters the drum by means of an air-sealed flop gate and gravity chute

Burner

Combustion zone flights

Showering flights

Conditioning flights

(a)

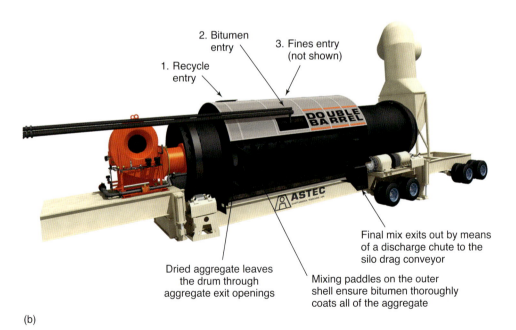

2. Bitumen entry

3. Fines entry (not shown)

1. Recycle entry

Final mix exits out by means of a discharge chute to the silo drag conveyor

Dried aggregate leaves the drum through aggregate exit openings

Mixing paddles on the outer shell ensure bitumen thoroughly coats all of the aggregate

(b)

Figure 14.12 The internal structure of a double drum, showing (a) the inner drum drying zone and (b) the outer drum mixing zone. (Courtesy of Astec Inc.)

Figure 14.13 A ContiMix continuous mix plant with a capacity of 320 tonne/h situated inside a building in Lyon, France. (Courtesy of Ammann)

before the mixing zone. The discharged hot aggregates can then be fed to a conventional asphalt batch plant. This type of arrangement is termed a 'convertible' or 'combination' plant, and provides versatility to the customer where a batch and a continuous plant are required.

The European development of the drum mix plant is the continuous mix plant. The drum mix rotating cylinder is replaced by a standard conventional asphalt batch plant dryer and a continuous mixer. The continuous mixer is an extended paddle mixer with a weir plate at one end. The weir plate can be adjustable to control the mixing time, and the mixer shaft speeds can also be varied, again to control the mixing time. The bitumen, imported filler, reclaimed filler and RAP are added to the continuous paddle mixer. The continuous mixer is usually vented to the dryer to incinerate the fumes. The maintenance and operational costs of a continuous mix plant are higher than for a counter flow drum mix or double drum plant, as a paddle mixer is incorporated in the design. A continuous mix plant is shown in Figure 14.13.

The counter flow drum mix, double drum and continuous mix plants enable high temperature mixtures and the utilisation of high percentages of RAP for asphalt production in a continuous plant. The benefits are

- they are economical for long production runs of one material specification
- they can be used to manufacture all European (BS EN 13108:2006 (BSI, 2006a)) and US individual state DOT mix specifications, and therefore worldwide mixture specifications
- high capacities are readily achieved
- high temperature mixtures can be produced
- maintenance costs are relatively low

- they are easily adapted to a mobile design
- a high percentage of RAP can be added to the mixture
- fumes or blue smoke are incinerated by the burner.

Although the counter flow drum mix, double drum and continuous mix plants overcome two of the main shortcomings of the drum mix plant, it is still necessary that the feed is composed of pre-graded, single-sized aggregates, otherwise the final mixed material product may not meet relevant specifications. The limitations are

- pre-graded, single-sized feed materials are essential, otherwise mixture quality cannot be guaranteed
- there is wastage of heated material at the beginning and end of production
- small quantities are uneconomical
- a mixed material system is essential on higher capacity plants.

14.3 Plant capacities

There are two main constraints on the capacity of an asphalt plant: the drying capacity and the mixing capacity (Hunter, 2000). Both these process constraints are complex. Factors affecting the drying capacity are

- altitude
- ambient air temperature
- type and grading of feed material
- feed material temperature
- moisture content of the feed material
- moisture content of the discharge material (usually less than 0.5%)
- dryer discharge temperature
- exhaust gas temperature from the dryer
- angle of inclination of the dryer
- calorific value of the fuel
- available exhaust air volume
- lifter pattern inside the dryer
- burner efficiency.

Factors affecting the mixing capacity to ensure a homogeneous mixture are

- mixture recipe
- mixer live zone (if applicable)
- paddle tip speed (if applicable)
- size of paddle tips and arms (if applicable)
- mixer arm configuration (if applicable)
- mixing zone lifter configuration (if applicable)
- method of feeding ingredients

- temperature of mixed materials
- mixing time.

Either the drying process or the mixing process determines the capacity of an asphalt plant. Both must be considered separately, and the capacities compared. The lower restricting capacity is then selected as the overall plant capacity.

The capacity of a batch plant is easy to determine, as the drying and mixing processes are separate. In drum mix, counter flow drum mix and double drum plants, the drying and mixing processes are combined, and therefore more detailed consideration is required. The constraints due to drying in a drum mix plants and its variants are similar to those for a conventional asphalt plant dryer. The mixing time in a drum mix type plant can be controlled by adjusting the throughput, altering the drum angle, adjusting the position of the bitumen injection pipe within the mixing zone, altering the design of the mixer lifters, or altering the speed of the rotating cylinder. However, in reality, once these variables have been set during commissioning, or at the design stage, they are not usually altered. The constraint due to the mixing process in a drum mix type plant is, therefore, usually determined from firsthand operational experience with the equipment used.

The following information is required to determine the size of an asphalt batch or continuous plant

- maximum plant capacity
- type of feed material
- average moisture content of the feed material
- types of mixtures to be produced
- temperature of mixtures
- special constraints (e.g. minimum mixing time).

14.4 Additive systems

There are many additive systems that can be incorporated in an asphalt production plant, either batch or continuous. However, an additive ingredient is usually easier to control in the batch process.

Asphalts can be produced in many different colours. The most common coloured additive is red, although blue, green, gold and clear mixtures are also available. The different coloured mixtures can be achieved by adding a pigment, in bag form, direct into the paddle mixer, or by adding the pigment into the bitumen supply.

Rubber can be added to the mixture in several formats. Latex can be added by means of a pump into the paddle mixer, or into the mixing zone in a drum mix type plant. Rubber granulate can be added, by weight, directly into the mixer in a batch plant, or through the RAP collar in a continuous plant.

Fibres are also required in some stone mastic asphalts and thin surface course systems. The fibres can be in loose form, which are small fibre 'strings' or pellets. The pellets can also be impregnated with bitumen. The fibres can be introduced in bulk form or as bags. The use of bulk fibres requires the inclusion of silos with associated handling equipment, and the bagged fibres require the use of bag splitters. The fibres are added to the mixture by volume or weight, directly into the paddle mixer.

14.5 Cold mix plants

Cold mix plants can be of batch or continuous design, although the majority in use now are continuous plants. The term 'cold mix plant' refers to the aggregates being fed into the plant, which are not heated. The bitumen and the bitumen emulsions added to the mixture are heated. The equipment is therefore very basic, as there are no dryers, burners, dust collectors, grading screens, hot stone bins or weigh hoppers, and mixed material storage is optional. The simplicity of the cold mix plants results in low level arrangements, which lend themselves to being mobile or skid mounted. A continuous cold mix plant is shown in Figure 14.14.

Cold mix plants can be used to produce cold mix asphalt with foamed bitumen, cold mix asphalt with emulsion, and hydraulically bound mixtures

Figure 14.14 A continuous cold mix asphalt plant with capacity of up to 400 tonne/h. (Courtesy of Ammann)

for use as a base. Foamed bitumen is created by spraying water under pressure through nozzles into the hot bitumen. The water expands when the pressure is released, resulting in a fine bitumen foam that is used to coat the cold aggregate feed (for an overview of the literature on asphalt mixing plant, see Ammann Schweiz AG, 2014).

In a continuous cold mix plant, pre-graded single size aggregates are fed into the plant using pre-calibrated, dedicated variable speed belt feeders. The feeders are set at the percentage required in the finished mixture specification, and the aggregates are then fed onto a belt weigh conveyor. Prior to the weigh section on the conveyor, the cement, hydrated lime or imported filler is added by way of a silo and a loss-in-weight system onto the aggregate feed. The belt weigher is therefore weighing the total of the wet weight of the aggregates and the dry filler weight. The aggregates and filler are discharged into a continuous paddle mixer. This is an extended paddle mixer with a weir plate fitted at the discharge end. The signal from the belt weigher controls the metering of the bitumen and/or water or bitumen emulsion feed into the continuous mixer through a spray bar. The cold mixture discharges directly into trucks, by means of a conveyor into trucks, or into an optional cold mix surge hopper. Direct truck loading involves the control system being set to produce the specific quantity required in the truck.

A cold mix batch plant has a paddle mixer with a bottom discharge instead of a continuous mixer, and a bitumen weigh hopper instead of a bitumen metering system. Each batch is made separately, whereas in a cold mix continuous plant each truck load or surge bin capacity is specifically set for production. Continual discharge into trucks from the surge hopper would equate to continuous plant production. Capacities of up to 400 tonne/h can easily be achieved in cold mix plants.

RAP can be directly added in the cold mix plants, either into the continuous mixer or the paddle mixer.

Cold mix plants are becoming increasingly popular in some parts of the world for the production of bitumen bonded bases and hydraulically bound mixtures, where only aggregates, cement and water are used. The initial cost of the plant is low, as are the operational and maintenance costs, because no heating or dust collection is required, and therefore the cost of the cold mix is economical. However, pre-graded, single size aggregates must be used to feed the plant, and the performance of the finished cold mix product is limited. Further development of these cold mixes is currently being undertaken. The benefits of cold mix plants are

- low initial cost of equipment
- low maintenance and operational costs

- low cost of the finished mixture
- reduced environmental impact, as no heating and drying of aggregates is necessary
- simple low level design
- easy to adapt to a skid mounted or mobile design
- high capacities are readily achieved
- small batches are possible
- no material wastage
- mixed material storage is not essential
- RAP can be added.

The limitations of cold mix plants are

- pre-graded, single size feed materials are essential, otherwise mixture quality cannot be guaranteed
- restricted performance of the finished cold mixture product.

14.6 Warm mix plants

The need to reduce the 'carbon footprint' when producing asphalt has led to the development of warm mix asphalt. Warm mix asphalt is categorised as being produced and mixed at temperatures roughly between 100°C and 140°C (EAPA, 2010; Nicholls *et al.*, 2013). Warm asphalt mixtures are produced in several ways

- the incorporation of chemical additives, such as amines, or the addition of waxes to reduce the bitumen viscosity (Hauguel, 2013; Walsh, 2013)
- the introduction of foamed bitumen, or the addition of moisture releasing additives, such as zeolites to the mixture (Nicholls *et al.*, 2013).

Whichever method is adopted the asphalt production plant can be adapted to accommodate the changes. Existing production plants can also easily be retrofitted with the necessary equipment.

The most common method used to produce warm mix asphalt is by the introduction of foamed bitumen. Foamed bitumen is created by injecting cold water into the flow of bitumen, which is usually at a temperature of 150°C (for more information on foamed bitumen, see section 22.1). The water is heated above boiling point in the hot bitumen and some of it evaporates, and the result is a mixture of steam, water and bitumen, which is then mixed with the low temperature aggregates. The foaming of the bitumen enables the aggregates to be coated by the bitumen at a lower temperature.

The quantity of water added is small when making foamed bitumen; for example, 1.25–2% of water by weight of the total bitumen content. The water is added to the process using a positive displacement pump and the quantity is controlled by a flow meter.

The benefits of producing warm mix asphalts compared with traditional hot asphalts are considerable

- lower manufacturing costs
- lower fuel consumption
- reduced carbon dioxide emissions to the atmosphere
- reduced fumes and odours at the production plant
- reduced fumes and odours at the laying site, and therefore a safer working environment for paving crews.

The major advantage with warm mix asphalts is the significant reduction in emissions. Research has shown that for every 11°C reduction in mixing temperature the emissions to the atmosphere are halved (Hauguel, 2013; Walsh, 2013). This is a substantial reduction in the carbon footprint for the asphalt production plant considering that the existing equipment can still be used.

14.7 Asphalt plant control

The controls on all asphalt plants are now fully automatic, usually with a PLC (programmable logic controller) to control the process and a PC (personal computer) to act as the interface with the operator. Plants are equipped with manual control mainly for maintenance purposes. The direction now being taken is to use a PC for manual control, instead of individual push buttons to start/stop each motor. The PC is then augmented by a backup PC, which is feasible due to the relatively low cost of a PC compared with the push buttons and associated wiring.

A PC can provide a dynamic image of the plant layout, with the separate components of the plant changing colour depending on whether or not they are active, and the flow of material through the plant is simulated using line graphics and interactive animated symbols. The image will also show the weights, tonnages and temperatures throughout the plant, so that the operator can see at a glance the overall plant performance. A dynamic image for a conventional asphalt batch plant is shown in Figure 14.15.

The control system has several functions

- plant operation
- fault diagnostics
- recipe storage
- production data
- daily production schedule
- operator interface

Figure 14.15 A dynamic display screen for an asphalt batch plant.
(Courtesy of Control Net Solutions)

- printing, storing and transferring production data
- quality assurance.

The control system detects and confirms whether the mixture has been produced in automatic or manual mode. This is important for quality assurance purposes. In fact, if the control system is in automatic mode, the correct data have been input, the correct aggregate sizes and bitumen grades have been fed into the plant, and no faults have been recorded, then the control system will ensure that the asphalt mixture is within specification. The plant operator just needs to observe the process, and intervention is only required if there is a problem. The quality of the mixture is therefore not dependent on the operator.

Process faults such as starved feeders, stopped motors, broken chains, tare weight faults, sticking doors, sticking valves etc. can be monitored by the control system and alarms raised when there is a problem. This reduces downtime and improves the overall plant efficiency. The control system can also be set to stop the plant if temperatures or weights are not within pre-set parameters, thus providing quality assurance.

The information from the production schedule can be printed, stored and/or transferred to another PC, which may be in the laboratory, at the weighbridge or in the accounts department.

Each aspect of the asphalt plant production process is or can be controlled and monitored automatically using the flow rates, the weights being processed, the levels in the hoppers, bins and silos, and the temperatures throughout the plant.

The variable speed volumetric belt feeders have a turndown ratio of 20 : 1 to control the flow of aggregates to the plant. The feeders are calibrated during plant commissioning to achieve the required capacities. The feeders can therefore be set to operate at between 5% and 100% capacity to ensure optimum aggregate feed to the plant with minimum material and heat wastage, providing the hoppers are filled with the correct size aggregates. The individual belt feeder speed is set in the controls to match the mixture recipe percentage of the particular aggregate being handled. Low level indicators can be fitted to the feed hoppers, with elevated warning lights above the hoppers to alert the shovel driver of the need to charge that particular aggregate hopper. No flow indicators can also be fitted to the feeders to warn the operator that the hopper is either empty or a blockage has occurred. A no flow condition can either activate an alert, or after a time delay the plant can be automatically stopped in a controlled sequence.

The burner on a batch or continuous plant can have a turndown ratio of up to 10 : 1. This means that for a plant with a maximum burner flow rate of 3000 l/h the minimum flow rate can be reduced to 300 l/h. This enables an economic use of fuel for a whole range of different flow rates and various discharge temperatures. The turndown ratios are achieved by variable speed fuel pumps and variable speed blowers fitted to the burner, the motor speeds and settings are individually controlled by the PLC. The discharge temperature of the material leaving the dryer, or the drum mix is measured by a non-contact infrared pyrometer, which is used to alter the burner fuel and air settings in a closed loop control.

The volume of exhaust air from the dryer, or drum mix is controlled by a pressure sensor mounted in the burner end box. The volume of exhaust air is increased or decreased to maintain a set pressure at the burner in order to sustain combustion and remove the products of combustion. The exhaust air is varied by employing a variable speed exhaust fan.

The levels of the materials within the hot stone bins and filler silos are monitored by the use of continuous level indicators within each hot stone bin compartment and in each filler silo. The hot stone bin level indicators provide the operator with trending information on the filling and discharge rate of each hot stone bin compartment. The operator can then adjust the feeders according to this information if the levels are not within the mixture specification requirements. This enables the plant to perform more economically with reduced overflow from the hot bins. The feeders are not automatically controlled from the hot bin level indicators because the levels are a trend and only accurate to within 10% of the actual level. In addition, and more importantly, the feed aggregates from each feed hopper will be distributed into several hot bins and the specification of these aggregates can be inconsistent, so automatic control is not feasible. The levels in the filler silos are for information only; however, they provide useful data instead of just 'full' or 'empty'.

The accuracy of the batch plant weigh gear for the aggregate, bitumen and filler ingredients is consistently repeatable to within ±0.3%. Weighing the bitumen relative to the actual weight of aggregate and filler recorded in the weigh hoppers enables the bitumen, which is the most expensive ingredient in the mixture, to have a recovered accuracy of ±0.1% of the specification target value. Self-learning in-flights from the hot stone bins discharge to provide accurate aggregate weights, and alternative hot bin selection are features incorporated in batch plant control systems to help sustain plant capacity while still maintaining accuracy. The load cells on the belt weigher in continuous plants have guaranteed accuracies of ±1%.

Temperatures are monitored throughout the plant prior to the asphalt being mixed. Thermocouple probes are located in the hot stone bins and the bitumen ring main to ensure that the temperatures are within the specification limits before mixing takes place. Non-contact infrared pyrometers are also positioned under the paddle mixer on batch plants and at the drum mix discharge to record the actual mixture temperature. Pyrometers can also be fitted at each mixed material silo discharge point to record the temperatures of the asphalt in the trucks.

The mixed material storage silos can be fitted with load cells or strain gauges to determine the weight in each silo, and to offer an accurate automatic load out facility to trucks, thus preventing wastage and saving time from excessive visits to the weighbridge. Alternatively, weighbridges can also be mounted underneath the mixed material silos to provide an automatic load out system; however, this does not record the weight in each silo.

14.8 Sampling

Testing of samples taken from an asphalt plant should only be carried out by skilled and experienced personnel using well maintained and calibrated test equipment (Forde, 2009). It is impractical to test every tonne produced, but measures must be taken to ensure that the aggregate grading and the bitumen content are in accordance with the mixture specification. Therefore, samples are taken at predetermined intervals, making sure that the samples are representative of the whole production. Recommendations about minimum sampling frequencies are provided in BS EN 13108-21:2006 (BSI, 2006b), and care must be taken to ensure that the sampling frequency provides evidence that the plant is under control.

In the UK, methods of sampling and achieving representative samples are given in PD 6682-2:2009 + A1:2013 (BSI, 2009/2013) for the incoming aggregates, sands and fillers to the asphalt plant, and are described in PD 6692:2006 (BSI, 2006c) for the asphalts being produced. These are useful UK guidelines on the use of BS EN 13043:2002 (BSI, 2002) and the various BS EN 12697 testing standards (BSI, 2005–2012), respectively.

Materials being delivered to the asphalt plant should also be sampled, including the aggregates, bitumens and fillers. If the asphalt plant is connected directly to the crushing plant storage bins then part of this control will already be in place.

Samples can also be taken from within the production process. On batch plants, sampling facilities are usually located beneath each hot stone bin compartment. Alternatively, these samples can be taken directly from the paddle mixer discharge. These samples confirm the aggregate grading within each hot stone bin compartment, helping to warn of material contamination or screen mesh breakages. Sampling points are also incorporated in the reclaimed and imported filler discharge points, and usually for each bitumen tank. These individual samples should be taken at regular intervals to help control the aggregate grading and to confirm that the correct bitumens and fillers are in use.

During the calibration and commissioning of asphalt plants, or if troubleshooting, samples are taken from the paddle mixer, drum mix discharge and the mixed material discharge points. The samples from the paddle mixer or drum mix confirm whether a homogeneous mixture has been produced. The samples from the mixed material discharge indicate whether the material has been stored successfully, or if segregation, excessive loss of temperature or some other deterioration has occurred. During full production, the samples are usually taken from the truck.

Sampling and testing in the laboratory takes time, and asphalt plants producing at high capacities can result in a large quantity of asphalt being laid on a highway or airport runway before the results are available. For this reason, small production runs and tests are usually carried out prior to large production batches being undertaken. Sampling of asphalt production plants is essential to show compliance with the European standard for asphalts, and is usually carried out in accordance with components of BS EN 12697 (BSI, 2005–2012). The positive results of sampling ensure confidence in the performance of an asphalt plant, and provide quality control of the production process. Sampling on an asphalt plant is essential, and must be performed regularly.

References

Ammann Schweiz AG (2014) See http://www.ammann-group.com (accessed 21/09/2014).

Astec Inc. (2014) See http://www.astecinc.com (accessed 21/09/2014).

BSI (British Standards Institution) (1999) BS EN 536:1999. Road construction machines, asphalt mixing plants, safety requirements. BSI, London, UK.

BSI (2002) BS EN 13043:2002. Aggregates for bituminous mixtures and surface treatments for roads, airfields and other trafficked areas. BSI, London, UK.

BSI (2005–2012) BS EN 12697:2012. Bituminous mixtures. Test methods for hot mix asphalt. BSI, London, UK.

BSI (2006a) BS EN 13108:2006. Bituminous mixtures – Material specifications. BSI, London, UK.

BSI (2006b) BS EN 13108-21:2006. Bituminous mixtures – Material specifications, factory production control. BSI, London, UK.

BSI (2006c) PD 6692:2006. Asphalt. Guidance on the use of BS EN 12697. BSI, London, UK.

BSI (2009/2013) PD 6682-2:2009 + A1:2013. Aggregates. Aggregates for bituminous mixtures and surface treatments for roads, airfields and other trafficked areas. Guidance on the use of BS EN 13043. BSI, London, UK.

EAPA (European Asphalt Pavement Association) (2010) *Position Paper: The Use of Warm Mix Asphalt.* EAPA, Brussels, Belgium.

Forde M (2009) *ICE Manual of Construction Materials.* Thomas Telford, London, UK.

Hauguel S (2013) Presentation to the Irish Branch of IAT. Edinburgh, UK, p. 14.

Hunter RN (2000) *Asphalts in Road Construction.* Thomas Telford, London, UK.

Jones L (1986) Recent developments in coating plant technology. *Quarry Management* **13(10)**: 25–30.

Nicholls JC, Bailey H, Ghazireh N and Day DH (2013) *Specification for Low Temperature Asphalt Mixtures.* TRL Ltd., Crowthorne, UK, PPR666.

US Patent (1931) US 1836754A. Hepburn, Donald Mcknight, 15 December, 1931.

Walsh T (2013) Irish Branch report. *Asphalt Professional* **58**: 14.

Chapter 15

Transport, laying and compaction of asphalts

Jonathan Core
Divisional Manager, Jean Lefebvre (UK) Ltd. UK

15.1 Transport

It is important that heat loss during transportation is minimised. Accordingly, delivery vehicles must have a body constructed with an acceptable insulating material on all surfaces in contact with the asphalt, together with a means of covering the top of the asphalt in a manner that minimises heat loss. This can be achieved using either a mechanical sheeting system, or a sheet positioned under the asphalt coupled with a waterproof sheet over the top of the hot asphalt. A typical modern delivery vehicle used for transporting asphalts is shown in Figure 15.1. Journey times and standing time on site should be as short as possible, and this is particularly important during inclement weather.

Care should be exercised during the loading of delivery vehicles at the asphalt production plant to ensure that opportunities for cold spots are minimised. This, in turn, will reduce the risk of inadequate compaction and early failure of the asphalt.

Any releasing agent used in the wagon body to assist discharge should be such that it will not contaminate or adversely affect the performance of the asphalt. The amount of releasing agent used should be kept to a minimum.

15.2 Preparatory works

Asphalts should only be laid on surfaces free from detritus and deleterious material. Prior to installation, the underlying course should be prepared to produce a stable surface of appropriate profile on which the asphalt can be placed. Any weak areas of the substrate should be strengthened, and major variations in profile corrected. Crazing and narrow cracks in the substrate are acceptable; however, wide cracks should be sealed using an appropriate treatment. Badly cracked areas should be removed completely.

The Shell Bitumen Handbook, Sixth Edition
ISBN 978-0-7277-5837-8
Shell International Petroleum Company Ltd: All rights reserved
http://dx.doi.org/10.1680/tsbh.58378.413

413

Figure 15.1 Typical asphalt delivery vehicle. (Courtesy of Eurovia UK Limited)

Where improvements are required to the receiving layer, the addition of a regulating course or cold milling may be necessary to achieve the designed profile. Cold milling (Figure 15.2), also described as 'planing', should be undertaken using a suitable milling machine that can complete the work to the required depth both effectively and efficiently. Depending on their size, modern cold milling machines can remove asphalt at depths of between 0 and 350 mm, and at operating widths of between 0.35 and 3.8 m. The cut edges should be left neat, vertical and in straight lines. The existing ironwork should not be disturbed by the milling process. Where necessary, surfacing in the vicinity of ironwork and in small or irregular areas should

Figure 15.2 Cold milling. (Courtesy of Eurovia UK Limited)

be cut out using pneumatic tools or other suitable methods and removed. Immediately after milling, surplus materials should be removed using a small loading shovel or back hoe loader, and the milled surface mechanically swept to remove all dust and loose debris. The finished surface should have a uniform texture.

Any water ponding on the receiving substrate must be removed.

Where asphalt is to be installed on existing concrete pavements with defective joints, the joints should be made good by cleaning out and refilling with a suitable joint treatment. This treatment should finish flush with the surface, and should be of a type that will not adversely affect the overlying asphalt.

Prior to the installation of the asphalt, the vertical faces of any access chamber covers, gullies, kerbs, channels and similar projections against which the asphalt is to abut should be cleaned and painted with a uniform coating of hot applied paving grade bitumen, cold applied thixotropic bitumen emulsion of similar grade, or polymer modified bitumen emulsion bond coat. The choice of material to be used will be dictated by the job specification. Many clients understandably believe that there is no substitute for paving grade bitumen.

15.3 Application of bond coats

It was explained in Chapter 9 that a tack coat is a bituminous emulsion used to promote adhesion of contiguous layers. It was further explained that bond coats perform the same function but are proprietary products manufactured with polymer modified bitumen. Compared to a tack coat, a suitable bond coat is much more likely to enhance adhesion of contiguous layers. It is for this reason that, in the UK, bond coats are invariably specified rather than the traditional tack coats as per the Note to Clause 3.2 of BS 594987:2010, which states: 'While tack coats have traditionally been used, they are no longer regarded as best practice' (BSI, 2010b).

The bond/tack coat should be hot applied either by using equipment separate from the paver (Figure 15.3), in which case it should be permitted to break (i.e. to turn from brown to black in colour) prior to paving, or by using a paver with an integrated spray bar (sometimes described as an 'integral paver'), in which case breaking occurs on contact with the overlying hot asphalt. Breaking is essential because lorries reversing into the paver hopper and the paver itself will lift the bond coat on their tyres.

Before spraying is started, the surface should be free of all loose material and standing water, which would affect the adhesion between the layers.

Rates of spread for bond coats or tack coats are typically found within the applicable specification and/or proprietary product guidelines. In the

Figure 15.3 Application of a bond coat. (Courtesy of Eurovia UK Limited)

UK, it is currently specified in Clause 5.5.2 of BS 594987:2010 (BSI, 2010b) as 0.35 kg/m^2 of residual binder. This is currently being reviewed with a recommendation that it should be reduced to 0.20 kg/m^2. Selection of an appropriate tack coat or bond coat will be dictated by the nature of the substrate. The aim is to leave an even coating of bitumen on the receiving layer. Permeable (dry and/or open) surfaces and hydraulically bound substrates may require a higher rate of spread than that required for asphaltic substrates.

Care should be taken to avoid over application or ponding, as this could result in a degree of slippage or instability in the overlying layer and, in some circumstances when overlaying with an open material, bleeding. Consequently, any emulsion accumulating in hollows should be dispersed by brushing.

In the UK, until 2006, the inclusion of a tack or a bond coat was normally only required below surface courses on areas of overlay or at tie-ins to existing pavements. However, the 2006 edition of the UK pavement design standard HD 26, which is still current, states that 'a tack or bond coat is required between all layers' (Highways Agency *et al.*, 2006). In recent years, there has been some discussion in the UK about the role played by tack coats and bond coats in aiding efficient pressure dissipation across asphalt layer interfaces. Although there have been no studies that have provided proof that a lack of bond may, of itself, cause a pavement to fail, there is no doubt that the application of a tack coat or a bond coat at all asphalt to asphalt interfaces is sound practice (Nicholls *et al.*, 2008).

15.4 Delivery and discharge

Asphalt deliveries should be coordinated primarily to avoid interruption to the laying operation but also to avoid any unnecessary waiting on site. Any such delay may result in the delivered asphalt cooling to a point at which workability becomes an issue and, in turn, have an adverse effect on the long term performance of the pavement.

Where the paver cannot be operated in a continuous mode, consideration should be given to the possibility of material at the extremes of the hopper and at the end of loads, where cooler material is often found, being discarded. Cool material can result in inadequate compaction and subsequent premature failures, such as ravelling. This type of failure can also occur when material is dropped from delivery vehicles in front of the paver and is not removed, as it will cool rapidly on the cold substrate before being overlaid with asphalt from the paver. Delivery vehicles should be completely empty before withdrawing from the hopper of the paver to avoid the tailgate dragging asphalt off the hopper onto the area in front of the paver.

15.4.1 Material transfer vehicle

In many countries, material transfer vehicles, often colloquially known as 'shuttle buggies', are used in tandem with a paver or pavers to remix the asphalt immediately prior to installation (Figure 15.4). This reduces the incidence of material segregation and temperature variations caused by subjecting the material to long haul distances or excessive waiting times.

Material transfer vehicles, in conjunction with a paver hopper insert, can provide additional receiving capacity of the delivered asphalt, with the result

Figure 15.4 Material transfer vehicle. (Courtesy of Eurovia SA)

that the waiting times of delivery vehicles on site are reduced and continuous paving operations are achievable. Additionally, during material discharge, the material transfer vehicle serves as a buffer between the truck and paver, eliminating any bumping and resulting surface regularity issues.

15.5 Pavers

Pavers can install the full range of asphalts as well as granular, hydraulically bound and cement bound materials. They can lay at depths up to 400 mm at widths up to 16 m. Pavers are fitted with powerful engines that provide traction through crawler tracks (Figure 15.5) or pneumatic tyres (Figure 15.6).

Crawler tracked pavers provide a large contact area with the substrate, allowing them to achieve a high tractive effort, and making them ideal for installing materials on difficult terrain or when paving at large widths.

Wheeled pavers are much easier to transport than tracked pavers, and, with a travel speed of around 20 km/h, moving them between sites in close proximity is also an option. Compared to a tracked paver, a wheeled paver has excellent manoeuvrability, a typical turning radius being 6.5 m.

Conventional pavers are capable of laying asphalt on the inclines typically found on the UK road network up to a gradient of 20° (36%). For specialist applications, such as dam walls or the steep banks of a racing circuit, a conventional paver will require modifications to lay asphalt either vertically

Figure 15.5 Tracked paver with integral spray bar. (Courtesy of Eurovia UK Limited)

Figure 15.6 Wheeled paver. (Courtesy of Eurovia UK Limited)

or horizontally on the incline. Alternatively, specialist 'slope' pavers can be used to lay asphalt vertically on an incline of up to 32° (62.5%).

After deposition into the paver hopper at the front of the machine, slat conveyors transfer the material through a tunnel in the paver to the augers, located in front of the screed. The augers distribute the material evenly in front of the screed. The width of the augers can be adjusted to ensure uniform compaction of the material by the screed at all times.

15.5.1 Screed

The screed is the main difference distinguishing a paver from other construction machinery capable of laying asphalt (e.g. a grader). Pivoting from tow points on each side of the paver, the screed generally 'floats' on the uncompacted material during the paving process. The screed alters with every change in the balance of forces (e.g. as a result of greater resistance from the paving material caused by variations in its temperature and/or inconsistent material properties/composition). The layer thickness should only change as a result of an alteration in the 'angle of attack' of the screed or an adjustment of the height of the screed tow points, which are located on each side of the paver. By operating in this manner, irregularities in the layer being overlaid are reduced in each subsequent layer.

All screed components that are in contact with the asphalt should be heated to approximately 90°C before the operation starts. Failure to do so may result in the asphalt sticking to the underside of the screed, resulting in the formation of strips and irregular surface texture in the laid material. In addition, the floating behaviour of the screed may be affected, resulting in irregular layer thickness.

The screed is capable of laying a wide range of transverse profile shapes in the paved layer. Adjustments allow for a positive crown profile up to +5% and a negative crown down to −2.5%. 'M' and 'W' profiles are also possible (so called because these are profiles that loosely take an 'M' or 'W' shape in cross section).

The screed also houses the compaction system, which pre-compacts the asphalt to the greatest possible extent and consequently reduces the influence of layer thickness on the amount of subsequent compaction by the roller(s). Different pre-compaction systems are available.

15.5.2 Tamping screed

Driven by an eccentric shaft, the tamper bar employs a vertical high amplitude movement at comparatively low frequencies, typically between 300 and 1800 revs/min, to assist the flow of material under the screed as well as providing optimum pre-compaction. Correct set up of the tamper to suit the layer thickness and paving speed is crucial in ensuring optimum installation. Two factors determine the rate of compaction: tamper speed and the length of the tamper stroke. Both can be set up precisely, and adjusted to match perfectly the flow of mixture, type of mixture and layer thickness.

The tamper speed can be regulated: the higher the tamper speed, the greater the compaction achieved. In addition, the length of the tamper stroke should be adjusted depending on the thickness of the layer to be paved. For thin layers or ultra-thin overlays, a short tamper stroke of 2 mm is recommended, along with a very low tamper speed. For medium layer thickness, typically binder courses and surface courses, best results are achieved with a stroke length of 4 mm. When laying thick layers (i.e. those in excess of 120 mm), a maximum tamper stroke of 7 mm will provide optimum compaction.

15.5.3 Vibrating screed

Vibrators acting on screed plates and the screed frame cause them to vibrate. This movement reduces the frictional resistance between the asphalt particles, and thus promotes particle realignment and consequently aids compaction. Like the tampers, the speed of the vibrators can be regulated, the vibrating intensity rises with an increase in vibrator speed.

Typically, for thin layers (i.e. up to 30 mm) or for asphalts having a small nominal aggregate size (up to 6 mm), a low vibrator speed between 1000 and 1200 revs/min is required. For medium layer thickness in the range 30–100 mm, best results are achieved using a medium vibrator speed of between 1200 and 2100 revs/min. Thick layers (i.e. those in excess of 100 mm thick) require a high vibrator speed. Depending on the type of screed, the maximum vibrator speed can be up to 3000 revs/min.

15.5.4 Combination screed

Combination screeds provide the versatility of both tamping and vibratory pre-compaction, either independently or in combined mode. The additional screed weight from housing both systems also contributes to higher levels of pre-compaction being achieved. Most modern pavers are fitted with a combination screed.

15.5.5 High compaction screed

When laying thick asphalt layers, a high compaction screed may be considered. Compared to a conventional screed, these types of screed contain double tamper and vibration systems, which add to the weight of the unit. This results in higher levels of pre-compaction being achieved.

Some manufacturers also incorporate pressure bars into the rear of these types of screed. These operate at between 40 and 130 bar depending on the depth of the asphalt layer, and force the mixture down until it cannot be compressed further and the maximum density is achieved (Figure 15.7).

Figure 15.7 High compaction screed. (Courtesy of Joseph Vögele AG)

15.7 Machine laying

During paving, when the position of the screed tow points are stationary, an equilibrium of forces resulting from a combination of paving speed, screed weight and tamper speed is established. If any of these parameters alter, there will be an immediate effect on the floating behaviour of the screed. Accordingly, machine laying should be planned so that the asphalt layer can be installed continuously, in order to produce an even and compact surface to the required width, thickness, profile and camber or crossfall without causing defects or irregularities such as segregation, dragging and/or burning. The topography of the site should be taken into consideration when selecting the plant required to undertake the work.

It is essential that the paver is set up correctly to lay the particular asphalt being laid. The flow gates and/or the material handling conveyor should be adjusted to deliver an adequate supply of material and to provide and maintain a constant head of material in front of the screed.

The auger height and speed should be adjusted to suit the laying depths and the required rate of delivery. Typically, the lower edge of the auger flight should be set at a height above the receiving layer that is equivalent to approximately five times the nominal aggregate size of the asphalt being placed. If the auger height is excessive, the head of material in front of the screed may not extend to the outer extremities of the screed. If the auger height is inadequate, the pre-compacted material may have an open appearance, which may cause compaction deficiencies and subsequent durability issues. Extension augers should be fitted if the width of the screed is increased. This is necessary to minimise material segregation.

The screed should be checked for true alignment across its width prior to installation. Particular attention should be paid to the adjustment of the screed compaction unit, to ensure it pre-compacts the material sufficiently and does not cause any damage to the material.

The finished surface should be continuously inspected as it is laid, and any defects rectified immediately and before any rolling takes place. A finished asphalt surface that looks good (i.e. homogeneous and properly compacted both longitudinally and laterally) is much more likely to perform properly than one of variable appearance. There should be no scattering back by hand of additional asphalt on machine-laid work where it has an open appearance. The reason for the laid asphalt appearing open should be ascertained and addressed.

It is particularly important that asphalt in the vicinity of longitudinal joints (such joints are sometimes called 'rip joints') is laid properly and compacted adequately. Longitudinal joint failure is probably the most common defect in asphalt highways.

15.7.1 Head of material

The height of the asphalt in front of the screed is called the head of material. The head of material should be constant across the full paving width, and it should remain so as the asphalt is laid. A constant head of material is a precondition for perfect floating behaviour of the screed.

Figure 15.11 illustrates the importance of ensuring that the head of material is correct.

(a) If the head of material is too high, then the result is a force tending to lift the screed. If the screed lifts, then the laid thickness will be greater than that required.
(b) If the head of material is too low, then the result is a force tending to push the screed down. If the screed moves down, then the laid thickness will be less than that required.
(c) This shows the screed in equilibrium, with the result that the thickness of mat laid is the required value.

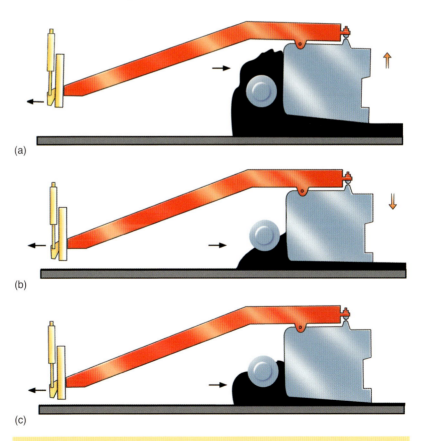

(a)

(b)

(c)

Figure 15.11 The importance of the head of material

The possibility of the head of material being incorrect can be reduced by the use of strike off and limiting plates for the auger tunnel, and sensors to regulate the auger feed.

15.7.2 Paving speed

Ideally, the paver should be operated at a constant speed, as any variations may result in surface irregularities. Stoppages can also cause profile irregularities and localised imperfections within the finished asphalt layer. After a stoppage, a paver screed tends to rise when work restarts. This creates a bump in the finished surface. To counter this tendency, modern pavers are fitted with a device called a screed stop or screed freeze.

The paving speed also determines the impact of the screed's pre-compaction system on the asphalt layer, so the speed should be set at a value that ensures optimum pre-compaction. To minimise the possibility of irregularities occurring, a small angle of attack should be adopted.

Material supply logistics and the capability of the compaction plant should also be considered when determining the paving speed.

15.7.3 Material segregation

Segregation of asphalts is the separation of the larger particles of coated aggregate in the mixture (i.e. the coarser fractions in the material). A compacted asphalt has to be homogeneous if it is to perform as expected. Segregation tends to be prevalent in asphalts of larger nominal sizes, which is why 40 mm nominal size asphalts were dropped from the UK asphalt production standard BS 4987-1:2003 (BSI, 2003). With 40 mm mixtures, it was always a concern that segregation would occur during loading and transportation. In the case of the latter, this was due to vibration of the load as it was transported. If segregation occurs in the hopper of the paver, the solution is to ensure the conveyors are covered with homogeneous material when the hopper sides are folded in (hopper sides are folded in from time to time during laying by the paver operator to move asphalt at the sides of the hopper into the area where the slats will convey the material through the tunnel to the augers).

If segregation occurs in front of the screed, it may be possible to rectify the issue by adjusting the auger height. According to the *Vögele Booklet on Paving* (Joseph Vögele AG, 2012), if, after adjustment, there is no improvement, additional smaller or different auger blades may be required. In these circumstances, the auger needs to rotate more quickly or more continuously, resulting in the asphalt being mixed more effectively.

If segregation occurs around the centre of the auger box (the auger box is the volume around the vicinity of the auger), the screed should be moved further back in order to increase the head of the material in front of the screed. Additionally, the height of the auger should be increased, and one or two auger blades in the central area of the auger box should be turned around to cause a reverse flow, which should ensure that all the asphalt fractions are conveyed behind the auger in the vicinity of its centre.

15.8 Hot-on-hot paving

Two layer construction of asphalt pavements by hot-on-hot paving is an efficient method of constructing durable roads. This method consists of simultaneously installing the binder course and surface course layers hot-on-hot (i.e. in a single pass), with both layers being compacted together. This approach results in perfectly bonded layers with a strong interlock between the courses, without the use of a bond coat or tack coat (Figures 15.12 and 15.13).

The process requires simultaneous supply of both materials, usually from two asphalt plants. On site, the materials are alternately fed into separate hoppers by a material transfer vehicle before being individually transferred and installed by two floating screeds running one behind the other. A high compaction screed is used to install the binder course, resulting in densities of 98% being achieved. Due to this high level of pre-compaction on the binder course, only a medium weight tandem roller is required to achieve the final density of the installed layers.

15.9 Hand laying

Hand laying is normally only permissible where machine laying is impractical or where small quantities of asphalt are being laid. Hand laying will

Figure 15.12 InLine Paving® system. (Courtesy of Joseph Vögele AG)

Figure 15.13 Compactasphalt™ system. (Courtesy of Eurovia Services GmbH)

never achieve the quality of finished pavement produced using a paver. Thus hand laying should be adopted only where there is no other option. When areas of handwork are unavoidable, installation should be undertaken quickly and efficiently, with the compaction plant being deployed as promptly as is safely possible.

The asphalt should be spread in a loose layer of uniform thickness and even texture, and thoroughly compacted immediately. Every precaution should be taken to minimise segregation and to avoid contamination. An inspection should be carried out after the initial rolling, with any rugous areas made good by carefully applying small quantities of hot material, commonly described as 'spotting up'.

15.9.1 Patching and repairs
As is always the case, the asphalt should continue to be protected from heat loss and the effects of adverse weather conditions. The material should be taken directly from the delivery vehicle, which should be kept sheeted for as long as possible. It should, preferably, be protected in a portable hot box.

As a means of ensuring reasonable riding quality (specifications often describe this aspect as 'surface regularity') after patching operations, the surface of any compacted patch should be flush with or slightly higher than the surrounding material. The finished surface should not be left below the surrounding surface level, as ponding may occur, and such depressions are more noticeable to vehicle drivers. (A slight bump upwards in the direction of travel is always less apparent to vehicle occupants than a drop down.)

15.10 Layer thickness and surface regularity

The nominal thickness of an asphalt layer can usually be found within the relevant transport, laying and compaction specifications. Typically, this is equivalent to multiplying the nominal aggregate size of the asphalt by around 2.5 or 3. However, there are exceptions to this rule. In the UK, the nominal and minimum compacted layer thickness for asphalts is specified in Clause 6.6 of BS 594987:2010 (BSI, 2010b).

Excessive thickness can lead to instability, causing surface regularity and level control problems. As a result, poor deformation resistance can be a potential long term issue in binder courses and surface courses of thicker pavements (total asphalt ≥ 180 mm).

Insufficient thickness can lead to aggregate damage (fracturing) within the asphalt, making the layer difficult to compact. As a result, the laid material may have poor durability. Such difficulties are often encountered when tying into an existing surface at the ends of an area of new construction or maintenance where the depth of asphalt laid falls below what is an acceptable thickness if the material is to be properly compacted. The solution is to ensure that the existing surface is milled out to an adequate depth to avoid this possibility.

Surface regularity (often described as 'rideability') can affect the durability of an asphalt pavement. The finished road profile is extremely important, because it is the main factor controlling ride quality, and hence user perception of the road condition. Indeed, many clients will judge the quality of a finished pavement solely on the quality of the ride. A sudden change in profile over a short length will significantly reduce the ride quality of a pavement. The profile of a road can also influence vehicle interaction. When a vehicle is moving along a road, unevenness will cause the vehicle to move up and down. In turn, this will cause a dynamic variation in the loading on the pavement, above and below their static values. This dynamic effect of road unevenness may accelerate the deterioration of the pavement, and its effect should not be underestimated.

The accuracy and quality of the surface regularity can be improved by using mechanical and electronic levelling devices, which are available for many modern pavers. These vary in design, and include devices described as contactable sensors, single cell sensors, sonic grade non-contacting sensors, variable length multiplex skis, averaging beams and laser receivers. All these systems detect level changes in the receiving layer and, depending on the readings, adjust the depth of asphalt laid by the paver accordingly. Each type of device can be used alone, on either side of the paver, or on both sides of the paver simultaneously (Figure 15.14). Another option for larger projects is the use of three dimensional positioning systems. These use a global

Figure 15.14 Averaging beams fitted with multi-cell sonic sensors. (Courtesy of Eurovia SA)

positioning system and specific digital design data for the site and automatically control the gradient, slope, paving width and direction of the paver.

15.11 Joints

Pavement durability can be extended by minimising the number of longitudinal and transverse joints. Thus, careful planning of the paving works should always be undertaken. Where possible, longitudinal joints should not be located within wheel tracks: wheel tracks are typically centred around 0.9 and 3 m from the nearside edge of the carriageway. In the UK, the wheel track zones are defined in Clause 903.21 of the *Specification for Highway Works* (Highways Agency *et al.*, 2009), which states that they 'shall be taken to be between 0.5 m and 1.1 m and between 2.55 m and 3.15 m from the centre of the nearside lane markings for each traffic lane (or, in the absence of lane markings, lane edges)'.

Transverse joints typically occur as a consequence of interrupted paving work. Such joints create a localised weakness within the pavement, resulting in a reduced service life as well as having a detrimental effect on ride quality. Accordingly, everything should be done to keep transverse joints to a minimum. On multi-layer pavements, the joints between the individual layers should be offset generally by 300 mm, or whatever is practicable if the applicable specification allows, from the joints in the course below. This minimises the possibility of water ingress throughout the pavement depth. It should always be remembered that water and pavements simply do not mix, and every possible step must be taken to prevent water getting into the pavement.

This is why the faces of kerbs should be painted, preferably with 40/60 penetration grade bitumen, although cold-applied joint paint or polymer modified emulsion may be permissible. This is essential to prevent rainwater, shed as a result of the camber, crossfall or superelevation of the carriageway into the channel, running down the face of the kerb and into the pavement.

Where joints have to be formed, it is imperative that care and consideration is given to their formation and compaction.

All transverse joints that occur when tying into adjacent surfacing or the day joint in new surfacing should be cut using a floor saw to provide a good vertical edge. The exposed edge should be thoroughly cleaned, and all loose material removed and discarded, before being sealed with an application of hot bitumen, cold applied joint paint or polymer modified emulsion, which should coat the entirety of the exposed vertical face. As above, many specifications understandably stipulate hot bitumen for joints. This bitumen seal is intended to ensure that the interface between adjacent layers is not permeable. For the same reason, transverse joints in base and binder course layers should be sealed, preferably with hot bitumen.

It is deemed best practice to form all exposed longitudinal joints. Indeed, many specifications will require this in future. This is most readily undertaken

Figure 15.15 Formation of a longitudinal joint using a chamfered cutting wheel. (Courtesy of Hamm AG)

Figure 15.16 Formation of a longitudinal joint using a cutting wheel. (Courtesy of Hamm AG)

by either using a chamfered cutting wheel, attached to the roller to form an edge at an angle of 45–60° (Figure 15.15). Alternatively, the longitudinal joint can be formed by cutting back the edge of the layer using a cutting wheel attached to the roller (Figure 15.16). This forms a near vertical face that exposes the full thickness of the layer.

Once the longitudinal joint has been formed, as with transverse joints the exposed edge should be thoroughly cleaned, and all loose material removed and discarded, before being sealed with an application of hot paving grade bitumen, cold-applied joint paint or polymer modified emulsion. It is important that the coat is applied to the entire exposed face. On base and binder course layers, the sealing of the joint at the surface with an additional application of a bitumen seal is also essential. As before, the use of properly applied penetration grade bitumen, either 40/60 or 100/150, is the most effective way of ensuring a waterproof seal.

It is not uncommon in car parks, including those at shopping centres, for joints to be left devoid of any bitumen seal. This is poor practice. All asphalt joints must be properly constructed if the asphalt is to last as long as possible, and this applies as much to a car park as it does to a highway.

When installing new asphalt layers adjacent to existing cold asphalt, the new layer must be thicker (surcharge) than the adjacent surface by the reduction in

Figure 15.17 Overlapping screed plate. (Courtesy of Joseph Vögele AG)

thickness that will occur as a result of the hot asphalt being compacted. This is essential to avoid the barrel of the roller being supported by the cold asphalt, resulting in the hot asphalt being compacted to a lesser degree than is required. In addition, the end plate of the screed should overlap the cold asphalt layer by 20–30 mm. Subsequently, the overlapped material should be manually pushed back into the area of the joint on the new asphalt layer prior to compaction by the roller. This procedure should not require any additional material, and keeps the required level of handwork to a minimum. This reduces the risk of material segregation and consequently improves the level of compaction at the joint (Figure 15.17). If the overlap is too narrow, then poor compaction at the joint will result. Alternatively, if the overlap is too wide, the paver will ride on the existing asphalt, causing damage to the cold material and resulting in poor compaction of the adjacent hot asphalt at the joint. This is a very common occurrence in asphalt surfacing.

Laying with two or more pavers working in echelon produces a hot-to-hot integral bond between the paved asphalt materials without the need to form a longitudinal joint (Figure 15.18). Elimination of the longitudinal joint is a very attractive proposition. However, as can be seen in Figure 15.18, the pavers cannot operate side by side. In these circumstances, the distance between the individual pavers should be minimised so that the joint face is still sufficiently hot to form a perfect bond with the adjacent asphalt.

Another way of hot jointing surface courses is to reheat the edge of the cold lane and bring it up to a plastic state using a joint heater prior to the new,

433

Figure 15.18 Paving in echelon. (Courtesy of Eurovia SA)

adjacent hot asphalt being laid. The heater must raise the temperature of the full depth of the surface course to within the specified range of minimum and maximum rolling temperatures for a width of not less than 75 mm. Care must be taken to ensure that the heated asphalt is not damaged in the process, as this would result in premature ageing.

15.12 Chipping of hot rolled asphalt surface course

In order for the finished surface to have an acceptable level of skid resistance, hot rolled asphalt surface course mixtures containing 35% or less nominal coarse aggregate have coated chippings rolled into them. This is achieved by the addition of 14/20 or 8/14 mm aggregate coated with not less than 1.5% by mass of 30/45 or 40/60 penetration grade bitumen. These are placed at a specified rate of spread on top of the newly laid asphalt, and then rolled into the asphalt using the compaction plant. Chippings of size 14/20 mm always have a superior appearance and result in less plucking than 8/14 mm chippings, and thus 14/20 mm chippings should be used wherever circumstances permit.

After initial compaction by the paver or, in the case of hand application, prior to the first pass of the roller, the installed asphalt surface should be covered with a uniform layer of coated chippings. Where possible, this operation should be undertaken using a mechanical chipping spreader (Figure 15.19). The chipped surface is then rolled to embed the chippings and compact the material. When spreading chippings on roads, an

Figure 15.19 Applying chippings to a hot-rolled asphalt surface course. (Courtesy of Eurovia UK Limited)

unchipped channel not less than 150 mm wide should be left adjacent to the kerb to facilitate the rapid flow of surface water into the gullies.

The degree of embedment and adhesion of the coated chippings is adversely affected by surface chilling of the asphalt layer. This occurs rapidly in windy conditions, even more so when combined with a low ambient temperature. Accordingly, all steps should be taken to ensure that embedment of the chippings and compaction of the layer is completed before the material reaches the relevant minimum rolling temperature.

15.13 Installing gussasphalt

Gussasphalt is a type of mastic asphalt, and consists of aggregate and high proportions of bitumen and limestone filler. The aggregate takes the form of sand or crushed rock fines. Typically, it contains between 6.8% and 7.5% binder. Once placed, the finished layer is impervious.

Gussasphalt is delivered in mobile cookers that maintain the material temperature between 200 and 260 °C, while also agitating the mixture in order to retain its workability and stability. Differences in the temperature and thickness of installed layers significantly affect the quality of the finished layer, so uniformity in both aspects is essential.

The paver, or as it is more typically known the 'finisher' (Figure 15.20), is fitted with a powerful engine that provides traction through crawler tracks,

Figure 15.20 Gussasphalt finisher. (Courtesy of Eurovia Services GmbH)

rail wheels or pneumatic tyres. Pavers are available in varying operational widths, typically between 1.0 and 14.5 m, and consist of a heated non-vibrating screed bar and a material distributor. Typical laying speeds range between 0.25 and 3.5 m/min.

Gussasphalt is typically laid at a depth between 20 and 40 mm. Due to its low viscosity, it must be laid between two restrained edges. These can be in the form of an adjacent existing layer of gussasphalt or other asphalt, a pre-constructed strip of gussasphalt or rails on which some finishers run. As a result, the restraining edges determine the laying depth.

The gussasphalt is deposited onto the substrate directly in front of the finisher. Subsequently, the finisher distributes the material across the width of a heated blade, which levels the surface to the pre-determined depth by travelling over it. The low viscosity of gussasphalt makes it self-levelling, so no vibration is required.

When installed as a surface course, gussasphalt is immediately gritted, either manually or mechanically, with aggregate coated with bitumen. The grit is subsequently embedded in the surface using a roller.

In common with other asphalts, gussasphalt should not be installed on a wet substrate, as the high installation temperatures and the impervious properties of the material trap the moisture beneath the layer, causing bubbles to appear on the surface.

15.14 Compaction

In a surfacing operation there is a myriad of activities that are undertaken before, during and after the asphalt has been laid. These range from cutting joints to the final rolling. Each of these elements is important in producing a finished road surface that will give many years of satisfactory service. However, compaction is perhaps the single most important activity in dictating the lifespan of a pavement. Accordingly, it is very important that all personnel involved ensure that compaction is carried out effectively and efficiently.

Compaction is the action of removing air from an asphalt, and is effected by rollers traversing the material, usually several times (called *passes*). Asphalt must be compacted in such a way that the particle displacement reduces the void content of the asphalt layer. Asphalts that are properly compacted will dissipate loads from traffic more quickly because compaction increases the stiffness of the mixture. High levels of compaction also reduce the rate of ageing of the binder, and, in doing so, prolong the life of the asphalt.

Compaction can be undertaken using two methods, or a combination of both

- static compaction
- dynamic compaction.

Static compaction uses the weight of the roller to exert downward pressure on the uncompacted material. Weight only operates vertically. This pressure causes the internal friction in the asphalt to be overcome, resulting in the mineral particles displacing air in the mixture and, as a consequence, reducing the void content.

The depth of the layer affected by static compaction is comparatively shallow. The higher the static linear load, the greater the depth to which static compaction has an effect. However, caution is warranted, as excessive linear loading may subsequently result in cracking and distortion in the layer being compacted.

Dynamic compaction penetrates to a much greater depth than static compaction. It is a more efficient compaction mode than static compaction. Eccentric weights cause the roller drum to vibrate, resulting in vibrations being transferred to the individual particles in the asphalt. This overcomes the frictional resistance between the particles, resulting in particle displacement. The effect of the vibrations together with the static load of the roller means that very high compaction can be achieved.

Some roller manufacturers now offer rollers that are capable of a particular variant of vibratory mode described as 'oscillation'. When a roller operates in oscillatory mode, the drum is in contact with the asphalt at all times, unlike normal vibratory rollers. Manufacturers claim that oscillation rolling only produces about 10% of the vibration-related stresses on the surrounding environment compared to those typically generated by a standard vibrating roller. This attribute makes their use more appropriate in situations where close control is required (e.g. in the vicinity of structures).

The amplitude and frequency at which rollers operate is important when compacting materials, including asphalt, in vibratory mode. The frequency is the rate at which the drum impacts the asphalt and is measured in hertz (Hz). The amplitude is the distance the drum moves from its axis, and is usually measured in millimetres. In the case of asphalt, the frequency would typically range from 50 to 70 Hz, while the amplitude will often be between 0.25 and 0.8 mm (Hamm AG, 2008). These parameters may be varied during compaction as the mixture densifies. It is important that the values of these parameters are correct in order to avoid problems that range from under-compaction to over-compaction, with the latter resulting in aggregate particles being crushed. Both under-compaction and over-compaction may well result in the asphalt having a shorter service life.

15.14.1 Types of rollers
The following roller types are typically used in compacting asphalts

- three wheeled static
- pneumatic tyred
- tandem.

Three wheeled rollers (Figure 15.21) compact asphalt layers through high static loading. The basis of this is a combination of their heavy operating weight, typically between 10.6 and 13.2 tonnes and the narrow width of the steel drums. These rollers are particularly suitable for levelling the asphalt surface and operating on sites where dynamic compaction may cause damage to the surrounding environment.

Pneumatic tyred rollers (Figure 15.22) achieve high levels of compaction through a combination of vertical loading from their heavy operating weight, typically between 9.5 and 30 tonnes, and a kneading effect caused by the wheels and, more specifically, their tyre pressure. This can be adjusted, within a range of 2 to 8 bar, to provide an even tyre-to-surface contact area. Curved tyre profiles should be avoided. These rollers are typically used in combination with a steel drum roller on asphalt possessing low stability properties, the latter being used purely to obtain an acceptable surface evenness on the compacted asphalt.

Tandem rollers (Figure 15.23) are the most common type of roller used for the compaction of asphalt. They comprise two steel drums, and are available

Figure 15.21 Three wheeled roller. (Courtesy of Eurovia UK Limited)

with operating weights of between 1.5 and 18 tonnes and corresponding working widths of between 0.8 and 2.14 m. They are fitted with either vibratory or oscillatory compaction modes, as well as having the capacity to operate as a deadweight roller. Additional options of articulated or pivot steering and all-wheel drives make these roller types very versatile.

Figure 15.22 Pneumatic tyred roller. (Courtesy of Eurovia SA)

Figure 15.23 Tandem roller equipped with oscillation. (Courtesy of Eurovia UK Limited)

Combination rollers (Figure 15.24) have a row of rubber wheels mounted on one axle and a smooth drum mounted on the other. As a consequence, these rollers combine the advantages of both a pneumatic tyred and tandem roller in one machine.

Figure 15.24 Combination roller. (Courtesy of Hamm AG)

15.14.2 Intelligent rollers

Compaction quality can be monitored and controlled more effectively by using rollers equipped with sensors. These sensors count each roller pass, plot the position of each pass and measure the surface temperature of the asphalt. All these data are available to the operator during compaction, and can be downloaded on completion of the work.

The advantages of intelligent rollers are

- uniform compaction
- detection of soft/weak spots
- optimising the number of passes
- efficient compaction
- potential cost saving.

15.14.3 Compaction of machine laid asphalt

The compaction of each layer should be carried out to a predetermined rolling pattern until the required density is achieved, and, in the case of surface courses, the layer has a homogeneous appearance.

As well as the size and weight of the roller(s) being sufficient to undertake the work effectively, the topography of the site should also be taken into consideration, so that the work is completed efficiently and safely. Smaller rollers should be considered for sites with restricted access, limited working area, restricted width of surfacing or where the underlying construction will not support a heavier roller.

The number of rollers available for use on a particular site should be such that, while travelling at a low but steady speed, sufficient passes can be made in order to compact the asphalt adequately in the time available. It is important to ensure that the frequency and amplitude of vibration or oscillation, along with the speed of travel of the rollers, are correctly matched to layer thickness and mixture composition.

Compaction should start without delay, and the rollers should operate as close as is safely possible to the paver. When using a tandem roller, compaction typically starts with an initial static compaction pass before the main compaction process is undertaken with vibration or oscillation. Care should be taken, as excessive vibration on a cooled material may fracture the aggregate. The final pass is typically undertaken using static compaction. To ensure that there are no uncompacted areas, the rolled tracks should overlap by at least 150 mm. Unsupported edges cannot be compacted. Accordingly, the edge should be either cut-back, thus removing the inadequately compacted material, or the edge should be supported during compaction. This can be achieved using a temporary restraint or an edge compactor fitted to the side of the roller, which compacts the asphalt leaving the edge sloping at 45–60° (see Figure 15.15).

There are many variables that affect the time available for compaction. One important parameter is the thickness of the layer being compacted. Figure 15.25 shows, in effect, the time available for compaction of two layers (20 and 150 mm thick). If the minimum temperature at which compaction can be effected is 80°C, then compaction has to be completed within 3 minutes of the 20 mm layer being placed, while compaction can still be effective after 30 minutes in the thicker layer. Thus, layer thickness is a very important factor in determining the time available for compaction (Hunter, 1986).

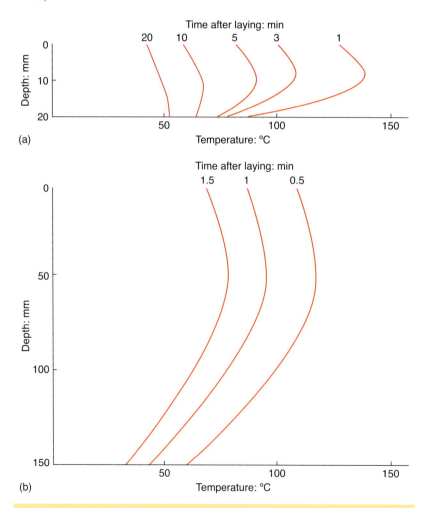

Figure 15.25 Cooling time in asphalt of (a) 20 mm thickness and (b) 150 mm thickness. (Courtesy of Dr Robert N Hunter)

Another crucial parameter is the speed of wind acting on the surface, as wind takes heat by convection from the hot asphalt. Convection losses are much more significant than the air temperature. The safest option is always to compact as quickly as possible. If the material moves unduly because it is too hot to be compacted, then the rollers can wait until the material has cooled sufficiently to permit compaction without damaging the surface. However, once a material has cooled to a point where further compactive effort is ineffective, nothing can be done to correct that shortcoming.

Due to the number of variables involved (up to 22), it is impossible to generalise about the best combination of rolling and roller pattern to use in all scenarios. There is no substitute for proper training and experience, combined with advice and support from roller manufacturers.

15.14.4 Compaction of hand laid and patching work
In areas where it is impractical to operate a large roller, such as footpaths, restricted access sites and limited working areas, either a 2.5 tonne static deadweight roller or vibrating roller having a minimum static weight of 750 kg would typically be used. The procedure adopted should closely replicate that used for larger scale work. However, more attention may be required with regard to material and rolling temperatures.

15.15 Application of grit to surface courses
Some contracts stipulate that the finished surface course should be gritted to improve initial skid resistance. This involves accurately applying a grit, which may be uncoated or coated with bitumen, at a specified rate of spread to the surface by way of a rear-mounted spreader fitted to a tandem roller. Grit application is undertaken after an initial compaction pass by the roller. The rolling pattern should, as far as is practicable, provide a single application of grit to the full width, with no overlap.

The rate of spread for the grit is typically no less than 600 g/m^2. The rate of spread should be automatically adjusted to take account of the roller speed, and should stop automatically when the roller halts or reverses.

Once the asphalt surface has reached ambient temperature, any surplus grit should be carefully removed using a mechanical sweeper prior to the application of road markings and/or being opened to traffic.

15.16 Opening to traffic
In order to avoid damage to newly laid sections of asphalt, it should not be overlaid or opened to traffic until the surface temperature has fallen to 40°C.

15.17 Specification and quality control

Compaction should be continuously assessed using an indirect density gauge, with readings taken at regular intervals in alternate wheel tracks. The gauge should be initially calibrated, and subsequently rechecked and updated based on correlations between gauge readings and core densities at the same locations. For each location, the in situ void content can be determined using the bulk density from the gauge reading compared with the maximum theoretical value ascertained in a laboratory. The average in situ void content should be calculated from any six consecutive indirect gauge readings.

Tests on pairs of cores can also be used to determine the achieved density. Cores have the additional advantage that they can be used to assess the void content at points below which the density gauge is effective. Cores are typically taken from each wheel track. In addition, a pair can be taken at a location centred 100 mm from the final joint position at any unsupported edge.

Other tests include longitudinal regularity and transverse regularity. On roads with crossfalls, this can be determined by measuring the depth of the depressions under a 3 m straight edge placed parallel to or at right angles to the centre line of the carriageway.

The most common method of determining the in situ texture depth of surface course materials is by means of the volumetric patch test, which involves taking ten individual measurements on a diagonal line across the lane width at a spacing of 5 m (BSI, 2010a). Although many UK authorities rely heavily on this parameter, published work suggests that such faith may well be misplaced (Hunter, 1996).

References

BSI (British Standards Institution) (2003) BS 4987-1:2003. Coated macadam (asphalt concrete) for roads and other paved areas. Specification for constituent materials and for mixtures. BSI, London, UK.

BSI (2010a) BS EN 13036-1:2010. Road and airfield surface characteristics. Test methods. Measurement of pavement surface macrotexture depth using a volumetric patch technique. BSI, London, UK.

BSI (2010b) BS 594987:2010. Asphalt for roads and other paved areas. Specification for transport, laying, compaction and type testing protocols. BSI, London, UK.

Hamm AG (2008) *Compaction in Asphalt Construction and Earthworks*. Hamm AG, Tirschenreuth, Germany.

Highways Agency, Scottish Executive, Welsh Assembly Government and Department for Regional Development Northern Ireland (2006) *Design Manual for Roads and Bridges*. Volume 7, *Pavement Design and Maintenance*. Section 2,

Pavement Design and Construction. Part 3, *Pavement Design*. The Stationery Office, London, UK, HD 26/06.

Highways Agency, Scottish Executive, Welsh Assembly Government and the Department for Regional Development Northern Ireland (2009) *Manual of Contract Documents for Highway Works*. Volume 1, *Specification for Highway Works*. The Stationery Office, London, UK.

Hunter RN (1986) The cooling of bituminous materials during laying. *Journal of the Institute of Asphalt Technology* **38**: 19–26.

Hunter RN (1996) A review of the measurement of the skidding resistance of roads with particular emphasis on new surfacings. *The Asphalt Yearbook*. IAT, Edinburgh, UK.

Joseph Vögele AG (2012) *Vögele Booklet on Paving*. Joseph Vögele, Ludwigshafen, Germany.

Nicholls JC, McHale MJ and Griffiths RD (2008) *Best Practice Guide for Durability of Asphalt Pavements*. TRL, Crowthorne, UK, Road Note RN42.

Chapter 16

Laboratory testing of asphalts

Dr Vincent Guwe
Solution Centre Manager,
The Shell Co. of Thailand Ltd.
Thailand

Xu Liting
Senior Application Specialist,
Shell China Ltd. China

Dr Jia Lu
Senior Application Specialist,
Shell China Ltd. China

The properties of asphalts need to be known for a variety of reasons, including performance evaluation, mixture or pavement designs, and production and/or construction specification compliance. In situ testing of material properties in full scale trial sections or in-service pavements is impractical or uneconomical in most cases, so engineers generally have to rely on laboratory testing to characterise or predict material properties. In addition, testing may also be necessary to ensure that specified requirements are met.

Laboratory tests should reproduce the anticipated in situ conditions as closely as possible (i.e. in terms of temperature, loading time, stress conditions, degree of compaction etc.). However, in situ conditions may be the subject of change, and selection of appropriate testing conditions may be difficult. Figure 16.1 (Pell, 1988) shows a simplified model depicting a representation of a pavement element, and shows the stresses to which it is subjected when a wheel load approaches. In practice, the stresses are applied three dimensionally: horizontal and shear stresses occur in planes that are perpendicular to those shown in Figure 16.1. As the wheel passes over the element, these stresses change with time, and this is shown in Figure 16.2 (Pell, 1988). Given the difficulty of reproducing such complex stress regimes accurately in the laboratory, simplified tests have been introduced that can reproduce certain aspects of the in situ behaviour. Such tests are also used to correlate the laboratory mixture design with in situ performance in relation

The Shell Bitumen Handbook, Sixth Edition
ISBN 978-0-7277-5837-8
http://dx.doi.org/10.1680/tsbh.58378.447

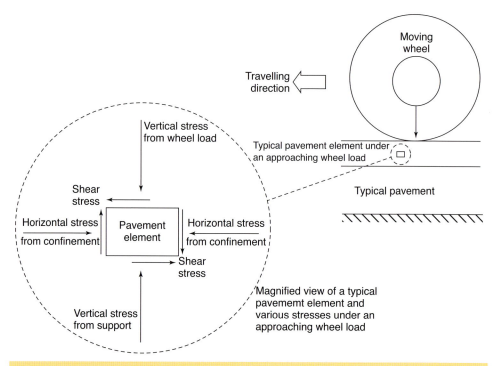

Figure 16.1 Stresses induced on a typical pavement element under an approaching wheel load

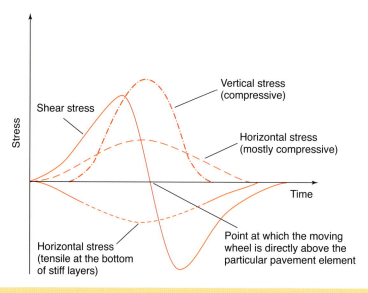

Figure 16.2 Stress state changes induced by a moving wheel load

to aspects such as moisture susceptibility, resistance to deformation, stiffness, fatigue and tensile strength.

Many laboratory tests have been proposed over the years. Efforts to standardise particular tests for routine usage are ongoing, and over time these have been improved by advances in equipment technology that measures material behaviour more accurately. In general, laboratory tests can be divided into the following three groups

- Fundamental
  - static creep test
  - repeated load test
  - dynamic stiffness and fatigue tests
  - indirect tensile tests.
- Simulative
  - wheel tracking tests
  - gyratory compaction
  - durability
  - cracking.
- Empirical
  - Marshall
  - indirect tensile strength.

16.1 Fundamental tests
16.1.1 Static creep test
In the 1970s, Shell developed a simple creep test whereby a static uniaxial compressive load was applied to unconfined cylindrical samples of asphalt to assess the resistance to deformation of the material (Hills, 1973). This test gained wide acceptance due to the ease of specimen preparation, the simplicity of the test procedure and the low cost of test equipment. The only requirements for the test specimen are that it should be prismatic, with flat and parallel ends normal to the axis of the specimen, while the test procedure involves the application of a constant axial stress of up to 500 kPa to a 100 mm dia test specimen for up to 10 000 s at a constant temperature, and measurement of the resultant deformation, as described in a now withdrawn British Standard (BSI, 1995).

The obvious limitation of the simple creep test is that the static loading mechanism does not simulate the dynamic traffic loading conditions to which asphalts are subjected in service, and it was found that rut prediction based on this test underestimated rut depths measured in trial pavements (Hills *et al.*, 1974). Furthermore, static loading could not capture the improved performance of binder modifiers that enhanced the elastic recovery properties of an

449

asphalt, whereas, in contrast, this could be demonstrated under repeated loading (Valkering *et al.*, 1990).

16.1.2 Repeated load tests

Repeated loading more closely simulates actual traffic loading. In the repeated load axial test, an unconfined test specimen is subjected to repeated block pulse loadings of 1 s duration separated by 1 s duration rest periods for up to 10 000 pulses, as described in a now withdrawn British Standard (BSI, 1996). A European Standard, EN 12697-25:2005 (BSI, 2005b), improved the position with the application of a confining pressure to the test specimen. This allows test conditions to reproduce more closely the stress conditions in an in-service pavement, and has the advantage that the applied loading is closer to the levels predicted in the pavement.

EN 12697-25:2005 (BSI, 2005b) describes two methods to achieve the confinement. In the first method, the cylindrical test specimen of 150 mm diameter is larger than the loading platens of 100 mm diameter, so the confinement is provided by the material along the outer cylindrical surface of the test specimen. In the second method, the confining pressure is exerted by a gaseous or liquid medium held in a triaxial cell, which makes the test equipment more expensive and test procedures more complex. A cyclic block pulse load can be used for both methods, and an alternative option to apply a sinusoidal compression (i.e. haversine, which is always positive) is also included for the second method with the triaxial cell. The results are typically reported as a cumulative strain curve, as shown in Figure 16.3. The load repetition number at which there is a turning point or an inflection point between the secondary flow and tertiary flow stages is identified as the flow number, which can be correlated with rutting potential.

In the USA, the National Cooperative Highway Research Program (NCHRP) developed a simple performance procedure using compact servo-hydraulic test equipment with a built in test chamber capable of controlling temperature and confining pressure (Bonaquist, 2008a, 2008b, 2011; Bonaquist *et al.*, 2003; Witczak *et al.*, 2002). Renamed the asphalt mixture performance tester (AMPT), this equipment (Figure 16.4) can be used to evaluate the rut resistance of an asphalt by subjecting a cylindrical test specimen of 100 mm diameter and 150 mm height to a repeated haversine axial compressive load pulse of 0.1 s every 1.0 s at a specified test temperature to find the flow number that can serve as the performance parameter to complement the Superpave mix design method (AASHTO, 2013b; FHWA, 2013).

450

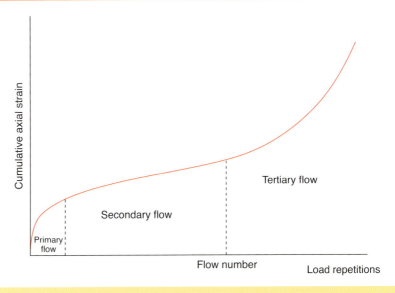

Figure 16.3 Typical cumulative creep curve

Figure 16.4 Asphalt mixture performance tester set up

16.1.3 *Dynamic tests*

Dynamic tests are more complex than repeated load tests in terms of the loading cycles and frequencies – dynamic tests apply repeated cyclic loads (usually a haversine wave) over a range of frequencies. This necessitates more accurate load application and deformation measuring systems.

Several dynamic test methodologies have been developed to determine stiffness and fatigue resistance. Bending tests using beams or cantilevers subjected to repeated applications of load have been used since the 1970s (Bonnaure *et al.*, 1977; Brown, 1983; Cooper and Pell, 1974; Monismith *et al.*, 1985). In such tests, the maximum stress occurs at a point on the surface of the specimen, and its calculation, using the standard beam bending formula, depends on the assumption of linear elasticity. The European Standards (BSI, 2012a, 2012b) have brought together various dynamic fatigue resistance and stiffness test methodologies, including the flexural bending tests as shown in Table 16.1.

In the two-point cantilever bending test, trapezoidal and square prismatic sample types are both included in the European standards (BSI, 2012a, 2012b), and different maximum aggregate sizes of the asphalts are also considered in the test sample dimensions. The base end of the sample is glued, and thus held rigidly while a sinusoidal load is applied at the head. For stiffness modulus measurements, the applied force or deflection at defined temperatures and frequencies should only cause up to 50 microstrain in the most heavily stressed part of the test sample so as to remain within the linear range. The load is applied for a minimum duration of 30 s and a maximum duration of 2 min, while the force, deflection and phase angles are measured and recorded over the last 10 s of the test. The test can be repeated for at least four temperatures at 10°C intervals and three

Table 16.1 Flexural bending test methods for fatigue resistance and stiffness (ASTM, 2010; BSI, 2012a, 2012b)

Test mode		Fatigue resistance[a]	Stiffness[a]
Two-point bending		✓ EN 12697-24	✓ EN 12697-26
Three-point bending		✓ EN 12697-24	✓ EN 12697-26
Four-point bending		✓ EN 12697-24 ✓ ASTM D7460-10	✓ EN 12697-26

[a] EN 12697-24 (BSI, 2012a); EN 12697-26 (BSI, 2012b); ASTM D7460-10 (ASTM, 2010)

frequencies at each temperature, in order to determine the master curve. For fatigue measurements, the European standards specify loading under constant displacement or strain for the trapezoidal samples, with at least one-third of the test set of 18 specimens able to reach one million test cycles, while the prismatic samples are tested under constant stress to a displacement of 280 μm. The French mixture design method has five levels that are chosen according to the type of asphalt, loadings and intended use, with additional tests required as the level increases. The two highest levels require fundamental tests for stiffness and fatigue using the two-point cantilever bending test (Delorme *et al.*, 2007). Figure 16.5 illustrates the two-point cantilever bending test set up with a trapezoidal sample.

For the three-point and four-point bending tests, a beam is subjected to periodic bending through vertical movements in the central load point(s), while the vertical positions of the two end points are kept fixed. Free rotation and horizontal movement are allowed at all load and reaction points in order to prevent the development of horizontal and torque stresses that can affect the behaviour of the material during the test. The sample width and height should be at least three times the maximum aggregate size of the asphalt, while the effective sample length should be at least six times the width or height. The European standard, EN 12697-24:2012 (BSI,

Figure 16.5 Two-point bending test with a trapezoidal sample

2012a), has a more specific set of sample dimensions (300 mm length and 50 mm width and height for the three-point bending fatigue test), while the American Society for Testing and Materials (ASTM, 2010) specifies a 380 mm long by 50 mm thick by 63 mm wide sample for the four-point bending fatigue test. The initial stiffness modulus is typically determined at the 50th or 100th cycle (by ASTM D7460-10 (ASTM, 2010) and EN 12697-24:2012 (BSI, 2012a), respectively) of the applied sinusoidal force. For fatigue, EN 12697-24:2012 specifies loading under the constant strain mode for the three-point bending test, while the four-point bending test can be in either the constant strain or constant stress mode. The ASTM method is carried out under the constant strain mode. Figures 16.6 and 16.7 illustrate the three-point and four-point bending test set ups, respectively.

The various fatigue tests are carried out until 'failure' occurs in the test specimen. 'Failure' can be an arbitrary end point, not where the test specimen literally fails. In a constant-strain test, the sample is usually deemed to have 'failed' when the load required to maintain that level of strain has fallen to 50% of its initial value. Because of the scatter of test results associated with fatigue testing, it is normal to test several specimens at each stress or strain level, and to plot the results plotted as stress or strain against cycles

Figure 16.6 Three-point bending test with a beam sample

Figure 16.7 Four-point bending test with a beam sample

to failure on a log–log graph, as shown in Figure 16.8 (Read, 1996). In EN 12697-24:2012, a minimum of 18 samples should be tested, generally over at least three selected levels of loading, while ASTM D7460-10 requires nine replicate samples for a complete fatigue curve.

The AMPT described in section 16.1.2 can also be used to apply controlled sinusoidal compressive stresses to determine the dynamic modulus. The test is undertaken at multiple temperatures and loading frequencies because these factors have a direct impact on material stiffness – the dynamic modulus values decrease with higher temperatures and lower loading frequencies. The multiple test results are then used to develop master curves to be input into mechanical–empirical pavement design methods. The American Association of State Highway and Transportation Officials (AASHTO) has

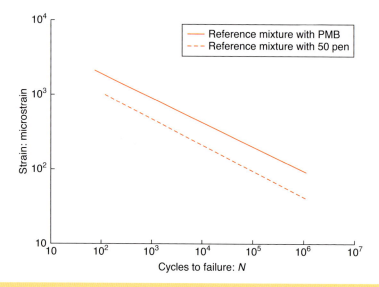

Figure 16.8 Typical fatigue test results

a provisional standard test method, TP 79-13 (AASHTO, 2013b), and a standard practice, PP 61-13 (AASHTO, 2013a), for determining the dynamic modulus and developing the master curve, respectively, while work is in progress to develop and standardise a direct tension test using the AMPT to assess fatigue and top down cracking resistance (Christensen and Bonaquist, 2009; Hou et al., 2010).

16.1.4 Indirect tensile tests

In addition to the uniaxial and triaxial compression tests and flexural bending tests described in the previous sections, indirect tensile tests have also been developed to evaluate the fundamental stiffness and fatigue properties. In such tests, a repeated loading is applied in the vertical diametrical plane of a cylindrical specimen using a loading strip, as shown in Figure 16.9. This vertical loading produces both a vertical compressive stress and a horizontal tensile stress on cylinders of the specimen. The magnitudes of the stresses vary across the cylinder, as shown in Figure 16.10, but are at a maximum in the centre of the specimen.

The situation, as depicted in Figure 16.10, enables the calculation of strain based on the following assumptions

- the specimen is subjected to plane stress conditions ($\sigma_z = 0$)
- the material behaves in a linear elastic manner
- the material is homogeneous

Figure 16.9 Indirect tensile stiffness modulus test set up

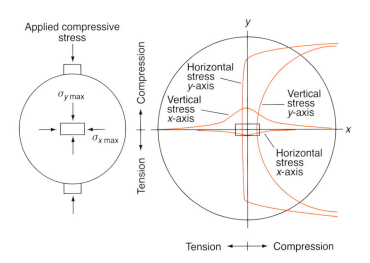

Figure 16.10 Stress distributions in the indirect tensile test mode

- the material behaves in an isotropic manner
- Poisson's ratio for the material is known
- the force is applied in the vertical diametrical plane.

16.1.4.1 Indirect tensile stiffness modulus

The indirect tensile stiffness modulus (ITSM) test, as defined in the European standard EN 12697-26:2012 (BSI, 2012b), is a simple test that can be completed quickly. The operator selects the target horizontal deformation and a target load pulse rise time (the time from the start of load application to the peak load). The force applied to the specimen is then automatically calculated by a computer, and a number of conditioning pulses are applied to the specimen. These conditioning pulses are used to make any minor adjustments to the magnitude of the force needed to generate the specified horizontal deformation and to seat the loading strips correctly on the specimen. Once the conditioning pulses have been completed, the system applies five load pulses. This generates an indirect movement on the horizontal diameter and, as the diameter of the specimen is known beforehand, the strain can be calculated. As the cross-sectional area of the specimen is also known and the force applied is measured, the applied stress can be calculated. Thus, as the stress and the strain are now known, the stiffness modulus of the material can be calculated.

Standard test conditions and requirements for the ITSM test are

- peak horizontal deformation of at least 5 µm
- rise time 124 ms – equivalent to a frequency of 1.33 Hz
- specimen diameter 80, 100, 120, 150 or 200 mm
- specimen thickness between 30 and 75 mm.

As material stiffness depends on the test frequency and temperature, EN 12697-26:2012 (BSI, 2012b) also defines a cyclic indirect tension test to be done at multiple temperatures and loading frequencies. A sinusoidal compressive loading is applied, without rest periods, under the following test conditions

- initial horizontal strain of up to 0.1%
- load frequency range 0.1–10 Hz
- specimen diameter 100 or 150 mm
- specimen thickness 40–90 mm
- minimum of four test temperatures: −10, 0, 10 and 20°C.

16.1.4.2 Indirect tensile fatigue test

The indirect tensile fatigue test (ITFT) uses a repeated controlled stress pulse to damage the specimen, and the accumulation of horizontal deformation against the number of load pulses is continually measured and recorded at preselected intervals.

Standard test conditions and requirements for the ITFT according to the European standard, EN 12697-24:2012 (BSI, 2012a), are

- initial strain range 100–400 μm/m
- loading amplitude: 250 kPa
- failure criterion – obvious vertical cracks or when the dynamic tensile strain increases to twice its initial value.

16.2 Simulative tests

As the stress conditions in a pavement loaded by a rolling wheel are extremely complex and cannot be replicated in a laboratory test on a sample of asphalt with any precision, simulative tests have been used to compare the performance of different materials.

16.2.1 Wheel tracking tests

Wheel tracking tests are based on the general operating procedure of tracking a load repeatedly over an asphalt sample to simulate in-service behaviour. The equipment ranges from large full scale pavement test facilities such as those at the UK's Transportation Research Laboratory (TRL), France's Laboratoire Central des Ponts et Chaussées (LCPC) and the USA's National Centre for Asphalt Technology (NCAT), to mid-scale pavement test facilities to the smaller laboratory scale wheel tracking machines (Figure 16.11). Full scale and mid-scale facilities can be used to test entire pavement structures to provide inputs for mechanistic pavement analyses, whereas laboratory tests are not mechanistic but do seem to simulate field behaviour (Brown *et al.*, 2001) so they can be employed to evaluate the relative resistance of particular asphalts to deformation.

The laboratory wheel tracking devices shown in Figure 16.11 have different sets of testing parameters and conditions. The European standard, EN 12697-22:2007 (BSI, 2007a), has brought together several wheel tracking test methodologies covering the Hamburg wheel tracking device (HWTD), the French rutting tester (FRT) and the UK wheel tracking machine WTM), while AASHTO also has test standards for the HWTD, T 324-11 (AASHTO, 2011b), and the asphalt pavement analyzer (APA), T 340-10 (AASHTO, 2010), another wheel tracking device more commonly used in the USA. Table 16.2 shows the test parameters commonly used for these four devices.

The test temperatures and initial air void contents have the greatest effects on test results (Cooley *et al.*, 2000) – rutting increases with higher test temperatures and air void contents. In relation to the former, all four test

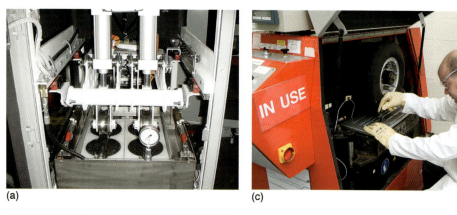

(a)

(c)

(b)

(d)

Figure 16.11 Laboratory scale wheel tracking devices: (a) asphalt pavement analyzer Jr (Courtesy of Pavement Technology, Inc.); (b) Hamburg wheel tracking device; (c) French rutting tester; (d) UK wheel tracking machine

devices are capable of testing at a range of higher temperatures (around 60°C), at which rut resistance is the critical performance requirement for asphalts in pavements. For the latter, the initial voids contents are commonly set at 7%, which is taken as the typical as-constructed condition, or 4% to assess actual rutting due to shear failure. The specimen density and aggregate orientation are influenced by the specimen types, sizes and preparation methods. The two common specimen types (beam/slab and cylinder) provide different rut depths but generally similar mixture ranking. With beam/slab specimens, roller compaction better simulates field practice compared to vibratory or kneading plate compaction. With cylindrical

Table 16.2 Laboratory-scale wheel tracking devices and common test parameters

Device	APA	HWTD	FRT	UK WTM
Test temperature: °C	US PG high temperature level, commonly 58 or 64	50 or 60	60	45 or 60
Specimen size: mm	150 diameter × 75 thickness cylinder, or 300 length × 125 width × 75 thickness beam	150 diameter × 38 to 100 thickness cylinder, or 320 length × 260 width × 38 to 100 thickness slab	500 length × 180 width × 50 or 100 thickness beam	200 dia. cores of various thickness, or 300 length × 300 width × 50 thickness slab
Compaction method	Gyratory compactor for cylinder specimen, vibratory compactor for beam specimen, 7% voids	Gyratory compactor for cylinder specimen, kneading plate compactor for slab specimen, recommended 7% voids	Roller compactor, compacted, void content depends on mixture type, at low and high ends of field as-compacted range	Roller compactor for slab specimen
Type of wheel	Stiff rubber hose under concave steel wheel	Solid treadless rubber tyre (EN) or solid steel wheel (AASHTO)	Pneumatic tyre	Solid treadless rubber tyre, rectangular cross-section
Wheel dimensions: mm	29.5 width (external diameter of rubber hose)	200 diameter × 50 width	400 diameter × 80 width	200 diameter × 50 width
Applied load: N	445	700 (EN) or 705 (AASHTO)	5000	700
Number of cycles	8000	10 000	30 000	1000
Test frequency: cycles/minute	60	26.5	60	26.5
Test time: minutes	135	380	500	38
Average test speed: m/s	0.60	0.20	0.82	0.20

specimens, gyratory compaction is usually used to achieve lower voids contents more easily.

The loading methods, magnitudes, effective stress applied and tracking speeds will also significantly affect the results. The FRT wheel is the largest (400 mm diameter), and is fitted with an 80 mm wide pneumatic tyre that contributes to higher surface shear and which simulates actual field conditions. The APA attempts to simulate the tyre pressure effect with a metal wheel running over a pressurised hose to transmit the load, but its representation of the pneumatic rubber tyres used in real life is limited, as are the solid rubber tyres of the UK machine and the HWTD's steel wheel. The steel wheel on the HWTD does not deform, resulting in a relatively high 0.73 MPa contact stress, and the HWTD test is also usually conducted with test samples submerged under water to assess moisture sensitivity, further increasing its severity.

The common test load applied for the FRT (5000 N) is significantly higher than that used with the other devices, but this is offset by its wider wheels and relatively high average tracking speed of 0.82 m/s. The APA has the same test frequency of 1 Hz but its average tracking speed is a lower 0.60 m/s. The average tracking speeds of the HWTD and UK machine are notably lower, and hence more severe than those of the other two devices.

Due to the variation in testing parameters and conditions, the different devices produce different results. However, comparative tests with full scale accelerated test facilities have shown that they are generally capable of assessing and ranking relative mixture performances similar to those in the field, with no clear superiority of any particular equipment. The various devices are also able to differentiate between binders of different grades (Cooley et al., 2000). As such, these devices are useful tools to highlight the contributions to improved mixture rut resistance when using modified binders.

Various transportation agencies have applied particular devices to set rut performance specifications. This requires tests to be conducted with local materials and mixtures to develop the reference results to compare against, as well as the performance criteria to be met. It should also be noted that, beyond the most basic level of the French mixture design method, a minimum rut resistance performance using the FRT is required for most mixtures (Delorme et al., 2007).

In China, the wheel tracking test equipment, sample sizes and test conditions are similar to the UK wheel tracking machine, with some modifications, such as test duration (60 versus 45 min) and analysis method – the results are analysed as the dynamic stability (DS) value calculated using the following formula, with higher DS values representing better rut resistance at 60°C

$$DS = \frac{N_{15}}{D_{60} - D_{45}}$$

where N_{15} is the number of loading passes after 15 min (i.e. 15 min × 42 passes/min), and $D_{60} - D_{45}$ is the change in rut depth in the last 15 min of the test.

16.2.2 Gyratory compaction

The gyratory compactor (Figure 16.12) simulates, to a reasonable degree, the compaction that actually takes place in service. The test procedure consists of placing a sample of hot asphalt in a cylindrical mould and applying a static pressure of a controlled magnitude over a round metallic insert covering the entire upper surface of the sample. The mould is then gyrated through a small angle in order to allow the aggregate particles to reorientate

Figure 16.12 Gyratory compactor

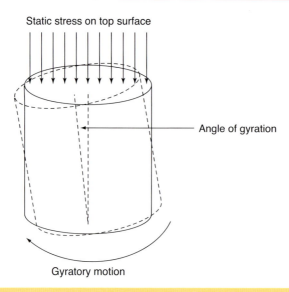

Static stress on top surface

Angle of gyration

Gyratory motion

Figure 16.13 Schematic diagram of gyratory motion

themselves under the loading. A schematic diagram of the gyratory motion is shown in Figure 16.13.

After compaction, the volumetric properties of the asphalt can be assessed against the mixture design specifications. This is a basic mixture design requirement in France (Delorme *et al.*, 2007) and in the Superpave mixture design method used in the USA (Asphalt Institute, 2001). The test conditions commonly used are a vertical stress level of 600 kPa and a gyration speed of 30 revs/min, but the angle of gyration used in France is 1.0° (Delorme *et al.*, 2007), whereas the Superpave gyratory compactor adopts a standard of 1.25° (ASTM, 2009) in order to achieve the higher level of densification of 4% design air voids (Harman *et al.*, 2002).

The gyratory compactor also allows a measure of the compactability of a mixture to be assessed by monitoring the vertical movement of the loading ram and the number of gyrations of the mould. Hence, if one mixture takes fewer gyrations in comparison to another mixture for the same vertical movement of the loading ram, then the former would be said to be more easily compacted.

Additionally, slab compaction using rollers and/or kneading plates is becoming more widespread, and is specified for the preparation of test specimens for certain tests such as wheel tracking (see Table 16.2). The European standard, EN 12697-33:2007, has brought together several methodologies for test specimen preparation by roller compactor (BSI, 2007b). A photograph of this equipment is shown in Figure 16.14.

Figure 16.14 Slab compactor

16.2.3 Durability testing

The durability of asphalts can be defined as the ability of the mixtures to resist the effects of water, ageing and temperature variations, for a given traffic loading without significant deterioration for an extended period of time. Accordingly, for durability evaluation, there is a need to be able to age artificially and/or to simulate the effect of water damage on compacted asphalt samples, as opposed to merely the components of the mixtures. A European standard, EN 12697-45:2012 (BSI, 2012d), specifies a saturation ageing tensile stiffness conditioning regime to age asphalts in the presence of water, while the comparison of the indirect tensile strength (AASHTO, 2011a) before and after a water-conditioning regime is one of the methods commonly used to assess moisture susceptibility. This test and other durability related moisture resistance tests are covered in section 19.4.4.

16.2.4 Low temperature cracking

The cracking of asphalt pavements in winter, or thermal cracking, is a common mode of distress in asphalt pavements in cold climates. While the asphalt binder is the main determinant in such failures, a European standard, EN 12697-46:2012 (BSI, 2012e), specifies several uniaxial tension tests with different stress–strain regimes to assess the resistance of asphalts to such cracking.

- In the uniaxial tension stress test, a specimen is pulled at a constant strain rate until failure while the temperature is kept constant. This yields the maximum stress (tensile strength) and the corresponding tensile failure strain at the test temperature.
- In the thermal stress restrained specimen test, as the test temperature is decreased at a constant rate, a test specimen is restrained from shrinking (as the name suggests), inducing an increase in the level of cryogenic stress in the test specimen. This yields the results of the progression of the cryogenic stress over the temperature and, finally, the failure stress at the failure temperature.
- In the relaxation test, a spontaneous strain is applied to the test specimen and held constant. Over time, relaxation causes the tension stress to decrease. The remaining tension stress and time of relaxation are monitored.
- In the tensile creep test, the test temperature is kept constant while subjecting the test specimen to a constant tension stress. The progression of the resultant strain is measured, and the stress is later withdrawn at a given time. From the measured strain data, the elastic and viscous properties of the asphalt are interpreted.
- In the uniaxial cyclic tension stress test, a cyclic tensile stress is applied as a sinusoidal stress to simulate the dynamic loading condition by traffic in combination with a constant stress, which acts as a surrogate for the cryogenic stress. The strain response is monitored as the test progresses, while the stiffness is also recorded until fatigue failure occurs. The resultant number of load cycles to failure is reported.

In China, for the cracking resistance performance of asphalts at low temperature, a bending beam test is carried out as shown schematically in Figure 16.15. A beam specimen of 250 mm length (L) by 30 mm breadth (b) and 35 mm height (h) is subjected to a vertical load applied at its centre at a rate of 50 mm/min at $-10°C$ until it fails. The load and vertical displacement (d) are recorded, and their values at failure are used to calculate the maximum tensile strain using the following equation

$$\varepsilon_b = \frac{6hd}{L^2}$$

16.2.5 Resistance to crack propagation
The fundamental fatigue tests described in sections 16.1.3 and 16.1.4 cover the initiation of cracking as the first phase of the failure mechanism during dynamic loading. The second part involves the propagation of the

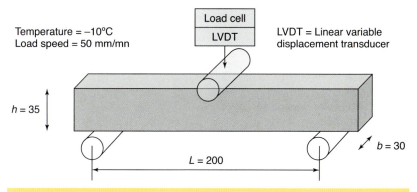

Figure 16.15 Low temperature bending beam test

cracks, and the resistance of asphalts to such crack propagation can be measured using the semi-circular bending test method described in a European standard, EN 12697-44:2010 (BSI, 2010). As the name suggests, a half cylinder test piece, with a 0.35 mm wide by 10 mm deep notch specially cut into the middle of the specimen, is loaded in three-point bending in such a way that its base is subjected to a tensile stress (Figure 16.16). A load is then applied at a rate of 5 mm/min until the test sample fails. The force and vertical displacement are recorded to calculate the maximum strain, stress at failure and fracture toughness.

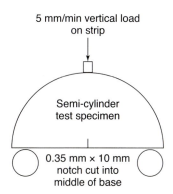

Figure 16.16 Semi circular bending test set up

16.3 Empirical

Empirical tests do not provide results such as fundamental material stresses and strains, but many have been widely used historically. One example, the Marshall test, is still used in many countries now, even though the complex stress system set up in the material is quite unrelated to the actual pavement condition in situ under traffic loading, and the tests do not provide basic information on the stress–strain characteristics of the material being tested.

16.3.1 The Marshall test

The concepts of the Marshall test were developed by Bruce Marshall, formerly bituminous engineer with the Mississippi State Highway Department. In 1948, the US Corps of Engineers improved and added certain features to Marshall's test procedure, and ultimately developed mixture design criteria (Asphalt Institute, 1997). Since 1948, the test has been adopted by organisations and government departments in many countries, sometimes with modifications either to the procedure or to the interpretation of the results.

In the ASTM and European standards (ASTM, 2006; BSI, 2012c), the Marshall test entails the manufacture of cylindrical specimens 102 mm in diameter by 64 mm high using a standard compaction hammer (Figure 16.17) and a cylindrical mould. The specimens are tested for their resistance to deformation at 60°C under a loading jig moving vertically at a constant rate of 50 mm/min. The jaws of the loading rig confine the majority, but not all, of the circumference of the specimen, the top and bottom of the cylinder being unconfined (see Figure 16.17). Thus, the stress distribution in the specimen during testing is extremely complex. Two properties are determined: the maximum load carried by the specimen before failure ('Marshall stability'), and the amount of deformation of the specimen before failure occurred ('Marshall flow'). The ratio of stability to flow is known as the 'Marshall quotient'.

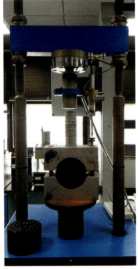

Figure 16.17 Marshall impact compactor and loading jig

Although the Marshall test is very widely used, it is important to recognise its limitations. Research at the University of Nottingham (Brown *et al.*, 1982) comparing the mechanical properties of various mixtures, using repeated load triaxial tests, triaxial creep tests, uniaxial unconfined creep tests and Marshall tests, suggested that the Marshall test was a poor measure of resistance to deformation and did not rank mixtures in order of their deformation resistance. In that study, the fundamental repeated-load triaxial tests gave more realistic results.

16.3.2 Indirect tensile strength test

In this test, the asphalt specimen is loaded diametrically, as shown in Figure 16.18, under a constant compression rate of 50 mm/min until it breaks (BSI, 2003; ASTM, 2012b). The indirect tensile strength, taken as the highest stress derived from the peak load applied at the break point, can be used as an indicator of cracking or rutting potential. The comparison of the indirect tensile strength before and after a water conditioning regime is

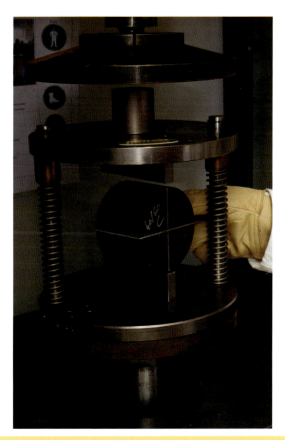

Figure 16.18 Indirect tensile strength test

also commonly used to assess moisture susceptibility (see section 19.4.4 for more information on this subject).

16.4 Determination of recovered bitumen properties

In addition to routine production control tests, there are a number of procedures available to assist highway engineers in the investigation of carriageway defects. One useful test is to recover the bitumen from the mixture and determine its properties, usually the penetration and softening point.

In this test, a sample of the asphalt is soaked in a solvent, such as trichloroethylene or dichloromethane (methylene chloride), to remove the bitumen from the aggregate into solution. The bitumen/solvent solution is separated from the fine mineral matter by filtration and centrifuging. The solvent is then evaporated under controlled conditions, using the apparatus shown in Figure 16.19, ensuring that all of the solvent is removed without

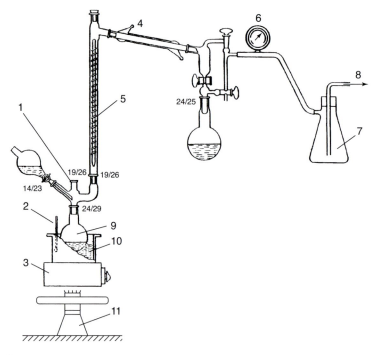

Key

1 Fit stirrer/CO_2 tube here
2 Thermometer with bulb opposite bottom of flask
3 Enclosed electrical heater
4 Condenser
5 Fractioning column
6 Vacuum gauge
7 Water trap
8 Pump
9 500 ml bottom flask
10 Oil bath
11 Jack

Figure 16.19 Distillation apparatus used for recovery of bitumen (BSI, 2005a)

Figure 16.20 Rotary evaporator

taking away any of the lighter components of the bitumen. When the recovery procedure has been completed, the penetration and softening point of the recovered bitumen can be determined using the standard tests. The method is described in detail in the European standard, EN 12697-4:2005 (BSI, 2005a). Alternatively, the solvent can be removed using a rotary evaporator (Figure 16.20) (ASTM, 2012a; BSI, 2013), which has the advantage of removing the solvent very quickly from the solution.

The test is very operator sensitive, and deviation from the standard method will almost certainly result in misleading data (Wadelin, 1982). It should also be noted that the extraction at high temperatures may cause the bitumen to harden, whereas residual solvent that remains after recovery may lead to some degree of softening, and these two opposite effects are usually taken to cancel each other out approximately (BSI, 2005a, 2013). For these reasons, caution should be exercised when using the test for contractual purposes. Notwithstanding, it is a very useful tool for investigating defects in service. With sensible interpretation of the test results, the bitumen recovery method can be used to determine if the bitumen in the mixture is abnormally hard or soft.

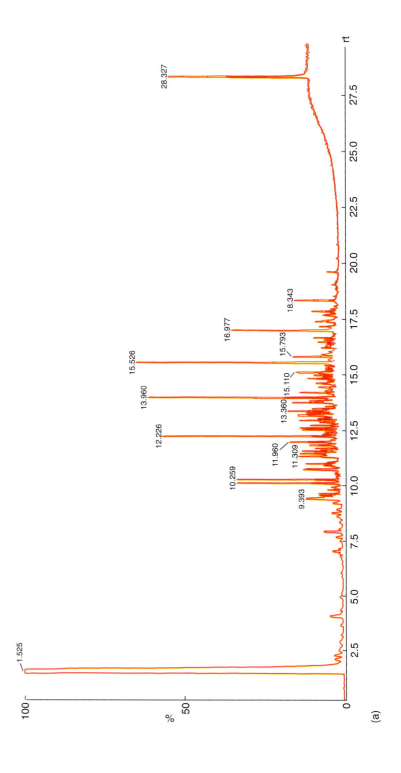

(a)

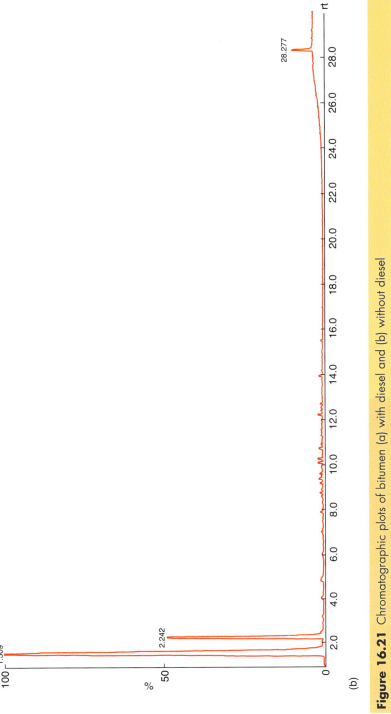

(b)

Figure 16.21 Chromatographic plots of bitumen (a) with diesel and (b) without diesel

In addition to the determination of the penetration and the softening point, chromatographic examination of the recovered bitumen can be useful to detect the presence of any contaminants. Figure 16.21(a) and (b) shows chromatographic plots, with and without the presence of diesel, respectively, generated using a combination of gas chromatography and mass spectroscopy. The presence of the contaminant is very clear. Identification, however, is less precise, because diesel can weather, losing lighter fractions, and the composition can change between summer and winter. Although the technique is only qualitative and not quantitative, it is a useful tool for examining recovered bitumens that are abnormally soft.

References

AASHTO (American Association of State Highway and Transportation Officials) (2010) AASHTO T 340-10. Standard method of test for determining rutting susceptibility of hot mix asphalt (HMA) using the Asphalt Pavement Analyzer (APA). AASHTO, Washington, DC, USA.

AASHTO (2011a) AASHTO T 283-11. Standard method of test for resistance of compacted hot mix asphalt (HMA) to moisture-induced damage. AASHTO, Washington, DC, USA.

AASHTO (2011b) AASHTO T 324-11. Standard method of test for Hamburg wheel-track testing of compacted hot mix asphalt (HMA). AASHTO, Washington, DC, USA.

AASHTO (2013a) AASHTO PP 61-13. Standard practice for developing dynamic modulus master-curves for asphalt mixtures using the Asphalt Mixture Performance Tester (AMPT). AASHTO, Washington, DC, USA.

AASHTO (2013b) AASHTO TP 79-13. Standard method of test for determining the dynamic modulus and flow number for hot mix asphalt (HMA) using the Asphalt Mixture Performance Tester (AMPT). AASHTO, Washington, DC, USA.

ASTM (American Society for Testing and Materials) (2006) ASTM D6927-06. Standard test method for Marshall stability and flow of bituminous mixtures. ASTM, Philadelphia, PA, USA.

ASTM (2009) ASTM D6925-09. Standard test method for preparation and determination of the relative density of hot mix asphalt (HMA) specimens by means of the Superpave gyratory compactor. ASTM, Philadelphia, PA, USA.

ASTM (2010) ASTM D7460-10. Standard test method for determining fatigue failure of compacted asphalt concrete subjected to repeated flexural bending. ASTM, Philadelphia, PA, USA.

ASTM (2012a) ASTM D5404-12. Standard practice for recovery of asphalt from solution using the rotary evaporator. ASTM, Philadelphia, PA, USA.

ASTM (2012b) ASTM D6931-12. Standard test method for indirect tensile (IDT) strength of bituminous mixtures. ASTM, Philadelphia, PA, USA.

Asphalt Institute (1997) *Mix Design Methods for Asphalt Concrete and Other Hot-Mix Types (MS-2)*. Asphalt Institute, Lexington, KY, USA.

Asphalt Institute (2001) *Superpave Mix Design (SP-2)*. Asphalt Institute, Lexington, KY, USA.

Bonaquist R (2008a) *Refining the Simple Performance Tester for Use in Routine Practice.* Transportation Research Board, Washington, DC, USA, NCHRP Report 614.

Bonaquist R (2008b) *Ruggedness Testing of the Dynamic Modulus and Flow Number Tests with the Simple Performance Tester.* Transportation Research Board, Washington, DC, USA, NCHRP Report 629.

Bonaquist R (2011) *Precision of the Dynamic Modulus and Flow Number Tests Conducted with the Asphalt Mixture Performance Tester.* Transportation Research Board, Washington, DC, USA, NCHRP Report 702.

Bonaquist RF, Christensen DW and Stump W (2003) *Simple Performance Tester for Superpave Mix Design: First Article Development and Evaluation.* Transportation Research Board, Washington, DC, USA, NCHRP Report 513.

Bonnaure F, Gest G, Gravois A and Uge PA (1977) A new method of predicting the stiffness of asphalt paving mixtures. *Journal of the Association of Asphalt Paving Technologist* **46**: 64–104.

Brown ER, Kandhal PS and Zhang J (2001) *Performance Testing for Hot Mix Asphalt.* National Center for Asphalt Technology, Auburn, AL, USA, NCAT Report 01-05.

Brown SF (1983) Practical mechanical tests for the design and control of asphaltic mixes. *Proceedings of RILEM 3rd International Symposium on the Testing of Hydrocarbon Binders and Materials, Belgrade, Serbia.*

Brown SF, Cooper KE and Pooley GR (1982) Mechanical properties of bituminous materials for pavement design. *Proceedings of the Eurobitume Symposium, Cannes, France.*

BSI (1995) BS 598-111:1995. Sampling and examination of bituminous mixtures for roads and other paved areas, Method for determination of resistance to deformation of bituminous mixtures subject to unconfined uniaxial loading. BSI, London, UK.

BSI (1996) DD 226:1996. Method for determining resistance to permanent deformation of bituminous mixtures subject to unconfined dynamic loading. BSI, London, UK.

BSI (2003) BS EN 12697-23:2003. Bituminous mixtures – Test methods for hot mix asphalt – Part 23: Determination of the indirect tensile strength of bituminous specimens. BSI, London, UK.

BSI (2005a) BS EN 12697-4:2005. Bituminous mixtures – Test methods for hot mix asphalt – Part 4: Binder recovery: fractionating column. BSI, London, UK.

BSI (2005b) BS EN 12697-25:2005. Bituminous mixtures – Test methods for hot mix asphalt – Part 25: Cyclic compression test. BSI, London, UK.

BSI (2007a) BS EN 12697-22:2007. Bituminous mixtures – Test methods for hot mix asphalt – Part 22: Wheel tracking. BSI, London, UK.

BSI (2007b) BS EN 12697-33:2007. Bituminous mixtures – Test methods for hot mix asphalt – Part 33: Specimen prepared by roller compactor. BSI, London, UK.

BSI (2010) BS EN 12697-44:2010. Bituminous mixtures – Test methods for hot mix asphalt – Part 44: Crack propagation by semi-circular bending test. BSI, London, UK.

BSI (2012a) BS EN 12697-24:2012. Bituminous mixtures – Test methods for hot mix asphalt – Part 24: Resistance to fatigue. BSI, London, UK.

BSI (2012b) BS EN 12697-26:2012. Bituminous mixtures – Test methods for hot mix asphalt – Part 26: Stiffness. BSI, London, UK.

BSI (2012c) BS EN 12697-34:2012. Bituminous mixtures – Test methods for hot mix asphalt – Part 34: Marshall test. BSI, London, UK.

BSI (2012d) BS EN 12697-45:2012. Bituminous mixtures – Test methods for hot mix asphalt – Part 45: Saturation ageing tensile stiffness (SATS) conditioning test. BSI, London, UK.

BSI (2012e) BS EN 12697-46:2012. Bituminous mixtures – Test methods for hot mix asphalt – Part 46: Low temperature cracking and properties by uniaxial tension tests. BSI, London, UK.

BSI (2013) BS EN 12697-3:2013. Bituminous mixtures – Test methods for hot mix asphalt – Part 3: Binder recovery: rotary evaporator. BSI, London, UK.

Christensen DW and Bonaquist R (2009) Analysis of HMA fatigue data using the concepts of reduced loading cycles and endurance limits. *Journal of the Association of Asphalt Paving Technologist* **78**: 335–370.

Cooley Jr LA, Kandhal PS, Buchanan MS, Fee F and Epps A (2000) *Loaded Wheel Testers in the United States: State of the Art*. National Center for Asphalt Technology, Auburn, AL, USA. NCAT Report 00-04.

Cooper KE and Pell PS (1974) *The Effect of Mix Variables on the Fatigue Strength of Bituminous Materials*. Transport and Road Research Laboratory (TRRL), Crowthorne, UK. Laboratory Report 633.

Delorme JL, de la Roche C and Wendling L (2007) *LCPC Bituminous Mixtures Design Guide*. Laboratoire Central des Ponts et Chaussées, Paris, France.

FHWA (Federal Highway Administration) (2013) *Asphalt Mixture Performance Tester (AMPT)*. FHWA, Washington, DC, USA, Report FHWA-HIF-13-005.

Harman T, Bukowski JR, Moutier F, Huber G and McGennis R (2002) The history and future challenges of gyratory compaction 1939 to 2001. *Transportation Research Record* **1789**: 200–207.

Hills JF (1973) The creep of asphalt mixes. *Journal of the Institute of Petroleum* **59(570)**: 247–262.

Hills JF, Brien D and Van de Loo PJ (1974) *The correlation of creep and rutting*. Institute of Petroleum, London, UK, Paper IP 74-001.

Hou T, Underwood BS and Kim R (2010) Fatigue performance prediction of North Carolina mixtures using the simplified viscoelastic continuum damage model. *Journal of the Association of Asphalt Paving Technologist* **79**: 35–80.

Monismith CL, Epps JA and Finn FN (1985) Improved asphalt mix design. *Journal of the Association of Asphalt Paving Technologist* **54**: 347–391.

Pell PS (1988) Bituminous pavements: materials, design and evaluation. Residential course at the University of Nottingham. Section E, Laboratory test methods, E1–E17. University of Nottingham, Nottingham, UK.

Read JM (1996) *Fatigue Cracking of Bituminous Paving Mixtures*. PhD thesis, University of Nottingham, Nottingham, UK.

Valkering CP, Lancon DJL, de Hilster E and Stoker DA (1990) Rutting resistance of asphalt mixes containing non-conventional and polymer modified binders. *Journal of the Association of Asphalt Paving Technologist* **59**: 590–609.

Wadelin FA (1982) Workshop on binder recovery from asphalt and coated macadam. *Journal of the Institute of Asphalt Technology* **32**: 28–42.

Witczak MW, Kaloush K, Pellinen T, El-Basyouny M and von Quintus H (2002) *Simple Performance Test for Superpave Mix Design.* Transportation Research Board, Washington, DC, USA, NCHRP Report 465.

Chapter 17

Properties of asphalts

Jeyan Vasudevan
Senior Application Specialist,
Shell Malaysia Trading Sdn
Bhd. Malaysia

Dr Jia Lu
Senior Application Specialist,
Shell China Ltd. China

Dr Sridhar Raju
Product Researcher,
Shell India Markets Pvt. Ltd.
India

The analytical design of asphalt pavements involves consideration of two aspects of material properties. These are

- the load/deformation or stress/strain characteristics used to analyse critical stresses and strains in the structure
- the performance characteristics of the materials that show the mode, or modes, of failure.

The two principal structural distress modes are cracking and deformation. The former applies to bound materials only in the pavement, while the latter is valid for all materials in the pavement, both bound and unbound.

17.1 Stiffness of asphalts

As discussed previously, asphalts behave visco-elastically (i.e. they respond to loading in both an elastic and a viscous manner). The proportion of each depends on the time of loading and on the temperature at which the load was applied. The complexity of this behaviour is increased by the proportion of the components of the mixture, the bitumen being responsible for the visco-elastic properties, while the mineral skeleton influences elastic and plastic properties. Mixture components and compositions can be extremely diverse, which makes prediction of the properties of a particular mixture difficult.

Asphalt stiffness can be divided into elastic stiffness, which dominates under conditions of low temperatures or short loading times, and viscous stiffness, which dominates at high temperatures or long loading times. The former is used to calculate critical strains in the structure in analytical design. The latter

The Shell Bitumen Handbook, Sixth Edition
ISBN 978-0-7277-5837-8
http://dx.doi.org/10.1680/tsbh.58378.479

is used to assess the resistance of the material to deformation. It has also been shown that, at intermediate temperatures, when the stiffness has both an elastic and a viscous component, the stiffness is stress dependent (Read, 1996). High stresses result in lower stiffness, and low stresses result in higher stiffness, making the assessment of performance even more difficult. However, this stress dependency is less important than the effects of both the time of loading and temperature.

Stiffness at a particular temperature and time of loading can be measured by a variety of methods

- bending or vibration tests on a beam specimen
- direct uniaxial or triaxial tests on cylindrical specimens
- indirect tensile tests on cylindrical specimens
- dynamic loading on the asphalt mixture performance tester.

Identical samples from the same mixture will return different modulus values, as the results depend on the measuring criteria as described in Chapter 16.

Different types of loading can be used in the tests but for elastic stiffness of materials that carry traffic, sinusoidal or pseudo-sinusoidal repeated loading at high frequency is most appropriate (Raithby and Sterling, 1972).

17.1.1 The prediction of asphalt stiffness

Asphalt stiffness can be measured quickly and easily using tests such as the indirect tensile stiffness modulus test (BSI, 2003). However, when testing is not feasible, such as in the design office, then the stiffness of a particular mixture at any temperature and time of loading can be estimated using empirical methods, to an accuracy that is acceptable for most purposes.

17.1.1.1 Shell method

In 1977, Shell produced a nomograph (Figure 17.1) for predicting the stiffness of asphalt (Bonnaure et al., 1977). The data required for this nomograph are

- the stiffness modulus of the bitumen (in Pa)
- the percentage volume of bitumen
- the percentage volume of the mineral aggregate.

The University of Nottingham has also developed a method for calculating mixture stiffness (Brown and Brunton, 1986) and the data required are

- the stiffness modulus of the bitumen (in Pa $\times$ 10^6)
- the voids in the mineral aggregate (VMA in %).

These two procedures can only be applied when the stiffness modulus of the bitumen exceeds 5 MPa (i.e. under high stiffness conditions appropriate to

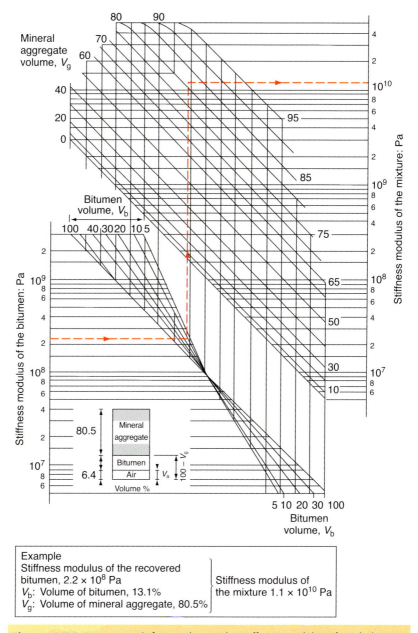

Figure 17.1 Nomograph for predicting the stiffness modulus of asphalt (Bonnaure *et al.*, 1977)

traffic when the response is predominantly elastic and, for the Nottingham method, values of VMA between 12% and 30%). These two methods assume that the grading, type and characteristics of the aggregate affect only the elastic stiffness of the mixture, as they influence the packing

481

characteristics of the aggregate and, thus, the state of compaction of the material.

17.1.1.2 Asphalt Institute method

Hwang and Witczak (1979) developed the DAMA computer program for the Asphalt Institute. They applied regression formulae to determine the dynamic modulus of asphalts (what is described in the USA as 'hot mix asphalt')

$$|E^*| = 100\,000 \times 10^{\beta_1}$$

$$\beta_1 = \beta_3 + 0.000\,005\beta_2 - 0.00189\beta_2 f^{-1.1}$$

$$\beta_2 = (\beta_4)^{0.5} + T^{\beta_5}$$

$$\beta_3 = 0.553\,833 + 0.028\,829(P_{200}f^{-0.1703}) - 0.034\,76V_a$$

$$\quad + 0.070\,377\lambda + 0.931\,757f^{-0.027\,74}$$

$$\beta_4 = 0.483V_b$$

$$\beta_5 = 1.3 + 0.498\,25\log f$$

where β_1 to β_5 are temporary constants, f is the load frequency (in Hz), T is the temperature (in °F), P_{200} is the percentage by weight of aggregate passing through a no. 200 sieve, V_a is the percentage volume of air voids, and λ is the bitumen viscosity at 70°F (21.1°C) in 10^6 poise.

Note: if insufficient viscosity data are available, the following equation may be used

$$\lambda = 29\,508.2\ (P77°F) - 2.1939$$

where P77°F is the penetration at 77°F (25°C) and V_b is the percentage volume of bitumen.

It can be seen that the factors considered by Hwang and Witczak (1979) for the Asphalt Institute model are very similar to those considered by Shell, with the following differences.

- The percentage of fines passing the no. 200 sieve is used in the Asphalt Institute method but not in the Shell method.
- The viscosity or penetration of bitumen utilised in the Shell method is derived from recovered binder from the mixture, while the Asphalt Institute method uses virgin bitumen.
- The temperature and viscosity of bitumen feature in the Asphalt Institute method, while the Shell method employs the penetration index.

17.1.1.3 Witczak equation for MEPDG

Witczak's equation is based on a non-linear regression analysis using the generalised reduced gradient optimisation approach. This model incorporates mixture volumetric and aggregate gradation, and is currently one of two options for level 3 analysis using the NCHRP 1-37A *Mechanistic–Empirical Pavement Design Guide* (MEPDG) program (Advanced Research Associates, 2004). The new *Mechanistic–Empirical Pavement Design Guide* adopted by the American Association of State Highway and Transportation Officials (AASHTO) represents a fundamental advance over the current 50 year old empirical pavement design procedures derived from the AASHTO road test. A hierarchical input data scheme has been implemented in the MEPDG to permit varying levels of sophistication for specifying material properties, ranging from laboratory measured values (level 1) to empirical correlations (level 2) to default values based on the prediction model (level 3).

In a study by Andrei *et al.* (1999), the original Witczak predictive equation was revised as follows

$$
\begin{aligned}
\log_{10}|E^*| \\
&= -1.249\,937 \\
&\quad + 0.029\,23P_{200} - 0.001\,767(P_{200})^2 - 0.002\,841P_4 \\
&\quad - 0.058\,09V_a - 0.082\,208\left(\frac{V_{beff}}{V_{beff} + V_a}\right) \\
&\quad + \frac{3.871\,977 - 0.0021P_4 + 0.003\,958P_{3/8} \quad -0.000\,017(P_{3/8})^2 + 0.005\,47P_{3/4}}{1 + \exp(-0.603\,313 - 0.313\,351\log f - 0.393\,532\,1\log\eta)}
\end{aligned}
$$

where P_{200} is the percentage of aggregate passing a no. 200 sieve, P_4 is the percentage of aggregate retained on a no. 4 sieve, $P_{3/8}$ is the percentage of aggregate retained on a $\frac{3}{8}$-in. (9.56 mm) sieve, $P_{3/4}$ is the percentage of aggregate retained on a $\frac{3}{4}$ in. (19.01 mm) sieve, V_a is the percentage of air voids (by volume of mixture), V_{beff} is the percentage of effective bitumen content (by volume of mixture), f is the loading frequency (Hz) and η is the binder viscosity at the temperature of interest (10^6 poise or 10^5 pa·s).

The limitations of Witczak's equation, acknowledged by Bari (2005), include relying on other models to translate the dynamic shear modulus into binder viscosity. Because the original Witczak equation is based on regression analysis, extrapolation beyond the calibration database should be restricted. Bari also mentions that there is limited volumetric influence (precision) when the model is compared to the Shell method. Dongre *et al.*

(2005) have also noted the need for improved sensitivity to volumetric values, such as the percentage of voids in mineral aggregate (VMA), the percentage of voids filled with bitumen (VFB), the bitumen percentage (V_b) and the percentage of air voids (V_a).

To include the dynamic shear modulus of the binder $|G^*|$ in the predictive model, Witczak reformulated the model to include the binder variable directly. The updated model is as follows

$$
\begin{aligned}
\log_{10}|E^*| \\
= -0.349 &+ 0.754(|G^*|_b^{-0.0052}) \\
&\times \Bigg[6.65 - 0.032P_{200} + 0.0027(P_{200})^2 + 0.011P_4 \\
&\quad - 0.0001(P_4)^2 + 0.006P_{3/8} - 0.000\,14(P_{3/8})^2 \\
&\quad - 0.08V_a - 1.06\left(\frac{V_{beff}}{V_{beff} + V_a}\right)\Bigg] \\
+ &\frac{2.558 + 0.032V_a + 0.713\left(\dfrac{V_{beff}}{V_{beff} + V_a}\right) + 0.124P_{3/8} - 0.0001(P_{3/8})^2 - 0.0098P_{3/4}}{1 + \exp(-0.7814 - 0.5785\log|G^*|_b + 0.8834\log\delta_b)}
\end{aligned}
$$

where $|G^*|_b$ is the dynamic shear modulus of asphalt binder (lb/in.^{2}) and δ_b is the binder phase angle associated with $|G^*|_b$ (degrees). This equation is one of two options for level 3 analysis in the most current NCHRP 1-40D MEPDG program.

Because some of the mixtures in this database do not contain $|G^*|_b$ data, the Cox–Mertz rule, using correction factors for the non-Newtonian behaviour (the relationship between the shear stress and the shear rate is different, and can even be time dependent) is used to calculate the dynamic shear modulus of the binder $|G^*|_b$ from A-VTS values, where A is the intercept of the temperature susceptibility relationship and VTS is the slope of temperature susceptibility relationship as follows

$$|G^*|_b = 0.0051 f_s \eta_{f_s,T}(\sin\delta_b)^{7.1542 - 0.4929f_s + 0.0211f_s^2}$$

$$\delta_b = 90 + (-7.3146 - 2.6162VTS)\log(f_{s_{f_s},T})$$

$$+ (0.1124 + 0.2029VTS)\log(f_{s_{f_s},T})^2$$

$$\log[\log(\eta_{f_s,T})] = 0.9699 f_s^{-0.0527} A + 0.9668 f_s^{-0.0527} \times VTS\log T_R$$

where f_s is the dynamic shear frequency (Hz), δ_b is the predicted binder phase angle (°), η_{fsT} is the viscosity of asphalt binder at a particular loading frequency (fs) and temperature (T) (centipoise (cP)) and T_R is the temperature using the Rankine scale (i.e. absolute temperature) (°R).

17.1.1.4 Hirsch model

Christensen et al. (2003) examined four different models based on the law of mixtures parallel model, and chose the model that incorporates the binder modulus, VMA, and voids filled with asphalt (VFA) because it provides accurate results in the simplest form. The other more complicated forms attempt to incorporate the modulus of the mastic or the film thickness, which are difficult parameters to measure. The suggested model for estimating the dynamic modulus of asphalt (again what would be described in the USA as 'hot mix asphalt') ($|E^*|$) is as follows

$$|E^*|_m = P_c\left[4\,200\,000\left(1 - \frac{VMA}{100}\right) + 3|G^*|_b\left(\frac{VFA \times VMA}{10\,000}\right)\right]$$

$$+ \frac{(1 - P_c)}{\frac{(1 - VMA/100)}{4\,200\,000} + \frac{VMA}{3|G^*|_b VFA}}$$

$$\phi = -21(\log P_c)^2 - 55 \log P_c$$

$$P_c = \frac{(20 + 3|G^*|_b VFA/VMA)^{0.58}}{650 + (3|G^*|_b VFA/VMA)^{0.58}}$$

where $|E^*|_m$ is the dynamic modulus of the asphalt (lb/in.^{2}), P_c is the aggregate contact volume (in.^{3}) and ϕ is the phase angle of the asphalt (°).

A strength of this model is the empirical phase angle equation, which is important for the inter-conversion of the dynamic modulus to the relaxation modulus or creep compliance. Weaknesses of the model include a lack of a strong dependence on volumetric parameters, particularly under conditions of low air void and voids filled with bitumen. Also, questions arise as to whether the dynamic shear modulus is capable of being altered to take account of the possible beneficial effects of modifiers (Al-Khateeb et al., 2006).

17.1.1.5 Al-Khateeb model

Based on their findings from the Hirsch model, Al-Khateeb et al. (2006) suggest the law of mixture parallel model as follows

$$|E^*|_m = 3\left(\frac{100 - VMA}{100}\right)\left[\frac{(90 + 10\,000(|G^*|_b/VMA))^{0.66}}{1100 + (900(|G^*|_b/VMA))^{0.66}}\right]|G^*|_g$$

where $|G^*|_g$ is the dynamic shear modulus of the asphalt binder at the glassy state (assumed to be 145 000 psi (999 050 kPa)).

Like the Hirsch model, this formulation is based on the law of mixtures for composite materials. In this model, the different material phases (aggregate, asphalt binder and air) are considered to exist in parallel. Therefore, this model is a simpler interpretation of the Hirsch model. The researchers note that their model addresses one of the primary shortcomings of the Hirsch model (namely, the inability of the Hirsch model to predict accurately the dynamic modulus of the mixture at low frequencies and high temperatures).

The strengths of this model include the improved prediction of high temperature and low frequency dynamic modulus. Weaknesses include a lack of model verification and the fact that the researchers who developed this model did so based on dynamic modulus values obtained from tests at higher than recommended strain amplitudes (200 microstrain versus the recommended maximum of 75–150 microstrain).

17.2 Deformation of asphalts

In order to determine the deformation characteristics of asphalt, the low stiffness response of the material (i.e. its response at high temperature or long loading times) must be analysed. When the stiffness of the bitumen is <0.5 MPa, mixture behaviour is much more complex than it is in the elastic zone (the area where the elastic behaviour of the mixture is more prominent than its viscous behaviour). Under these conditions, the stiffness of the mixture not only depends on the bitumen stiffness and the voids in the mixed aggregate and bitumen, but also on a variety of other factors. These include the aggregate grading, its shape, texture and degree of interlock, and the degree of compaction. This is illustrated in Figure 17.2.

The simplest test used to study the deformation behaviour of asphalts is the creep test. Figure 17.3(a) shows the deformation behaviour of a mixture tested at different temperatures. If these same results are plotted on a graph of mixture stiffness (S_{mix}) against bitumen stiffness (S_{bit}), the test results form a single continuous 'master curve' for the mixture, as shown in Figure 17.3(b). Thus, the effect of testing at different temperatures is combined in a single curve. Similarly, the effect of using different grades of bitumen or the application of different stress levels can also be combined.

Thus, the S_{mix} versus S_{bit} curve is a means of assessing the resistance to deformation that is independent of arbitrarily chosen test variables for a particular asphalt. For example, Figure 17.4 shows the deformation characteristics of two asphalts having identical aggregate grading but containing 5% and

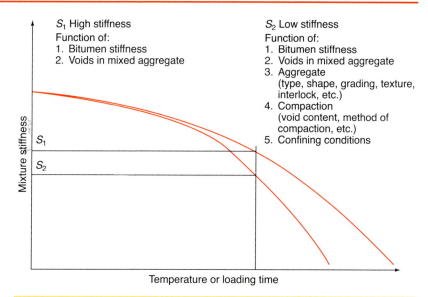

Figure 17.2 Mixture stiffness as a function of temperature or loading time (Brown, 1988)

11% bitumen, respectively. It can be seen that the stiffness of the leaner mixture levels out with decreasing bitumen stiffness. In contrast, the stiffness of the richer mixture continues to decrease with decreasing bitumen stiffness. Clearly, therefore, maintaining a high value of S_{mix} when S_{bit} is decreasing is a desirable characteristic for long term resistance to deformation. Similarly, Figure 17.5 demonstrates the effect of aggregate shape on asphalt having the same aggregate grading and bitumen content (Hills *et al.*, 1974). As would be expected, crushed aggregate increases the degree of aggregate interlock, and this results in an increase in the value of S_{mix} and higher resistance to deformation.

The resistance of mixtures to deformation can be determined by tests such as the unconfined creep test and the repeated load uniaxial or triaxial test, their results being plotted as S_{mix} versus S_{bit}. These tests reproduce stress conditions on the road more accurately and, as a result, are gaining widespread popularity.

17.2.1 The prediction of deformation

When attempting to calculate the amount of deformation on a highway, account has to be taken of a range of wheel load spectra, contact areas and pressures, lateral distribution of the wheel loads and ambient temperature gradients within the asphalt layers. Clearly, the situation is very complex.

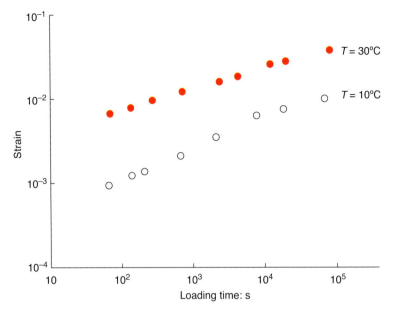

(a)

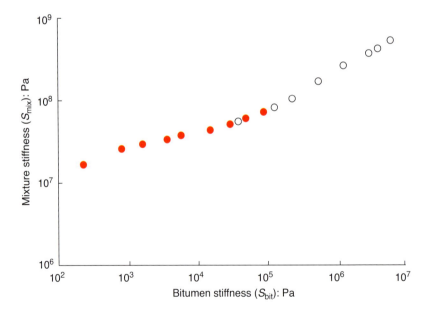

(b)

Figure 17.3 Results from a creep test (Hills *et al.*, 1974): (a) effect of temperature on deformation; (b) master curve for a particular mixture

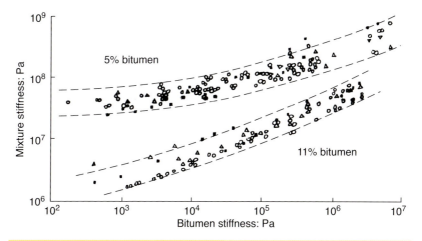

Figure 17.4 The effect of bitumen content on creep properties (Shell, 1978)

17.2.1.1 Shell method

The final step in the *Shell Pavement Design Manual* (Claessen *et al.*, 1977; Shell, 1978) when layer thicknesses have been determined is to predict the rut depth from

$$\text{Rut depth} = C_m \times h \times \sigma_{ave} / S_{mix}$$

where C_m is the empirical correction factor, h is the thickness of the layer, σ_{ave} is the average stress, and S_{mix} is the stiffness of the mixture.

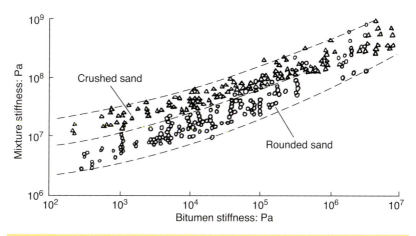

Figure 17.5 The effect of aggregate shape on creep properties (Hills, 1973)

However, as temperature gradients occur within the asphalt layers and different mixture types are often used in the surface course, binder course and base, the total asphalt thickness must be subdivided and each layer considered individually. The rut depth is the sum of the deformation in each of these layers.

17.2.1.2 VESYS method

VESYS is a linear computer program that can be used to determine the deformation based on resilient modulus (FHWA, 1978). The VESYS method of prediction of rut depth is based on an assumption that permanent strain is proportional to resilient strain.

$$\varepsilon_p(N) = \mu \varepsilon N^{-\alpha}$$

where $\varepsilon_p(N)$ is the permanent or plastic strain due to a single load application at the Nth application, ε is the elastic or resilient strain at the 200th repetition, N is the load application number, μ is the constant between permanent and elastic strain, and α is the rate of decrease in deformation.

The total deformation can be obtained by integrating the earlier equation

$$\varepsilon_p = \int_0^N \varepsilon_p \mu \varepsilon(N) \, dN = \varepsilon \mu \left(\frac{N^{1-\alpha}}{1-\alpha} \right)$$

Converting the above equation into a log–log relationship gives the following relationship

$$\log \varepsilon_p = \log \left(\frac{\varepsilon \mu}{(1-\alpha)} \right) + (1-\alpha) \log N$$

where the slope of the plot is $S = 1 - \alpha$ (i.e. the rate of decrease of deformation $\alpha = 1 - S$.

The intercept $I = \varepsilon \mu / (1 - \alpha)$ or constant $\mu = IS/\varepsilon$.

VESYS also has an option to determine the deformation parameter of a layered pavement.

17.2.1.3 MEPDG rutting model

Similar to the stiffness prediction model, the new MEPDG also includes a rutting prediction model. The analysis using the rutting model was simplified, and separate rutting at the asphalt concrete surface course was used, in contrast to the earlier version that assessed deformation of the full depth of the asphalt layers. The current version of the MEPDG model (National

Cooperative Highway Research Program, 2004) for rutting of asphalt concrete is based on the following relationship

$$\frac{\varepsilon_p}{\varepsilon_r} = k_1 10^{-3.35412} T^{1.5606} N^{0.4791}$$

where ε_p is the plastic strain, ε_r is the elastic strain, T is the temperature (°F), N is the number of load applications, and k_1 is a constant that depends on the depth at which the elastic strain is calculated based on the equation

$$k_1 = (C_1 + C_2 \times depth) \times 0.328196^{depth}$$

$$C_1 = -0.1039 h_{ac}^2 + 2.4868 h_{ac} - 17.342$$

$$C_2 = -0.0172 h_{ac}^2 - 1.7331 h_{ac} - 27.428$$

in which 'depth' is the depth to the point of strain calculation and h_{ac} is the thickness of the asphalt layer. The total rutting in the asphalt layer is calculated by integrating the calculated plastic strain over the thickness of the layer.

17.3 Fatigue characteristics of asphalts

Fatigue can be defined as: 'The phenomenon of fracture under repeated or fluctuating stress having a maximum value generally less than the tensile strength of the material' (Pell, 1988). However, this has been generally accepted as referring to tensile strains induced by traffic loading, and, because other means of generating tensile strains in a pavement exist, a better definition may be: 'Fatigue in bituminous pavements is the phenomenon of cracking. It consists of two main phases, crack initiation and crack propagation, and is caused by tensile strains generated in the pavement by not only traffic loading but also temperature variations and construction practices' (Read, 1996).

Under traffic loading, the layers in a flexible pavement are subjected to flexing that is virtually continuous. The size of the strains is dependent on the overall stiffness and nature of the pavement construction, but analysis confirmed by in situ measurements has suggested that tensile strains of the order of 30×10^{-6} to 200×10^{-6} for a standard wheel load occur. Under these conditions, the possibility of fatigue cracking exists (Pell, 1988).

As described in section 16.1.3, dynamic bending tests are normally used to measure the fatigue characteristics of asphalts. An example of the constant-stress fatigue life characteristics for the same mixture at different temperatures is shown in Figure 17.6. The lines are essentially parallel, and show longer fatigue lives at lower temperatures. If the tests are carried out at a different frequency, the result is similar. Thus, fatigue lives are longer at higher frequencies. The fatigue results could also be presented in terms

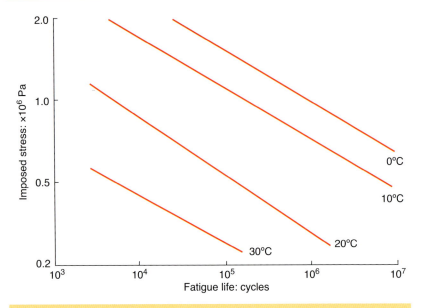

Figure 17.6 Typical fatigue lines – stress criterion (Pell, 1988)

of initial strain, as shown in Figure 17.7. When the criterion of failure is strain, the temperature and time of loading will affect the mixture stiffness. This failure mechanism or effect is known as the 'strain criterion' (Pell and Taylor, 1969).

The general relationship defining the fatigue life, in terms of initial tensile strain, is

$$N_f = c(1/\varepsilon_t)^m$$

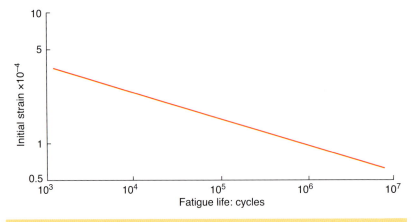

Figure 17.7 Fatigue lines from Figure 17.6 – strain criterion (Pell, 1988)

where N_f is the number of load applications to initiate a fatigue crack, ε_t is the maximum value of applied tensile strain, c and m are factors depending on the composition and properties of the mixture, and m is the slope of the strain/fatigue life line.

However, the strain criterion does not account for the differences in fatigue characteristics predicted using controlled stress and controlled strain experiments. Accordingly, researchers have tried to find alternative failure criteria that take account of the differences in fatigue characteristics.

It has been suggested by Van Dijk and others (Van Dijk, 1975; Van Dijk and Visser, 1977; Van Dijk et al., 1972) that the differences in the fatigue life of asphalt determined under conditions of stress and strain control can be explained by the dissipated-energy concept. This is the amount of energy that is lost from the system, due to fatigue damage, per cycle summed for the entire life of the specimen. Van Dijk and others stated that, for a given mixture, the relationship between dissipated energy and the number of load repetitions to failure is valid, independent of testing method and temperature. This work has been progressed by Himeno et al. (1987), who developed the dissipated-energy concept for three dimensional stress conditions and applied it to the failure of an asphaltic layer in a pavement, and by Rowe (1993), who has shown that dissipated energy can be used to predict the life to crack initiation accurately. The dissipated-energy concept shows considerable promise as a failure criterion that encompasses all the variables of fatigue.

The fatigue characteristics of a mixture can be influenced significantly by its composition. Mixture stiffness is also influenced by the composition of the asphalt, and distinguishing between the two is important. For example, adding filler or reducing the void content will increase mixture stiffness, resulting in increased fatigue life at a given level of stress, as the resulting strain is smaller. In other words, a point has been reached lower down on the strain–fatigue life line. At a particular strain level, however, it has been found that the entire line moves to the right (Pell and Cooper, 1975) (i.e. the fatigue life is improved if, for example, the volume of bitumen is increased). Basic fatigue performance is influenced by other mixture variables only insofar as these affect the volume of bitumen in the mixture. If rounded aggregate is used, for instance, a denser mixture with a lower void content will result, and this will increase the relative volume of bitumen, and thus improve the fatigue performance.

17.3.1 The prediction of fatigue deformation

Traditionally, the establishment of the fatigue line for a particular mixture involved specialised testing equipment. A simpler procedure to predict fatigue performance with sufficient accuracy for pavement design purposes was clearly needed.

17.3.1.1 Shell method

A method for predicting the fatigue life of asphalts was developed by Shell (Bonnaure *et al.*, 1980) using the nomograph shown in Figure 17.8. The required data are

- the percentage volume of bitumen
- the penetration index of the bitumen
- the stiffness modulus of the mixture (in Pa)
- the initial strain level.

The equations used to develop the nomographs are as follows.

For constant stress tests

$$N_f = [0.0252PI - 0.00126PI(V_b) + 0.006\,73V_b - 0.0167]^5 \varepsilon_t^{-5} S_m^{-1.4}$$

For constant strain tests

$$N_f = [0.17PI - 0.0085PI(V_b) + 0.0454V_b - 0.112]^5 \varepsilon_t^{-5} S_m^{-1.8}$$

Other prediction techniques based on some or all of the above input data (Asphalt Institute, 1981; Cooper and Pell, 1974) have also been developed, but these empirical techniques are unable to deal with material developments such as polymer modified mixtures, as they are based on historical data. This has led to the usage of other types of fatigue measurement as described in detail in Chapter 16.

The application of laboratory determined fatigue lives to predict actual pavement performance in practice is a complex problem that is likely to yield conservative results. This happens because simple continuous cycles of loading neglect the beneficial effects of rest periods that occur in practice between axle loads. Longer lives are also likely in practice because of the lateral distribution of wheel loads in the wheel track, and the fact that a degree of crack propagation will occur before the performance of the pavement is adversely affected. As a result of these problems, laboratory fatigue lives have to be calibrated to correlate with actual pavement performance, and the calibration factor is likely to depend on environmental and loading conditions. However, these factors have been incorporated in a prediction technique (Read, 1996) for use in analytical pavement design, based around work carried out at the University of Nottingham.

17.3.1.2 Asphalt Institute method

The Asphalt Institute fatigue equation (Asphalt Institute, 1969) is based on constant stress criteria only.

$$N_f = 0.004\,32C\varepsilon_t^{-3.291}|E^*|^{-0.854}$$

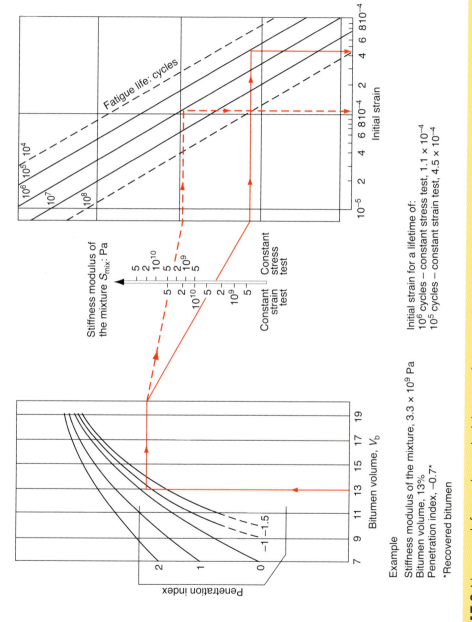

Figure 17.8 Nomograph for predicting the laboratory fatigue performance of asphalt (Bonnaure *et al.*, 1980)

495

where C is a correction factor, where $C = 10\,M$

$$M = 4.84 \left[\frac{V_b}{V_a + V_b)} \right] - 0.69$$

where V_b is the percentage binder content and V_a is the percentage of air voids.

17.3.1.3 MEPDG method

Besides providing a stiffness and rutting prediction model, the new *Mechanistic–Empirical Pavement Design Guide* includes a fatigue prediction model (Advanced Research Associates, 2004). The MEPDG fatigue model uses the modified Asphalt Institute method with an additional correction factor for the layers. This fatigue equation predicts the number of load repetitions to fatigue cracking failure.

$$N_f = 0.00432 k_1 C \left(\frac{1}{\varepsilon_t} \right)^{3.9492} \left(\frac{1}{E} \right)^{1.287}$$

where ε_t is the tensile strain at the bottom of the asphalt surface layer, and E is the elastic modulus of the asphalt layer.

k_1 and C are parameters calculated using the following equations

$$k_1 = \frac{1}{0.000398 + \dfrac{0.003602}{1 + e^{11.02 - 3.49 h_\alpha}}}$$

$$C = 10^M$$

$$M = 4.84 \left(\frac{V_b}{V_a + V_b} - 0.69 \right)$$

where V_a is the percentage air voids and V_b is the percentage binder content.

17.4 Failure theories for cracking in asphalt layers

In the process of designing a structural member, the designer has to ensure that the member under consideration does not fail under service conditions. As empirical failure methods are adopted in the design of flexible pavements, there is little emphasis on research related to the applicability of failure theories to asphaltic layers. Failure theories represent a fundamental approach for evaluating the susceptibility to failure of a material under given loading conditions. Under uniaxial loading, failure occurs when the applied stress reaches the tensile strength of the material. However, under complex, three dimensional stress fields, the situation is more complex, as both normal and shear stresses of varying magnitude may be present. There are five main failure theories, and these are discussed briefly in the following paragraphs. These theories are named after the principal researcher involved in the study.

17.4.1 Maximum principal stress theory (Rankine)

According to this theory, the maximum principal stress in the material determines failure, regardless of the value of the other two principal stresses, as long as they are algebraically similar. This theory is not supported by experimental results (Raju, 2008). It is considered to be reasonably satisfactory for brittle materials that do not fail by yielding. In uniaxial tension or compression modes, failure occurs when the maximum principal stress at any point reaches a value equal to the tensile or compressive elastic limit or yield strength of the material obtained from the uniaxial test. Thus, if σ_1, σ_2 and σ_3 are the principal stresses at a point and $\sigma_1 > \sigma_2 > \sigma_3$ and σ_y is the yield stress for the material under a uniaxial test, then failure occurs when $\sigma_1 \geq \sigma_y$.

17.4.2 Maximum shearing stress theory (Tresca)

Observations made in the course of extrusion tests on the flow of soft metals through orifices lent support to the assumption that the plastic state in such metals is created when the maximum shear stress just reaches the value of the resistance of the metal against shear. Assuming $\sigma_1 > \sigma_2 > \sigma_3$, yielding, according to this theory, occurs when the maximum shearing stress reaches a critical value. The maximum shearing stress theory is accepted as being fairly well justified for ductile materials.

If $\sigma_1 > \sigma_2 > \sigma_3$ are the three principal stresses at a point, failure occurs when

$$\tau_{max} = \frac{\sigma_1 - \sigma_3}{2} \geq \frac{\sigma_y}{2}$$

where $\sigma_y/2$ is the shear stress at yield point in the uniaxial test (Raju, 2008).

17.4.3 Maximum principal strain theory (Saint Venant)

Maximum principal strain theory is analogous to maximum principal stress theory, but in this case failure is expected when the maximum principal strain exceeds the tensile yield strain, or the minimum principal strain exceeds the compressive yield strain (Raju, 2008).

$$|\sigma_1 - \mu(\sigma_2 + \sigma_3)| = \sigma_{yt} \text{ or } |\sigma_3 - \mu(\sigma_1 + \sigma_2)| = \sigma_{yc}$$

where σ_2 is the intermediate principal stress. This failure theory has found limited use in the design of thick walled cylinders (Raju, 2008).

17.4.4 Maximum distortion energy failure theory (von Mises)

The maximum distortion energy failure theory states that a material will fail when, at any point in the material, the distortion energy per unit volume reaches the distortion energy occurring in a uniaxial tension test at failure (Raju, 2008). The failure criteria in terms of normal as well as principal stresses are

$$(\sigma_x - \sigma_y)^2 + (\sigma_y - \sigma_z)^2 + (\sigma_z - \sigma_x)^2 + 6(\tau_{xy}^2 + \tau_{yz}^2 + \tau_{zx}^2) = \sigma_{yp}^2$$

where σ_x, σ_y and σ_z are the normal stresses in the x, y and z directions, and τ_{xy}, τ_{yz} and τ_{zx} are the shearing stresses on the xy, yz and zx planes.

This can also be stated in terms of principal stresses.

$$(\sigma_1 - \sigma_2)^2 + (\sigma_2 - \sigma_3)^2 + (\sigma_3 - \sigma_1)^2 = 2\sigma_{yp}^2$$

where σ_1, σ_2 and σ_3 are the principal stresses, and σ_{yp} is the yield stress.

The maximum distortion energy theory predicts a failure envelope (as defined when plotted using the principal stresses as coordinates) that is very similar to that predicted by maximum shear stress theory, and it is widely applied to failure in ductile materials (Raju, 2008).

17.4.5 Octahedral shearing stress theory

According to this theory, the critical quantity is the shearing stress on the octahedral plane. The plane that is equally inclined to all the three principal axes is called the 'octahedral plane'. The octahedral shear stress theory is, in essence, a different way of stating the von Mises theory. Octahedral shear stress is given by the following equation (Raju, 2008)

$$\tau_{oct} = \tfrac{1}{3}[(\sigma_x - \sigma_y)^2 + (\sigma_y - \sigma_z)^2 + (\sigma_z - \sigma_x)^2 + 6(\tau_{xy}^2 + \tau_{yz}^2 + \tau_{zx}^2)]^{1/2}$$

where τ_{oct} is the octahedral shear stress and the other variables are as the von Mises theory. Combining this relationship with the equation gives the failure criterion for the maximum octahedral shear stress theory (Raju, 2008)

$$\tau_{oct} = 0.47\sigma_{yt}$$

Because the maximum distortion energy theory and the octahedral shear stress theory are equivalent, they are applicable to similar situations – triaxial loading conditions in ductile materials.

Figure 17.9 shows failure envelopes obtained by maximum shear stress and von Mises (octahedral shear stress) theories. Ameri-Gaznon and Little (Raju, 2008) applied octahedral shear stress theory to the design of asphalt concrete overlays.

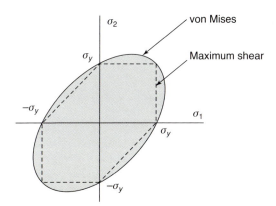

Figure 17.9 Failure envelope

References

Advanced Research Associates (2004) *2002 Design Guide: Design of New and Rehabilitated Pavement Structures*. National Cooperative Highway Research Program, National Research Council, Washington, DC, USA, Project NCHRP 1-37A.

Al-Khateeb G, Shenoy A, Gibson N and Harman T (2006) A new simplistic model for dynamic modulus predictions of asphalt paving mixtures. *Proceedings of the Association of Asphalt Paving Technologists Annual Meeting, Savannah, GA, USA*, Paper Preprint CD.

Andrei D, Witczak MW and Mirza MW (1999) *Development of a Revised Predictive Model for the Dynamic (Complex) Modulus of Asphalt Mixtures*. Department of Civil Engineering, University of Maryland, College Park, MD, USA, Inter Team Technical Report prepared for the NCHRP 1-37A Project.

Asphalt Institute (1969) *Research and development of the Asphalt Institute's Thickness Design Manual (MS-1)*, 9th edn. Asphalt Institute, Lexington, KY, USA, Research Report 82-2.

Asphalt Institute (1981) *Thickness Design, Asphalt Pavements for Highways & Streets*. Asphalt Institute, Lexington, KY, USA, Manual Series No. 1 (MS-1).

Bari J (2005) *Development of a New Revised Version of the Witczak E* Predictive Models for Hot Mix Asphalt Mixtures*. PhD thesis. Arizona State University, Phoenix, AZ, USA.

Bonnaure F, Gest G, Gravois A and Uge PA (1977) A new method of predicting the stiffness of asphalt paving mixtures. *Proceedings of the Association of Asphalt Paving Technologists* **46**: 64–104.

Bonnaure F, Gravois A and Udron J (1980) A new method for predicting the fatigue life of bituminous mixes. *Proceedings of the Association of Asphalt Paving Technologists* **49**: 499–529.

Brown SF (1988) Bituminous pavements: materials, design and evaluation. In: *Bituminous Materials: Elastic Stiffness and Permanent Deformation*. University of Nottingham, Nottingham, UK, Residential course at the University of Nottingham, Section 1, pp. 1–13.

Brown SF and Brunton JM (1986) *An Introduction to the Analytical Design of Bituminous Pavements*, 3rd edn. University of Nottingham, Nottingham, UK.

BSI (British Standards Institution) (2003) BS EN 12697-23:2003. Bituminous mixtures – Test methods for hot mix asphalt – Part 23: Determination of the indirect tensile strength of bituminous specimens. BSI, London, UK.

Christensen JRDW, Pellinen TK and Bonaquist RF (2003) Hirsch model for estimating the modulus of asphalt concrete. *Proceedings of the Journal of the Association of Asphalt Paving Technologists* **72**: 97–121.

Claessen AIM, Edwards JM, Sommer P and Uge PC (1977) Asphalt pavement design – the Shell method. *Proceedings of the 4th International Conference on the Structural Design of Asphalt Pavements* **1**: 39–74.

Cooper KE and Pell PS (1974) *The Effect of Mix Variables on the Fatigue Strength of Bituminous Materials*. Transport and Road Research Laboratory, Crowthorne, UK, Laboratory Report 633.

Dongre R, Myers L, D'Angelo J, Paugh C and Gudimettla J (2005) Field evaluation of Witczak and Hirsch models for predicting dynamic modulus of hot mix asphalt. *Proceedings of the Journal of the Association of Asphalt Paving Technologists* **74**: 381–442.

FHWA (Federal Highways Association) (1978) *Predictive Design Procedures, VESYS User's Manual*. FHWA, Washington, DC, USA, Report FHWA-RD-77-154.

Hills JF (1973) The creep of asphalt mixes. *Journal of the Institute of Petroleum* **59(570)**: 247–262.

Hills JF, Brien D and Van De Loo PJ (1974) The correlation of rutting and creep tests on asphalt mixes. *Journal of the Institute of Petroleum* IP74-001.

Himeno K, Watanabe T and Maruyama T (1987) Estimation of fatigue life of asphalt pavements. *Proceedings of the 6th International Conference on Structural Design of Asphalt Pavements, Ann Arbor, MI, USA*, pp. 272–288.

Hwang D and Witczak MW (1979) *Program DAMA (Chevron), User's Manual*. Department of Civil Engineering, University of Maryland, College Park, MD, USA.

National Cooperative Highway Research Program (2004) *Mechanistic–Empirical Pavement Design Guide*. Transportation Research Board, Washington DC, USA, NCHRP Project 1-37A.

Pell PS (1988) Bituminous pavements: materials, design and evaluation. In: *Bituminous Materials: Fatigue Cracking*. University of Nottingham, Nottingham, UK residential course, Section J, pp. J1–J13.

Pell PS and Cooper KE (1975) The effect of testing and mix variables on the fatigue performance of bituminous materials. *Proceedings of the Association of Asphalt Paving Technologists* **44**: 1–37.

Pell PS and Taylor IF (1969) Asphaltic road materials in fatigue. *Proceedings of the Association of the Asphalt Paving Technologists* **38**: 371–422.

Raithby KD and Sterling AB (1972) *Some effects of loading history on the fatigue performance of rolled asphalt*. TRL, Crowthorne, UK, Laboratory Report 496.

Raju S (2008) *Analysis of Top-Down Cracking in Bituminous Pavements*. PhD thesis. Indian Institute of Technology, Karagpur, India.

Read JM (1996) *Fatigue Cracking of Bituminous Paving Mixtures*. PhD thesis, University of Nottingham, Nottingham, UK.

Rowe GM (1993) Performance of asphalt mixtures in the trapezoidal fatigue test. *Proceedings of the Association of Asphalt Paving Technologists* **62**: 344–384.

Shell (Shell International Petroleum Co. Ltd.) (1978) *The Shell Pavement Design Manual*. Shell Bitumen, London, UK.

Van Dijk W (1975) Practical fatigue characterization of bituminous mixes. *Proceedings of the Association of the Asphalt Paving Technologists* **44**: 38–74.

Van Dijk W and Visser, W (1977) The energy approach to fatigue for pavement design. *Proceedings of the Association of the Asphalt Paving Technologists* **46**: 1–41.

Van Dijk W, Moreaud H, Quediville A and Uge P (1972) The fatigue of bitumen and bituminous mixes. *Proceedings of the 3rd International Conference on Structural Design of Asphalt Pavements*, London, UK, pp. 354–366.

Chapter 18

Influence of binder properties on the performance of asphalts

Dr Shifa Xu
Regional Technology
Manager,
Shell China Ltd. China

Kathy Wang
QA/QC Manager,
Shell China Ltd. China

Andrew Wayira
Global Product Portfolio
Specialist,
Shell Markets (Middle East)
Ltd. UAE

Dr Jia Lu
Senior Application Specialist,
Shell China Ltd. China

This chapter considers the relationship between bitumen properties and pavement performance, and a number of failure mechanisms, with particular emphasis on the role that is played by the binder.

18.1 Introduction

The performance of asphalts, in both the short and the long term, is influenced by the binder properties, aggregate and grading characteristics, as well as the composition, with the binder playing a critical role.

Despite the wide range of applications to which asphalts are put and the substantial variations in weather and loading to which they are subjected, the vast majority of asphalt pavements perform well for many years. However, failures such as rutting, cracking, fretting, fatting up etc. do occur. An understanding of the mechanisms that cause asphalt pavements to fail is important if designers, contractors and producers are to employ specifications, manufacturing techniques, equipment and methods that will minimise the possibility of defects occurring.

The performance of asphalts in service is influenced significantly by the rheological (or mechanical) properties and, to a lesser extent, the chemical constitution of the binder. These factors are, in turn, influenced by changes due to the effects of air, temperature and water on the binder.

The Shell Bitumen Handbook, Sixth Edition
ISBN 978-0-7277-5837-8
Shell International Petroleum Company Ltd: All rights reserved
http://dx.doi.org/10.1680/tsbh.58378.503

The chemical constitution of the binder is particularly important at the road surface because it influences the rate of oxidation and, as a result, how rapidly the binder is eroded by traffic.

There are, of course, many other factors influencing behaviour, including the nature of the aggregate, mixture composition, binder content (i.e. binder film thickness), degree of compaction, etc. – all of which influence long term durability.

Binders are visco-elastic materials, and their behaviour varies from purely viscous to wholly elastic depending on loading time and temperature. During the mixing and compaction of asphalts and at high service temperatures, the properties can be considered in terms of viscosity, but for most service conditions, binders behave visco-elastically, and their properties can be considered in relation to their stiffness modulus.

The rheological requirements for binders during mixing, compaction and in service are illustrated in Figure 18.1, and the critical requirements are summarised in Table 18.1.

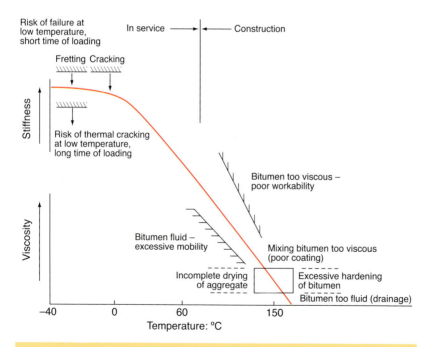

Figure 18.1 Properties of penetration grade bitumens during construction and subsequently in service

Table 18.1 Engineering requirements for binders during application and in service (Dormon, 1969)

	Condition		Significant property of the binder in the mix
	Temperature: °C	Time of loading: s	
Behaviour during application			
Mixing	High (>100°C)	–	Viscosity, approximately 0.2 Pa·s
Laying	High	–	Viscosity
Compaction	High	–	Viscosity, maximum 30 Pa·s
In-service deformation	High road temperature (>30°C)	Long (>10^2)	Maximum viscosity determined by penetration and the softening point of the binder
Fatting up	High road temperature (>30°C)	Long (>10^2)	
Cracking – traffic stresses	Low road temperature	Short	
Cracking – thermal stresses	Low road temperature	Long	
Fretting	Low road temperature	Short	

As the traditional methods of characterising bitumens (i.e. penetration, ductility and softening points) are empirical, they cannot be used to predict how an asphalt or an asphalt pavement will perform in service.

Some researchers have, in the past, used penetration, softening point and viscosity data to assess the complex properties of bitumen. Values of parameters such as penetration indexes (PIs) and penetration-viscosity numbers (Anderson et al., 1983; McLeod, 1972) were believed capable of reflecting the effect of temperature on rheological behaviour. However, it is now known that these parameters cannot accurately describe the effect of time of loading and temperature on the stress–strain response of bitumen.

In the USA, the Strategic Highway Research Program (SHRP) was initiated in the 1990s because specifications based on penetration value and viscosity could not be used to describe asphalt pavement performance fully . One of the major objectives of the SHRP project was to develop test methods that could be used to characterise the performance related physical properties of bitumen binders, so that a performance related binder specification could be developed based on rational relationships between bitumen properties and pavement performance.

The SHRP asphalt research program addressed the major distress modes encountered in asphalt concrete pavements

- rutting
- low temperature cracking
- fatigue cracking.

The parameters considered against these distress modes in the SHRP programme (Petersen *et al.*, 1994; Kennedy *et al.*, 1990) are

- for rutting – the ratio of the complex modulus (G^*) and the sine of the phase angle (sin δ) measured on the original binder at the maximum pavement temperature
- for low temperature cracking – creep stiffness and the *m* value, which is the slope of the master stiffness curve at 60 s, measured on the residue from the pressure ageing vessel (PAV) test at the minimum pavement temperature plus 10°C
- for fatigue cracking – the product of the complex modulus (G^*) and the sine of the phase angle (sin δ) measured on the residue aged in the PAV at an intermediate pavement temperature.

More detailed information about the SHRP programme can be found in sections 5.7, 7.2.3.2, 12.1 and 12.4.5.

In Europe, highway engineers faced similar challenges in trying to ascertain which binder properties dictated the performance of asphalts and pavement performance. In 1995, Eurobitume hosted a workshop on rheology, and subsequently held an international workshop on performance-related properties of bituminous binders in 1999. The workshop identified key properties of binders associated with pavement performance, as listed in Table 18.2 (Eurobitume, 2002).

In 2000, CEN Technical Committee 336, Bituminous Binders, was assigned the task of drafting the second generation of European standards for bituminous binders. The aim of these standards was to define the relevant binder properties affecting the performance of asphalts, surfacing dressings and other road or industrial applications.

In 2012, Eurobitume published an updated position report to provide the views of the bitumen industry on the development of performance-related bitumen specifications. It proposed bitumen test methods appropriate for

Table 18.2 Binder properties linked to asphalt performance (Eurobitume, 2002)

Binder properties	Performance requirements for pavement/mix
Rheological property at elevated service temperature	Resistance to deformation
Ageing behaviour: short term and long term	Resistance to surface cracking due to binder ageing
Rheological property: complex modulus	Structural strength
Combination of rheological and failure properties	Resistance to low temperature cracking
Failure property	Resistance to fatigue cracking
Viscosity vs temperature, storage stability	Manufacturing and laying

performance-based binder specification at some time in the future (Eurobitume, 2012). This initiative focused on the properties that Eurobitume considered to be directly related to pavement performance, namely

- mechanical properties of the binder
- bearing capacity
- fatigue cracking
- resistance to fretting
- resistance to low temperature thermal cracking.

The background to the relevance of these parameters is as follows.

- *Mechanical properties of binder.* The most commonly used properties are rheological parameters such as viscosity, stiffness, phase angle and deformation energy, because these parameters change dramatically depending on test conditions such as temperature, frequency/loading time, stress/strain level, number of loading cycles etc. By selecting sets of test conditions, it is possible to simulate the loading actually applied on the binder in the pavement and, in such circumstances, the response of the binder will provide an indication of its performance in the pavement.
- *Bearing capacity.* Bearing capacity is related to the stiffness of the pavement structure. The stiffness of the pavement is determined by the stiffness of the individual constituents of the asphalt. Therefore, bearing capacity is related to binder stiffness.
- *Fatigue cracking.* Fatigue cracking is the result of the accumulation of damage due to a large number of loading cycles. Even though there have been many attempts to relate the fatigue behaviour of bituminous binders to a simple rheological parameter, no clear relationship has yet been established. There are currently no harmonised test methods for determining the fatigue behaviour of bituminous binders. However, some laboratories are developing different methods. RILEM (the International Union of Laboratories and Experts in Construction Materials, Systems and Structures (Réunion Internationale des Laboratoires et Experts des Matériaux, systèmes de construction et ouvrages) is evaluating the use of a dynamic shear rheometer (DSR) to determine the fatigue behaviour of binders. IFSTTAR (the French Institute of Science and Technology for Transport, Spatial Planning, Development and Networks (Institut français des sciences et technologies des transports, de l'aménagement et des réseaux)). IFSTTAR is evaluating a tension–compression fatigue test based on diabolo shaped specimens. The University of Catalunya in Barcelona is also evaluating a tension–compression fatigue test with cylindrical specimens.

- *Resistance to fretting.* Resistance to fretting is related to the cohesive properties of the binder, which can be determined by measuring the deformation energy using the tensile test or force ductility test.
- *Resistance to low temperature thermal cracking.* The resistance of binders to low temperature cracking can be evaluated using different methods: stiffness at low temperatures, bending beam rheometer test, and fracture toughness and tensile test.

Eurobitume (2012) concluded that 'A new specification is required only for rheologically 'complex' bitumens such as polymer modified and hard paving grade bitumen. Rheologically 'simple' bitumen that meets the EN 12591 specification does not need new specifications as EN 12591 is considered adequately related to performance.' EN 12591 is, of course, the European standard for paving grade bitumens (BSI, 2009).

18.2 The influence of binder properties during construction
18.2.1 Mixing and transport
During mixing, the dried hot aggregate has to be coated by the hot binder in a relatively short mixing time (typically 30–60 s). While the mixing temperature must be sufficiently high to allow rapid distribution of the binder on the aggregate, the use of the minimum mixing time at the lowest temperature possible is advocated. The higher the mixing temperature, the greater the tendency of the binder exposed in thin films on the surface of the aggregate to oxidise. This is illustrated in Figure 18.2, where an increase of 5.5°C in the mixing temperature, for a mixing time of 30 s, results in an increase of 1°C in the softening point of the binder (Whiteoak and Fordyce, 1989). There are, therefore, upper and lower limits of mixing temperatures. If asphalt concretes are mixed too hot, drainage of binder from the aggregate may occur during hot storage or transport to site, leading to variations in binder content. In such circumstances, the filler serves an important function. As it is generally added to the mixture after coating of the aggregate has been completed, the filler 'stabilises' or increases the apparent viscosity of the binder, reducing binder drainage. In thin surface course systems, stone mastic asphalts and porous asphalts, binder drainage can be prevented by the addition of fibres to the mixture, or by using a polymer modified binder.

These different considerations combine to give an optimum binder viscosity of 0.2 Pa·s at the mixing temperature. It has been shown (Jacobs, 1981) that the temperature required to achieve a viscosity of 0.2 Pa·s can be crudely estimated by simply adding 110°C to the softening point of the binder (Figure 18.3). The disadvantage of this method is that it does not take account of the PI of the binder (Figure 18.4). A more precise estimate can be determined using both the penetration and softening point of the binder on

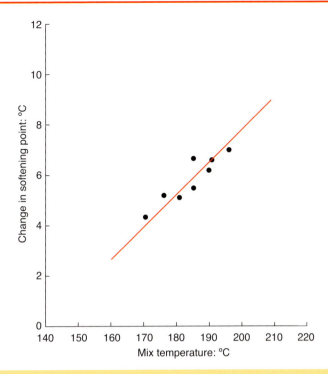

Figure 18.2 Relationship between the temperature of the mixture and the change in softening point (Whiteoak and Fordyce, 1989)

the binder test data chart. The various conditions described above are summarised in Figure 18.4.

When materials are being laid at low ambient temperatures, or if haulage over long distances is necessary, mixing temperatures are often increased to offset these factors. However, increasing the mixing temperature will tend to accelerate the rate at which the binder oxidises, and this increases the viscosity of the binder. Thus, a significant proportion of the reduction in viscosity achieved by increasing the mixing temperature may be lost because of additional oxidation of the binder. If asphalts are transported in properly sheeted, well-insulated vehicles, the loss in temperature is very low, about 2°C/h.

18.2.2 Laying and compaction

Discharging the asphalt into a paver and spreading the material on to a substrate will reduce the temperature by about 20–30°C. The temperature loss will be dependent on a number of factors, including the thickness of the layer, ambient temperature, wind speed and the temperature of the substrate on which the new material is being placed. The two most crucial factors

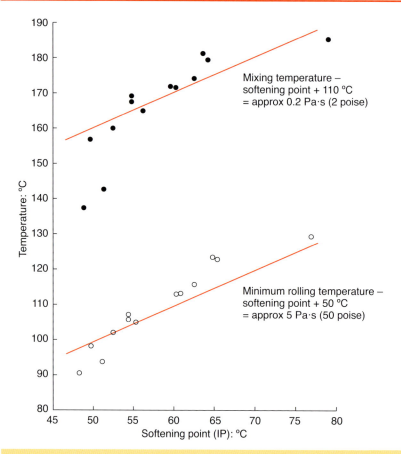

Figure 18.3 Relationship between the softening point of the bitumen and equiviscous temperatures for mixing and rolling of a hot rolled asphalt surface course

influencing the cooling of the layer are wind speed and layer thickness (Daines, 1985).

Once the asphalt has been laid, it must be sufficiently workable to enable the material to be satisfactorily compacted with the available equipment. For effective compaction to take place, the viscosity of the binder should be between 5 and 30 Pa·s. At viscosities lower than 5 Pa·s, the material will probably be too mobile to compact, and at viscosities greater than 30 Pa·s the material will be too stiff to allow further compaction. The minimum rolling temperature can be estimated by adding 50°C to the softening point of the binder (Jacobs, 1981) (see Figure 18.3).

The slope of the viscosity–temperature relationship, or PI, in this temperature region is particularly important because it determines the temperature range over which the viscosity of the binder remains at a suitable level for compaction. Figure 18.4 clearly shows that binders with a high temperature susceptibility (i.e. those with a low PI) have a much narrower temperature 'window' within which satisfactory compaction of the material can be achieved.

It should be noted that the principles and observations discussed above apply to traditional hot asphalts but may not apply to warm asphalts due to these materials containing a range of additives not found in traditional asphalts.

18.3 The influence of binder properties on the performance of asphalts in service

Once an asphalt has been successfully manufactured, laid and compacted, its behaviour in service (i.e. its low and high temperature stability, moisture sensitivity, durability etc.) can be forensically analysed by considering the circumstances that prevailed while the material was in service and the relevance of those circumstances to specific categories of pavement defect.

18.3.1 Cracking

Cracking of pavements is a complex phenomenon that can be caused by several factors. It is associated with stresses induced in the asphalt layers by wheel loads, temperature changes or a combination of the two. Furthermore, the volume of binder in the mixture and its rheological behaviour have a major bearing on the susceptibility of the asphalt to cracking. An asphalt, by virtue of the binder it contains, displays visco-elastic behaviour. If an asphalt test specimen is strained to a predetermined point and the amount of strain held constant, a stress will be induced. Depending on temperature, this stress will dissipate more or less quickly. This process is called 'relaxation'. At high temperatures, the viscous component dominates, and total stress relaxation may take a few minutes. At very low temperatures, relaxation can take many hours or even days.

Cracking occurs when the tensile stress and related strain induced by traffic and/or temperature changes exceed the breaking strength or breaking strain of the mixture. At elevated temperatures, stress relaxation will prevent these stresses reaching a level that can cause cracking. On the other hand, at low temperatures, the tensile condition will persist and, therefore, pavement cracking is more likely.

It is also recognised that binder in an asphalt ages during its service life, and this is called 'curing'. This results in a progressive increase in the stiffness modulus of the asphalt, together with a reduction in its stress relaxation capability. This will further increase the likelihood of the pavement cracking.

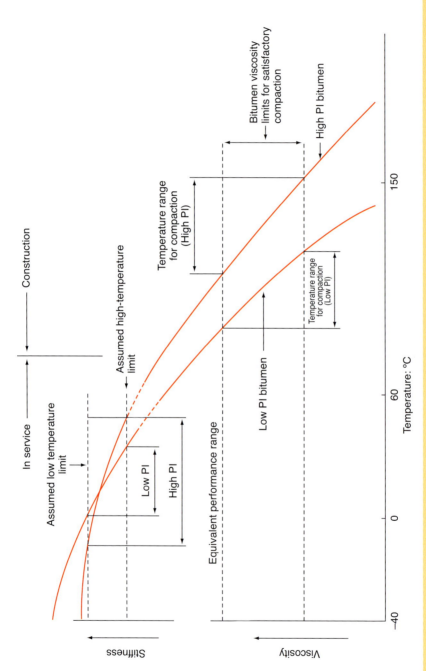

Figure 18.4 The influence of the penetration index (PI) on behaviour during construction and subsequent performance in service

During mixing and laying, the penetration of binder generally reduces to about 70% of the pre-mixed value. This is known as 'short term ageing'. This ageing continues during its service life, albeit at a lower rate, and is known as 'long term ageing'. The most dominant of these mechanisms is oxidative hardening, and this is influenced by the thickness of the binder film, the air voids in the mixture and the temperatures experienced.

Pavement cracking can take several forms. The most frequent are

- longitudinal cracking, which occurs generally in the wheel track
- transverse cracking, which can be in any area and is not necessarily associated with the wheel path
- alligator or crocodile cracking, where longitudinal and transverse cracks link up to form a network of cracks
- reflective cracking, which results from an underlying defect (typically joints in concrete layers or cracks in hydraulically stabilised bases).

Traditionally, pavement design has only considered load-associated cracking, in which cracks are initiated at the underside of the asphalt base caused by repeated pavement flexure under traffic. This has led to the design of pavements of greater thickness to withstand the higher flow rates predicted for future commercial traffic. A programme of research carried out at the Transport Research Laboratory (TRL) (Nunn *et al.*, 1997) to investigate how these thicker asphalt pavements (thicker pavements are those having at least 180 mm of total asphalt in the base, binder course and surface course – see Figure A3 in TRL 250 (Nunn *et al.*, 1997)) were performing revealed that cracks in thick asphalt pavements invariably initiate at the surface and propagate downwards. It should be noted that such cracking in thicker pavements was confined to the surfacing (i.e. in the surface course and binder course only). Similar observations have been made by others (Schorak and Van Dommelen, 1995; Uhlmeyer *et al.*, 2000).

18.3.1.1 Thermal cracking

Cracking that results from extreme cold is generally referred to as low temperature cracking, whereas cracking that develops from thermal cycling is normally referred to as thermal fatigue cracking. Thermal cracking will occur when the binder becomes too stiff to withstand the thermally induced stress, and it is related to the coefficient of thermal expansion and the relaxation characteristics of the mixture. Both these properties are related to the nature of the binder, and the risk of thermal cracking increases with the age of the pavement because the binder hardens as a result of oxidation or time-dependent physical hardening.

Two different thermal cracking mechanisms can occur. At low pavement temperatures, transverse cracks that run the full depth of the pavement can suddenly appear. Pavement temperatures generally have to fall below about −30°C to induce this form of cracking. Accordingly, documented cases of low temperature cracking in the UK are extremely rare.

Under milder conditions, cracks may develop at a slower rate, taking several seasons to propagate through the asphalt layers. This form of cracking initiates at the surface and propagates relatively slowly with each thermal cycle. This is generally described as 'thermal fatigue cracking'.

The general mechanism responsible for these two forms of cracking is considered to be similar. The main differences are as follows.

- Low temperature cracking is a single event phenomenon that is the result of the full depth of asphalt being put into thermal tension under conditions where stress relaxation cannot occur.
- Thermal fatigue cracking is more dependent on the properties of the surface course, and cracks first have to initiate at the surface and propagate through the surface course before they affect the lower asphalt layers.

18.3.1.2 Low temperature cracking

The mechanism of low temperature cracking is illustrated in Figure 18.5. The asphalt layer is subjected to a tensile stress distribution with depth. These stresses are caused by the contraction of the asphalt as it cools, and they are a function of the temperature change and the relaxation characteristics, stiffness modulus and coefficient of expansion of the asphalt. They are not necessarily uniform, because pavement temperature can vary with depth. These stresses can, potentially, cause a crack to propagate down from the surface.

The coefficient of thermal expansion of the binder is an order of magnitude higher (i.e. ten times higher) than that of the aggregate in the mixture. A rough estimate for the thermal expansion coefficient of the mixture can be determined using the equation (European Commission, 1999)

$$a_m = \frac{a_b v_b + a_a v_a}{v_b + v_a}$$

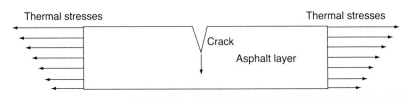

Figure 18.5 Thermal cracking mechanism

where a_m is the coefficient of expansion of the mixture, v_b is the proportion of the binder by volume, v_a is the proportion of the aggregate by volume, a_b is the coefficient of thermal expansion of the binder, and a_a is the coefficient of thermal expansion of the aggregate.

The coefficient of volumetric expansion of bitumen is approximately $6 \times 10^{-4}/°C$. Typically, the linear coefficient of thermal expansion of an asphalt is between 2 and $3 \times 10^{-5}/°C$.

As an asphalt becomes colder, its tensile strength (β_z) initially increases and then begins to decrease as a result of micro-cracking in the binder matrix. This is caused by the differential contraction that results from the large difference between the coefficients of thermal expansion of the aggregate and binder. These fractures can be detected as acoustic events, using a sensitive microphone to record their increasing occurrence with falling temperature (Valkering and Jongeneel, 1991).

At the same time, thermal stress (σ) builds up as the material loses its relaxation ability and, at some point, the thermal stress will exceed the strength of the material. This defines the probable low temperature fracture temperature (spontaneous cracking). The difference between the tensile strength and the low temperature stress is known as the tensile strength reserve $(\Delta\beta_z)$, and it is this reserve that is available to accommodate additional superimposed stresses (e.g. traffic-induced stresses). This is illustrated in Figure 18.6.

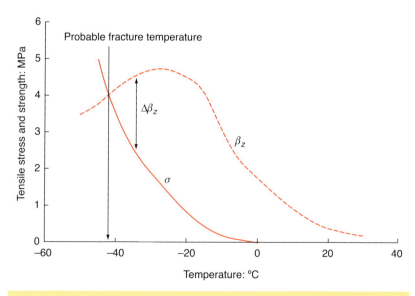

Figure 18.6 Low temperature stress (σ) and tensile strength of asphalt (β_z) as a function of temperature

The curves shown in Figure 18.6 can be derived using the thermal stress restrained specimen test (TSRST) and the isothermal direct tensile test over a range of low temperatures (Jung and Vinson, 1992). Although expensive and time consuming, these tests are considered (King et al., 1993) to offer a reliable means of predicting the temperature at which a pavement will crack due to excessive thermal stresses. The results are reasonably reproducible, and the relative ranking of materials is consistent with field performance. For example, for asphalts manufactured using 60/80 pen bitumen, the TSRST tested cracking temperature is about −30°C (Fan et al., 2012).

Various studies of low temperature cracking have generally concluded that, in order to reduce cracking, the binder stiffness must not exceed some defined limit at the coldest pavement temperature (Bahia, 1991).

The bending beam rheometer (BBR) can be used to produce a master curve for the binder stiffness as a function of loading time and temperature. This curve is used to predict the temperature at which the binder has a stiffness of 300 MPa at a loading time of 60 s (Anderson et al., 1994).

In 1996, a research project was undertaken (Kandhal et al., 1996) to verify whether the BBR test results could have predicted the low temperature cracking of six AC-20 bitumens (ASTM, 2014). One bitumen sample was found to have had a stiffness exceeding 300 MPa at the minimum design temperature but it had not cracked during its 7 year service life.

Marasteanu (2004) investigated the relationship between the stiffness and the m value calculated from the BBR experimental data and the development of thermal stresses in asphalt pavements. The analysis showed that the benefit of a high m value for low temperature performance is not obvious. But it was also pointed out that

- in climates where the temperature stays at reasonably low values for extended periods of time, higher m value binders may perform better as they allow more relaxation to occur
- in climates characterised by extremely cold temperatures, higher m values result in worse performance, as thermal stresses develop faster and can result in the occurrence of cracking before relaxation takes place.

Jung and Vinson (1992) concluded that the TSRST provided an excellent indicator of low temperature cracking. Furthermore, King et al. (1993) demonstrated that results obtained using the TSRST correlated very well with the temperature prediction using the BBR ($R_2 = 0.96$). However, later work (Anderson et al., 2001) suggests that fracture toughness should be included in the applicable criteria when a polymer modified binder is used.

Low temperature cracking is of particular concern in Canada, North America, northern and eastern Europe and northern Asia, where temperatures can fall as low as $-40°C$. To counter this problem, softer binders are often used. It is recognised that the risk of low temperature cracking is related primarily to the properties of the binder, with viscosity and temperature susceptibility of the binder being the most important properties. The risk increases as the hardness of the binder rises. Variations in the grading and type of aggregate have very little effect, and increasing the binder content reduces only slightly the susceptibility of the mixture to thermal cracking.

18.3.1.3 Thermal fatigue and load-associated surface cracking

Cracking in the surface of asphalt pavements is relatively common. These cracks can be transverse or longitudinal, and are usually located in the wheel tracks. Longitudinal wheel track cracking has often been regarded as evidence of conventional fatigue, in which cracks are assumed to have initiated at the bottom of the base and then propagated to the surface. However, investigation by coring thick asphalt pavements has invariably found that either the cracks partially penetrate the thickness of asphalt or, if the crack is full depth, the propagation is downwards rather than upwards. A typical example of this type of cracking in a major UK motorway is shown in Figure 18.7. This crack extended for several hundred metres. Coring showed that only at the most seriously cracked locations were there cracks that had progressed further than the top 100 mm of the asphalt (i.e. cracking was restricted to the surface course and binder course only). None of the cracks had penetrated the full thickness of the asphalt layers.

Investigations of a number of relatively thick UK motorways showed that this behaviour was typical. However, crack investigations by coring in the USA (Uhlmeyer et al., 2000) and Holland (Schorak and van Dommelen, 1995) have shown that, in fully flexible pavements with less than 160 mm of asphalt, the cracking is likely to be full depth. Above this thickness, an increasing proportion of the pavements have cracks that are confined to the asphalt surfacing (i.e. the binder course and the surface course). In the Dutch study, all pavements less than 160 mm thick had full depth cracking, and all pavements with 300 mm or more of asphalt had cracks that were confined to the top 100 mm of asphalt.

Penetration tests on the binder recovered from hot rolled asphalt surface courses in the UK study showed somewhat lower values in cracked areas compared with uncracked lengths. However, binder penetration is not always a good indicator of the susceptibility of the surface course to surface cracking, and it is evident that more detailed studies will be required to understand the loading, material and environmental factors that determine the initiation and propagation of cracks.

Figure 18.7 Longitudinal crack in the wheel track

Surface cracking is not always longitudinal: transverse or block cracking can occur. This is not confined to wheel tracks, and it can occur at any location across the carriageway. Figure 18.8 shows an example of transverse cracks in a UK motorway that had carried traffic totalling some 30 msa (million standard axles) since it was laid 24 years previously. During this time, it had not been resurfaced or overlaid. The penetration of recovered binder from the surfacing was 15 dmm, and it was concluded that the cracking was not traffic related but due to brittle surfacing resulting from aged binder. As with longitudinal cracking, transverse cracks generally only penetrated to a depth of about 100 mm.

The phenomenon of surface cracking has received relatively little attention. However, there are a considerable number of observations of surface cracking of flexible pavements that span all climatic regions, from cold to tropical regions. This suggests that higher temperatures associated with the warmer

Figure 18.8 Transverse cracks in the M32

season cause binder to harden with age, and this reduces its capacity to withstand the thermal stresses generated during the cooler nights, particularly during the coldest season.

With all types of surface cracking, age hardening of the binder in the surface course will play a part. As stated above, this hardening over time progressively reduces the ability of the surface course to withstand the thermal and traffic generated stresses at the surface. Binder rheological studies (Nunn et al., 1997) have shown that the binder in the top few millimetres of the surface course becomes particularly hard with age.

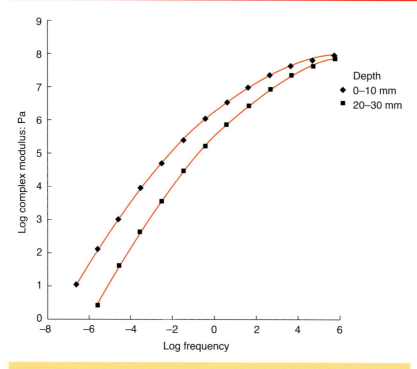

Figure 18.9 DSR master curves at 25°C, illustrating increased binder ageing near the surface

Figure 18.9 shows an example of the results obtained using a DSR on binder extracted from the top 10 mm of the surface course and from between 20 and 30 mm below the surface.

Figure 18.9 shows that the binder close to the surface is substantially harder than binder that is deeper in the layer. Investigations of four sites that were 18 to 24 years old indicated that the penetration of the binder recovered from the top 10 mm of surface course was approximately 50% of that obtained from the lower layer. This aged, and hence hard, upper skin of the surface course is considered to be a major factor in the initiation of surface cracks.

18.3.1.4 Modelling surface cracking

The mechanism of surface cracking is complex and, to date, there is no completely satisfactory explanation of this phenomenon. It is now being recognised that tyre stresses can induce a tensile condition at the pavement surface and this can initiate a crack. De Beer *et al.* (1997) have shown that the tyre can transmit non-uniform horizontal and vertical stresses to the pavement, and pavement stresses in the locality of the tyre need to be taken into account in modelling crack behaviour. Finite element modelling predicts that a tensile condition can be created in the surface course close to the edge of

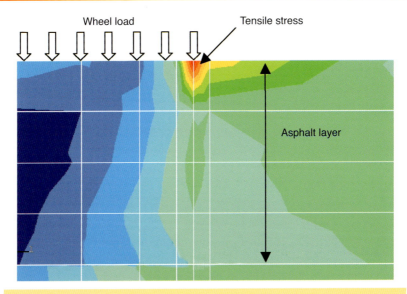

Figure 18.10 Pavement stresses induced by tyre loading

the tyre as shown in Figure 18.10. Similar conditions are also predicted near the edge of tyre treads (Myers and Roque, 2001).

These near-tyre tensile stresses, shown in red in Figure 18.10, extend only approximately 10 mm into the asphalt. Consequently, they may initiate a longitudinal crack in the wheel track, but another mechanism is required to account for crack propagation to any depth. There is increasing conviction that these surface cracks are propagated by thermal stresses.

Groenendijk (1998) concluded that the ageing of asphalt concrete at the surface combined with the non-uniform contact stress could result in critical tensile stress on the surface.

Traditionally, it was considered that there were two categories of cracking: top down and bottom up. As the names suggest, top down cracking starts at the surface, with bottom up cracking beginning at the underside of the base. Nunn *et al.* (1997) discovered that in thicker pavements cracking was likely to start at the surface and was found only in the surfacing (i.e. the surface course and the binder course) – there was no cracking in the base.

There is also a view that top down cracking is caused by the shear stress/ strain in the surface course rather than by tensile stress. Bensalem *et al.* (2000) concluded that the shear strains on the vertical plane play a dominant role in top down cracking. In a research report from Michigan State University, Svasdisant (2003) pointed out that the cracks caused by shear failure along a vertical plane should incline at an angle from the vertical plane. Wang *et al.* (2003) suggested that both tensile type and shear type cracking could initiate top down cracking. Using strength theory and a finite element method to analyse stress/strain in asphalt pavements under non-uniformly distributed tyre loads, Jia *et al.* (2008) stated that top down cracking could be caused by the combination of tensile stress and shear stress.

Thermal fatigue is the mechanism responsible for the development of transverse surface cracks, with crack propagation resulting from the cyclic, daytime thermal stresses. However, transverse rolling cracks, sometimes seen at the time of compaction, are likely to play a major part in the initiation of transverse cracks.

The models predict (Nesnas and Merrill, 2002) that, for a thick asphalt pavement, a thermally propagated crack will stabilise at some point and will not propagate through the full thickness of the asphalt. On the other hand, the prediction for a thin pavement is that the rate of crack propagation will accelerate as the remaining thickness of asphalt is subjected to greater thermal and traffic loading.

18.3.1.5 Fatigue cracking

Fatigue cracking has received more attention from the research community than any other deterioration mechanism. Fatigue is the phenomenon of cracking under the repeated application of a stress that is less than the tensile strength of the material.

When a wheel load passes over a point in an asphalt pavement, the pavement flexes and a tensile strain is induced at the underside of the base layer. Continuous flexure and relaxation over many years carries with it the possibility of fatigue cracks initiating at the underside of the asphalt base and propagating upwards.

Fatigue is a major component of all modern analytical pavement design methods. These methods use a simple shift factor, or calibration factor, to relate the laboratory-determined fatigue life to the performance of the material in the road. The conditions and the behaviour of material in service are much more complex (Thrower, 1979) than those used in laboratory studies. For example, the nature of the stress system and long term physico-chemical changes in the binder are not considered.

The results of studies on the fatigue performance of the asphalt indicated that asphalt fatigue life was a function of the initial loss of stiffness in sinusoidal loading. Thus, it seemed logical to use the binder loss stiffness as the parameter to control mixture fatigue, as fatigue is attributable to cracking in the binder phase of the mixture. Therefore, SHRP uses the product of the complex modulus (G^*) and the sine of the phase angle ($\sin \delta$) to measure the fatigue cracking (Deacon *et al.*, 1997). This product is known as the SHRP fatigue parameter or fatigue parameter.

The fatigue parameter ($G^*\sin \delta$) is measured after a relatively small number of load cycles, and therefore cannot reflect the non-linear behaviour of a binder. Many experts agree that fatigue damage resistance cannot be predicted from linear visco-elastic properties alone, as $G^*\sin \delta$ (Bahia *et al.*, 2010). A number of studies have shown that there is a poor correlation between the linear visco-elastic $G^*\sin \delta$ and mixture fatigue performance (Stuart and Mogawer, 2002; Tsai *et al.*, 2005). In NCHRP Report 459 (Bahia *et al.*, 2001), an analysis method was outlined that uses the ratio of dissipated energy of the binder to replace the SHRP fatigue parameter to quantify the amount of fatigue damage in an asphalt.

The development of the long-life pavement concept (Nunn *et al.*, 1997) has demonstrated that long term changes in the binder properties can result in an increased fatigue life of the pavement. Also, in the USA, where this concept was subsequently developed, it has been suggested that, at low strain amplitudes, asphalt has a fatigue endurance limit (Newcombe *et al.*, 2002).

The fatigue resistance of an asphalt is especially sensitive to binder volume. The simplest means of increasing the predicted fatigue life of the pavement is to construct the pavement using a binder-rich lower asphalt layer (Harm and Lippert, 2002).

18.3.1.6 Reflective cracking

A composite pavement consists of a continuously laid cementitious base under asphalt surfacing, or an existing cementitious pavement overlaid with asphalt. A regular pattern of thermally induced transverse cracks can appear in the cementitious base soon after laying, and these cracks begin to appear as reflection cracks in the asphalt surfacing several years later, as shown in Figure 18.11.

For some composite pavements, particularly those in which an existing cementitious pavement has been overlaid with asphalt and carries heavy traffic, as observed in China on the Hanyi expressway (Liao and Chen, 2010), these cracks were caused either by the crack in the cementitious layer opening and closing as a result of thermal expansion and contraction, or by a flexing and shearing action caused by wheel loads passing over the crack. These mechanisms will induce a stress concentration in the asphalt immediately above the crack, and this causes a crack in the asphalt to initiate and propagate towards the surface.

However, for some other composite pavements, particularly those consisting of a continuously laid cementitious base under asphalt surfacing, as

Figure 18.11 Reflection cracks on a motorway with a composite pavement

observed in the UK, extensive coring of in-service roads has shown that reflection cracks often start at the surface of the road and propagate downwards to meet an existing crack or joint in the underlying concrete layer (Figure 18.12). Furthermore, this study (Nunn, 1989) has shown that environmental effects, rather than traffic loading, cause reflection cracks in as-laid composite pavements. When the pavement is new, the surface course is ductile enough to withstand thermally induced stresses but, as the pavement ages, it will progressively lose this capability. The study showed that the occurrence of reflection cracking in as-laid pavements correlated with the amount of strain that the surface course could accommodate before cracking. This reduced with age, and it is related to the type and volume of binder used.

In their early stages of development, the cracks are not considered to present a structural problem. However, once they propagate through the asphalt layer, water infiltration and the pumping action of the traffic will weaken the foundation layers. At the same time, the load transfer across the slab will deteriorate, and under these conditions the cementitious slabs will move

Figure 18.12 Core of reflection crack initiating from surface

under heavy traffic, and further cracking, spalling and general deterioration will result. The reflection cracking of a strengthening overlay over a crack of this nature will be dependent on traffic-induced forces and the severity of the crack.

Modelling reflection cracking in as-laid composite pavements has generally treated the asphalt as a passive layer that has to respond to movements in the concrete layer. These models either assume that the thermal opening and closing of the crack in the cement bound layer or shearing caused by traffic will induce a high stress concentration in the asphalt immediately above the crack, causing it to initiate and propagate upwards. These models have not considered that

- the thermal coefficient of expansion of the asphalt surfacing is several times greater than that of concrete
- greater daytime temperature changes occur close to the surface
- age hardening of the binder is more severe close to the surface.

525

Modelling by Bensalem (2001) takes into consideration all these factors and uses the finite element technique to model the behaviour of the asphalt and cement bound layers as a single system, as illustrated in Figure 18.13(a). Figure 18.13(b) shows the thermally induced stress contours predicted using this model. The model predicts that the highest thermal stresses will occur at the surface immediately above the crack in the cement bound base. Furthermore, it predicts that a compressive stress condition will exist at the underside of the asphalt layer adjacent to the crack in the cement bound layer. This implies that it is not possible for a crack to propagate upwards. This model

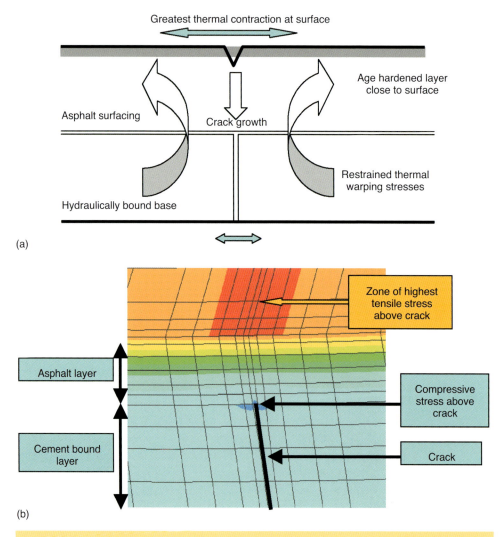

(a)

(b)

Figure 18.13 (a) Schematic model of surface-initiated reflection cracking; (b) predicted thermally induced stress contours

provides a qualitative explanation for the observed manner in which cracks initiate at the surface and propagate downwards.

However, for existing cement pavements overlaid with asphalt, the void under the cement concrete slab will cause additional deformation and change the stress/strain status in the pavement structure. Therefore, the modelling needs to take into account the supporting condition beneath the cement concrete slab.

18.3.2 Deformation

Deformation (often called 'rutting') may be restricted to one or more of the asphalt layers, or it may extend throughout the entire pavement and into the subgrade. It is classified as wearing, post-compaction and shearing rutting. In practice, however, the majority of deformation defects result from plastic deformation of the asphalt layers, mostly in the surface course, or surface course and binder course. Severe rutting can occur under moving or stationary traffic, and particularly under the high shearing stresses imposed by braking, accelerating, turning, overloaded traffic or long and steep sloping sections of road (Figure 18.14). The primary internal factor influencing plastic deformation is the composition of the mixture, while the primary external factors are stress and temperature. Plastic deformation is greatest at high service temperatures, for which 70°C may be taken as a maximum

Figure 18.14 Severe rutting in a long and steep sloping section after an extremely hot summer and overloading

in situ temperature. At such temperatures, the cumulative effect of repeated loadings of short duration will be determined by the binder viscosity. It has been estimated that during the long hot summer experienced in the UK during 1976, deformation in the wheel tracks of hot rolled asphalt surface courses was between two and four times the rate of an average UK summer.

Bitumen rheology is one of the most critical factors influencing the propensity of an asphalt to rut. The significance of the PI of the binder was confirmed by an extensive full scale road trial carried out on the Colnbrook bypass. Binders of widely differing rheological characteristics were employed in hot rolled asphalt surface course mixtures used in this project. Figure 18.15 shows the relationship between rut depth on the road after 8 years and the PI of the binder. The advantage of using higher PI binders is evident.

Work in several laboratories has shown that deformation occurring in a wheel tracking test can be predicted using the uniaxial unconfined creep compression test (Van de Loo, 1974, 1976). Creep curves of mixture stiffness (S_{mix}) plotted against binder stiffness (S_{bit}) can be used to compare different mixtures and to predict their relative deformations if subjected to the same temperature and loading conditions.

The shape of a creep curve for a typical hot rolled asphalt mixture surface course is such that it can be approximated for most of its length by a straight line having a slope of 0.25. Thus

$$S_{mix} = k(S_{bit})^{0.25}$$

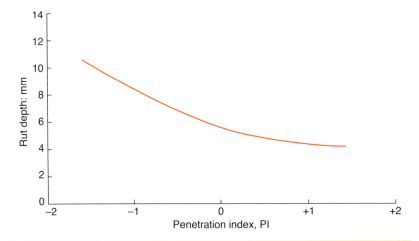

Figure 18.15 Rut depth as a function of in situ penetration index (PI) measured on the Colnbrook bypass after 8 years in service (Dormon, 1969)

For two mixtures having S_{bit} values in a known ratio,

$$S_{mix} \text{ ratio} = (S_{bit} \text{ ratio})^{0.25}$$

For asphalts in similar road structures subjected to the same traffic

$$\text{Deformaton} = k(1/S_{mix})$$

For two mixtures having S_{mix} values in a known ratio,

$$\text{Deformation ratio} = 1/S_{mix} \text{ ratio}$$

The above relationships can be used to compare the effect of changing the penetration and/or the PI of the binder.

Calculations made on this basis show that if the deformation of a 50 pen binder is regarded as unity, a mixture manufactured using a 30 pen binder subjected to the same temperature and loading regime will deform by 0.75 units. This assumes that the loading time for determination of viscosity and stiffness are the same. Conversely, if a 100 pen binder is used, 1.5 units of deformation will occur (Table 18.3(a)).

Increasing the PI of the binder significantly improves the resistance to deformation. For example, at 40°C a binder with a penetration of 40 and a PI of 0.5 has a viscosity of 4×10^4 Pa·s, whereas a binder with the same penetration but having a PI of +2.0 has a viscosity of 6×10^5 Pa·s at 40°C. This gives a factor of 15 in the viscosity, and hence in the stiffness of the two binders. Thus, applying the 0.25 power formula stated above gives an increase of approximately 2 in mixture stiffness. Accordingly, the deformation using a mixture made with a binder having a PI of +2.0 would be half of one made with a standard binder (Table 18.3(b)). Figure 18.16 shows the above theoretical relationship for relative deformation plotted as a function of PI. This clearly shows that the theoretical relationship is supported by both simulative laboratory wheel tracking tests and full scale road trials.

Correlation of binder properties with both Marshall and wheel tracking tests (Jacobs, 1981) shows that the relationship between binder penetration and Marshall stability is poor, whereas the softening point relates very well to both Marshall stability and deformation in the wheel tracking test (Figure 18.17).

Table 18.3 (a) Relative deformation of mixtures at 40°C, effect of binder penetration

Binder penetration	PI	Viscosity at 40°C: Pa·s	Relative S_{bit}	Relative S_{mix}	Relative deformation
100	0.8	6×10^3	0.2	0.67	1.5
60	0.6	18×10^3	0.6	0.88	1.1
50	0.5	30×10^3	1	1	1
40	0.5	40×10^3	1.3	1.07	0.94
30	0.4	90×10^3	3.0	1.32	0.75

Table 18.3 (b) Relative deformation of mixtures at 40°C, effect of binder penetration index

Binder penetration	PI	Viscosity at 40°C, Pa·s	Relative S_{bit}	Relative S_{mix}	Relative deformation
40	−0.5	4×10^4	1	1	1
40	+0.5	1.5×10^5	3.8	1.41	0.71
40	+2.0	6×10^5	15	1.97	0.50

For every 5°C increase in softening point, the Marshall stability increased by over 1.3 kN (see Figure 18.17 (b)) and the wheel tracking rate almost halved (see Figure 18.17 (c)).

The SHRP programme has provided a binder parameter, designated the 'rutting factor' (the ratio of the complex modulus and the sine of the phase angle, i.e. $G^*/\sin \delta$) representing a measure of the high temperature stiffness of the binder. Laboratory research had found that the binder parameter, $G^*/\sin \delta$, was related to the depth of rutting in wheel tracking results. An earlier indication of the suitability of $G^*/\sin \delta$ as the key specification criterion for rutting was contained in the paper by Petersen et al. (1994).

Mitchell et al. (2004) have used the Federal Highway Administration's two accelerated loading facility machines to validate the rutting factors of Superpave binder properties. However, no overall relationship between $G^*/\sin \delta$ and pavement rut depth was found.

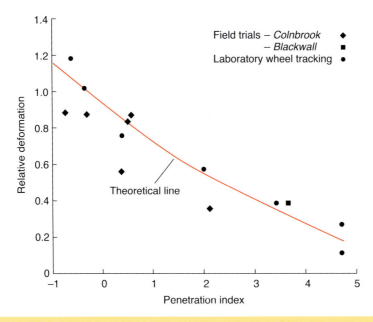

Figure 18.16 Relative deformation as a function of penetration index

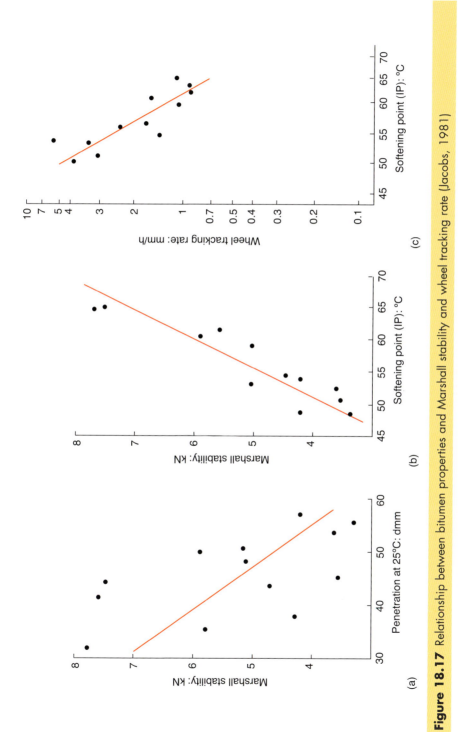

Figure 18.17 Relationship between bitumen properties and Marshall stability and wheel tracking rate (Jacobs, 1981)

18.3.2.1 The zero shear viscosity concept

In terms of deformation, it is thought that the binder contribution to this mode of distress is primarily associated with viscosity. However, identifying the relevant viscosity in terms of shear, shear rate and stress is a complex problem, and a single quantifiable regime would have to be established for assessment purposes. Some pavement design methods are based on the assumption that the material response to traffic loading is within the linear regime, and to remain consistent with this premise the corresponding linear viscosity of the binder should be examined. This is the zero shear viscosity, η_0, sometimes termed 'Newtonian viscosity' as stress is proportional to shear. In this regime, stress and strain are linearly proportional to each other, so the stiffness modulus is independent of stress or strain level, and the resultant viscosity, η_0, is independent of the shear rate.

By measuring binder properties within the linear regime, conclusions can be made as to the likelihood of a binder contributing to or mitigating the rutting process. Zero shear viscosities can be measured using a dynamic shear rheometer, which also permits the elastic and viscous components to be identified. The higher the value of zero shear viscosity, the lower the influence of the binder in relation to deformation. Research has shown that the useful upper limit on zero shear viscosity of unaged binder is 10^5 Pa·s, above which further resistance to deformation gives diminishing returns (Phillips and Robertus, 1996). For binders that exhibit zero shear viscosities below this value, the relative value of the zero shear viscosity coupled with the elastic component of the binder can be used to interpret the deformation behaviour of the binder.

18.3.2.2 Multiple stress creep recovery

Another test, called the multiple stress creep recovery test (MSCR), has been suggested, and of particular interest is the resistance to deformation at high temperature under the repetitive action of heavy 'traffic' loadings. This test was proposed by D'Angelo et al. (2007) in the USA, and features in ASTM standard D7405 published in 2008 (ASTM, 2008). The MSCR test is a standard rheological test based on the repeated creep recovery test (D'Angelo et al., 2007).

In the MSCR test, a specimen is subjected to a constant load for a fixed time period, t_1. The resultant deformation is measured as a function of time and applied stress (τ). At a time t_1 (in the MSCR test $t_1 = 1$ s), the specimen is allowed to recover at zero load for a fixed time period, t_2 (in the MSCR test $t_2 = 9$ s). The time dependent and stress dependent recoverable shear deformation or strain is the difference between the maximum shear strain $\gamma(t_1)$ and the shear strain at $\gamma(t_2)$ with $t_2 > t_1$. The total recoverable deformation γ_r for $(t_2 - t_1)$ is a measure of the elasticity of the material (i.e. the mechanical energy stored in the sample during the creep phase). The non-recoverable part relates to the viscosity.

In addition, a property called 'non-recoverable compliance', J_{nr}, was also derived from the non-recovered strain normalised for the applied stress during the creep portion of the test ($J_{nr} = \gamma_{nr}/\tau$). According to a Federal Highway Administration study (FHWA, 2011), this J_{nr} value at a stress of 3.2 kPa has been shown to be an indicator of the rutting resistance of an asphalt to deformation under repeated load. The study by D'Angelo *et al.* (2007) showed that reducing J_{nr} by half typically reduced rutting by half. Therefore, the J_{nr} value obtained at a stress of 3.2 kPa, which is based on the response of binders at their performance grade temperature, will be used in the MSCR high temperature specification as an indicator of rutting resistance.

18.3.2.3 The 'softening point' of asphalt concept

Considering that the aggregate grading, binder content and design air voids are all important factors that will have an effect on the propensity of a particular asphalt to rut, any attempt to establish the direct relationship between binder parameters and pavement deformation must take account of these factors. Accordingly, any method for predicting rutting must investigate the temperature susceptibility of asphalt.

Xu and Ji (2006) have developed a new method to evaluate the anti-rutting performance of asphalt. A series of wheel tracking tests is carried out at different temperatures. Following this, the relationship of dynamic stability (the number of wheel loads necessary to cause a deformation of 1 mm) versus temperature is plotted. It was found that the whole plot can be divided into three stages as follows.

- Stage 1: as the temperature increases, the dynamic stability slowly decreases.
- Stage 2: as the temperature increases, the dynamic stability remains constant, with the maximum temperature in this stage being described as the 'softening point' of the asphalt.
- Stage 3: there is a dramatic increase in dynamic stability.

These stages are illustrated in Figure 18.18.

The explanation for this behaviour is as follows. During stage 1, as the temperature increases, the stiffness and viscosity of the binder drop quickly, which results in deformation of the asphalt. During stage 2, although the stiffness and viscosity of the binder are continually decreasing, the loading is carried by the mechanical interlock of the asphalt, and the dynamic stability remains constant. During stage 3, a further decrease in viscosity causes the binder to

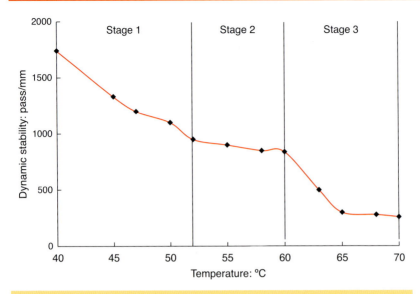

Figure 18.18 Illustration of dynamic stability versus temperature and asphalt softening point

act as a lubricant between the particles of aggregate, with a consequent loss of strength, resulting in greater deformation.

It was found that the 'softening point' of a conventional asphalt has good correlation with the softening point of the constituent bitumen, and is about 10°C higher than the softening point temperature of that bitumen. However, the 'softening point' of a styrene–butadiene–styrene (SBS) modified asphalt has poor correlation with the softening point of a binder but has a good correlation with the equivalent softening point T800 (the temperature at which the penetration of the binder is 800 dmm) and is also about 10°C higher than the equivalent softening point of binder T800.

The significance of the 'softening point' of an asphalt lies in its validity for selection of binder and asphalt to meet the requirement of deformation resistance according to the highest pavement design temperature. The advantage is that it takes into account binder, aggregate and grading characteristics when assessing the deformation resistance of asphalts, rather than only a binder property.

The research on the 'softening point' of asphalts has suggested that the grading of the asphalt is a key element in determining anti-rutting performance. An asphalt having a grading in which there is a high degree of stone-to-stone contact always exhibits high rutting resistance, as shown in Figure 18.19. Indeed, this substantiates the view long held by experienced pavement engineers that asphalt concrete mixtures generally have good

Figure 18.19 Stone-to-stone contact in an asphalt concrete

resistance to deformation, while hot rolled asphalts perform poorly in relation to rutting. This is why the use of a hot rolled asphalt binder course may cause concern due to its propensity to rut.

18.3.3 Fatting up

Fatting up (sometimes described as 'flushing up') is a migration of binder from the asphalt to the top of the pavement, resulting in a smooth, shiny surface that has poor resistance to skidding in wet weather. One reason why fatting up occurs is as a result of secondary compaction of the aggregate in the mixture by traffic. The void content is reduced, eventually squeezing binder from within the structure to the surface. This will be exacerbated if the binder content is too high, resulting in a thicker binder film, or if the void content, after completion of compaction, is too low (Youtcheff, 2002). Fatting up is most likely to occur at high service temperatures. Accordingly, increasing the softening point or viscosity of the binder at 60°C will limit this failure mechanism.

18.3.4 Fretting

Fretting is the progressive loss of interstitial fines from the road surface. It occurs when traffic stresses exceed the breaking strength of the asphalt itself or the asphalt mortar, depending on the nature of the mixture. Fretting is more likely to occur at low temperatures, and at short loading times when the stiffness of the binder is high.

535

The major factors influencing fretting are the binder content of the mixture and the degree of compaction. Loss of aggregate can be due either to the loss of adhesion between the aggregate and the binder or to brittle fracture of the binder film connecting particles of aggregate. The first condition should not arise if suitable aggregates and/or binders are selected. In the majority of cases, fretting is associated with a low degree of compaction or inadequate binder content. The choice of binder grade may control the degree of compaction that can be achieved. This is especially true if operations are being conducted in adverse weather conditions or if compaction is effected at temperatures below which the asphalt can be properly compacted. The higher the penetration of the binder or the lower the stiffness at low temperatures, the greater will be the resistance to fretting.

Fretting is also often associated with wet weather. It is well known that many types of asphalt defect are caused or increased in severity by the presence of water. Water penetrating the binder–aggregate mixture causes moisture damage: early loss of strength and durability due to loss of adhesion between the binder and the aggregates or the loss of cohesion in the binder.

Over many years, researchers have established that moisture damage, one of the major causes of pavement failure, is affected by many factors. More recent studies are those by Petersen (2002), Thomas *et al.* (2002), the Pennsylvania Transportation Institute (2002) and Caro *et al.* (2010). These and other studies have established that the parameters relevant to fretting defects are

- the chemistry of the binder and binder additives
- the rheology of the binder
- the chemistry of the binder–aggregate interaction
- the asphalt film thickness
- the aggregate surface chemistry and shape characteristics
- the surface energy
- the air void distribution and permeability
- the climate.

The most significant of the above factors are the binder chemistry, the binder rheology, the aggregate surface chemistry and the physical properties of the aggregate.

A recent study has addressed the effect of water vapour on asphalt pavement materials and found it to be significant. It was found that water vapour from subgrade soils migrates upwards into the pavement. This water vapour accumulates in the pavement at a rapid rate and reaches near-saturated vapour pressure within a period of 6 months. Moreover, wetting processes in the pavement layer brought about by sub-surface water vapour diffusion occurs day and night. Therefore, the water presence attributable to vapour diffusion in the asphalt surface layer is one of the major water movement mechanisms in pavement, and it greatly accelerates the deterioration of the asphalt (Tong *et al.*, 2013).

18.3.4.1 The effect of ageing
The ageing of the mortar is attributed to the effects of oxygen, ultraviolet radiation and temperature. The effects of ageing of asphalts can be classified in two major groups

- short term ageing, which involves the loss of volatiles and oxidation of the bitumen during production, transportation, laying and compaction of the asphalt
- long term ageing, which is predominantly the continuation of the oxidation process during the service life of the asphalt pavement.

The ageing process is influenced by the type of asphalt: for example, a higher rate of ageing is expected in a porous asphalt compared to a dense asphalt (Roberts *et al.*, 1996).

In asphalt surface courses, the bituminous mortar is subjected to shear, tension and/or compression, so the strength and relaxation behaviour of the bitumen are important. However, the effect of ageing increases the stiffness of the asphalt. This ageing also results in a change in creep and relaxation properties of the bitumen, resulting in higher susceptibility to fracture due to brittleness at low temperatures and a decrease in healing potential at intermediate temperatures (Yero, 2012).

Ageing changes the rheological, mechanical and chemical properties of the binder. The changes in the properties of the binder, and therefore the asphalt, can be quantified by means of ageing indicators. Ageing indicators commonly used in practice are either empirical parameters describing changes in binder properties, including changes in the visco-elastic behaviour of the binder, or a measure of the change in the chemical composition of the binder, indicating a change in colloidal structure after ageing (Hagos, 2008).

18.3.4.2 The effect of water
Water infiltration has a negative effect on the material characteristics of the individual components of the asphalt, and damages the bond between the components, which may result in ravelling of the surface. Adhesive failure is largely attributed to damage caused by the effect of water, and cohesive failure is caused when the stress levels in the binder/mortar are exceeded.

The introduction of water into an asphalt increases the chance of cohesive failure due to the 'softening' of the binder. Further studies have shown that the water damage phenomenon is an effect of advective transport (a phenomenon causing the mastic films in an asphalt to be washed away when the material is subjected to a moving flow of water) and diffusion, leading to mastic–aggregate interface failure and dispersion of the mastic. The analyses showed that water damage in asphalts is highly dependent on the adsorption characteristics of the aggregate–mastic system, as well as the diffusion and dispersion characteristics of the mastic. It is deduced that, by minimising the susceptibility of the mastic, the damage to the mastic will be solely due to advective transport, and damage due to dispersion phenomena will then be limited (Kringos and Scarpas, 2005).

18.3.4.3 Fretting of asphalt concrete surface courses
The following two different forms of fretting are identifiable in asphalt concrete surface courses.

- *Superficial fretting*. This involves the loss of interstitial fines from the road surface. The material subsequently closes up under traffic without further deterioration or detrimental effect on performance.
- *Severe fretting*. If superficial fretting does not close up under traffic, the loss of interstitial fines results in a reduction of the level of mechanical interlock that is likely to lead to the loss of coarse aggregate on the surface and a substantial reduction in internal cohesion due to the ingress of water. The presence of water breaks down the adhesive bond between the binder and the aggregate resulting in stripping of the binder (i.e. a physical separation of the binder from the aggregate). In addition, the uncoated fines and water form a slurry that is pumped through the material, as a result of pore pressures induced by moving traffic, abrading the coarse aggregate and exacerbating the problem. This will eventually lead to collapse of the material and a lack of internal stability, often visible on the road surface as slight rutting or as potholes that are bowl-shaped holes of various sizes.

Some research results have suggested that adhesive failures and thin film cohesive failures need to be differentiated. Designing binders with improved adhesion properties and/or improved cohesion properties could be an essential step in improving the resistance of the aggregate–binder system to moisture damage (Kanitpong and Bahia, 2003).

A number of procedures and recommendations to eliminate or minimise moisture damage have been formulated. One of those procedures involves treating the asphalt with an anti-stripping agent such as hydrated lime or other commercially available anti-stripping additives (Kennedy and Ping, 1991).

In addition, other studies have identified a range of approaches to this type of defect (Bagampadde *et al.*, 2006; Hefer *et al.*, 2005), namely

- theory of (weak) boundary layers
- mechanical theory
- electrostatic theory
- chemical bonding theory
- thermodynamic theory.

It is hoped that better use of additives, higher testing frequencies and improved testing protocols will result in better in-service performance.

18.3.4.4 Fretting of hot rolled asphalt surface courses
In 1984, a survey was carried out by the British Aggregate Construction Materials Industries (BACMI, now the MPA – Mineral Products Association) (White, 1989) as part of a wider survey into chipping loss by the Institution of Highways and Transportation (Johnson and Salt, 1985a,1985b). The BACMI survey showed that over half of the locations where chipping loss had occurred were high stress low speed sites such as roundabouts and junctions, where a minimum 1.5 mm texture depth had been needlessly specified.

Fretting of a hot rolled asphalt surface course may well follow a failure to embed the precoated chippings properly. Failure to achieve adequate chipping embedment usually occurs as a result of one or more of the following factors

- the asphalt is unworkable (i.e. too stiff)
- the temperature of the asphalt is too low
- the asphalt cools quickly due to poor weather conditions during laying
- excessive application of precoated chippings
- inadequate compaction of the material (see Figure 18.20).

One method of treating patches where the chippings are inadequately embedded is to reheat the surface using an infrared heater and, when the asphalt is sufficiently hot, embed the chippings further into the asphalt. Used judiciously, this technique should not significantly harden the binder except at the surface. This hardened surface material will be eroded by traffic over time to expose the aggregate in the asphalt, a situation that will not threaten the long term integrity of the surface. However, this technique is often difficult to apply, requires great care and is not always successful. The best way of avoiding such problems is to ensure proper chipping embedment when the surface course is laid.

Some chipping loss may occur as a result of a failure to achieve or maintain bond between the precoated chippings and the asphalt. It has been found that the binder on precoated chippings can be 'coked' and made

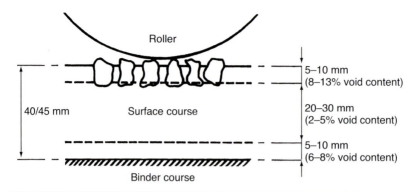

Figure 18.20 Compaction of a hot rolled asphalt surface course under adverse weather conditions

non-adhesive by storage of the chippings in large stockpiles immediately after manufacture. To avoid this possibility, chippings should be stored no higher than 1 m. The adhesivity of the binder coating can be checked using the hot sand test (BSI, 2005).

If the chippings have plucked out of the asphalt, the asphalt mortar is exposed to the rigours of the prevailing traffic. It is likely that this material will be poorly compacted and, as a result, it will fret rapidly.

18.3.5 Ravelling

Ravelling differs from the two fretting modes described above in that it involves plucking out of the asphalt itself (i.e. not just interstitially). It occurs when individual aggregate particles move under the action of traffic. If the tensile stress (induced in the binder as a result of the movement) exceeds the breaking strength of the binder, cohesive fracture of the binder will occur, and the aggregate particles will be detached from the road surface. Thus, ravelling is most likely to occur at low temperatures and at short loading times, when the stiffness of the binder is high. Obviously, a high binder content (and preferably a bitumen appropriately modified) will increase the ravelling resistance of the asphalt under high frequency traffic stress. A 1977 survey carried out by the Asphalt and Coated Macadam Association (ACMA, subsequently BACMI, then the QPA and currently the MPA) established that the use of higher binder contents and softer grades of binder will limit these fretting occurrences, while binder hardening may also lead to fretting.

Ravelling in asphalt concrete pavements depends very strongly on the bond between bitumen and the aggregate. Some studies found that the chemistry of both the binder and aggregate surface determine the degree of water sensitivity of the binder–aggregate bond. If the bond is poor, adhesion failure could occur at the binder–aggregate interface, leading to stripping and ravelling in asphalt pavements (Kandhal and Rickards, 2001). A methodology has been developed to establish a relationship between moisture content and the reduction in strength of the mastic–aggregate bond (Copeland et al., 2006).

Recent studies have shown that the adhesive and cohesive characteristics of asphalt surface courses are significant factors in determining whether the asphalt may be susceptible to ravelling (Kanitpong and Bahia, 2003).

The adhesive characteristics are those that are responsible for the adhesion between aggregates and the bituminous mortar, while the cohesive characteristics are those that are responsible for the cohesion in the bituminous mortar. Because of the very nature of asphalt surface courses, the cohesive and adhesive characteristics are strongly influenced by the effects of ageing and water.

18.3.6 *Skid resistance*

The formal UK definition of skid resistance is 'the frictional properties of the road surface measured using a specified device under standardised conditions' (Highways Agency et al., 2004). A less technical definition would be a measure of the ease with which a vehicle stops when the brake is applied. The higher the skid resistance, the shorter the distance a vehicle takes to stop. Improving the skid resistance of a road is often the reason for surface dressing.

Maintaining adequate levels of skid resistance reduces accidents, meaning fewer injuries and deaths in road accidents. In countries where skid resistance standards apply, a loss of skid resistance is likely to be the most common reason for maintenance treatment.

The skid resistance of a road is determined by the microtexture of the aggregate and the macrotexture of the road surface itself. Means of measuring skid resistance are discussed in section 21.1.2.4.

The microtexture, or degree of polish of the exposed aggregate, is primarily dependent on the petrographic characteristics of the aggregate and the intensity of traffic. The macrotexture of the road surface is dependent not only on the type of surface course but also on the characteristics of the binder used in the surface course.

As mentioned above, restoration of an adequate level of skid resistance on an existing road may be achieved by surface dressing. It can also be effected by replacing the surface course. Whatever maintenance measure is chosen, one of the functions of the binder in the surface dressing or surface course is to provide adequate resistance to the forces between the aggregate particles in the chosen system under conditions of compression, shear and tension. Accordingly, the binder should have high cohesive strength and an excellent ability to adhere to the aggregate, in addition to possessing good visco-elastic properties. Binder durability is also required to ensure retention of adequate rheological, cohesive and adhesive properties during the service life of any replacement system.

An 'ideal' asphalt pavement is supposed to be one that possesses exceptional performance, exhibiting minimal surface distress and having a smooth surface over the design period (Von Quintus, 2009). Good materials selection, design procedures and construction practices are essential for ideal asphalt pavements. Eliminating segregation and moisture damage, achieving good compaction (high density and low air voids), constructing good longitudinal joints and maintaining adequate bond between pavement layers improves pavement performance, increasing the service life and reducing maintenance costs.

References

Anderson DA, Dukatz EL and Rosenberger JL (1983) Properties of asphalt cement and asphaltic concrete. *Proceedings of the Association of Asphalt Paving Technologists* **52**: 291–324.

Anderson D, Christensen D, Bahia H *et al.* (1994) *Binder Characterization and Evaluation*. Volume 3, *Physical Characterization and Evaluation*. Strategic Highway Research Program, National Research Council, Washington, DC, USA, SHRP A-369.

Anderson D, Champion-Lapalu L, Marasteanu M *et al.* (2001) Low temperature thermal cracking of asphalt mixture binders as ranked by strength and fracture properties. *Transportation Research Board 80th Annual Meeting, Washington*. Transportation Research Board, Washington, DC, USA.

ASTM (American Society for Testing and Materials) (2008) D7405 Standard test method for multiple stress creep and recovery (MSCR) of asphalt binder using a dynamic shear rheometer. ASTM, West Conshohocken, PA, USA.

ASTM (2014) http://www.astm.org/DATABASE.CART/HISTORICAL/D3381-09.htm (accessed 08/12/14).

Bagampadde U, Isacsson U and Kiggundu BM (2006) Impact of bitumen and aggregate composition on stripping in bituminous mixtures. *Materials and structures* **39**: 303–315.

Bahia HU (1991) *Low Temperature Physical Hardening of Asphalt Mixture Cements*. Pennsylvania State University, University Park, PA, USA. Research Report.

Bahia HU, Hanson DI, Zeng M *et al.* (2001) *Characterization of Modified Asphalt Binders in Superpave Mix Design*. Transportation Research Board, National Research Council, Washington, DC, USA, NCHRP Report 459.

Bahia HU, Wen HF and Johnson CM (2010) Developments in intermediate temperature binder fatigue specifications. In *Development in Asphalt Binder*

Specification. Transportation Research Board, Washington, DC, USA, Transportation Research Circular E-C127.

Bensalem A (2001) Private communication. TRL Ltd, Crowthorne, UK.

Bensalem A, Broen AJ, Nunn ME, Merrill DB and Lloyd WG (2000) Finite element modeling of fully flexible pavements: surface cracking and wheel interaction. Finite Element Modeling of Pavement Structures: *Proceedings of the Second International Symposium on 3D Finite Element for Pavement Analysis, Design, and Research, Charleston, West Virginia*. University of West Virginia, Morgantown, VV, USA.

BSI (British Standards Institution) (2005) BS 598-108:2005. Sampling and examination of bituminous mixtures for roads and other paved areas. Methods for determination of the condition of the binder on coated chippings and for measurement of the rate of spread of coated chippings. BSI, London, UK.

BSI (2009) BS EN 12591:2009. Bitumen and bituminous binders – Specifications for paving grade bitumens. BSI, London, UK.

Caro S, Masad E, Bhasin A, Little D and Sanchez-Silva M (2010) Probabilistic modeling of the effect of air voids on the mechanical performance of asphalt mixtures subjected to moisture diffusion. *Proceedings of the Technical Sessions AAPT* **79**: 221–252.

Copeland A, Kringos N, Scarpas A and Youtcheff J (2006) Determination of bond strength as a function of moisture content at the aggregate–mastic interface. *Proceedings of 10th Conference of the International Society for Asphalt Pavement (ISAP)*, Quebec, Canada, pp. 711–719.

Daines ME (1985) *Cooling of Bituminous Layers and Time Available for Their Compaction*. TRL, Crowthorne, UK, Research Report 4.

D'Angelo J, Robert K, Dongre R, Stephens K and Zanzotto L (2007) Revision of the Superpave high temperature binder specification: the multiple stress recovery test. *Journal of the Association of Asphalt Paving Technologists* **76**: 126–116.

Deacon JA, Harvey JT, Tayebali A and Monismith CL (1997) Influence of binder loss modulus on the fatigue performance. *Journal of the Association of Asphalt Paving Technologists* **66**: 633–685.

De Beer M, Fisher C and Jooste F (1997) Determination of pneumatic tyre/pavement interface contact stresses under moving loads and some effects on pavements with thin asphalt mixture surfacing layers. *Proceedings of the 8th International Conference on Asphalt Mixture Pavements, Seattle, Washington, USA*.

Dormon GM (1969) *Some Observations on the Properties of Binder and their Relation to Performance in Practice and Specifications*. Shell International Petroleum Company Ltd, London, UK, Construction Service Report.

Eurobitume (2002) *Future Specification System for Bituminous Paving Binders*. European Bitumen Association, Brussels, Belgium, Eurobitume Position Paper D/2002/7512/07.

Eurobitume (2012) *Performance Related Specification for Bituminous Binder*. European Bitumen Association, Brussels, Belgium, Eurobitume Position Paper D/2012/7512/25.

European Commission (1999) *COST 333 – Development of New Bituminous Pavement Design Method. Final Report of the Action*. European Commission, Brussels, Belgium.

Fan XL, Ji J, Suo Z *et al.* (2012) Evaluation of low temperature cracking of asphalt mixtures containing seam. *Journal of China Road and Traffic Engineering* **12**: 211–217.

FHWA (Federal Highway Administration) (2011) *FHWA Techbrief 2011: The Multiple Stress Creep Recovery (MSCR) Procedure*. Office of Pavement Technology, Federal Highway Administration, Washington, DC, USA, FHWA-HIF-11-038.

Groenendijk J (1998) *Accelerated Testing and Surface Cracking of Asphaltic Concrete Pavements*. PhD thesis, Delft University of Technology, Delft, The Netherlands.

Hagos ET (2008) *The Effect of Aging on Binder Properties of Porous Asphalt Concrete*. Thesis, Delft University of Technology, Delft, The Netherlands.

Harm E and Lippert D (2002) Developing Illinois' extended life hot mix asphalt mixture pavement specifications. *Transportation Research Board 81st Annual Meeting*. Transportation Research Board, Washington, DC, USA.

Hefer AW, Little DN and Lytton RL (2005) A synthesis of theories and mechanisms of bitumen–aggregate adhesion including recent advances in quantifying the effects of water. *Proceedings of the Technical Sessions AAPT* **74**: 139–195.

Highways Agency, Scottish Executive, Welsh Assembly Government and the Department for Regional Development (2004) *Design Manual for Roads and Bridges*, Volume 7, *Pavement Design and Maintenance*, Part 1, HD 28/04, *Skid Resistance*. The Stationery Office, London, UK.

Jacobs FA (1981) *Hot Rolled Asphalt: Effect of Binder Properties on Resistance to Deformation*. TRL, Crowthorne, UK, Laboratory Report 1003.

Jia L, Sun LJ, Hu XD and Hu X (2008) Strength theory, a better explanation for top-down cracking in asphalt pavement. In *Pavement Cracking*. Taylor & Francis, London, UK.

Johnson WM and Salt GF (1985a) Loss of chippings from rolled asphalt wearing courses. Paper presented at the *Institution of Highways and Transportation National Conference, Keele University*. Institution of Highways and Transportation, London, UK.

Johnson WM and Salt GF (1985b) Loss of chippings from rolled asphalt mixture wearing courses. *Highways and Transportation* **32**: 22–27.

Jung DH and Vinson TS (1992) *Final Report on Test Selection – Low Temperature Cracking*. Strategic Highway Research Program, National Research Council, Washington, DC, USA.

Kandhal PS and Rickards IJ (2001) Premature failure of asphalt overlays from stripping: case histories. *Proceedings of the Technical Sessions AAPT* **70**: 301–351.

Kandhal PS, Dongre R and Malone MS (1996) Prediction of low temperature cracking using Superpave binder specifications. *NCAT Report. 96-02*. Paper presented at the Annual Meeting of the Association of Asphalt Paving Technologists, Baltimore, MD, USA.

Kanitpong K and Bahia HU (2003) Role of adhesion and thin film tackiness of asphalt binders in moisture damage of HMA. *Proceedings of the Technical Sessions AAPT* **72**: 502–528.

Kennedy TW and Ping VV (1991) An evaluation of effectiveness of anti-stripping additives in protecting asphalt mixtures from moisture damage. *Proceedings of the Technical Sessions AAPT* **60**: 230–263.

Kennedy TW, Cominsky RJ, Harrigan ET and Leahy RB (1990) SHRP-A/UWP-90-007, The SHRP Asphalt Research Program: 1990 Strategic Planning Document. Strategic Highway Research Program, National Research Council, Washington, DC, USA.

King GN, King HW, Harders O, Arand W and Planche P-P (1993) Influence of asphalt mixture grade and polymer concentration on the low temperature performance of polymer modified asphalt mixture. *Journal of the Association of Asphalt Paving Technologists* **62**: 1–22.

Kringos N and Scarpas A (2005) Raveling of asphaltic mixes due to water damage, computational identification of controlling parameters. *Journal of the Transportation Research Board* **1929**: 79–87.

Liao WD and Chen SF (2010) *PMB Stress Absorbing Interlayer's Material Properties and Mechanical Behavior*. Science Press, Marrickville, NSW, Australia.

McLeod NW (1972) A four year survey of low temperature transverse pavement cracking on three Ontario test roads. *Journal of the Association of Asphalt Paving Technologists* **41**: 424–493.

Marasteanu MO (2004) The role of bending beam rheometer parameters in thermal stress calculations. *Journal of the Transportation Research Board* **1875**: 9–13.

Mitchell T, Stuart K, Qi XCH et al. (2004) ALF testing for development of improved Superpave binder specification. *Proceedings of the 2nd International Conference on Accelerated Pavement Testing (ICAPT), Minneapolis, MN, USA*.

Myers AL and Roque R (2001) *Evaluation of Top-Down Cracking in Thick Asphalt Mixture Pavements and Implications for Pavement Design*. Washington. Transportation Research Board, Washington, DC, USA, Transport Research Circular 503.

Nesnas N and Merrill DM (2002) Development of advanced models for the understanding of deterioration in long-life pavements. Workshop on Modelling Materials and Structures. *Proceedings of the 6th International Conference on the Bearing Capacity of Roads and Airfields, Lisbon, Portugal*.

Newcombe DE, Huddleston IJ and Bruncher M (2002) US perspective on design and construction of perpetual asphalt mixture pavements. *Proceedings of the 9th International Conference on Asphalt Mixture Pavements, Copenhagen, Denmark*.

Nunn ME (1989) An investigation of reflection cracking in composite pavements. *Proceedings of RILEM Conference on Reflective Cracking, Liege, Belgium*.

Nunn ME, Brown A, Weston D and Nicholls JC (1997) *Design of Long-Life Flexible Pavements for Heavy Traffic*. TRL, Crowthorne, UK, Report 250.

Pennsylvania Transportation Institute (2002) Improved conditioning procedure for predicting the moisture susceptibility of HMA pavements. *Moisture Damage Symposium, Laramie, Wyoming, USA*. Transportation Research Board, Washington, DC, USA.

Petersen JC (2002) Chemistry of the asphalt–aggregate interaction. *Moisture Damage Symposium, Laramie, Wyoming, USA*. Transportation Research Board, Washington, DC, USA.

Petersen JC, Robertson RE, Branthaver JF et al. (1994) *Binder Characterization and Evaluation*. Volume 1. Strategic Highway Research Program, National Research Council, Washington, DC, USA, SHRP-A-367.

Phillips M and Robertus C (1996) Binder rheology and asphalt mixture pavement permanent deformation: the zero shear viscosity concept. *Proceedings of the Eurasphalt and Eurobitume Congress, Strasbourg, France*, Session 5, paper 134.

Roberts FL, Kandhal PS, Brown R et al. (1996) *Hot Mix Asphalt Materials, Mixture Design, and Construction*. NAPA Research and Education Foundation, Lanham, MD, USA.

Schorak N and Van Dommelen A (1995) Analysis of the structural behaviour of asphalt mixture concrete pavements in SHRP-NL test sections. *International Conference: SHRP and Traffic Safety, Prague, Czech Republic*.

Stuart KD and Mogawer WS (2002) Validation of the Superpave asphalt binder fatigue cracking parameter using the FHWA's accelerated loading facility. *Journal of the Association of Asphalt Paving Technologists* **71**: 116–146.

Svasdisant T (2003) *Analysis of Top-Down Cracking in Rubblized and Flexible Pavements*. PhD thesis, Michigan State University, East Lansing, MI, USA.

Thomas K, McKay J and Branthaver J (2002) Asphalt chemistry and its relationship to moisture damage. *Moisture Damage Symposium, Laramie, Wyoming, USA*. Transportation Research Board, Washington, DC, USA.

Thrower EN (1979) *A Parametric Study of a Fatigue Prediction Model for Bituminous Pavements*. TRL, Crowthorne, UK, Laboratory Report 892.

Tong YW, Luo R and Lytton RL (2013) Modeling water vapor diffusion in pavement and its influence on fatigue crack growth of fine aggregate mixture. *Transportation Research Record* No. 2373 *Asphalt Materials and Mixtures* **4**: 71–80.

Tsai B-W, Monismith CL, Dunning M et al. (2005) Influence of asphalt binder properties on the fatigue performance of asphalt concrete pavements. *Journal of the Association of Asphalt Paving Technologists*. **74**: 733–789.

Uhlmeyer JF, Pierce LM, Willoughby K and Mahoney JP (2000) Top-down cracking in Washington State asphalt mixture concrete wearing courses. *Transportation Research Board 79th Annual Meeting, Washington*. Transportation Research Board, Washington, DC, USA.

Valkering CP and Jongeneel DJ (1991) Acoustic emission for evaluating the relative performance of asphalt mixture mixes under thermal loading conditions. *Journal of the Association of Asphalt Paving Technologists* **60**: 160–187.

Van de Loo PJ (1974) Creep testing, a simple tool to judge asphalt mix stability. *Proceedings of the Association of Asphalt Paving Technologists*. **43**: 253–285.

Van de Loo PJ (1976) Practical approach to the predication of rutting in asphalt pavements: the steel method. *Transportation Research Record* **616**: 15–21.

Von Quintus HL (2009) Performance characteristics of the ideal asphalt pavement. *Proceedings of the Technical Sessions AAPT* **78**: 941–968.

Wang LB, Myers LA, Mohammad LN and Fu YR (2003) Micromechanics study on top-down cracking. *Transportation Research Record* **2853**: 121–131.

White MJ (1989) *Flexible Pavements and Bituminous Materials*, Section G. *Bituminous Materials – Special Requirements, Problems and Processes*. Residential course at Newcastle University, Newcastle, UK, pp. G1–G16.

Whiteoak CD and Fordyce D (1989) Asphalt mixture workability, its measurement, and how it can be modified. *Shell Binder Review* **64**: 14–17.

Xu SF and Ji J (2006) An investigation of the critical temperature of asphalt mixture representing its permanent deformation resistance. *Journal of Highways* **5**: 176–179.

Yero SA (2012) *The Influence of Short-Term Ageing on Bituminous Properties.* Thesis, University OF Technology Malaysia, Johor, Malaysia.

Youtcheff J (2002) Moisture sensitivity. *Moisture Damage Symposium, Laramie, Wyoming, USA.* Transportation Research Board, Washington, DC, USA.

Chapter 19

Adhesion of bitumen and moisture damage in asphalts

Dr Richard Taylor
Global Technical
Development Manager,
Shell International Petroleum
Co. Ltd. UK

Prof. Gordon Airey
University of Nottingham. UK

The primary function of bitumen in an asphalt is to act as an adhesive. Bitumen is required either to bind aggregate particles together or to provide a bond between particles and an existing surface. When used in an asphalt, bitumen binds together the aggregate, sand and filler that comprise the mixture. In a pavement, the bitumen needs sufficient cohesive and adhesive strength to ensure minimum loss of aggregate under the shear stresses induced by traffic. Although the incidence of premature failure attributed to adhesion is relatively rare, failures, when they occur, may involve considerable expense. The need to ensure adhesion between the aggregate and the bitumen is, therefore, very important.

The adhesion of bitumen to most types of dry and clean aggregate presents few problems. However, aggregate is easily wetted by water, the presence of which can result in unexpected problems. These may occur at any time during the life of an asphalt, from the initial coating of the aggregate during the mixing process to its time in service, when it has to maintain an adequate bond between the bitumen and aggregate under traffic conditions. Damage to asphalts in the presence of water is a complex phenomenon and has been the subject of research over many decades. Although it can be described and quantified to an extent using thermodynamic first principles, in practice moisture damage is a complex phenomenon and is commonly assessed using simulative tests in the laboratory.

Measurement of adhesion is notoriously difficult to achieve in a reproducible manner, and many test methods have been employed to measure the combined forces of adhesion and cohesion, often referred to as 'adhesive strength'. Several laboratory procedures have been developed to assess the dry strength of asphalts and their adhesive strength in the presence of water, typically termed 'moisture (induced) damage'.

The Shell Bitumen Handbook, Sixth Edition
ISBN 978-0-7277-5837-8
Shell International Petroleum Company Ltd: All rights reserved
http://dx.doi.org/10.1680/tsbh.58378.549

The aim of this chapter is to introduce the fundamentals of adhesion and cohesion, draw attention to aggregate–bitumen adhesion and the ways in which it can be assessed, and how it may be possible to reduce the possibility of premature failures in service resulting from a lack of adhesion.

19.1 Thermodynamic principles of adhesion, cohesion and surface free energy

A relationship exists between Gibbs free energy (the energy associated with a chemical reaction that can be used to do work), work of adhesion and surface energy, which provides a fundamental definition of adhesion and cohesion.

In terms of cohesion, the total work expended per unit of surface area in forming two surfaces is equal to twice the surface energy per unit of surface area

$$W^c = 2\gamma \tag{19.1}$$

where W^c is the work of cohesion and γ is the surface energy.

When two dissimilar materials form an interface by being in intimate contact this is known as adhesion, and the work of adhesion (W^a) can be defined by means of the Dupré (1869) equation as

$$W^a_{12} = \gamma_1 + \gamma_2 - \gamma_{12} \tag{19.2}$$

where γ_i is the surface energy of the ith material and γ_{12} is the interfacial energy between the two materials in contact.

When a drop of liquid is placed on a clean smooth horizontal surface, it either spreads over the solid surface or takes the shape of a drop with a finite contact angle between the solid and liquid phases, as shown in Figure 19.1. Contact angles are commonly used to measure the surface energy of solids based on the relationship between contact angles, surface energy of solids, wetting of solids and thermodynamic considerations.

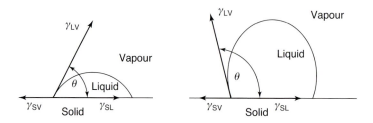

Figure 19.1 The three phase boundary of a liquid drop on a solid surface in vapour

The properties of the solid–liquid, solid–vapour and liquid–vapour interfaces can normally be described as a three phase boundary. Young proposed an equation to obtain surface tension from the contact angle formed when a drop of liquid is placed on a perfectly smooth, rigid solid (Young, 1805). In the contact angle experiment presented in Figure 19.1, the contact angle θ is defined as the angle between the solid–liquid interface and the tangent of the liquid–vapour interface. If the liquid under consideration is water, the schematic diagram on the left hand side would be an example of a hydrophilic (water loving) surface, as is evident from the drop that tends to spread over the solid due to a favourable interaction between the interfacial forces. The contact angle on the right hand side, however, is greater than 90°, indicating that water does not spread readily over this surface, and this is thus a hydrophobic (water hating) surface.

The surface energy γ of a solid or liquid can be defined as the reversible work required to create a unit area of new surface. Various theories can be used to describe the interaction or intermolecular forces that contribute to the surface energy of a material (Fowkes, 1962; Good, 1966; Owens and Wendt, 1969; Van Oss et al., 1988). For bitumen and aggregate, the three component acid–base theory, or Good–Van Oss–Chaudhury theory, is most commonly used with the total surface energy being expressed as (Good and Van Oss, 1991)

$$\gamma = \gamma^{LW} + 2\sqrt{\gamma^+ \gamma^-} \tag{19.3}$$

where γ is the total surface energy, γ^{LW} is the Lifshitz–van der Waals component of the surface energy, γ^+ is the Lewis acid component of surface interaction, and γ^- is the Lewis base component of surface interaction.

When the system is in equilibrium, the relationship between the contact angle and the surface energies of the solid and liquid is given by Young's equation as

$$\gamma_{SV} = \gamma_{SL} + \gamma_{LV} \cos \theta \tag{19.4}$$

where γ_{SV} is the surface free energy of the solid in equilibrium with the vapour, γ_{SL} is the surface–liquid interfacial free energy, and γ_{LV} is the surface free energy (surface tension) of the liquid in equilibrium with the vapour.

Equations (19.3) and (19.4) can, therefore, be combined to form the Young–Dupré equation, which is the starting point for any method that utilises contact angles to obtain surface energies by relating the contact angle to the work of adhesion.

$$W_{SL} = \gamma_{LV}(1 + \cos \theta) \tag{19.5}$$

where W_{SL} is the work of adhesion between the solid and the liquid.

The work of adhesion between two materials, such as bitumen and aggregate, can be calculated by performing surface energy calculations on both materials and then using Equation (19.2). If the value of adhesion is positive, it means that the two phases of the material tend to bind together, with a lower magnitude dictating the likely mode of failure.

It is also very useful to know the surface energy of adhesion in the presence of water when considering moisture damage. For the general case, the work of adhesion for two different materials in contact within a third medium, W_{132}, can be explained by the following equation

$$W_{132} = \gamma_{13} + \gamma_{23} - \gamma_{12} \tag{19.6}$$

Using the process proposed by Van Oss and colleagues (Good and Van Oss, 1991; Van Oss et al., 1988), Equation (19.7) can then be used to calculate the adhesion between bitumen and aggregate in the presence of water, where the subscripts 1, 2 and 3 represent bitumen, aggregate and water, respectively.

$$\begin{aligned}
W_{132} = {} & 2\gamma_3^{LW} + 2\sqrt{\gamma_1^{LW}\gamma_2^{LW}} - 2\sqrt{\gamma_1^{LW}\gamma_3^{LW}} - 2\sqrt{\gamma_2^{LW}\gamma_3^{LW}} \\
& + 4\sqrt{\gamma_3^+\gamma_3^-} - 2\sqrt{\gamma_3^+}\left(\sqrt{\gamma_1^-} + \sqrt{\gamma_2^-}\right) - 2\sqrt{\gamma_3^-}\left(\sqrt{\gamma_1^+} + \sqrt{\gamma_2^+}\right) \\
& + 2\sqrt{\gamma_1^+\gamma_2^-} + 2\sqrt{\gamma_1^-\gamma_2^+}
\end{aligned} \tag{19.7}$$

The two bond energy parameters in the dry condition (Equation (19.2)) and for the bitumen–aggregate system in water (Equation (19.7)) can be used to predict the moisture sensitivity of asphalts.

19.2 Factors affecting bitumen–aggregate adhesion

In practice, many factors (in addition to thermodynamics) influence the bitumen–aggregate bond; the importance of aggregate mineralogical composition has been recognised for many years (Saal, 1933; Winterkorn, 1936). Failure of the aggregate–bitumen bond is commonly referred to as 'stripping'.

One of the main factors is the type of aggregate. This has a considerable influence on bitumen adhesion due to differences in the degree of affinity for bitumen. The vast majority of aggregates are classified as 'hydrophilic' (water loving) or 'oleophobic' (oil hating). Aggregates with high silicon oxide content (e.g. quartz and granite (i.e. acidic rocks)) are generally more difficult to coat with bitumen than are basic rocks such as basalt and limestone. The majority of adhesive failures have been associated with siliceous aggregates such as granites, rhyolites, quartzites, cherts, etc. The fact that satisfactory performance is achieved with these same aggregates and that failures occur using aggregates that have good resistance to stripping (e.g. limestone) emphasises the complexity of bitumen–aggregate adhesion, and raises the possibility that other factors may play a role in the failure.

Other factors affecting the initial adhesion and subsequent bond are the surface texture of the aggregate, the presence of dust on the aggregate and, to a lesser extent, the degree of acidity of the water in contact with the interface. It is generally agreed that rougher aggregate surfaces have better adhesion characteristics. However, a balance is required between wetting of the aggregate (smooth surfaces being more easily wetted) and rougher surfaces, which hold the bitumen more tenaciously once wetting has been achieved.

Physio-mechanical adsorption of bitumen into the aggregate depends on several factors, including the total volume of permeable pore space, the size of the pore openings, and the viscosity and surface tension (surface energy) of the bitumen (Thelen, 1958). The presence of a fine microstructure of pores, voids and micro-cracks can bring about an enormous increase in the adsorptive surface available to the bitumen. It has also been shown (Plancher et al., 1977; Scott, 1978) that fractions of the bitumen are strongly adsorbed in the aggregate surface to a depth of approximately 180 Å $(18 \times 10^{-9} \, \text{m})$.

It has been suggested that the good mechanical bond achieved on a rough aggregate can be more important than the aggregate mineralogy in maintaining bitumen–aggregate adhesion (Lee and Nicholas, 1957). The properties of the bitumen are also important in the acquisition and subsequent retention of the bitumen–aggregate bond. In particular, the viscosity of the bitumen during coating and in service, polarity and constitution all influence the adhesion characteristics. However, it is the nature of the aggregate that is, by far, the most dominant factor influencing bitumen–aggregate adhesion.

Table 19.1 summarises the main factors that influence bitumen–aggregate adhesion. It is considered that approximately 80% of these factors are controllable during production and construction.

19.3 Disbonding mechanisms in asphalts

Many studies have been carried out to determine the mechanism of bitumen disbonding in asphalts (Asphalt Institute, 1981; Hughes et al., 1960; Taylor and Khosla, 1983). There are two main methods by which the bitumen–aggregate system may fail: adhesive and cohesive mechanisms. If the aggregate is clean and dry and the mixture is effectively impermeable, the mode of failure will be cohesive. However, in the presence of water, the failure mode will almost certainly be due to a loss of adhesion caused by stripping of the

Table 19.1 Material properties and external factors that can affect the bitumen–aggregate bond

Aggregate properties	Bitumen properties	Mixture properties	External factors
Mineralogy	Rheology	Void content	Rainfall
Surface texture	Electrical polarity	Permeability	Humidity
Porosity	Constitution	Bitumen content	Water pH
Dust	Surface free energy	Bitumen film thickness	Presence of salts
Durability		Filler type	Temperature
Surface area		Aggregate grading	Temperature cycling
Surface free energy		Type of mixture	Traffic
Absorption			Design
Moisture content			Workmanship
Shape			Drainage
Weathering			

bitumen from the aggregate surface. Several mechanisms of disbonding are possible, and these are discussed below.

The description above is an oversimplification, and the loss of adhesion and the effects of moisture damage are often described as a mixture of adhesive and cohesive damage. The situation is further complicated by scalar effects, in which cohesion of a bitumen–filler mastic is described, whereas such a mastic is a function itself of adhesive strength and physical factors such as bitumen absorption by the filler.

19.3.1 *Displacement*

Displacement theory relates to the thermodynamic equilibrium of the three-phase bitumen–aggregate–water system. If water is introduced at a bitumen–aggregate interface, consideration of the surface energies that are involved shows that the bitumen will retract along the surface of the aggregate.

19.3.2 *Detachment*

Detachment occurs when a thin film of water or dust separates the bitumen and aggregate, with no obvious break in the surface of the bitumen film being apparent. Although the bitumen film completely encapsulates the aggregate particle, no adhesive bond exists, and the bitumen can easily be peeled from the aggregate surface. This process may be reversible (i.e. if the water is removed, the bitumen may re-adhere to the aggregate). A second mechanism of disbonding must occur to allow the ingress of water between the bitumen and the aggregate.

19.3.3 *Film rupture*

Film rupture may occur despite the fact that the bitumen fully coats the aggregate. At sharp edges or asperities on the aggregate surface, where the bitumen film is thinnest, it has been shown (Hughes *et al.*, 1960) that water

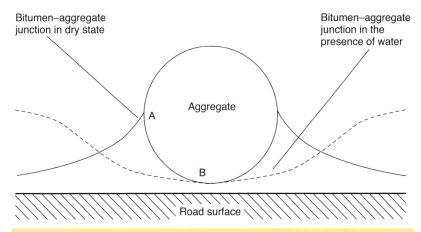

Figure 19.2 Retraction of the bitumen–water interface in the presence of water

can penetrate through the film to reach the surface of the aggregate. This movement of water to the aggregate surface may occur with the water in either a vapour or liquid form. Once this process has started, it is possible for the water to spread between the bitumen and aggregate surface to produce a detached film of bitumen.

The speed with which the water can penetrate and detach the bitumen film will depend on the viscosity of the bitumen, the nature of the aggregate surface, the thickness of the bitumen film and the presence of filler and other components such as surface active agents. Once significant detachment of the bitumen film from the aggregate has occurred, stresses imposed by traffic will readily rupture the film, and the bitumen will retract, exposing water-covered aggregate. The displacement theory (Blott *et al.*, 1954) relates to the thermodynamic equilibrium of the three phase bitumen–aggregate–water system. If water is introduced at a bitumen–aggregate interface, then consideration of the surface energies that are involved shows that the bitumen will retract along the surface of the aggregate. Figure 19.2 shows an aggregate particle embedded in a bitumen film, with point A representing the equilibrium+ contact position when the system is dry. When in contact with water, the equilibrium point shifts, and the new interface moves or retracts over the surface to point B. This new equilibrium position has a contact angle that will depend on the type and viscosity of the bitumen used.

19.3.4 Blistering and pitting

If the temperature of a flexible pavement increases, the viscosity of the bitumen within that pavement will reduce. If this is associated with a recent rainfall, the bitumen may creep up the edges of water droplets to form a blister, as shown in Figure 19.3. If the temperature continues to increase, the

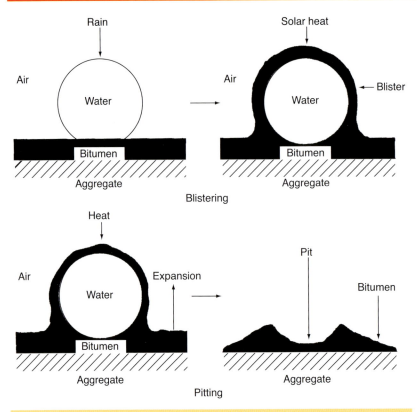

Figure 19.3 Formation of blisters and pits in an asphalt

blister will expand, leaving a hollow or a pit, which may allow water to access the surface of the aggregate (Hughes *et al.*, 1960; Thelen, 1958).

19.3.5 *Hydraulic scouring*
Hydraulic scouring or pumping occurs in the surface course, and is caused by the action of vehicle tyres on a saturated pavement surface (i.e. water is forced into surface voids in front of the vehicle tyre). On passing, the action of the tyre sucks up this water, thereby inducing a compression–tension cycle in these surface voids, which may result in disbonding of the bitumen from the aggregate. Suspended dust and silt in the water can act as an abrasive and can accelerate disbonding.

19.3.6 *Pore pressure*
This type of disbonding mechanism is most important in open or poorly compacted asphalts, where it is possible for water to be trapped as the material is compacted by traffic. Once the asphalt becomes effectively impermeable, subsequent trafficking induces a pore water pressure (Lee and Nicholas,

1957). This creates channels around the bitumen–aggregate interface, leading to loss of bond. Higher temperatures acting on the entrapped water result in expansive stresses accelerating water migration and disbonding. Low temperatures may also lead to the formation of ice, which is equally destructive.

19.4 Methods of measuring and assessing adhesion between bitumen and aggregates and moisture damage

Given the potential for premature failure due to adhesion related problems and the number of factors at play, the need for predictive laboratory tests is self-evident. A number of different types of test have been developed to compare combinations of aggregate, bitumen and water. However, the problem with many of these methods is a lack of information relating the laboratory prediction to performance in practice. Several techniques have been developed, and these can be categorised broadly as set out below.

- Surface free energy and work of adhesion calculations
  - surface energy measurements of bitumen
  - surface energy measurements of aggregate
  - calculation of the work of adhesion with and without the presence of water.
- Bitumen–substrate tests
  - peel tests
  - pull off tests.
- Simulative measurements on bitumen–aggregate combinations
  - coating and immersion tests
  - adsorption tests.
- Simulative measurements on asphalts
  - immersion mechanical tests
  - immersion trafficking tests.

Each of the above is discussed in the following sections.

19.4.1 Surface free energy and work of adhesion calculations

The surface energy of solids cannot be measured directly. Indirect methods are therefore used, and the surface energies of solids (including bitumen as well as aggregates) are inferred from the known surface energy values of several liquids or gases. Several techniques can be used, including contact angle measurement methods and vapour sorption techniques.

19.4.1.1 Surface energy measurements of bitumen

The Wilhelmy plate can be used to determine the surface energy of bitumen by measuring the contact angle between bitumen and a liquid, based on the kinetic force equilibrium when a very thin plate is immersed or withdrawn

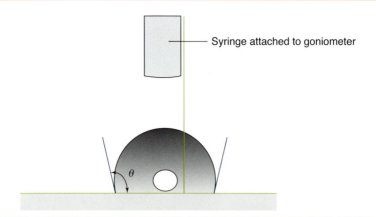

Figure 19.4 A droplet of liquid on a bitumen substrate during contact angle measurement using a goniometer

from a liquid solvent at a very slow and constant speed (Adamson and Gast, 1997). The dynamic contact angle between bitumen and a solvent liquid measured during the immersing process is called the 'advancing contact angle'; while the dynamic contact angle during the withdrawal process is called the 'receding contact angle'. The advancing contact angle, which is a wetting process, is associated with the fracture healing process, while the receding angle is associated with the fracture mechanism. The total surface energy and its components of bitumen can be calculated from these advancing and receding contact angles. The surface energy calculated from the advancing contact angles is called the 'surface energy of wetting', while the surface energy computed from the receding contact angles is called the 'surface energy of dewetting'.

Alternatively, bitumen surface free energy can be derived using contact angles measured using a goniometer. In this technique, bitumen is applied evenly to a glass slide and drops of different liquids of known surface free energy are placed onto the surface, and the contact angle measured. Figure 19.4 shows a droplet at rest during a contact angle measurement using a goniometer.

Typical values for surface free energy vary by test method and the choice of solvent but typically fall within the range 25–50 mN/m^2, some example values (measured using the goniometer method and presented using the Van Oss approach (Van Oss et al., 1988)) are given in Table 19.2.

19.4.1.2 Surface energy measurements of aggregate

A goniometer can be used to measure the surface energy of aggregate but the measurement presents some difficulties. The samples need to be non-porous, and this can be problematic with some aggregates. Specimens are

Table 19.2 Typical measured values of surface free energy for various bitumen grades

	Surface free energy, γ: mN/m^2				
	Lifshitz–van der Waals component of surface energy, γ^{LW}	Acid component of surface energy, γ^+	Basic component of surface energy, γ^-	Acid–base component of surface energy, γ^{AB}	Total surface energy, γ^T
70/100 pen A	38.90	1.00	3.50	3.74	42.64
70/100 pen B	26.29	0.00	0.57	0.00	26.29
40/60 pen A	33.35	0.85	1.90	2.54	35.89
40/60 pen B	24.62	0.00	0.59	0.00	24.62
10/20 pen A	34.66	0.00	0.59	0.00	34.66
10/20 pen B	27.25	0.02	0.34	0.16	27.41
SBS PMB A	31.90	0.14	3.60	1.42	33.32
SBS PMB B	36.40	3.10	4.90	7.79	44.19

required to be flat and low in texture, and this requires cutting and often polishing of aggregates in order to carry out the contact angle measurements. This introduces issues around isotropy in the aggregate and also the impact of polishing, and thus deviation from the true nature of the aggregate.

The surface energies of aggregates can be measured in their natural form using various vapour sorption techniques such as dynamic vapour sorption systems and universal sorption methods (Bhasin, 2006; Li, 1997). The basis of surface energy determination in vapour sorption techniques for high energy surfaces such as is found in aggregates is the use of the gas adsorption characteristics of selected solvents, the surface energies of which are known indirectly to measure the surface energies of the aggregate. This method can accommodate the particular shape, size, mineralogy and surface roughness of aggregates.

Again, typical values for surface free energy vary by test method and choice of solvent, and aggregates show a much wider range of values (50–200 mN/m^2) than does bitumen, reflecting the wide range of mineralogies found in rocks suitable for aggregate production. Values for some typical aggregate types measured using the vapour sorption technique and adopting the Van Oss approach are presented in Table 19.3.

19.4.1.3 Calculation of the work of adhesion with and without the presence of water

Surface free energy measurements for the different types of both bitumen and aggregate and their bond energies provide an improved understanding of the material properties that influence the moisture sensitivity of asphalts but, due to the complex nature of damage in asphalts, cannot provide a full account of the likelihood of moisture damage alone. Similarly, conventional moisture-sensitive asphalt tests are also able to compare mechanical property

559

Table 19.3 Typical measured values of surface free energy for various aggregate types

Material	Surface energy components: mN/m^2				
	Lifshitz–van der Waals component of surface energy, γ^{LW}	Acid component of surface energy, γ^+	Basic component of surface energy, γ^-	Acid–base component of surface energy, γ^{AB}	Total surface energy, γ^T
Limestone A	68.72	99.34	14.49	75.88	144.60
Limestone B	45.00	8.00	214.00	82.75	127.75
Basalt	51.68	14.12	59.82	58.13	109.81
Granite	56.55	1.58	1.90	3.47	60.02
Gritstone	40.00	5.00	131.00	51.19	91.19

results for dry and moisture conditioned specimens, but they are unable to quantify the causes of good or poor performance beyond comparative tests.

Measurement of the surface free energies of bitumens and aggregates allows the calculation of their interfacial work of adhesion and the reduction in free energy of the system (work of disbonding) when water displaces bitumen from the bitumen–aggregate interface. These two parameters can then be used to estimate the moisture sensitivity of asphalts based on the principles of thermodynamics and physical adhesion (see section 19.1).

For an asphalt to be durable and less sensitive to moisture, it is desirable that the work of adhesion W_{12} between the bitumen and the aggregate be as high as possible. In addition to the work of adhesion, the greater the magnitude of work of debonding when water displaces bitumen from the bitumen–aggregate interface, W_{132}, the greater will be the thermodynamic potential that drives moisture damage. Therefore, it is desirable that the magnitude of work of disbonding be as small as possible.

19.4.2 Bitumen–substrate tests

19.4.2.1 Peel tests

A peel test is a well established and standardised methodology in which a rigid substrate and a flexible substrate are separated and the force required to achieve the separation under set conditions is measured to provide insights into the adhesive strength between the two materials. This approach can be further extended to investigate the influence of conditioning in the presence of water. Various geometries and loading regimes can be adopted, and these have become standardised in many regions (ASTM D1876-08 and ASTM D6862-11 (ASTM, 2008, 2011)). The methods are employed to measure the bonded strength of composites, joints, laminates and related materials.

Recent studies (Blackman et al., 2013) have examined the possibility of adapting peel tests to bitumen–aggregate combinations and a range of

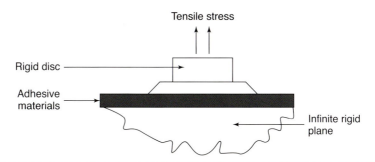

Figure 19.5 The pull off test

bitumen–aggregate combinations with adhesion promoters. The studies show that adhesion promoters could be successfully applied to bitumen–aggregate combinations.

19.4.2.2 Pull off tests
A pull off test is widely used to measure the mechanical tensile strength of films, paints and other coatings. Generally, a pull off test is conducted by measuring the minimum tensile stress necessary to detach or fracture the coatings of adhesive materials in a direction perpendicular to the substrate. A typical arrangement is shown in Figure 19.5.

The limpet pull off test (Craig, 1991) was developed to measure, quantitatively, the bond strength between the aggregate of a surface dressing and the underlying surface. The test uses a limpet apparatus that was originally developed for measuring the tensile strength of concrete. A 50 mm dia. metal plate is fixed to the road surface and the maximum load to achieve pull off determined. The method may be used both in the laboratory and on the actual road surface.

Bitumen adhesion has been assessed using different types of pull off test. The Instron pull off test (Craig, 1991) uses an Instron tensile apparatus to extract aggregate test specimens from containers of bitumen under controlled laboratory conditions. Test variables such as rock type, dust coatings, test temperature, rate of loading, bitumen type etc., have been shown to alter the results. An example of the data obtained for four different rock types with increasing dust contents is shown in Figure 19.6, and demonstrates how the maximum stress during testing varied for the different aggregates.

19.4.3 Simulative measurements on bitumen–aggregate combinations
19.4.3.1 Coating and immersion tests
This type of testing attempts to assess adhesion between aggregate and bitumen when water is also present. For example, in the immersion tray test

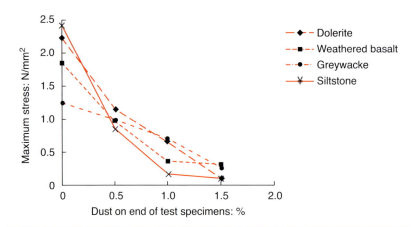

Figure 19.6 Instron pull off data showing the effect of dust content for four different aggregates

(Nicholls, 2002) aggregate chippings are applied to a tray of bitumen covered with a layer of water. By careful examination of the chippings, it may be possible to determine whether surface active agents improve adhesion under wet conditions. Typically, aggregate is coated with bitumen, immersed in water under controlled conditions and the effect of stripping determined after a period of time. The various methods differ in the type of specimen used, the conditions under which the sample is immersed in water and the method that is used to assess the degree of stripping. This is the simplest type of test, and consists of aggregate being coated with bitumen and then being immersed in water. The degree of stripping is estimated by a visual inspection after a period of time. For example, in the total water immersion test, 14 mm single size chippings are coated with a known quantity of bitumen. The coated aggregate is then immersed in distilled water at 25°C for 48 h. The percentage of bitumen stripped off the aggregate is assessed visually.

A static immersion test was standardised (AASHTO T 182-84 (AASHTO, 2002, ASTM D1664 (ASTM, 1985)) and involved coating 100 g of aggregate with bitumen, immersing it for 16–18 h in 400 ml of distilled water with a pH of 6–7, and visually estimating the total visible area of the coated aggregate as above or below 95%. The visual assessment is made while the mixture is still immersed in the water.

The fundamental problem with such methods is their subjective nature and resultant poor reproducibility. An experienced operator may be capable of ranking the aggregates in relation to their performance in situ, but it must be recognised that, in some cases, an aggregate with good laboratory performance may occasionally perform poorly on the road and those with poor static immersion test results may perform satisfactorily in practice. In summary, although such methods may indicate which combinations of aggregate and bitumen show degrees of water sensitivity, it is doubtful that the long term potential of stripping is adequately addressed, and the static immersion test was discontinued as an ASTM standard in 1993.

A dynamic immersion test is very similar to the static immersion test but the sample is agitated mechanically by shaking, stirring or kneading. Again, the degree of stripping is estimated visually, together with a subjective judgement of whether the mixture remains cohesive or separates into individual particles of aggregate. The reproducibility of this type of test is also very poor. There are several variants of this technique. The rolling bottle (flask) method (BS EN 12697-11 (BSI, 2012a)) is used to determine the adhesion between single size aggregate and bitumen. In the test, 200 spherical aggregate particles (6.3–8 mm) are coated with a 0.1 mm thick binder film. The coated aggregate is then placed in a 250 ml flask containing deionised cold water and a glass rod, and rotated at 40 revs/min for 3 days. The amount of retained bitumen is then visually determined.

The boiling water test (ASTM D3625 (ASTM, 2012)) involves placing a 200–300 g sample of coated aggregate (single size aggregate, or aggregate graded to design specifications) in boiling water (500 ml distilled water) for 1–10 min. For the 10 min test, the mixture is stirred three times with a glass rod while it is being boiled. After boiling, the mixture is dried and the amount of bitumen loss is determined by visual assessment. The test is very subjective and is known to provide inconsistent results in terms of identifying water-sensitive mixtures. In addition, the test only reflects the loss of adhesion, and does not address loss of cohesion.

In the majority of these tests, the coated aggregate is immersed in water. However, salt based or de-icing solutions and those containing fuel oils may also be used. Certain types of test may tend to coat an aggregate in the presence of water. In such cases, the degree of coating obtained is used as an index of adhesivity.

In this type of test, aggregate coated in bitumen is boiled in solutions containing various concentrations of sodium carbonate. The strength of sodium carbonate solution in which stripping is first observed is used as a measure of adhesivity. However, the artificial conditions of the test are unlikely to predict the likely performance on the road.

19.4.3.2 Adsorption tests
The Strategic Highway Research Program (SHRP) in the USA investigated the effect of moisture damage as one of six major distress areas requiring further investigation. The result was the net adsorption test method (Curtis *et al.*,

1993). This test combined a fundamental measurement of bitumen–aggregate adhesion with a measure of moisture sensitivity. The method was able to match an aggregate and a bitumen to give optimum performance conditions.

In the test, a stock solution of bitumen is first prepared by adding 1 g of bitumen to 1000 ml of toluene. The aggregate is crushed to produce 50 g test samples of 5 mm to dust grading, and then 140 ml of the stock solution is added to a conical flask containing the graded aggregate placed on a rotating table to agitate the mix. After 6 h, a 4 ml sample is removed and diluted to 25 ml using toluene. The absorbance is measured using a spectrophotometer at 410 nm.

The reading is compared with the one obtained for the stock solution, and the percentage initial adsorption calculated. Distilled water (2 ml) is then added to the flask, and the flask is agitated overnight. Another 4 ml is removed and assessed as before. The value obtained at this stage is the percentage net adsorption, which should be lower than the percentage initial adsorption value, as some of the bitumen will have been removed from the aggregate due to the introduction of moisture into the system.

19.4.4 Simulative measurements on asphalts
19.4.4.1 Immersion mechanical tests
Immersion mechanical tests involve the measurement of a change in a mechanical property of compacted asphalt after it has been immersed in water. Thus, the ratio of the property after immersion divided by the initial property is an indirect measure of stripping (usually expressed as a percentage or index).

Different types of mechanical properties can be measured, and these include shear strength, flexural strength and compressive strength. The two most common are the AASHTO T 283 (AASHTO, 2011) method sometimes referred to as the 'modified Lottman procedure', and the 'retained Marshall stability test'. Indexes based on retained stiffness are also employed, and these have the advantage of the index measurement being non-destructive, allowing for repeated measurements on the same specimen over time and under different conditions. Common approaches to immersion mechanical testing for asphalts are discussed below.

One of the most globally used procedures is the modified Lottman procedure, which combines features of both the Lottman (1978, 1982) and Tunnicliff and Root (1984) procedures. The Lottman procedure attempts to achieve a 100% saturation level in its specimens, while the Tunnicliff and Root procedure attempts to control the level of saturation between 55% and 80%. Concern that over-saturation induces damage in specimens that is not associated with moisture damage, but rather with the over-saturation of the specimen, meant that for the modified Lottman procedure the degree of saturation was decreased to between 60% and 80%. As the saturation level achieved by partial vacuum is primarily responsive to the magnitude of the vacuum and relatively independent of the length of time, this reduced saturation was achieved by lowering the partial vacuum from 600 to 508 mmHg.

Moisture damage tests based on Marshall stability are still in widespread use. There are a number of different versions of each type of test. In the version used by Shell, at least eight Marshall specimens are manufactured using a prescribed aggregate type, aggregate grading, bitumen content and void content. Four specimens are then tested using the standard Marshall method to give a standard stability value. The four remaining specimens are vacuum treated under water at a temperature of 0–1°C to saturate the pore volume of the mixture with water. The specimens are then stored in a water bath at 60°C for 48 h, after which their Marshall stability is determined. The ratio of the Marshall stability of the soaked specimens to the standard Marshall stability is termed the 'retained Marshall stability'. A retained Marshall stability value of >75% is usually regarded as acceptable.

The principle of the retained stiffness test is similar to that of the retained Marshall test. The major difference is that the test specimens are assessed using the indirect tensile stiffness modulus test (BS EN 12697-26:2012 (BSI, 2012b)), which is carried out using a Nottingham asphalt tester (NAT) or similar device. This has the additional benefit that the method is non-destructive, allowing the same test specimens to be used after soaking. A number of versions of the method exist in which the soaking period and water temperature may be varied.

The saturation ageing tensile stiffness (SATS) test (BS EN 12697-45:2012 (BSI, 2012c)) is the first procedure of its kind that combines both ageing and water-damage mechanisms (intended to simulate the conditions to which some asphalt pavements are subjected in service) within a single laboratory test protocol. The conditions used in the SATS procedure were selected in order to reproduce in the laboratory the moisture damage observed in the field, as demonstrated by a decrease in stiffness modulus for particular asphalts as detailed by Airey et al. (2005). The procedure has been found to reproduce successfully the moisture damage observed in asphalts in the field, and to distinguish between poorly performing materials and alternative asphalts incorporating aggregate with good durability track records (Airey and Choi, 2002). The results obtained from the SATS moisture conditioning procedure tend to rank asphalts in terms of moisture sensitivity in the same order as the AASHTO T 283 (AASHTO, 2011) procedure, although the relative performance of a mixture containing a moisture sensitive aggregate is usually significantly lower in the SATS test (Airey *et al.*, 2005).

The standard SATS procedure involves conditioning five pre-saturated specimens simultaneously in a pressure vessel under 0.5 MPa air pressure at a temperature of 85°C for a period of 24 h. This conditioning is followed by a cooling period of 24 h before the air pressure is released and the vessel opened to remove the specimens for stiffness testing. The pressure vessel used can hold five nominally identical specimens (100 mm diameter and 60 mm thick) in a custom made specimen tray.

19.4.4.2 Immersion trafficking tests

A major problem with most types of adhesion test is that they do not consider the effect of trafficking in causing stripping. A method that considers this factor is the immersion wheel tracking test (Mathews and Colwill, 1962) shown in Figure 19.7. In this test, a specimen is immersed in a water bath and traversed by a loaded reciprocating solid rubber tyre. In the standard method, three specimens that have been compacted in moulds are tracked using a 20 kg load at 25 cycles/min at a water temperature of 40°C until failure occurs.

The development of the rut that forms is measured until stripping starts to occur. This is typically marked by a steep increase in the rut depth and

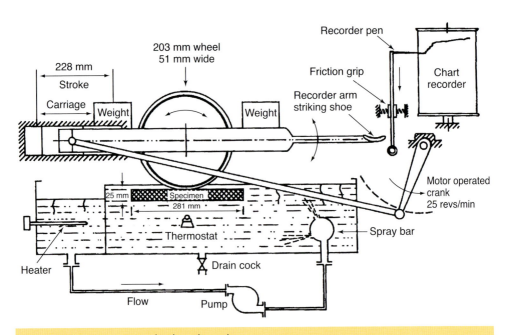

Figure 19.7 Immersion wheel tracking device

surface ravelling of the test specimen. Good correlation has been shown to exist for stripping failures of heavily trafficked roads. It has been found that factors such as aggregate shape, aggregate interlock, bitumen viscosity and sample preparation affect failure times.

The wet wheel track test is now being reconsidered as a means of assessing asphalts such as porous asphalts and high stone content, thin surface course systems. The method is easily adaptable in which temperature, immersion medium and loading may be varied. The test may be performed for a specific time or until failure occurs.

A locked wheel method was developed to assess the likelihood of ravelling due to the presence of moisture (McKibben, 1987). The only alteration to the equipment involved the provision of a locking ratchet that allowed the wheel to travel freely in one direction but also to be locked in position as it is dragged back across the specimen surface on the remainder of the cycle. Surface ravelling has been related to grading, degree of compaction, bitumen and aggregate characteristics. This method is, however, very aggressive, and may be most suitable for highly stressed trafficking locations.

Similar to the immersion wheel tracker described above, the Hamburg wheel tracking device consists of a pair of samples subjected to wheel loading under water. A sample is typically 260 mm wide, 320 mm long and 40 mm thick. The samples are submerged in water at 50°C, although the temperature can vary from 25°C to 70°C, and subjected to a steel wheel (47 mm wide) with a load of 705 N. The wheel makes 50 passes/min over each sample, with a maximum velocity of 0.34 m/s in the centre of the sample. Testing is undertaken to 20 000 passes or until a 20 mm deep rut develops.

The onset of stripping is termed the 'stripping inflection point' and relates to the intersection of the creep slope and the stripping slope where there is a sharp increase in the rate of permanent deformation. Both the stripping slope and the stripping inflection point are related to the moisture damage of the sample. Based on an evaluation of 20 pavements, Aschenbrener (1995) found an excellent correlation between the Hamburg wheel tracking device and pavements of known in-service performance.

19.5 Improving bitumen–aggregate adhesion

Typically, the adhesion of bitumen to aggregate is not a problem. However, in the presence of water, unexpected adhesion related problems may occur. There are a number of traditional methods used to reduce the likelihood of this happening (i.e. using higher viscosity bitumen, hydrated lime or surface active agents that improve the bond between the bitumen and the aggregate). While modifying bitumen viscosity may be easily achieved, this may result in workability and compaction related problems, particularly for thin surface course systems that use high stone content asphalts.

19.5.1 Hydrated lime

The use of 1–3% of hydrated lime as part of the filler content has traditionally been used as an anti-stripping agent. It has been suggested (Plancher *et al.*, 1976) that the hydrated lime reacts with the carboxyl acids present in the bitumen and allows other carbonyl groups such as ketones to attach themselves to the aggregate surface. These ketones are not as easily removed by water as the acids, and so the mixtures are less susceptible to stripping.

It has also been suggested (Ishai and Craus, 1977) that a hydrated lime solution will result if water is present at the bitumen–aggregate interface. The calcium ions in this solution cause the surface of the aggregate to become basic. The electrochemical balance forces the water away from the aggregate and into an emulsion in the bitumen. The balance will then force attachment to the hydrophobic surface of the aggregate.

19.5.2 Anti-stripping agents

Bitumen–aggregate adhesion may be improved by the addition of chemical additives. These act in two main ways

- they may change the interfacial conditions between the aggregate and the bitumen so that the bitumen preferentially wets the aggregate, which improves adhesion
- they may improve the adhesive bond between the aggregate and the bitumen, thus increasing the long term resistance to bitumen detachment due to water.

Typically, 0.1–1.0% of fatty amines is the main type of additive used to improve adhesion. It is believed that the amine groups are attracted to the surface of an aggregate, while the fatty groups remain in the bitumen. The result is an ionically bonded cross-link between aggregate and bitumen. However, these additives may be relatively unstable at bitumen storage temperatures, and can become deactivated. It is also possible that a given additive will not improve the adhesion of all aggregate types (i.e. they may be rock type specific). There is also the issue that, while they may improve initial adhesion, they may have no or limited long term effect.

It is recommended that, prior to their use, laboratory testing should be carried out to optimise the type and amount of additive used for a given bitumen–aggregate combination. This may be undertaken using the wide range of the test methods already mentioned in this chapter. For example, Table 19.4 shows data obtained using the net adsorption test for two different

Table 19.4 Net and initial adsorption results with different binders

Aggregate type	Initial adsorption: %		Net adsorption: %	
	100 pen bitumen	100 pen bitumen with 0.5% adhesion agent	100 pen bitumen	100 pen bitumen with 0.5% adhesion agent
Quartz dolerite	52.2	63.1	46.4	49.6
Greywacke	64.6	67.6	54.3	50.4

aggregates and a 70/100 pen bitumen with and without 0.5% adhesion agent. It can be seen that, while the adhesion agent increased initial adsorption by 10.9% for the quartz dolerite, the adsorption only improved by 3% for the greywacke. In terms of improving moisture sensitivity, there was a small improvement for the quartz dolerite, and no improvement for the greywacke. This example shows how aggregate properties influence attempts at improving adhesion. For the greywacke, the addition of adhesion agents essentially did little to improve performance.

References

Adamson AW and Gast AP (1997) *Physical Chemistry of Surfaces*, 6th edn. John Wiley & Sons, New York, USA, pp. 18–40.

Airey GD and Choi YK (2002) State of the art report on moisture sensitivity test methods for bituminous pavement materials. *Road Materials and Pavement Design* **3(4)**: 355–372.

Airey GD, Choi YK, Collop AC, Moore AJV and Elliott RC (2005) Combined laboratory ageing/moisture sensitivity assessment of high modulus base asphalt mixtures (with discussion). *Journal of the Association of Asphalt Paving Technologists* **74**: 307–346.

Aschenbrener T (1995) Evaluation of Hamburg wheel-tracking device to predict moisture damage in hot mix asphalt. *Transportation Research Record* **1492**: 193–201.

Asphalt Institute (1981) *Cause and Prevention of Stripping in Asphaltic Pavements*. Asphalt Institute, Lexington, KY, USA, Educational Series No. 10.

ASTM (American Society for Testing and Materials) (1985) ASTM D1664-80. Test method for coating and stripping of bitumen aggregate mixtures. ASTM International 1985 (Withdrawn 1992). ASTM, West Conshohocken, PA, USA.

ASTM (2008) ASTM D1876-08. Standard test method for peel resistance of adhesives (t-peel test). ASTM, West Conshohocken, PA, USA.

ASTM (2011) ASTM D6862-11. Standard test method for 90 degree peel resistance of adhesives. ASTM, West Conshohocken, PA, USA.

ASTM (2012) ASTM D3625. Standard practice for effect of water on bituminous-coated aggregate using boiling water. ASTM, West Conshohocken, PA, USA.

AASHTO (American Association of State and Highway Transportation Officials) (2002) AASHTO T 182-84. Standard method of test for coating and stripping of bitumen–aggregate mixtures. ASTM, Washington, DC, USA.

AASHTO (2011) AASHTO T 283. Standard method of test for resistance of compacted asphalt mixtures to moisture-induced damage. American Association of State and Highway Transportation Officials, Washington, DC, USA.

Bhasin A (2006) *Development of Methods to Quantify Bitumen–Aggregate Adhesion and Loss of Adhesion due to Water*, Department of Civil Engineering, Texas A&M University, College Station, TX, USA. PhD Thesis.

Blackman BRK, Cui S, Kinloch AJ and Taylor AC (2013) The development of a novel test method to assess the durability of asphalt road–pavement materials. *International Journal of Adhesion and Adhesives* **42(1)**: 1–10.

Blott JFT, Lamb DR and Pordes O (1954) Weathering and adhesion in relation to the surface dressing of roads with bituminous binder. In *Adhesion and Adhesives: Fundamentals and Practice*. Society of Chemical Industry, London, UK.

BSI (British Standards Institution) (2012a) BS EN 12697-11:2012. Test methods for hot mix asphalt Determination of the affinity between aggregate and bitumen. BSI, London, UK.

BSI (2012b) BS EN 12697-26:2012. Bituminous mixtures. Test methods for hot mix asphalt. Stiffness. BSI, London, UK.

BSI (2012c) BS EN 12697-45:2012. The saturation ageing tensile stiffness (SATS) conditioning test. BSI, London, UK.

Craig C (1991) *A Study of the Characteristics and Role of Aggregate Dust on the Performance of Bituminous Materials*. PhD Thesis, University of Ulster, Newtownabbey, UK.

Curtis CW, Ensley K and Epps J (1993) *Fundamental Properties of Asphalt–Aggregate Interactions Including Adhesion and Adsorption*. Strategic Highway Research Program, National Research Council, Washington, DC, USA, Report SHRP-A-341.

Dupré MA (1869) *Théorie Mécanique de la Chaleur*. Gauthier-Villars, Paris, France (in French).

Fowkes FM (1962) Determination of interfacial tensions, contact angles, and dispersion forces in surfaces by assuming additivity of intermolecular interactions in surfaces. *Journal of Physical Chemistry* **66**: 382.

Good RJ (1966) Intermolecular and interatomic forces. In *Treatise on Adhesion and Adhesives*, Volume 1, *Theory* (Patrick RL (ed.)). Marcel Dekker, New York, USA, pp. 10–65.

Good RJ and Van Oss CJ (1991) The modern theory of contact angles and the hydrogen bond component of surface energies. In *Modern Approach to Wettability: Theory and Applications* (Schrader ME and Loeb G (eds)). Plenum Press, New York, USA, pp. 1–27.

Hughes RI, Lamb DR and Pordes O (1960) Adhesion of bitumen macadam. *Journal of Applied Chemistry* **180**: 433–440.

Ishai I and Craus J (1977) Effect of the filler on aggregate–bitumen adhesion properties in bituminous mixtures. *Proceedings of the Association of Paving Technologists* **46**: 228–257.

Lee AR and Nicholas JR (1957) The properties of asphaltic bitumen in relation to its use in road construction. *Journal of the Institute of Petroleum* **43**: 235–246.

Li W (1997) *The Measurement of Surface Free Energy of Aggregates SHAP RB –*

Final Report of Cahn Balance Thermogravimetry Gas Adsorption Experiments. Chemical Engineering, Texas A&M University, College Station, TX, USA.

Lottman RP (1978) *Predicting Moisture-Induced Damage to Asphaltic Concrete.* Transportation Research Board, Washington, DC, USA, NCHRP Report 192.

Lottman RP (1982) Laboratory test method for predicting moisture-induced damage to asphalt concrete. *Transportation Research Record* **843**: 88–95.

McKibben DC (1987) *A Study of the Factors Affecting the Performance of Dense Bitumen Macadam Wearing Courses in Northern Ireland.* PhD Thesis, University of Ulster, Newtownabbey, UK.

Mathews DH and Colwill DM (1962) The immersion wheel-tracking test. *Journal of Applied Chemistry* **12(11)**: 505–509.

Nicholls JC (2002) *Design Guide for Road Surface Dressing*, 5th edn. TRL Ltd, Crowthorne, UK, Appendix D, Road Note 39.

Owens DK and Wendt RC (1969) Estimation of the surface free energy of polymers. *Journal of Applied Polymer Science* **13**: 1741–1747.

Plancher H, Green E and Paterson JC (1976) Reduction of oxidative hardening of asphalts by treatment with hydrated lime – a mechanistic study. *Proceedings of the Association of Asphalt Paving Technologists* **45**: 11–24.

Plancher H, Dorrene SM and Petersen JC (1977) Identification of chemical types strongly absorbed at the asphalt–aggregate interface and their displacement by water. *Proceedings of the Association of Asphalt Paving Technologists* **46**: 151–175.

Saal RNJ (1933) Adhesion of bitumen and tar to solid road building materials. *Bitumen* **3**: 101.

Scott JAN (1978) Adhesion and disbonding mechanisms of asphalt used in highway construction and maintenance. *Proceedings of the Association of Asphalt Paving Technologists* **47**: 19–48.

Taylor MA and Khosla NP (1983) Stripping of asphalt pavements: state of the art. *Transportation Research Record* **911**: 150–157.

Thelen E (1958) Surface energy and adhesion properties in asphalt/aggregate systems. *Highway Research Board Bulletin* **192**: 63–74.

Tunnicliff DG and Root RE (1984) *Use of Antistripping Additives in Asphaltic Concrete Mixtures.* TRB, Washington, DC, USA, NCHRP Report 274.

Winterkorn HF (1936) Surface chemical aspects of the bond formation between bituminous materials and mineral surfaces. *Proceedings of the Association of Asphalt Paving Technologists* **17**: 79–85.

Van Oss CJ, Chaudhury MK and Good RJ (1988) Interfacial Lifshitz–van der Waals and polar interactions in macroscopic systems *Chemical Reviews* **88**: 927–941.

Young T (1805) An essay on the cohesion of fluids. *Philosophical Transactions of the Royal Society, London* **95**: 65–87.

Chapter 20

Durability of bitumens and asphalts

Dr Nigel Preston
Bitumen Technical Manager,
Shell Australia Limited.
Australia

Lee O'Nions
Senior Application Specialist,
Shell UK Oil Products. UK

Bituminous road surfaces are expected to provide serviceability to the road user for many years before a maintenance intervention or rehabilitation is required. Indeed, there are undoubtedly examples of asphalt surfaces in many countries that are 40 or 50 years old and still provide a useable pavement in lightly trafficked environments. The term 'durability' is often used to describe the retention of the desirable engineering and serviceability characteristics over the lifetime of an asphalt, and is frequently included in the engineer's lexicon of performance attributes. Greater clarity of the term is helpful, and one definition of the durability of asphalts is as follows.

> The ability of the materials comprising the mixture to resist the effects of water, ageing and temperature variation, in the context of a given amount of traffic loading, without significant deterioration for an extended period. (Scholz, 1995)

While the integrity of the aggregates is an important factor in the long term performance of asphalts, it is primarily aspects associated with the ageing characteristics of bitumen that have the greatest influence on the durability of asphalts, surface dressings, chip seals and sprayed seals.

Quantification of bitumen durability has proved to be a difficult task, and methods of assessing this property tend to be indicative rather than precise. Long term studies have shown that, if an asphalt is to achieve its design life, it is important that the bitumen is not excessively hardened during hot storage, during the asphalt manufacturing process or in service in the road. Bitumen, in common with many organic substances, is affected by the presence of oxygen, ultraviolet radiation and changes in temperature. In bitumen, these external influences cause it to harden, resulting in a decrease in penetration, an increase in softening point and, usually, an increase in penetration index (PI). In recent years, the phenomenon of the hardening

The Shell Bitumen Handbook, Sixth Edition
ISBN 978-0-7277-5837-8
Shell International Petroleum Company Ltd: All rights reserved
http://dx.doi.org/10.1680/tsbh.58378.573

of bitumen and, hence, the hardening of an asphalt has been viewed as being beneficial in pavement layers as it increases the stiffness of the material and, therefore, the load spreading capabilities of the structure. This hardening is known as 'curing', and is believed to extend the life of a pavement (Scholz, 1995).

In the surface course, where the material is exposed to the environment, hardening of the bitumen can have a detrimental effect on its performance and can lead to fretting and/or cracking. This effect is still referred to as 'hardening', the term implying that the change in bitumen properties is detrimental to the service life of the surface course.

20.1 Bitumen hardening

The tendency for bitumen to harden under the influence of the atmosphere has been known and studied for many years. As many as 15 different factors that influence bitumen ageing have been identified, and these are detailed in Table 20.1.

The most important of these 15 mechanisms of bitumen hardening are

- oxidation
- volatilisation
- steric or physical factors
- exudation of oils.

Table 20.1 Mechanisms of bitumen ageing (Traxler, 1963)

Factors that influence bitumen ageing	Influenced by					Occurring	
	Time	Heat	Oxygen	Sunlight	Beta and gamma rays	At the surface	In the mixture
Oxidation (in dark)	✓	✓	✓			✓	
Photo-oxidation (direct light)	✓	✓	✓	✓		✓	
Volatilisation	✓	✓				✓	✓
Photo-oxidation (reflected light)	✓	✓	✓	✓		✓	
Photochemical (direct light)	✓	✓		✓		✓	
Photochemical (reflected light)	✓	✓		✓		✓	✓
Polymerisation	✓	✓				✓	✓
Steric or physical	✓					✓	✓
Exudation of oils	✓	✓				✓	
Changes by nuclear energy	✓	✓			✓	✓	✓
Action by water	✓	✓	✓	✓		✓	
Absorption by solid	✓	✓				✓	✓
Absorption of components at a solid surface	✓	✓				✓	
Chemical reactions	✓	✓				✓	✓
Microbiological deterioration	✓	✓	✓			✓	✓

20.1.1 Oxidation

This is considered to be the most important cause of hardening of bitumen in service.

Like many organic substances, bitumen is slowly oxidised when in contact with atmospheric oxygen. Polar groups containing oxygen are formed, and these tend to associate into micelles (an aggregate of molecules in a colloidal solution) of higher micellar weight, thereby increasing the viscosity of the bitumen. Reaction of oxygen with bitumen molecules causes the formation of carbonyl species, resulting in larger and more complex molecules that make the bitumen harder and less flexible. The degree of oxidation is highly dependent on the temperature, period of exposure and the thickness of the bitumen film. The rate of oxidation doubles for each 10°C increase in temperature above 100°C. Hardening due to oxidation has long been held to be the main cause of ageing, to the extent that other factors have been given scant consideration. However, it has been shown that, although other factors are generally less important than oxidation, they are measurable.

20.1.2 Loss of volatiles by evaporation

The evaporation of volatile components depends mainly on temperature and the exposure conditions. The rate of evaporation is controlled by the diffusion rate and the length or thickness of the diffusion path. Well compacted dense asphalt will have a slow diffusion process, whereas open grade asphalts and surface dressings will present more rapid evaporative conditions. However, paving grade bitumens are relatively involatile, and therefore the amount of hardening resulting from evaporation of volatiles is usually fairly small.

20.1.3 Steric or physical hardening

The phenomenon of steric hardening was first reported in 1944 by Traxler, when it was observed that bitumen samples stored at 25°C showed an increase in viscosity. This physical hardening is usually attributed to a combination of reorientation or restructuring of molecules within the bitumen and the slow crystallisation of waxes. The molecular reorientation that results in the hardening effect is reversible on the application of heat, with Traxler noting that raising the storage temperature to 70°C reversed the changed structure, with the bitumen sample returning to its original viscosity.

20.1.4 Loss of volatiles by exudation

Exudative hardening results from the movement of oily components that exude from the bitumen into the mineral aggregate (Van Gooswilligen *et al.*, 1989). It is a function of both the exudation tendency of the bitumen and the porosity of the aggregate.

20.2 Hardening of bitumen during storage, mixing and in service

The circumstances under which hardening occurs vary considerably. During storage, the bitumen is in bulk at a high temperature for a period of days or weeks. During mixing, hot storage, transport and laying, the bitumen is a thin film at high temperature for a relatively short period. In service, the bitumen is a thin film at a low or moderate temperature for a very long time. The degree of exposure to the air of asphalts in service is important and depends on the void content of the mixture. In dense, well-compacted mixtures, the amount of hardening is relatively small, while asphalts that have a more open constitution, such as porous asphalt, will undergo significant hardening. The degree of exposure to air for surface dressings or sprayed seals is potentially much higher than is the case with dense asphalt layers, and is influenced by the mosaic achieved by the aggregate and the absorptivity of the aggregates (Marais, 1974).

20.2.1 Hardening of bitumen in bulk storage

Very little hardening occurs when bitumen is stored in bulk at high temperature. Data from Shell terminals shows that 3–4 kt of paving grade bitumen stored at 150°C in a 7 kt tank shows virtually no change in viscosity or penetration over a 4 week storage period. This is because the surface area of the bitumen that is exposed to oxygen is very small in relation to the volume. However, if the bitumen is being circulated and is falling from the pipe entry at the top of the tank to the surface of the bitumen, significant hardening may occur. This arises because the surface area of the bitumen will be relatively large as it falls from the entry pipe, exposing it to the action of the oxygen. This effect can be minimised using the storage tank layout described in section 2.4.2 and shown in Figure 2.6.

20.2.2 Hardening of bitumen during mixing with aggregate

This is described as 'short term ageing', a term that is also applied to hardening that occurs during laying. During the mixing process, it is estimated that all the aggregate and filler is coated with a thin film of bitumen, usually between 5 and 15 μm thick. If the bitumen from 1 tonne of a dense asphalt concrete was spread at 10 μm thick, it would occupy an area of around 10 000 m^2, the equivalent of over one and a half average size football pitches. Thus, when bitumen is mixed with hot aggregate and spread into thin films in a paddle mixer, conditions are ideal for the occurrence of oxidation and the loss of volatile fractions within the bitumen. Hardening of bitumen during this process is well known, and is taken into account when selecting the grade of bitumen to be used. As a very rough approximation, during mixing with hot aggregate in a paddle mixer, the penetration of a paving grade bitumen falls by about 30%. However, the amount of

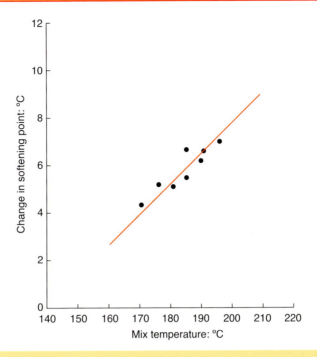

Figure 20.1 Relationship between the temperature of the mixture and change in softening point (Whiteoak and Fordyce, 1989)

hardening depends on a number of factors, such as temperature, duration of mixing and bitumen film thickness. Minimising the hardening during mixing depends on careful control of all these factors. Control of the temperature and the bitumen content are particularly critical. Figure 20.1 clearly shows increasing bitumen hardening, measured by higher values of softening point, as mixing temperatures are raised.

Similarly, Figure 20.2 shows that reducing the thickness of the bitumen film significantly increases the viscosity of the bitumen. The latter is measured by the ageing index, which is defined as the ratio of the viscosity of the aged bitumen, η_a, to the viscosity of the virgin bitumen, η_o.

Note that the ageing index is not a fundamentally defined parameter – it is usually a ratio of two values (e.g. viscosity, stiffness or penetration) measured at different times.

The type of mixer used also affects the amount of hardening during mixing. It has been recognised for some time that the amount of hardening in a drum mixer is often less than that which occurs in a conventional batch mixer (Haas, 1974). This is due to the presence of steam in the drum, which limits the availability of oxygen. However, the multiplicity of different designs of

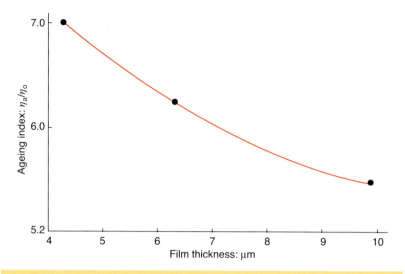

Figure 20.2 Effect of bitumen thickness on the ageing index (Griffin *et al.*, 1955)

drum mixers means that variation in the amount of hardening between different designs of plant is almost inevitable. Modern drum mix plants, notably those with counter flow drum mix and double drum configurations with very high mixing efficiencies, result in minimal hardening of bitumen compared with older drum mix plants. Notwithstanding, a study carried out by Shell Bitumen (1973) on two different drum mixers showed that, for equivalent mixing temperatures, the overall reduction in penetration and increase in softening point can be less than half of that which occurs in a conventional batch mixer.

20.2.3 Hardening of bitumen in asphalts during hot storage, transport and laying

Drum mixers, which produce large volumes of asphalts, require silo storage for the mixed material to account for peaks and troughs in demand. In such circumstances, mixtures are stored in hot silos, as well as in the delivery vehicle during transportation.

Some hardening of the bitumen will take place during hot storage, whether it is in a silo or in a truck. It was stated above that the amount of hardening will depend principally on the duration of exposure to oxygen, the thickness of the bitumen film and the temperature of the mixture. When a mixture is discharged into a storage silo, air enters with the mixture and some is trapped in the voids of the material. During the storage period, some of the oxygen in this entrained air will react with the bitumen. If no additional air enters the silo, oxidation of the bitumen will cease.

It is important that the entry and discharge gates are airtight and that there are no other openings where air can enter the silo. If the discharge gate is not airtight, the silo may behave like a chimney, drawing air in at the discharge gate (which exits the loading gate) and resulting in oxidation and cooling of the stored material. In addition, the silo should be as full as is practicable in order to minimise the amount of free air at the top of the silo. Air remaining at the top of the silo will react with the top surface of the material. This reaction forms carbon dioxide which, because it is heavier than air, tends to blanket the surface of the mixture, protecting it from further oxidation. In the USA, some silos have the facility to be pressurised with exhaust gases, containing no oxygen, from a burner. These exhaust gases purge the silo of entrained air and provide a slight positive pressure, preventing more air entering the silo.

Studies carried out in the USA (Brock, 1986) suggest that if oxidation in the silo is limited to that induced by entrained air, then little or no additional oxidation will occur during transportation and laying. It is hypothesised that this is because no significant quantity of fresh air is entrained in the mixture during discharge into the truck. Thus, little or no additional air is available for oxidation. In fact, it was observed that if the mixture was discharged directly from the paddle mixer into the delivery vehicle, the amount of hardening during transportation was very similar to that which occurs during silo storage.

If materials are being laid at low ambient temperatures or if the mixture has to be retained for a period in hot storage, there is a temptation to increase mixing temperatures to offset these two factors. However, increasing the mixing temperature will considerably accelerate the rate of bitumen oxidation, resulting in an increase in bitumen viscosity. Thus, a significant proportion of the reduction in viscosity achieved by increasing the mixing temperature will be lost because of additional oxidation of the bitumen, which may adversely affect the long term performance of the material.

20.2.4 Hardening of bitumen in the pavement and on the road

As explained above, a significant amount of bitumen hardening occurs during mixing and, to a lesser extent, during hot storage and transportation. However, hardening of the binder will continue in the pavement until some limiting value is reached. This behaviour is described as 'long term ageing' and is illustrated in Figure 20.3, which shows the ageing index of the bitumen after mixing, storage, transport, paving and subsequent service.

20.2.4.1 Bitumen hardening in asphalt surface courses

The main factor that influences bitumen hardening in asphalts, including surface courses, is the void content of the mixture. Table 20.2 shows the properties of bitumens recovered from three asphalt concretes after 15 years'

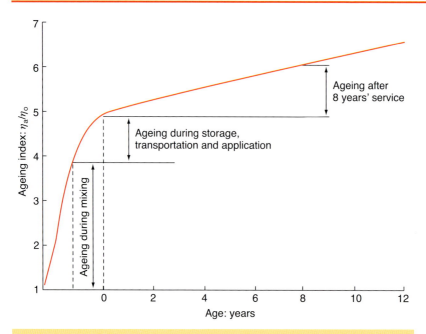

Figure 20.3 Ageing of bitumen during mixing, subsequently during storage, transportation and application and in service

Note that ageing index is not a fundamentally defined parameter – it is usually a ratio of two values (e.g. viscosity, stiffness, penetration) measured at different times

service. The bitumen recovered from the mixture with the lowest void content had hardened very little. However, where the void content was high, allowing constant ingress of air, substantial hardening had occurred. The PI of the material with the highest void content had increased appreciably. This limits the amount of stress relaxation that can occur, and may result in cracking of the compacted material. Similarly, Figure 20.4 shows the in situ bitumen properties of 5 year old asphalt concrete with void contents ranging from 3% to 12% (Lubbers, 1985). At void contents less than 5%, very little hardening occurred in service. However, at void contents greater than 9%, the in situ penetration fell from 70 dmm to less than 25 dmm.

The bitumen at the surface of the road hardens much more quickly than the bitumen in the bulk of the pavement. There are three reasons for this

- the existence of a constant supply of fresh oxygen
- the occasional incidence of high temperatures at the road surface
- the occurrence of photo-oxidation of the bitumen by ultraviolet radiation.

Photo-oxidation causes a skin, 4–5 µm thick, to be rapidly formed on the surface of the bitumen film. This is induced by natural ultraviolet radiation

Table 20.2 Hardening of bitumen in service (Edwards, 1973)

Road	A	B	C
Voids in mix: %	4	5	7
Properties after mixing and laying			
Softening point (IP): °C	64	63	66
Penetration at 25°C: dmm	33	33	30
Penetration index (Pen, SP)	+0.7	+0.7	+0.9
Stiffness (S_0): Pa (calc.)			
10^4 s, 25°C	1.4×10^3	1.4×10^3	2.5×10^3
10^4 s, 0°C	5.0×10^5	5.0×10^5	7.0×10^5
10^{-2} s, 25°C	2.5×10^7	3.0×10^7	3.0×10^7
10^{-2} s, 0°C	3.0×10^8	3.0×10^8	3.0×10^8
Properties after 15 years' service			
Softening point (IP): °C	68	76	88
Penetration at 25°C: dmm	24	15	11
Penetration index (Pen, SP)	+0.8	+1.1	+2.1
Stiffness (S_{15}): Pa (calc.)			
10^4 s, 25°C	4×10^3	20×10^3	150×10^3
10^4 s, 0°C	13×10^5	40×10^5	80×10^5
10^{-2} s, 25°C	4×10^7	7×10^7	8×10^7
10^{-2} s, 0°C	4×10^8	6×10^8	6×10^8
Ageing index (S_{15}/S_0)			
10^4 s, 25°C	2.8	14	60
10^4 s, 0°C	2.6	8	11
10^{-2} s, 25°C	1.6	2.3	2.7
10^{-2} s, 0°C	1.3	2.0	2.0

Note: Ageing index is not a fundamentally defined parameter. It is usually a ratio of two values (e.g. viscosity, stiffness, penetration) measured at different times.

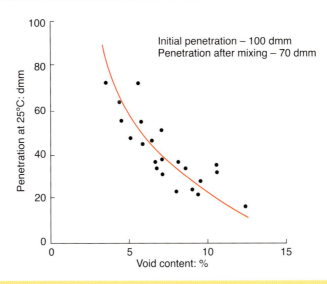

Figure 20.4 The effect of void content on the hardening of bitumen on the road (Lubbers, 1985)

- porous asphalt   >12 μm
- stone mastic asphalt   >6 μm.

This method of calculating bitumen film thickness determines surface area factors for the aggregate down to 75 μm. Clearly, some of the material passing this sieve will be substantially finer than 75 μm, particularly for materials in which limestone filler is added to the mixture. Therefore, for high filler content materials such as hot rolled asphalt, the bitumen film thickness will be lower than the value given above. However, if it is assumed that the bitumen and filler coat the coarse and fine aggregate, then the above values underestimate the binder film thickness.

20.2.4.2 Bitumen curing in asphalt bases

Contrary to the effects of bitumen hardening in the surface course, a gradual hardening of the main structural layers of the pavement appears to be beneficial, and is described as 'curing'. As with all hot mixed asphalts, the penetration of the bitumen in a base will harden by approximately 30% during the mixing and laying process, and, despite asphalt bases being locked inside the pavement construction and shielded from exposure to the environment, the penetration of the bitumen will continue to exhibit varying rates of hardening.

The Transport Research Laboratory (TRL) carried out a wide ranging investigation into the behaviour of asphalt pavements in service (Nunn *et al.*, 1997). This very important study noted in section 2.1 that

> One of the findings of this research was that changes occurring in asphalt over the life of the road are crucial in understanding its behaviour. These changes, which are referred to as curing, can help to explain why conventional mechanisms of deterioration do not occur and why, provided the road is constructed above a minimum threshold strength, it should have a very long, but indeterminate, structural life. The increase in the stiffness of the asphalt base causes the traffic-induced strains in the pavement structure, which control fatigue and structural deformation, to decrease with time. Therefore, a road will be more vulnerable to structural damage in its early life, before curing has increased the structural strength of the material. If the road is designed and constructed with sufficient strength to prevent structural damage in its early life, it has been found that curing doubles the stiffness of DBM roadbase [now base, of course] in the first few years in service and this will substantially improve the overall resistance of the pavement to fatigue and structural deformation. The improvement in the bearing capacity of the road, as determined by deflection measurements, provides confirmation of this improvement.

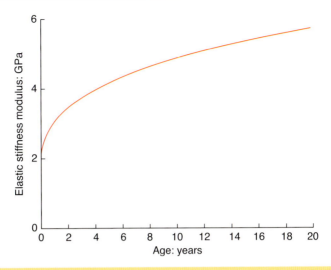

Figure 20.7 Typical curing curve for asphalt concrete base (Nunn *et al.*, 1997). (Reproduced by courtesy of the Transport Research Laboratory (TRL))

Figure C2 in TRL 250 (Nunn *et al.*, 1997) is reproduced here as Figure 20.7. As can be seen by examining the figure, the stiffness of an asphalt concrete base (the most common type of base in UK roads), which is initially 2 GPa at the time of laying, increases to around 4 GPa 4 years after laying. So, the stiffness has doubled in a period of 4 years. Clearly, the increase in the stiffness of asphalt bases over time will improve the structural competence of a road pavement, and has implications for the structural design of flexible pavements (see section 13.3).

20.2.4.3 Bitumen hardening in surface dressings

Surface dressings probably have the greatest exposure to environmental ageing mechanisms of all bituminous products. Sprayed surface treatments, known as 'surface dressings' in the UK, 'chip seals' in the USA and New Zealand, and 'sprayed seals' in Australia, will present a binder film with a depth of around 1.5 mm on the existing road surface on top of which the chippings will be spread. The mosaic achieved by the chippings will determine the extent of the dressing that is directly exposed to the air but the overall proximity of the surface dressing to the elements means that the binder is aggressively subjected to ageing factors.

Countries such as Australia that rely on sprayed seals for much of the rural road network have paid close attention to the rate of hardening of bitumen in seals. The Australian Road Research Board conducted investigations into the performance and longevity of sprayed seals, and proposed a relationship between binder ageing and seal life (Dickinson, 1984).

20.3 Bitumen ageing tests

It is clearly desirable that there should be laboratory tests that quantitatively determine the resistance of bitumens to hardening at the various stages during the production process. A number of tests already exist to measure the effect of heat and air on bitumen. The main aim of these tests is to identify bitumens that are too volatile or are too susceptible to oxidation to perform well in service (for more information on this testing see section 5.5.4).

The thin-film oven test (Lewis and Welborn, 1940) simulates practical conditions. In this test, the bitumen is stored at 163°C for 5 h in a layer 3.2 μm thick. It is claimed that in this test the amount of hardening that takes place is about the same as that which occurs in practice. However, diffusion in the bitumen film is also limited, and it is not possible to obtain homogeneous hardening or ageing. Accordingly, the test is far from ideal. This test was adopted initially by the American Society for Testing Materials (ASTM) in 1969 as method ASTM D1754 and has been modified since that time to include improvements (ASTM, 2009).

In 1963, the State of California Department of Public Works, Division of Highways, developed a test that more accurately simulates what happens to a bitumen during mixing. It is called the rolling thin-film oven test (RTFOT) (Hveem *et al.*, 1963). In this test, eight cylindrical glass containers each containing 35 g of bitumen are fixed in a vertically rotating shelf. During the test, the bitumen flows continuously around the inner surface of each container in a relatively thin film, with preheated air blown periodically into each glass jar. The test temperature is normally 163°C for a period of 75 min. The method ensures that all the bitumen is exposed to heat and air, and continuous movement ensures that no skin develops to protect the bitumen. A homogeneously aged material, similar to that which is produced during full scale mixing, is obtained. Clearly, the conditions in the test are not identical to those found in practice but experience has shown that the amount of hardening in the RTFOT correlates reasonably well with that observed in a conventional batch mixer. However, the mixing conditions in efficient modern drum mix plants are less aggressive, and bitumen hardening during asphalt production in such plants is less than predicted by the RTFOT.

The RTFOT was accepted in 1970 by the ASTM as method ASTM D2872-12e1 (ASTM, 2012) and was included as part of the European specification for paving grade bitumens in BS EN 12591:2000 (BSI, 2000) and is part of the Superpave specification used in the USA (AI, 1997).

(a)

(b)

Figure 20.8 Shell sliding plate viscometer: (a) general view; (b) sample being subjected to shear loading

Australia utilises the RTFOT apparatus artificially to age samples of bitumen to a specified apparent viscosity level (SAVL – taken to be 5.67 log Pa·s at 45°C). The SAVL is considered to equate to the viscosity of a bitumen in a sprayed seal at the end of its service life, and is measured using the Shell sliding plate viscometer (Figure 20.8) (Standards Australia, 1997a).

A correlation between the time taken for a bitumen to reach the SAVL and the in situ service life of a seal was proposed by the Australian Road Research Board (ARRB) from data obtained from bitumen recovered from aged sprayed seals. The assessment is called the 'durability test', and is applied by a number of road agencies in Australia as a control point on the suitability of bitumen for use in sprayed seals (Standards Australia, 1997a). Australia is one of the few countries where a specific test is employed to quantify the durability of bitumen (Standards Australia, 1997b).

Over the years there have been a number of attempts to simulate the long term ageing of bitumen in asphalts but this has proved to be extremely difficult because of the number of variables that affect binder ageing – void content, mixture type, aggregate type, etc. The Strategic Highway Research Program (SHRP) investigated accelerated ageing techniques and identified the pressure ageing vessel (PAV) (Anderson *et al.*, 1994) as the preferred apparatus to simulate long term ageing of bitumen. The US Superpave specification (AI, 1996) uses the RTFOT to simulate initial ageing, followed by ageing over 20 h at elevated temperature (90, 100 or 110°C) and pressure 2070 kPa in a PAV. After this ageing procedure, the residue is used for dynamic shear rheometry, bending beam rheometry and direct tension testing. The use of the PAV, using modified conditions, is currently being considered in Europe as a method for ageing bitumens in the laboratory. The artificial ageing of binders in the PAV to simulate ageing in situ has still to be fully validated but the technique is now widely accepted as a satisfactory approach.

References

AI (Asphalt Institute) (1996) *Superpave Mix Design*. AI, Lexington, KY, USA, Superpave Series No. 2 (SP-2).

AI (1997) *Superpave Performance Graded Asphalt Binder Specification and Testing*. AI, Lexington, KY, USA, Superpave Series No. 1, pp. 15–19.

Anderson DA, Christiansen DW, Bahia HU *et al.* (1994) *Binder characterization, vol. 3: Physical characterization, SHRP-A-369*. Strategic Highway Research Program, National Reseearch Council, Washington, DC, USA.

ASTM (American Society for Testing and Materials) (2009) D1754-09. Standard test method for effect of heat and air on asphaltic materials (thin-film oven test). ASTM, West Conshohocken, PA, USA.

ASTM (2012) D2872-12e1. Standard test method for effect of heat and air on a moving film of asphalt (rolling thin-film oven test). ASTM, West Conshohocken, PA, USA.

Blokker PC and van Hoorn H (1959) Durability of bitumen in theory and practice. *Proceedings of the 5th World Petroleum Congress*, New York, NY, USA, paper 27, p. 417.

Brock JD (1986) *Oxidation of Asphalt*. Astec Industries, Chattanooga, TN, USA, Technical Bulletin T-103.

BSI (British Standards Institution) (2000) BS EN 12591:2000. Bitumen and bituminous binders. Specifications for paving grade bitumens. BSI, London, UK.

Campen WH, Smith JR, Erickson LG and Mertz LR (1959) The relationship between voids, surface area, film thickness and stability in bituminous paving mixtures. *Proceedings of the Association of Asphalt Paving Technologists* **28**: 149.

Dickinson EJ (1984) *Bituminous Roads in Australia*. ARRB, Melbourne, Australia.

Dickinson EJ, Nichols JH and Boas-Traube S (1958) Physical factors affecting the absorption of oxygen by thin films of bituminous binders. *Journal of Applied Chemistry* **8**: 673.

Edwards JM (1973) Dense bituminous mixes. Paper presented at *Conference on Road Engineering in Asia and Australasia*, Kuala Lumpur, Malaysia.

Griffin RL, Miles TK and Penther CJ (1955) Microfilm durability test for asphalt. *Proceedings of the Association of Asphalt Paving Technologists* **24**: 31.

Haas S (1974) Drum-dryer mixes in North Dakota. *Proceedings of the Association of Asphalt Paving Technologists* **34**: 417.

Hveem FN, Zube E and Skog J (1963) Proposed new tests and specifications for paving grade asphalts. *Proceedings of the Association of Asphalt Paving Technologists* **32**: 271–327.

Lewis RH and Welborn JI (1940) Report on the properties of the residues of 50-60 and 85-100 penetration asphalts from oven tests and exposure. *Proceedings of the Association of Asphalt Paving Technologists* **11**: 86–157.

Lubbers HE (1985) *Bitumen in de weg- en waterbouw.* Nederlands Adviesbureau voor Bitumentoepassingen, Gouda, The Netherlands (in Dutch).

Marais CP (1974) Gap-graded asphalt surfacings – the South African scene. *Proceedings of the 2nd Conference on Asphalt Pavements for Southern Africa, Durban, South Africa.*

Nunn ME, Brown A, Weston D and Nicholls JC (1997) *Design of Long-Life Flexible Pavements.* TRL Ltd, Crowthorne, UK, TRL Report 250.

Scholz TV (1995) *Durability of Bituminous Paving Mixtures.* PhD Thesis, University of Nottingham, Nottingham, UK.

Shell Bitumen (1973) The rolling thin-film oven test. *Shell Bitumen Review* **42**: 18.

Standards Australia (1997a) AS/NZS 2341.13:1997 Methods of testing bitumen and related roadmaking products. Method 5: Determination of apparent viscosity by 'Shell' sliding plate micro-viscometer. Standards Australia, Sydney, NSW, Australia.

Standards Australia (1997b) AS/NZS 2341.13:1997 Methods of testing bitumen and related roadmaking products. Method 13: Long-term exposure to heat and air. Standards Australia, Sydney, NSW, Australia.

Traxler RN (1963) Durability of asphalt cements. *Proceedings of the Association of Asphalt Paving Technologists* **32**: 44–63.

Van Gooswilligen G, De Bats FTh and Harrison T (1989) Quality of paving grade bitumen – a practical approach in terms of functional tests. *Proceedings of the 4th Eurobitume Symposium, Madrid, Spain,* pp. 290–297.

Whiteoak CD and Fordyce D (1989) Asphalt mixture workability, its measurement, and how it can be modified. *Shell Binder Review* **64**: 14–17.

Chapter 21

Surface treatments

Qian PeiZhong
Application Advisor,
Shell China Ltd. China

Frits de Jonge
Technology Excellence
Manager, Shell Downstream
Services International B.V.
The Netherlands

Mario Jair
Technical Manager,
Shell Cia Argentina de
Petroleo SA. Argentina

21.1 Surface dressing

Surface dressing is a treatment undertaken on roads and other types of pavement. As a maintenance process it has three purposes

- to provide both texture and skid resistance to the surface
- to seal the road surface against ingress of water
- to arrest disintegration, and hence to extend the life, of the pavement and to assist sustainable development.

The aims listed above are taken from Road Note 39 (Roberts and Nicholls, 2008). Road Note 39 is the default design guide for UK surface dressing activities and would be used for the vast majority of surface dressing designs in the UK.

In addition to the purposes listed above, surface dressing can be used as a treatment to provide

- a distinctive colour to the road surface
- a more uniform appearance to a patched road.

However, surface dressing cannot restore good surface regularity to a deformed road, nor will it contribute to the structural strength of the pavement.

In a new road, adequate surface texture is designed into the running surface by specifying requirements for both aggregate properties and texture depth. However, during its service life, the surface becomes polished under the action of traffic, and the skid resistance will eventually fall below the acceptable

The Shell Bitumen Handbook, Sixth Edition
ISBN 978-0-7277-5837-8
Shell International Petroleum Company Ltd: All rights reserved
http://dx.doi.org/10.1680/tsbh.58378.591

591

minimum value. A major benefit of surface dressing is that the process restores skid resistance to a surface that has become smooth under traffic.

Surface dressing is the most extensively used form of surface treatment on roads. In Europe, the USA, China and Australia, it is widely used for preventive maintenance. In the UK, data suggest that the road lengths of surface dressing are more than 50% of total lengths of maintenance in England and Wales (DfT, 2014). In Australia, about 90% of paved roads are surface dressed (Holtrop, 2008). With care, surface dressing can be used on roads of all types, from a country lane carrying only an occasional vehicle to trunk roads carrying tens of thousands of vehicles a day. However, in some locations, particularly those where vehicles undertake sudden or sharp turning manoeuvres, surface dressing may not be the appropriate surface treatment. This is due to the relatively poor ability of the finished surface to resist tangential forces. In addition, the durability of a surface dressing does not match that of a road resurfaced with traditional asphalt. Some large cities, Hamburg for example, prohibit the use of surface dressings on their carriageways.

Basic surface dressing consists of spraying a film of binder onto the existing road surface, followed by the application of a layer of aggregate chippings as shown in Figure 21.1. The chippings are then rolled to promote contact between the chippings and the binder, and to initiate the formation of an interlocking mosaic. On carriageways having a low value of road hardness, rolling may also cause the chippings to start the process of embedment into

Figure 21.1 Surface dressing. (Courtesy of Tarstone Surfacing Limited)

the existing road surface. (Road hardness is a property that represents the resistance of an existing road surface to the embedment of chippings – see section 7.2.2 of Road Note 39 (Roberts and Nicholls, 2008).)

21.1.1 *The design of surface dressings*

The design of surface dressings primarily addresses the retention of surface texture. The aim is to design a system that takes account of the hardness of the existing road such that the aggregate is adequately embedded by rolling and, subsequently, traffic.

In the UK, Road Note 39 (Roberts and Nicholls, 2008) and the *Code of Practice for Surface Dressing* (RSTA, 2014a) are commonly used to provide guidance on the appropriate type of dressing, its design and execution. The principal steps per Road Note 39 are as follows

- select the size and type of dressing that is appropriate for the particular site and the time of application
- identify the chipping size and basic binder application rate and type of binder
- refine the binder application rate to match the source of chippings selected
- further refine the binder application rate on site immediately before work starts.

Road Note 39 is set out in such a way that the application rate of the binder is only obtainable after completion of the steps outlined above. Tables provide coefficients for application rates of binders for the full range of types of surface dressing (i.e. single, racked-in, double, sandwich and inverted double surface dressings). These coefficients are then adjusted by reference to tables that take into account the type of aggregate and its shape, the condition of the road surface, the gradient, the degree to which the site is shaded by trees and structures, and steps to be taken for sites where traffic volumes are exceedingly light. The coefficients are then translated into actual target application rates of the binder expressed in litres per square metre.

Many countries in Europe employ the CEN (European Committee for Standardization) standards. The CEN specifications for surface dressing are based on the UK's Road Note 39.

In the USA, individual states have their own specifications. Some states have published documents on the subject (e.g. the *Minnesota Seal Coat Handbook 2006* (Wood *et al.*, 2006) and the *Texas Seal Coat and Surface Treatment Manual* (TxDOT, 2010)). In addition, the Asphalt Institute (a US-based international trade association of bitumen producers, manufacturers and affiliated businesses) and ASTM International (the

American Society for Testing and Materials (ASTM)) also publish surface treatment standards (Asphalt Institute 2009; ASTM, 2010a).

Austroads is the association of Australian and New Zealand road transport and traffic authorities. It provides two technical reports for the design of binder rates of application and aggregate spread rates for sprayed seal

- AP-T68/06 (*Update of the Austroads Sprayed Seal Design Method*) (Austroads, 2006)
- AP-T236–13 (*Update of Double/Double Design for Austroads Sprayed Seal Design Method*) (Austroads, 2013).

Guidance on selection of sprayed-seal-type materials and sprayed-seal construction is provided in the *Guide to Pavement Technology*

- Part 3: *Pavement Surfacings* (AGPT03-09) (Austroads, 2009a)
- Part 4F: *Bituminous Binders* (AGPT04F-08) (Austroads, 2008)
- Part 4K: *Seals* (AGPT04K-09) (Austroads, 2009b)
- Part 8: *Pavement Construction* (AGPT08-09) (Austroads, 2009c).

In China, two industry standards dictate the requirements for type selection, materials and construction specifications for surface dressing. These standards are

- JTG D50-2006: *Specifications for Design of Highway Asphalt Pavement* (RIOH, 2006a)
- JTG F40-2004: *Technical Specifications for Construction of Highway Asphalt Pavements* (RIOH, 2004).

21.1.2 Factors that influence the design and performance of surface dressings

The principal factors that influence the design, performance and service life of a surface dressing are

- traffic volumes and speeds
- condition and hardness of the existing road surface
- size and other chipping characteristics
- surface texture and skid resistance of the existing road surface
- type of binder
- adhesion
- geometry of site, altitude, latitude and local circumstances
- time of year.

21.1.2.1 Traffic volumes and speeds

The degree of chipping embedment is a function of the quantity of vehicles that use the route being designed. Accordingly, it is necessary to establish the number of heavier vehicles that regularly use that route and the speed of such traffic. In the UK, the number of medium and heavy vehicles (i.e. those having an unladen weight exceeding 1.5 tonnes) currently travelling in the lane under consideration is normally used for design purposes according to Table 7.2.3 of Road Note 39 (Roberts and Nicholls, 2008).

21.1.2.2 Condition and hardness of the existing road surface

If the existing road surface is rutted, cracked or in need of patching, these defects must be corrected before surface dressing can be undertaken (RSTA, 2014b). Other surface conditions that require special consideration are those in which the binder has flushed up, resulting in a binder-rich surface and, conversely, surfaces that have become very dry, lacking in binder.

The hardness of the road surface affects the extent to which the applied chippings become embedded in the road surface during the life of the dressing. The choice of chipping size is directly related to the hardness of the existing road surface. The use of chippings that are too small will result in early embedment of the chippings in the surface, leading to a rapid loss of texture depth and, in the worst cases, 'fatting up' of the binder, which may cover the entire surface of the road. ('Fatting up' is where there is an excess of bitumen on the road surface, sometimes but less commonly described as 'flushing-up'.) The use of chippings that are too large may result in immediate failure of the treatment due to stripping of the binder from the aggregate under the applied stresses of the traffic, and can also lead to excessive surface texture and an increase in the noise generated between the tyre and the road surface. The hardness of the road surface also influences the rate of application of binder required for a given size of chipping. The rate of spread must be reduced where the road surface is soft in order to compensate for the greater embedment of chippings in the road surface under the action of traffic.

In the UK, the categories of hardness of the existing road surfaces for the purposes of surface dressing design are described in Road Note 39. Four graphs take account of the geographical location within the UK and the altitude of the site. The geographical location is not strictly related to latitude but also takes account of climatic conditions. The method of measuring hardness on the site is by use of a road hardness probe, which is described in BS 598-112 (BSI, 2004). Hardness measurements are taken in conjunction with road surface temperature readings. The standard specifies that road surface temperatures must be in the range 15–35°C.

Having identified the applicable surface temperature category and measured the depth of penetration and the surface temperature, the hardness category of the road can be established by reference to Figure 7.2.2 in Road Note 39 (Roberts and Nicholls, 2008), as shown in Figure 21.2.

In Australia, the ball penetration test is used for testing the pavement hardness. The test method is described in Austroads test method AG:PT/T251 (Austroads, 2010).

In the USA, some guides (TxDOT, 2010; Wood *et al.*, 2006) for surface dressing provide a surface correction factor based on the existing pavement condition.

21.1.2.3 Size and other chipping characteristics

The selection of the size and type of chipping should take into account the following factors.

- The size of chipping has to offset the gradual embedment in road surfaces of different hardness caused by traffic.
- Maintaining both microtexture and macrotexture during the life of the surface dressing by selecting a chipping of appropriate
  - size
  - polished stone value (PSV)
  - aggregate abrasion value (AAV).
- Relating the traffic category for each lane and the road surface hardness category to the size of chipping required.

In the UK, chippings for surface dressing should meet the requirements of BS EN 13043:2002 (BSI, 2002a). Usually, chippings have a nominal size of 2.8/6.3, 6.3/10 or 8/14 mm. Extreme caution should be exercised when using chippings larger than 8/14 mm nominal size due to the risk of premature failure. In China, aggregate requirements are given in JTG F40-2004 (RIOH, 2004), the nominal size is from 3/5 to 15/30 mm, and in Australia it is up to 20 mm. In the USA, the nominal size of the largest grade is normally 19 mm, although in some states it is 12.7 mm.

The presence of dust can delay or even prevent adhesion and this problem is particularly acute at low temperatures and with the smaller sizes of chipping. As well as being of nominal single size and dust free, surface dressing chippings need to satisfy other characteristics. They should

- not crush under the action of traffic or shatter on impact
- resist polishing under the action of traffic
- have an acceptable value of 'flakiness index' (the ratio of length to thickness).

Chippings are rarely cubic, and when they fall onto the binder film and are rolled they tend to lie on their longest dimension. If the chippings used for a

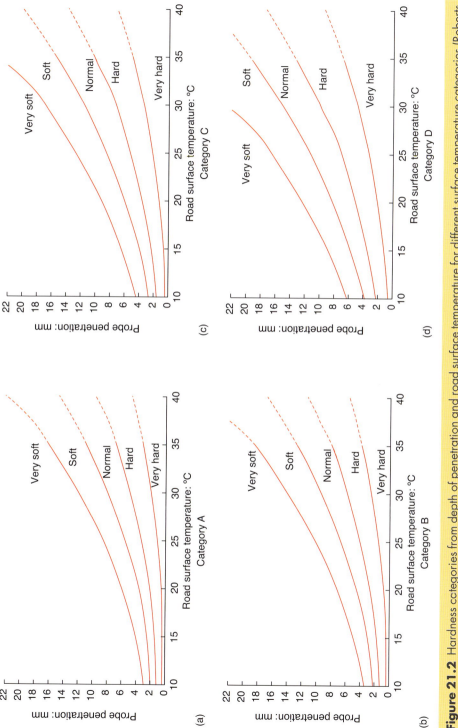

Figure 21.2 Hardness categories from depth of penetration and road surface temperature for different surface temperature categories (Roberts and Nicholls, 2008). (Reproduced by courtesy of the Transport Research Laboratory (TRL))

Table 21.1 Typical properties of surface dressing chippings

Property	Value
Los Angeles coefficient (LA)	LA_{30}
Flakiness index (FI)	FI_{20}
Polished stone value (PSV)	PSV_{56}[a]
Aggregate abrasion value (AAV)	AAV_{10}

[a] The required PSV value is usually stated in individual contracts. Because the PSV categories in BS EN 13043:2002 do not always use values that correspond with established UK practice, the standard allows an aggregate supplier to declare an intermediate value.

surface dressing have a large proportion of flaky chippings, less binder will be required to hold them in place, and the excess binder may 'flush-up' onto the surface of the dressing. Such a dressing will have a reduced texture and life. It is for this reason that the UK sets a maximum limit of flakiness for single size chippings for use in surface dressing. Typical requirements for surface dressing chippings are shown in Table 21.1.

Gritstones, basalts, quartzitic gravels and artificial aggregates such as slag constitute the majority of chippings used for surface dressing. However, not all aggregates with a high PSV are suitable for surface dressing due to them possessing insufficient strength. The PSV of the aggregate defines its skid resistance. In the UK, its determination is dictated by BS EN 1097-8:2009 (BSI, 2009) and the relationship between PSV, traffic and skid resistance is detailed in HD 36 (Highways Agency et al., 2006). Coated chippings have a thin film of bitumen applied at an asphalt plant. This bitumen film eliminates surface dust and promotes rapid adhesion to the bitumen. Coated chippings are used to improve adhesion with cut-back bitumen, particularly in the cooler conditions that occur at the extremes of the season. They should not be used with emulsions, as the shielding effect of the binder will delay 'breaking' of the emulsion (emulsions contain water, which evaporates after spraying; the process of 'breaking' is discussed in detail in section 21.1.2.5).

21.1.2.4 Surface texture and skid resistance
In the UK, two machines are used for monitoring the skid resistance of roads: the SCRIM (Figure 21.3) and the GripTester (Figure 21.4).

SCRIM is the acronym for the 'sideways-force coefficient routine investigation machine'. This machine was introduced into the UK in the 1970s. It provides a measure of the wet road skid resistance properties of a bound surface by measurement of a sideways-force coefficient (SFC) at a controlled speed.

Figure 21.3 SCRIM. (Courtesy of PTS Ltd.)

Figure 21.4 GripTester. (Courtesy of PTS Ltd.)

The GripTester was introduced in the 1980s, and is a braked wheel fixed-slip trailer. The resulting horizontal drag and the vertical load on the measuring wheel are continuously measured, and the coefficient of friction, known as the GripNumber, is calculated. The GripNumber can be converted to a SCRIM coefficient using the following relationship (Dunford, 2010)

$$SC = 0.89 \times GN$$

where SC is the SCRIM coefficient and GN is the GripNumber.

In China, besides the SCRIM, two other machines are used: one is the Mu-Meter (Figure 21.5) and the other is the dynamic friction tester (Figure 21.6). The Mu-Meter is manufactured in the UK, and the principle is the same as that employed in the SCRIM but, like the GripTester, the Mu-Meter is much smaller than the SCRIM. The dynamic friction tester is made in Japan and its use is described in ASTM E1911-09ae1 (ASTM, 2009a).

The values of both the SFC and the GripNumber vary with speed. Testing is normally carried out at 50 km/h. Corrections may be made to take account of variations in speed, but for these to be reliable it is necessary for the macrotexture of the surface to be known. Tight radii, such as occur at roundabouts, are normally tested at 20 km/h.

The SFC and GripNumber also vary with traffic flow, temperature and time of year. Network testing is restricted to May to September. Three tests are carried out at fairly similar intervals throughout the season. The average is defined as the 'mean summer SCRIM coefficient' (MSSC), and skid resistance is normally expressed in terms of the MSSC.

The skid resistance of a road surface is determined by two basic characteristics, the microtexture and macrotexture, as shown in Figure 21.7, which is

Figure 21.5 Mu-Meter. (Courtesy of Curtiss-Wright Corporation)

Figure 21.6 Dynamic friction tester. (Courtesy of the Transtec Group, Inc.)

Figure 2.1.2 in Road Note 39 (Roberts and Nicholls, 2008). Microtexture is the surface texture of the aggregate. A significant level of microtexture is necessary to enable vehicle tyres to penetrate thin films of water and thus achieve dry contact between the tyre and the aggregate on the carriageway. Macrotexture is the overall texture of the road surface. An adequate value of macrotexture is necessary to provide drainage channels for the removal of bulk water from the road surface.

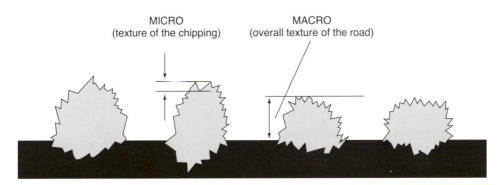

Figure 21.7 Microtexture and macrotexture (Roberts and Nicholls, 2008). (Reproduced by courtesy of the Transport Research Laboratory (TRL))

TRL Research Report 296 (Roe *et al.*, 1991) considered the relationship between accident frequency and surface texture on roads. It concluded that, on asphalt road surfaces with speed limits greater than 64 km/h, and possibly on roads with lower speed limits

- skidding and non-skidding accidents, in both wet and dry conditions, are fewer if the microtexture is coarse than if it is fine
- the texture level below which accident risk begins to increase is a sensor measured texture depth (SMTD, see below) of around 0.7 mm
- all major types of surfacing provide texture depths across the practical range
- macrotexture has a similar influence on accidents, whether they occur near hazards such as junctions or elsewhere.

Microtexture is the dominant factor in determining the level of skid resistance at speeds up to 50 km/h. Thereafter, macrotexture predominates, particularly in wet conditions. Traditionally, macrotexture has been measured using the volumetric patch technique (BSI, 2010).

Mature surface dressings in good condition typically have texture depths between 1.5 and 2.0 mm; double dressings usually have slightly lower values. Double surface dressings, however, usually stay above intervention levels, particularly on hard surfaces such as those surfaced with a hot rolled asphalt surface course.

Laser texture measurement (technically, SMTD) is a technique that tests areas much more quickly than is the case with volumetric patch testing. Accordingly, it is used as a screening method. The technique is illustrated in Figure 21.8, which is Figure 1 in TRL Research Report 143 (Roe *et al.*, 1988). This method is not directly comparable with measurements resulting from volumetric patch testing. Thus, when quoting texture depth values, the method used to measure the parameter should be stated.

It is the petrographic characteristics of the aggregate that largely influence its resistance to polishing, and this property is measured using the PSV test (BSI, 2009). A high PSV (say 65 or higher) indicates good resistance to polishing. The majority of road aggregates have a good microtexture prior to trafficking and, consequently, most road surfaces have a high skid resistance when new, particularly after the bitumen has worn off the aggregate. The time required for the bitumen to be abraded from the aggregate at the road surface depends on a number of factors, including the type of mixture and, in particular, the thickness of the film of bitumen on the particles of aggregate. Until the latter has been removed by traffic, the microtexture of the aggregate will not be fully exposed to vehicle tyres. Subsequently, the

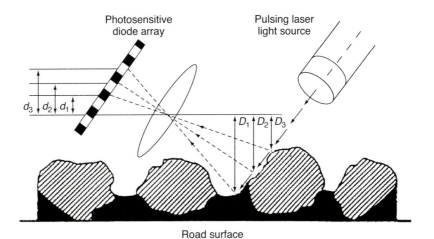

Photosensitive
diode array

Pulsing laser
light source

d_3 d_2 d_1

D_1 D_2 D_3

Road surface

Figure 21.8 Principle of the laser-based contactless sensor for measuring texture depth (Roe *et al.*, 1988). (Reproduced by courtesy of the Transport Research Laboratory (TRL))

exposed aggregate surfaces become polished and, within a year, the skid resistance normally falls to an equilibrium level.

The macrotexture is determined by the nominal size of aggregate used and the nature of the asphalt. However, resistance to abrasion is also important. It is expressed as the AAV. Aggregates with poor abrasion resistance, determined using the AAV test (BSI, 2009), will quickly be worn away, with a consequent loss of macrotexture. The lower the AAV, the greater the resistance to abrasion. High speed roads normally require an AAV of 14 or less. Surface dressings generally tend to last longer when the chippings have a lower AAV.

The rate at which skid resistance falls with increasing vehicle speed is influenced by the macrotexture. The TRL has shown that there is a correlation between the macrotexture, expressed in terms of average texture depth, and the skid resistance, measured by the percentage reduction in the braking-force coefficient between 130 and 50 km/h, as shown in Table 21.2, which is Table 4 in Laboratory Report 510 (Salt and Szatkowski, 1973). In the table, 'flexible' means fully asphaltic pavements, whereas 'concrete' means rigid, cementitious pavements.

The skid resistance necessary at any site will depend on the stresses likely to be induced when the road is in service. Sections of road with minor road junctions and with low traffic volumes are unlikely to require a high value of skid resistance, whereas sharp bends on roads carrying heavy volumes of traffic will require significantly higher skid resistance characteristics. It is

Table 21.2 The effect of macrotexture on the change in skid resistance with speed (Salt and Szatkowski, 1973)

Drop in skid resistance with speed change from 50 to 130 km/h: %	Average texture depth: mm	
	Flexible	Concrete[a]
0	2.0	0.8
10	1.5	0.7
20	1.0	0.5
30	0.5	0.4

[a] When textured predominantly transversely

important to note that roads carrying high volumes of traffic will require aggregates with higher PSVs than will roads where traffic levels are low or moderate.

21.1.2.5 Type of binder

The essential functions of the binder are to seal surface cracks and to provide the initial bond between the chippings and the road surface. The viscosity of the binder must be such that it can wet the chippings adequately at the time of application and prevent dislodgement of chippings when the road is opened to traffic. In addition, the bitumen should not become brittle during periods of prolonged low temperature, and should function effectively for the design life of the dressing.

The subsequent role of the binder is dependent on the traffic stresses applied to the surface dressing, from low stressed, lightly trafficked country roads to highly stressed sites carrying large volumes of traffic. Clearly, a range of binders is necessary if optimum performance is to be achieved under these widely differing circumstances.

The most commonly and widely used binders for surface dressing are

- conventional bitumen emulsion
- polymer modified bitumen
- polymer modified bitumen emulsion.

Other types are also used in some countries

- conventional bitumen
- conventional fluxed or cut-back bitumen
- polymer modified fluxed or cut-back bitumen
- epoxy resin thermoset binder.

Conventional bitumen emulsion

Emulsions used for surface dressing are normally rapid setting. In the EU, they should comply with BS EN 13808:2013 (BSI, 2013). In Australia, these are described in AS 1160-1996 (Standards Australia, 1996), and in JTG F40-2004 (RIOH, 2004) in China.

The relatively low viscosity of bitumen emulsions enables them to be sprayed at temperatures between 75°C and 85°C through either swirl or slot jets (jet types are discussed in section 21.1.4.1). There is the possibility that the binder will flow on road surfaces at the time of spraying. While this is helpful in some respects, migration of the binder from the centre to the side of cambered roads, or downhill on gradients, requires the application of chippings to the binder film as quickly as possible after spraying in order to counter any tendency to flow.

After spraying, the water migrates from the emulsion, leaving a film of bitumen, a process known as 'breaking'. This can take from as little as 10 min to several hours depending on ambient temperature, humidity, aggregate water absorption, wind speed and the extent to which the dressing is subject to slow moving traffic. Humidity is one of the most important of these factors, and the use of emulsions when humidity is above 80% should only be undertaken on very minor roads and where traffic speeds can be kept within the range 10–20 mph until the emulsion has fully broken. It is simple to check whether the emulsion has broken, as the colour of the emulsion changes from brown to black. In conditions of high humidity, work can continue with emulsions, particularly when the binder is polymer modified, by varying the type of dressing. Good results can be obtained with double and sandwich surface dressings. Precoated chippings should not be used with emulsions, as the coating delays breaking and provides no tangible benefit but carries additional cost.

Binder manufacturers are able to control the rate at which emulsions break within certain limits, but these have to be compatible with the storage, transport and application of the emulsion. On site, breaking can be accelerated by spraying the binder film with a mist of a breaking agent before the application of the chippings. However, this usually requires specially adapted spray bars and the fitting of storage tanks for the breaking agent.

The advantages of emulsion binders are

- low spraying temperature
- low risk of early bleeding
- may be used on higher viscosity base bitumen
- may be used on damp (not wet) surfaces
- do not require precoated chippings.

The disadvantages are

- low initial adhesion

- risk of brittle failure in the first winter (employing modified binders will reduce this risk considerably)
- skinning – the aggregates will not be wetted properly resulting in poor adhesion, together with the fact that the aggregate particles tend to sit on top of the emulsion layer rather than penetrating through it. The use of a sandwich or, preferably, a double surface dressing will reduce the film thickness and, thus, the possibility of skinning.

Skinning is the phenomenon that occurs when a thin layer at the surface of the emulsion breaks and forms a 'skin', or layer of bitumen, over the rest of the emulsion, which then stops it from breaking.

Polymer modified bitumen
The relentless increase in the volume of all types of vehicles on roads throughout Europe has led to the development of modified binders with improved adhesivity and substantially higher cohesive strengths than traditional fluxed/cut-back and emulsion binders. These characteristics are particularly important immediately after chippings have been applied to the binder film, and the use of a modified binder means that the chippings are less likely to be plucked from the road surface by passing traffic than would be the case with conventional unmodified binders. As well as greater adhesive and cohesive strength, modified binders perform satisfactorily over a wider temperature range than is the case with traditional binders. The result is that modified binders do not flush up as rapidly in hot conditions as traditional binders, and are not subject to the same degree of brittleness in winter conditions.

A high proportion of the European road network is lightly or moderately trafficked, and many of the roads can be surface dressed using bitumen emulsion or cut-back bitumen. However, enhanced treatment may be necessary on certain highly stressed sections, such as at sharp bends, road junctions and in shaded areas under trees and bridges and adjacent to trees or buildings. At these locations, it is probable that the same binder can be used, but the specification will change in respect of one or more of the following

- rate of application of binder
- chipping size and rate of spread
- double chip application
- double binder, double chip application.

The double chip techniques are widely practised in France on relatively heavily trafficked roads, and, by adopting appropriate procedures, the risk of flying chippings and consequential windscreen breakage can be reduced significantly.

As a further aid to consistent performance under variable weather conditions, a number of proprietary adhesion agents are available. These act in two basic ways. Some types (active) modify surface tension at the stone/binder interface to improve the preferential wetting of chippings by the binder in the presence of water. Other types (passive) improve the adhesion character- istics of the binder, resulting in greater resistance to the subsequent detach- ment of the binder film in the presence of water throughout the early life of the dressing.

Bearing these performance requirements in mind, it is possible to examine the ways in which the various polymers can be used with bitumen to modify its performance. The bitumen properties in which improvement would be sought through the addition of a modifier are as follows

- Reduced temperature susceptibility, which can be loosely defined as the extent to which a binder softens over a given temperature range. It is clearly desirable for a binder to exhibit a minimum variation in viscosity over a wide range of ambient temperatures.
- Cohesive strength is the property that enables the binder to hold chippings in place when they are subject to stress by traffic. This property can also be coupled with the elastic recovery properties of the binder, to enable a surface dressing to maintain its integrity even when it is subjected to high levels of strain.
- Adhesive power or 'tack' is an essential property of binders but it is difficult to define. The adhesion characteristics are particularly important during the early life of a surface dressing before any mechanical interlock or embedment of the chippings has taken place.

In addition, it is desirable for the modified binder to

- maintain its premium properties for long periods in storage and subsequently during its service life
- be physically and chemically stable during storage and spraying and at in-service temperatures
- be capable of being applied using conventional spraying equipment
- be cost effective.

Technically, the choice between an emulsion or cut-back formulation for modified binders is not easy, but health and safety concerns, and applicable regulations, often limit the use of cut-backs. In the UK, the use of cut-backs has dropped enormously because of safety concerns, cut-backs being applied at much higher temperatures than is the case with emulsions.

The development of polymer modified surface dressing binders to comple- ment conventional and epoxy resin systems offers highway engineers a range of binders suitable for all categories of site. Regardless of the improved

Table 21.3 Grades of polymer modified bitumen emulsions (Roberts and Nicholls, 2008)

UK Classification	Minimum peak cohesion: J/cm^2	Vialit pendulum class (BS EN 13808:2013)
Intermediate	1.0	4
Premium	1.2	3
Super premium	1.4	2

properties of a particular binder, it is the performance of the aggregate–binder system and application mode that will dictate the success or otherwise of the dressing.

Polymer modified bitumen emulsion
This class of materials can be produced using polymer or latex modified bitumen. Modifying with latex is normally done prior to or after emulsification. These materials have similar advantages to those of polymer modified bitumen. In the UK in 2004 about 80% of surface dressings were with polymer modified emulsions (Roberts and Nicholls, 2008). They should meet the requirements of BS EN 13808:2013 (BSI, 2013). There are three grades based on the minimum levels of peak cohesion of residual bitumen according to Table 5.3.4 in Road Note 39 (Roberts and Nicholls, 2008). The test method for measuring peak cohesion is described in BS EN 13588:2008 (BSI, 2008a), and the grades are shown in Table 21.3. Emulsions having a higher level of cohesion are appropriate for higher stress situations on the road.

Similar to conventional bitumen emulsion, surface dressing using polymer modified emulsion is also vulnerable during the early life of the dressing before the emulsion has fully broken. The rate at which breaking occurs is influenced by a number of factors, including emulsion formulation, weather and mechanical agitation during rolling and trafficking. Under certain conditions, the emulsion goes through a state during curing when the bitumen droplets have agglomerated but not coalesced, and the dressing is extremely vulnerable. The timing of this condition is variable, and is generally worsened by the presence of high polymer contents, which can cause skinning and retard breaking of the emulsion. It is essential that the emulsion has broken fully before uncontrolled traffic is allowed on the dressing. This can be a serious disadvantage on heavily trafficked roads or where work is executed under lane rental contracts. Chemical after-treatment sprays have been shown to be effective in promoting breaking.

Fluxed or cut-back bitumen
Fluxed or cut-back bitumen for surface dressing is bitumen that has been diluted with kerosene, usually 70/100 or 160/220 pen. Cut-back bitumens have benefits because of their immediate cohesive grip of the road and chippings,

but generally the residual binder properties are more variable due to the differences in the evolution of the diluent and its absorption into the road surface.

Cut-back bitumen is typically sprayed at temperatures in the range 130–170°C (compared with 75–85°C for conventional bitumen emulsion). During spraying, between 10% and 15% of the kerosene evaporates, and a further 50% dissipates from the surface dressing in the first few years after application. At the spraying temperatures, wetting of the road surface and the applied chippings is achieved rapidly. Cut-back bitumen, once applied to the road surface, cools quickly, and it is important to ensure that the chippings are applied as soon as possible after spraying to ensure satisfactory wetting of the chippings.

The trend in most countries is to move away from the use of traditional cut-back/fluxed bitumens where possible.

Epoxy resin thermoset binder
This group of binders is bitumen extended epoxy resin binders used with high PSV aggregates such as calcined bauxite. Epoxy resin based binders are used in high performance systems, such as Spraygrip, that are fully capable of resisting the stresses imposed by traffic on the most difficult sites (e.g. roundabouts or approaches to traffic lights and designated accident black spots). It is these systems that are often described as 'Shellgrip' but are more correctly generically described as 'high friction surfaces'. Binders used in such systems are classified as 'thermosetting', as the epoxy resin components cause the binder to cure by chemical action and harden, and it is not subsequently softened by high ambient temperatures or by the spillage of fuel. The dressing thus acts as an effective seal against the ingress of oil and fuel, which is particularly important on roundabouts, where spillages regularly occur. Very little embedment of the chippings in the road surface takes place with this type of binder, and the integrity of the surface dressing is largely a function of the cohesive strength of the binder.

The extended life of this binder justifies the use of a durable aggregate with an exceptionally high PSV. The initial cost of these surface dressing systems is high compared with systems based on conventional binders. However, their exceptional wear resistant properties and the ability to maintain the highest levels of skid resistance throughout their service lives make them a cost effective solution for very difficult sites by significantly reducing the number of skidding accidents. Statistics suggest that, in the UK, the average value of prevention per serious accident is far less than a fatal accident (DfT, 2013).

21.1.2.6 Adhesion
Some aggregates adhere more rapidly to binder than others, and it is wise to avoid problems by testing the compatibility of the selected binder and

chippings. When using cut-back binder, this can be checked using the total water immersion test. In this test, clean 8/14 mm chippings are totally coated with cut-back bitumen. After 30 min of curing at ambient temperature, the chippings are immersed in demineralised water at 40°C for 3 h. After soaking, the percentage of binder that is retained on the chips is assessed visually. The extent to which the base of the chipping has been coated with the binder after 10 min is then assessed visually. If the coated area is considered to be below 70%, serious consideration should be given to lightly precoating the chippings, as it is much easier to obtain a bond between the binder film and a light coating of binder on the chippings than is the case with uncoated chippings.

The amount of binder required to coat surface dressing chippings lightly varies according to the type of binder and the size of the chipping but it is typically around 1% by weight. A thick coating of binder (e.g. as in the precoated chippings often used with hot rolled asphalt surface course) will make the chippings sticky and will inhibit free flow through chipping machines. When lightly coated chippings are used, this free flow property is critical to a satisfactory surface dressing. Minor pinholes in the coating of the chippings are not detrimental. It is particularly important that the amount of filler is less than the specified level (usually $\leq$1% passing a 0.063 mm sieve).

Once the chippings have been lightly coated, it is important that they are not allowed to come into contact with dust, as this will impair the bond between the chippings and the applied binder film on the road surface.

21.1.2.7 Geometry of site, altitude and latitude and local circumstances
Sections of road that include sharp bends induce increased traffic stresses on a surface dressing. Three categories of bend are identified in Road Note 39 (Roberts and Nicholls, 2008) – under 100 m radius, 100–250 m radius and over 250 m radius. Gradients that are steeper than 10% will affect the rate at which the binder should be applied due to the fact that on the ascent side of the road traffic, loading will be present for longer than on the descent side. Typically, there is a difference of 0.2 l/m^2 in the rate of binder application between the ascent and the descent sides.

21.1.2.8 Time of year
This is discussed in section 21.1.6.

21.1.3 Types of surface dressing
The original concept of a normal single layer surface dressing has been developed over the years, and there are now a number of techniques available that vary according to the number of layers of chippings and the number

of applications of binder. Each of these techniques has its own particular advantages and associated cost, and it is quite feasible that along the length of a road a variety of techniques would be used depending on the features of particular sections. Figure 21.9 (a composite of Figures 2.2.2 to 2.2.6 in

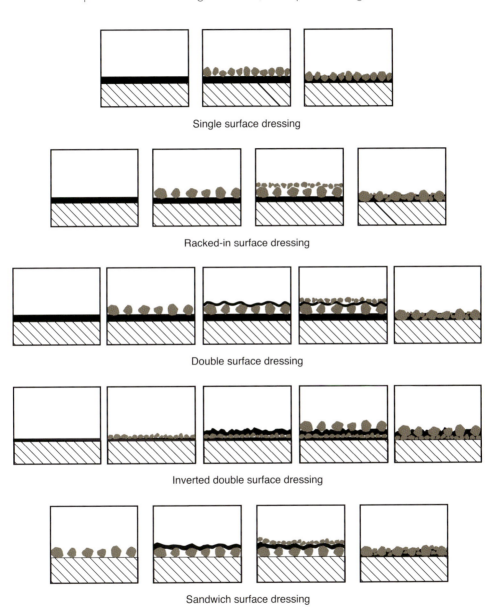

Single surface dressing

Racked-in surface dressing

Double surface dressing

Inverted double surface dressing

Sandwich surface dressing

Figure 21.9 Schematic representation of different types of surface dressing prior to embedment (Roberts and Nicholls, 2008). (Reproduced by courtesy of the Transport Research Laboratory (TRL))

Road Note 39 (Roberts and Nicholls, 2008)) diagrammatically depicts the different types of surface dressing that are available.

21.1.3.1 Pad coats

A pad coat consists of a single dressing with small chippings, and is applied to a road that has uneven surface hardness, possibly due to extensive patching by utilities. The reason for applying a pad coat is to produce a more uniform surface than would result by simply surface dressing the road. The chipping size for a pad coat is traditionally 2.8/6 mm, with a slight excess of chippings. Pad coats can also be used on very hard road surfaces (such as concrete or heavily chipped asphalts) to reduce the effective hardness of the surface, although using a racked-in system with a polymer modified binder is now the preferred option.

After laying and compaction by traffic, excess chippings should be removed before the road is opened to unrestricted traffic. Pad coats may be left for several months before the application of the main dressing, which may be either a single dressing or a racked-in system. Either system will embed in the pad coat and have immediate significant mechanical strength, reducing the risk of failure. All loose material should be swept from the surface of the road prior to the application of the final dressing.

21.1.3.2 Single surface dressing

The single surface dressing system is still the most widely used type of surface dressing. It consists of one application of binder followed by one application of single size chippings. Its advantages are that it has the least number of operations, uses the least amount of material and is sufficiently robust for minor roads and some main roads where excessive braking and acceleration are unlikely to occur and where speeds are unlikely to exceed 100 km/h.

21.1.3.3 Racked-in surface dressing

This type of dressing uses a single application of binder. A first application of 6.3/10 or 8/14 mm chippings (called the 'principal chippings') is spread at approximately 80–90% cover, which leaves a window of binder between the chippings. This window is filled with small chippings of size 2/4 mm (for a 10 mm principal chipping) or 2.8/6.3 mm (for a 14 mm principal chipping) to achieve high mechanical interlock (called a 'mosaic'). The rate of application of binder on this type of surface dressing is usually slightly higher than that which would be required for a single size dressing.

The initial surface texture and mechanical strength of a racked-in dressing are high. The configuration of the principal chippings is different from that which is achieved with a single dressing. The mosaic, referred to above, cannot be formed because the principal chippings are locked in place. In time, small chippings that are not in contact with binder are lost to the system without damaging vehicles or windscreens, resulting in increasing surface texture despite some coincident embedment of the larger chipping.

On a racked-in dressing, vehicle–tyre contact is principally on the large aggregate, with little contact on the small aggregate. Thus, the microtexture of the small aggregate is less critical, and consideration should be given to the use of cheaper lower PSV aggregates, such as limestone, for the small chippings.

The advantages of the racked-in system are

- virtual elimination of 'flying' 8/14 or 6.3/10 mm chippings, thereby reducing the incidence of broken car windscreens
- early stability of the dressing through good mechanical interlock
- good adhesion of larger size chippings
- a rugous surface texture with an initial texture depth exceeding 3.0 mm.

21.1.3.4 Double surface dressing
In this technique, two surface dressings are laid consecutively, the first consisting of 6.3/10 or 8/14 mm chippings and the second of 2.8/6.3 mm chippings. It is used on main roads as an alternative to the racked-in system to provide a mechanically strong dressing with a texture that is marginally less than a racked-in dressing, and therefore quieter. The method is also suitable for minor roads that have become very dry and lean in binder.

21.1.3.5 Inverted double surface dressing
In this system, two surface dressings are applied consecutively. The first uses 2.8/6.3 mm chippings, followed by a second application of 6.3/10 or 8/14 mm chippings. This technique is appropriate for use on minor concrete roads where chipping embedment does not occur and surface texture is not an important issue, or on sections of road that have been widely patched.

21.1.3.6 Sandwich surface dressing
Sandwich surface dressing is used in situations where the road surface condition is very rich in binder. The first layer of 8/14 or 6.3/10 mm chippings is spread onto the existing road surface before any binder is applied. The binder is then sprayed over these chippings, followed by an application of 2.8/6.3 mm chippings.

Sandwich dressings can be considered as double dressings in which the first binder film has already been applied. The degree of binder richness at the surface has to be sufficient to hold the first layer of chippings in place until the remainder of the operation has been completed. The sizes of the chippings must be chosen such that they are appropriate for the quantity of excess binder on the surface and the rate of application of the second coat of binder.

21.1.4 Surface dressing equipment

Improvements in the design of surface dressings and advances in binder technology and aggregate quality have been major contributory factors in improving the performance of surface dressings. Developments in application machinery in conjunction with these system developments have enabled the whole process to be improved in terms of the speed of operation, compliance with specifications and continued improvements in standards of safety.

21.1.4.1 Sprayers

In the UK, surface dressing sprayers must satisfy the requirements of BS 1707:1989 (BSI, 1989). This sets requirements for evenness of distribution of the spray bar, heat retention, capacity of heaters and other important issues.

In the UK, it has been traditional for spray bars to have swirl jets. In swirl jets, the binder passing through the jets swirls and forms a curtain in the shape of a cone. When using such jets it is advisable to use lower viscosity binders. However, the increased use of polymer modified binders has resulted in some sprayers being equipped with bars having slot jets. In slot jets, the binder travels through the jets in the shape of a flat fan. Slot jets can spray higher viscosity binders. On main roads, the need to undertake work quickly and with the minimum number of longitudinal joints has led to the use of expanding spray bars of widths up to 4.2 m. In the UK, the usual way of checking that the rate of spread across the bar is uniform and within 15% or better of the target rate has been by use of the depot tray test, which is detailed in BS 1707:1989 (BSI, 1989). The rate of spread is normally checked before the start of the surface dressing season.

Figure 21.10 shows a modern surface dressing sprayer. Programmable bitumen distributors are extremely sophisticated spray machines using two bars to achieve differential rates of spread across the width of the machine. Application rates can be changed for areas of patching, wheel tracks, binder rich fatty areas and binder lean areas. The rate of spread is automatically reduced by up to 30% in the wheel tracks, and can be increased between wheel tracks or in shaded areas. Both longitudinal and transverse binder distributions can be preprogrammed. Vehicle speed, binder temperature, spray bar width and application rates are controlled by an on-board computer.

Figure 21.10 A modern surface dressing sprayer

21.1.4.2 Chipping spreaders

Figure 21.11 shows an example of a self-propelled forward driven chipping spreader operating on a site in Estonia. These machines not only allow work to proceed rapidly but also spread the chippings more evenly and accurately at the required rate of spread. The chippings are released much closer to the

Figure 21.11 A modern chipping spreader. (Courtesy of the Phoenix Engineering Co Ltd.)

Figure 21.12 Combined spraying and chip spreading vehicle. (Courtesy of CTP Constructeur)

road surface, reducing the likelihood of chippings bouncing either elsewhere on the road surface or off the road surface altogether and exposing the binder.

The latest developments in self-propelled chipping spreaders include

- a four wheel drive for pulling heavy delivery vehicles up steep inclines without juddering
- a mechanism for breaking lumps of lightly coated chippings within the hopper
- methods for spraying additive onto the binder film
- incorporation of a pneumatic tyred roller to ensure that the chippings are pressed into the binder film at the earliest possible moment while it is still hot.

The latest equipment for spraying and chip spreading incorporates two functions within one vehicle. The chippings are spread, followed by the binder. This increases the efficiency of the operation. Figure 21.12 illustrates the latest equipment.

21.1.4.3 Rollers
Rubber covered steel drummed vibratory rollers (see Figure 21.13) are now regarded as the best means of establishing a close mosaic and ensuring

Figure 21.13 Rubber covered steel drummed vibratory roller. (Courtesy of Tarstone Surfacing Limited)

initial bond of the chipping to the binder film without crushing the carefully selected aggregate. Pneumatic tyred rollers are also used, and some rolling is still carried out using steel wheeled rollers. One of the risks of using steel wheeled rollers is that chippings can be crushed as a result of point loading. It should be noted that rolling is largely a preliminary process undertaken before the main stabilisation of the dressing by subsequent trafficking at slow speed.

21.1.4.4 Sweepers
It is necessary to remove surplus chippings before a surface dressing is opened to traffic travelling at uncontrolled speed. This is normally achieved with full width sweepers, which often have a full width sucking capability. Where sections of major roads have been closed to allow surface dressing to take place, sweepers are often used in echelon in order to ensure the rapid removal of surplus chippings. Sweeping that is properly organised and implemented will do much to alleviate the damage caused to surface dressings in their early life through chipping loss.

21.1.4.5 Traffic control and aftercare
Traffic control and aftercare are vital if the dressing is to be properly embedded. Ideally, when traffic is first allowed onto the new dressing, it should be behind a control vehicle that travels at a speed of about

15 km/h. If there are gaps in the traffic, it may be necessary to introduce additional control vehicles. The objectives should be to ensure that vehicles do not travel at more than 15 km/h on the new dressing and to prevent sharp braking and acceleration of these vehicles.

The length of time for which control is necessary will depend largely on the type of binder used and the prevailing weather conditions. It is likely to be longest where emulsions are being used and the weather conditions are humid. The use of wet or dusty chippings or the early onset of rain delays adhesion and necessitates a longer period of traffic control. Thus, ensuring that the chippings are clean, by washing if necessary, and that they comply with grading requirements will minimise the period necessary for traffic control and reduce disruption. Traffic passing slowly over a dressing immediately after completion creates a strong interlocking mosaic of chippings, a result only otherwise obtained by long periods of rolling.

21.1.5 *Types of surface dressing failure*
The majority of surface dressing failures fall into one of the following five categories

- loss of chippings immediately or soon after laying
- loss of chippings during the first winter
- loss of chippings in later years
- bleeding during the first hot summer
- fatting up in subsequent years.

The type and rate of application of the binder and size of chippings has a large influence on performance. Incorrect application rates are a frequent cause of premature failure. Every care should be taken to ensure that the selected application rates are maintained throughout the surface dressing operation.

Very early loss of chippings may be due to slow breaking in the case of emulsions or poor wetting in the case of cut-back bitumens. The use of cationic rapid setting emulsions with high binder contents (>70% binder) largely overcomes the former problem, and the use of adhesion agents and/or precoated chippings should negate the latter difficulty.

Many surface dressing failures only become apparent at the onset of the first prolonged frosts, when the binder is very stiff and brittle. These failures can normally be attributed to a combination of inadequate binder application rate, the use of too large a chipping or inadequate embedment of the chippings in the old road due to the surface being too hard and/or there being insufficient time between applying the dressing and the onset of cold weather. Inadequate binder application and/or the use of excessively large chippings will exacerbate this problem. Loss of chippings in the long term usually results from a combination of low durability binders and, again, poor embedment of chippings.

Bleeding may occur within a few weeks in dressings laid early in the season, or during the following summer in dressings that were laid late the previous year. It results from the use of binders that have a viscosity that is too low for high ambient temperatures, or from the use of binders that have high temperature susceptibility.

Fatting up is complex in nature and may result from one or more of the following factors

- application of excessive binder
- crushing of chippings
- absorption of dust by the binder.

Each of these factors is considered in more detail below.

21.1.5.1 Application of excessive binder
Gradual embedment of chippings in the road surface causes the relative rise of binder between the chippings to a point where the chippings disappear beneath the surface. This is largely dependent on the intensity of traffic, particularly the proportion of heavy commercial vehicles in the total traffic. The composition of the binder also affects the process. Cut-back bitumens that have a high solvent content can soften the old road surface and accelerate embedment.

21.1.5.2 Crushing of chippings
Most chippings will eventually split or crush under heavy traffic, leading to the loss from the dressing of many small fragments. The binder/chipping ratio therefore tends to increase, adding to the process of fatting up. A dilemma here is that many of the best aggregates for skid resistance often have a poor resistance to crushing.

21.1.5.3 Absorption of dust by the binder
Binders of high durability tend to absorb dust that falls on them. Soft binders can absorb large quantities of dust, thereby increasing the effective volume of the binder. This effect, coupled with chipping embedment, leads to the eventual loss of surface texture.

21.1.6 Surface dressing season
It is clear from the above that the success of a surface dressing operation may well depend on the weather conditions during and following the application of the dressing. Generally speaking, surface dressing is undertaken from

May to September in the UK. This period varies throughout Europe, and there are areas where it is much shorter. In the western isles of Scotland, the period during which ambient temperatures are high is relatively short, with the result that surface dressing using both conventional and polymer modified emulsions can only be carried out from mid-May until the end of June.

At the start of the surface dressing season, environmental factors are critical. The weather conditions during the early life of the dressing have a significant influence on performance. For example, high humidity can significantly delay the break of a surface dressing emulsion, and that, in turn, delays opening the road to traffic. Heavy rainfall immediately after a road has been surface dressed can have a detrimental effect on the bond between the binder and the aggregate, which can result in stripping of the binder from the aggregate.

In the first summer, it is important that some chipping embedment in the existing road surface occurs, as this will aid chipping retention during the first winter, when the binder is at its stiffest and most vulnerable to brittle failure. Therefore, the later in the season the dressing is applied the more likely it is that chipping loss will occur during the first winter, particularly where the traffic volumes are insufficient to promote initial embedment. This explains why, in areas such as the western isles of Scotland, the season finishes at the end of June rather than September.

21.1.7 Safety

Safety in dealing with bitumen is addressed in detail in Chapter 3. The following briefly highlights some of the potential hazards associated with the handling of surface dressing binders. However, this information is no substitute for the health and safety information available from individual suppliers, or the advice contained in the relevant codes of practice (Energy Institute, 2005; RSTA, 2014a).

21.1.7.1 Handling temperatures

Cut-back bitumens are commonly handled at temperatures that are well in excess of their flashpoints. The flashpoint is the temperature at which the vapour given off will burn in the presence of air and an ignition source. Accordingly, it is essential to exclude sources of ignition in the proximity of cut-back handling operations by displaying suitable safety notices.

Every operation should be carried out at a temperature that is as low as possible, to minimise the risks from burns, fumes, flammable atmosphere and fire. Such temperatures must always be less than the maximum values given in Table 21.4, which is part of Table B.1 in Part 11 of the *Bitumen Safety Code* (Energy Institute, 2005, 2011). Some bitumen emulsions are applied at

Table 21.4 Recommended handling temperatures for cut-back binders (Energy Institute, 2011)

Grade of cut-back binder: s	Temperature: °C		
	Minimum pumping[a]	Spraying[b]	Maximum safe handling[c]
50	65	150	160
100	70	160	170
200	80	170	180

[a] Based on a viscosity of 2 Pa·s.
[b] Based on a viscosity of 0.03 Pa·s.
[c] Based on generally satisfactory experience of the storage and handling of cut-back grades in contact with air. Subject to the avoidance of sources of ignition in the vicinity of tank vents and open air operations.

ambient temperatures but the majority of high bitumen content cationic emulsions are applied at temperatures up to 85°C.

21.1.7.2 Precautions, personal protective clothing and hygiene for personnel

All operatives should wear protective outer clothing when spraying emulsions or cut-back bitumen binders. This includes eye protection, clean overalls and impervious shoes and gloves to protect against splashes and avoid skin contact. In addition, operatives working near the spray bar should wear ori-nasal fume masks.

Although bitumen emulsions are applied at relatively low temperatures, the protection prescribed above is essential because emulsifiers are complex chemical compounds and prolonged contact may result in allergic reactions or other skin conditions. In addition, most surface dressing emulsions are highly acidic and, therefore, require the use of appropriate personal protective equipment (PPE). Barrier creams, applied to the skin prior to spraying, assist in subsequent cleaning should accidental contact occur but these are no substitute for gloves and other PPE. If any bitumen comes into contact with the skin, operatives should wash thoroughly as soon as is practicable and always before going to the toilet, eating or drinking.

21.2 Slurry surfacing/microsurfacing

Slurry surfacing (which includes slurry sealing) was first developed in the 1950s and was primarily used as a preservative treatment for airfield runways. The introduction of quick setting cationic emulsions and the development of polymer modified emulsions for thick film slurry applications, now termed microsurfacing, has broadened its use in highway maintenance. It was first developed in Germany in the late 1960s. Now, slurry surfacing using conventional bitumen emulsion is usually described as 'slurry seal'. Figure 21.14 illustrates slurry surfacing.

Figure 21.14 Slurry surfacing

The purpose of slurry surfacing is to

- seal surface cracks and voids against the ingress of air and water
- arrest disintegration and fretting of an existing surface
- fill minor surface depressions to provide a more regular surface
- give a paved area a more uniform surface appearance
- restore the skid resistance to a polished surface.

In addition, microsurfacing can also be used to fill areas where rutting has occurred.

21.2.1 Applications

21.2.1.1 Footway slurry

Usually, this is mechanically mixed and hand applied using a brush to provide surface texture. This protects the footway and is aesthetically pleasing. Modifiers can be used to give the material greater cohesive strength, and coloured slurries can be used to denote cycle tracks.

21.2.1.2 Thin carriageway slurries

These are usually laid 3 mm thick and are applied by a continuous flow machine. This type of slurry can be laid rapidly with minimal disruption to traffic. The texture depth that is achieved with this material is relatively low, so it is only suitable for sites carrying relatively slow moving traffic.

21.2.1.3 Thick carriageway slurries

These are normally single treatments using 0/6.3 mm aggregates blended with fast breaking polymer modified emulsions. They are more durable than thin slurries and are, therefore, suitable for more heavily trafficked locations.

21.2.1.4 Microsurfacing

Microsurfacing typically uses aggregates up to 6.3/10 mm and provides a new surface course. It can also be placed in two layers. The first layer regulates and re-profiles, including filling ruts, while the second layer provides a dense surface with reasonable texture.

21.2.1.5 Airfield slurries

Slurry surfacing has been used for runways and taxiways for many years. It is an ideal maintenance technique for airfields as the application is relatively rapid and the aggregates used are too small to constitute a significant foreign object damage hazard to jet engines.

Slurry surfacing only provides a thin veneer treatment to an existing paved area. Consequently, it does not add any significant strength to the road structure, nor will it be durable if laid on an inadequate substrate. Slurry seals are usually applied to lightly trafficked situations such as housing estate roads, footways, light airfield runways, car parks, small traffic highways etc. Microsurfacing can be used on heavily trafficked roads because it can provide higher texture depth.

Slurry surfacing can be laid on the surface dressing. It can reduce the noise generated by a surface dressing and inhibit the loss of chippings from a surface dressing. The system of surface dressing followed by slurry surfacing is described as 'cape seal'.

The use of fibre reinforcement is being developed, and it seems likely that, in the future, use of this technique will help overcome some of the problems of heavily cracked surfaces.

Pigmented slurry surfacings are available as proprietary products in a range of colours. The binder is synthesised, and the pigment and aggregate should be of the same colour. These surfacings are applied in a very thin layer, and are only suitable for pedestrian and lightly trafficked areas.

21.2.2 Materials

The materials used in slurry surfacing are bitumen emulsion, aggregate, mineral filler and water. A chemical additive may be used when the air temperature is much higher or lower than is usual for slurry surfacing application.

21.2.2.1 Aggregate

The aggregate normally used to manufacture asphalts can also be used in slurry surfacing. In China, basalt and diabase are commonly used in micro-surfacing. Other types of aggregate, such as granite, sandstone, limestone etc., can also be used.

The aggregate should meet the same requirements as would be specified for asphalts, but some characteristics warrant special attention.

- *Cleanliness*. The sand equivalent value indicates the aggregate cleanliness and can be obtained by tests described in BS EN 933-8:2012 (BSI, 2012) and ASTM D2419-09 (ASTM, 2009b). Aggregate usually contains some clays, which may require a shorter mix time for the slurry mixture and cause it to be difficult to spread. In addition, the presence of clay can result in long curing times and a material that has poor cohesion. Accordingly, aggregate cleanliness is very important. The higher the sand equivalent value, the cleaner the aggregate.
- *Surface activity*. Aggregates have a surface charge that plays a key role in dictating the nature of the reaction between aggregates and bitumen emulsions. If the surface charge is strong, the reaction is more intense. For freshly crushed aggregates, the amount of surface charge decreases with time, and much more quickly when the aggregates are wet. The type of mineral in the aggregate also has an effect on the surface charge.

The International Slurry Surfacing Association (ISSA) has defined three aggregate grades, type I, type II and type III, in a table in section 4.2.3 of ISSA A105 (ISSA, 2010a). The grading envelopes of these three categories are shown in Table 21.5. Materials having these gradings are widely used in the USA and China.

- *Type I*. The aggregate size is the smallest of the three gradings. It has a higher filler content so it will be capable of a degree of crack

Table 21.5 ISSA aggregate gradation envelopes

Sieve size: mm	Percentage by mass passing		
	Type I	Type II	Type III
9.5	–	100	100
4.75	100	90–100	70–90
2.36	90–100	65–90	45–70
1.18	65–90	45–70	28–50
0.6	40–65	30–50	19–34
0.3	25–42	18–30	12–25
0.15	15–30	10–21	7–18
0.075	10–20	5–15	5–15

penetration. It can be used to fill surface voids, cracks and correct minor to moderate surface defects. The surface texture with this grading is small because the aggregate is small, so it is normally used on lightly trafficked or low speed roads such as residential streets, rural roads etc. The main function of type I is to prevent pavements weathering.

- *Type II.* This aggregate grading can be used to fill surface voids and correct moderate surface distress, provide sealing and a durable surface. It can provide medium macrotexture and adapt to medium to heavily trafficked roads. It can also be placed on a flexible base or a stabilised base as a sealer prior to placement of the next layer.
- *Type III.* The aggregate size is the largest of the three gradings. As it contains more coarse particles, it can provide higher surface texture, so a surface paved with this material type has higher skid resistance. It can improve the surface durability because larger aggregate sizes increase the thickness of the pavement. This type of microsurfacing is appropriate for high speed or heavily trafficked roads. It can also be used for rut filling and minor surface reprofiling.

Normally, only types II and III are used for microsurfacing, with type I being used for slurry seal.

21.2.2.2 Emulsion

The emulsions used in slurry seal are traditionally slow setting and can be anionic or cationic. Slurry surfacing using quick setting cationic emulsion CQS-1h (ASTM D2397/D2397M-13 (ASTM, 2013)) can be opened to traffic more quickly than is the case with traditional slurry seals. Only polymer modified, quick setting cationic emulsions can be used in micro-surfacing. In China it is named BCR and defined in *Technical Guidelines for Micro-surfacing and Slurry Seal* (RIOH, 2006b). In Europe, the emulsions should comply with the requirements of BS EN 13808:2013 (BSI, 2013). Latex is mostly used as a polymer modifier for microsurfacing emulsions. Particular polymers can promote adhesion between emulsion and aggregate, and also increase the cohesion of the binder in service, with the result that the microsurfacing is not easily detached under heavy traffic.

21.2.2.3 Mineral filler

The mineral filler is usually Portland cement or hydrated lime, which is added to the mixture to control its consistency, setting and curing rate. Portland cement is normally used up to 2% of total aggregate for slurry seal and up to 3% for microsurfacing. The rate of hydrated lime is normally 0.3–0.5% of the weight of total aggregate. Mineral filler should be considered as part of the aggregate.

21.2.2.4 Water

Water used in slurry surfacing should be clean and free of salt, silt etc. The proportion of water influences the mix time and the consistency of the slurry mixture.

21.2.2.5 Additives

Additives are often used to extend the mix time of slurry mixtures when ambient temperatures are high. Sometimes, additives that accelerate curing are used when ambient temperatures are low. Additives are usually surfactants (substances that reduce surface tension).

21.2.3 Mixture design and specifications

The objectives of mixture design are to assess the compatibility of components and the consistency of the slurry mixture, and to determine the amount of each component. Appropriate consistency should result in the slurry surfacing having a uniform appearance and exhibiting suitable roughness.

21.2.3.1 Trial mixtures

Trial mixtures are used to determine

- the appropriate combination of aggregate and emulsion
- the type and content of mineral filler
- the approximate water content
- the type and content of additive, if appropriate.

In the trial, all preliminary materials are mixed, allowing the mix time to be assessed. During this process, the consistency, breaking and curing characteristics of the slurry mixture are checked and adjusted as necessary. The consistency of a slurry mixture with conventional slow setting emulsion can be measured using the applicable standard (e.g. ASTM D3910-11 (ASTM, 2011), BS EN 12274-3 (BSI, 2002b) or, in China, *Technical Guidelines for Micro-Surfacing and Slurry Seal* (RIOH, 2006b). The minimum mix time is 180 s for slurry seal and 120 s for microsurfacing. These minimum values are necessary to ensure that the slurry mixture can be successfully spread on the road surface.

21.2.3.2 Wet stripping

This test is used to determine the water resistance of the slurry mixture quickly and simply, and to screen the raw material. It is used in the USA and China. The test process may make reference to ISSA TB-114 (*Wet Stripping Test for Cured Slurry Seal Mix*) (ISSA, 2012).

21.2.3.3 Cohesion of mixture

The set and cure characteristics of slurries can be assessed by measuring the torsion resistance. For microsurfacing the torsion resistance should be up to

2.0 N m within 1 h. The test method is controlled by standards such as ASTM D6372-05 (ASTM, 2010b), BS EN 12274–4 (BSI, 2003a) or in China, *Technical Guidelines for Micro-Surfacing and Slurry Seal* (RIOH, 2006b).

21.2.3.4 Wet track abrasion

This test dictates the minimum binder content. The test measures the loss of weight of the slurry mixture specimen after having been soaked for 1 hour and abraded for a specified time by a rubber pad or wheel. The higher the binder content of the slurry mixture, the lower the abrasion loss of the specimen. If the soak time of the specimen is 6 days, the abrasion loss dictates the water resistance of the slurry mixture. The test method is controlled by standards such as ASTM D3910-11 (ASTM, 2011) and D6372-05 (ASTM, 2010b), BS EN 12274–5 (BSI, 2003b) or in China, *Technical Guidelines for Micro-Surfacing and Slurry Seal*. Figure 21.15 shows an example of a wet track abrasion machine.

When microsurfacing is used on heavily trafficked roads, two other tests are necessary.

■ *Loaded wheel test*. This test determines the maximum permissible binder content if bleeding of the surface under heavy traffic is to be avoided.

Figure 21.15 A wet track abrasion machine

Figure 21.16 A loaded wheel test machine

The test measures the amount of sand that adheres to the compacted microsurfacing specimen under a specified load. The higher the binder content, the greater is the amount of adhered sand. The test method is controlled by standards such as ASTM D6372-05 (ASTM, 2010b) or in China, *Technical Guidelines for Micro-Surfacing and Slurry Seal* (RIOH, 2006b). Figure 21.16 shows a loaded wheel test machine.

- *Aggregate filler–bitumen compatibility*. This test assesses the moisture sensitivity of the slurry mixture. Cylindrical specimens are soaked for 6 days after which the amount of water absorption is measured. Thereafter, the specimens are placed in the abrasion machine and the samples are abraded for 3 h. The loss of mass due to abrasion is measured. Finally, the specimens are placed in boiling water for 30 minutes after which they are weighed. The ratio of the mass of boiled specimens after being boiled to the mass before being boiled is an indication of the integrity of the samples. The detailed method may refer to ASTM D6372-05 (ASTM, 2010b) or, in China, *Technical Guidelines for Micro-Surfacing and Slurry Seal* (RIOH, 2006b). A further standard that may apply is the BSI test method set out in BS EN 12274-7 (BSI, 2006), which is similar to the ASTM D6372-05 and the method used in China.

21.2.3.5 Specifications

The International Slurry Surfacing Association has developed two guides and a series of technical bulletins. The two slurry surfacing guides are

- ISSA A105, *Recommended Performance Guidelines for Emulsified Asphalt Slurry Seal Surfaces* (ISSA, 2010a)

- ISSA A143, *Recommended Performance Guidelines for Polymer Modified Micro Surfacing* (ISSA, 2010b).

ASTM has two standard practice documents that address materials, design and construction of slurry seal and microsurfacing. They are

- ASTM D3910-11, Standard practices for design, testing, and construction of slurry seal (ASTM, 2011)
- ASTM D6372-05, Standard practice for design, testing, and construction of micro-surfacing (ASTM, 2010b).

CEN also developed two standards on slurry surfacing

- BS EN 12273, Slurry surfacing. Requirements (BSI, 2008b)
- BS EN 12274-1 to 12274-8, Slurry surfacing – Test methods, of which there are eight parts (BSI, 2002b, 2002c, 2002d, 2003a, 2003b, 2003c, 2005, 2006).

China published a guide on slurry surfacing in 2006 under the title *Technical Guidelines for Micro-Surfacing and Slurry Seal* (RIOH, 2006b).

21.2.4 Construction

The surface that is to be treated should be thoroughly cleaned before applying the slurry surfacing. Any loose material on the surface can cause the slurry surfacing to peel off. Any existing defects such as potholes, cracks etc. should be treated in advance of applying the slurry surfacing. A tack coat or bond coat may be required. Slurry surfacing cannot be applied at low temperatures or in wet conditions. Normally the air temperature should exceed 10°C.

A slurry/micropaver is a special continuous flow machine for laying slurry surfacing or microsurfacing. These machines contain an aggregate bin, a bitumen emulsion tank, a water tank, an additive tank, a mineral filler feed, a paddle mixer and a spreader box. All materials can be metered and transported into the paddle mixer, which feeds a spreader box towed behind the machine. These machines are capable of applying up to 8000 m^2/day. Figure 21.17 depicts a slurry/micropaver.

A 'scratch course' may be required for pavements having minor ruts or depressions. It is applied using the standard spreader box with a steel or stiff rubber screed. If the ruts are deep, they should be filled in advance using a rut filling box. Scratch courses and areas where ruts have been filled should be allowed to cure under traffic for at least 24 h before the surface course is applied. The surface course on flat pavements is applied using a standard spreader box. Figure 21.18 illustrates rut filling and paving a surface course.

Figure 21.17 Slurry/micropaver

Figure 21.18 Rut filling and paving a surface course

Figure 21.19 Rolling a slurry surface

It is very important that the consistency of a slurry mixture should be correct when it is applied. The water content has a significant effect on the consistency of the slurry. If the water content is too low, the slurry will be stiff, making uniform application difficult to achieve. A low water content may result in the slurry mixture breaking in the spreader box. If the slurry mixture contains too much water the surface texture of the finished surface will be reduced.

Light compaction with a pneumatic tyred roller may be required depending on the traffic condition. Rolling is applied as soon as the material has set sufficiently to support the compaction plant, and normally consists of two full passes. Figure 21.19 illustrates rolling a slurry surface.

21.2.5 Defects
Defects that must be avoided during installation include

- poor joints
- drag marks
- washboarding
- flush surface
- ravelling
- bleeding
- delamination.

Figure 21.20 Acceptable joints

The cause(s) may be due to shortcomings of operatives, the paving machine, raw materials, mixture design etc.

21.2.5.1 Poor joints
Poor joints can be caused by excessive overlap or areas left uncovered on longitudinal or transverse joints. Operatives should be careful to avoid these issues. Figure 21.20 shows acceptable joints.

21.2.5.2 Drag marks
Drag marks on a slurry surface are of two types. One is caused by oversized particles or broken slurry mixtures that have lodged under the strike off plate and left furrows on the surface as they are pulled along with the paving machine. The other is caused by the strike off plate having some small particles of aggregate or hardened slurry stuck to it. Small particles may stick to the strike off plate when the spreader box is lifted up off the road when paving has stopped. Another possibility is that some slurry has hardened and has stuck to the strike off plate during the application process of a slurry having insufficient mixing time. Figure 21.21 illustrates two types of drag marks. The drag marks caused by oversize particles can be repaired by judicious use of a squeegee.

Figure 21.21 Drag marks

21.2.5.3 Washboarding

Figure 21.22 illustrates the washboarding effect. It occurs when the slurry mixture is too stiff and the spreader box is set up incorrectly. This results in the spreader box bumping during installation and causing the washboarding effect. Washboarding can be eliminated by use of a secondary strike off plate. It can be avoided by adjusting the rubbers on the primary strike off plate, making the slurry mixture slower setting. In addition, adding weight to the back of the spreader box may also be effective.

Figure 21.22 Washboarding

Figure 21.23 Flush surface

21.2.5.4 Flush surface

If the slurry has too much water or emulsion, the total volume of water and emulsion is greater than the voids of the aggregate, and thus the aggregate is submerged by the surplus water. Figure 21.23 shows an example of a flush surface. Due to the emulsion being too hot, high ambient temperature or the use of dirty aggregate, it may be difficult to spread the slurry. In such circumstances operatives may be tempted to add more water, and this may result in a flush surface.

21.2.5.5 Ravelling

Ravelling occurs after opening to traffic. This can be due to the slurry curing more slowly than expected, possibly due to low ambient temperatures or high humidity. Either could result in the slurry having insufficient resistance to shear when exposed to traffic. In addition, aggregate that is not clean may also delay setting of the slurry. Ravelling may also occur as a result of excessive rainfall.

21.2.5.6 Bleeding

Bleeding may occur after reopening to traffic or during the first summer. The cause may be a high binder content, poor adhesion between aggregate and binder, or a softening point of the binder that is too low for heavy traffic.

Figure 21.24 Delamination of the slurry surface

21.2.5.7 Delamination

This is separation of the slurry and the layer underneath at the interface. Following any such debonding, the slurry may break up as a result of traffic. Figure 21.24 illustrates this defect. Such failures occur due to poor bond between the subsurface material and the slurry. This may arise because the subsurface is polluted by soil, oil etc., which has not been cleaned sufficiently before application of the slurry. Other reasons why this condition may occur are the existing surface is too old or has excessive ravelling, or the slurry itself may be too stiff. Finally, it has been suggested that the slurry and the sub-surface may contract at different rates during inordinately cold periods of weather, causing the bond between the two materials to break.

21.3 High friction surfaces

In 1965, a study carried out by the Greater London Council (GLC) suggested that 70% of all road accidents occurred at or within 15 m of conflict locations such as road junctions and pedestrian crossings (Hatherly and Lamb, 1970). This caused Les Hatherly, then the chief engineer of the GLC, to approach Shell, asking whether it would be possible to produce a high friction surfacing suitable for such sites. Shell developed a bitumen extended epoxy resin system and, following various successful road trials, the result was Shellgrip.

In the intervening period, alternative systems have been developed, including polyurethane resin, acrylic resin and thermoplastic rosin ester materials. A resin is generally a manufactured material or a natural secretion from certain plants, whereas rosin is a hard residue from the distillation of turpentine. A number of these systems are discussed below.

- *Bitumen extended epoxy resins*. These are produced by blending two components, one containing a resin and the other a hardener. When combined, they react chemically to form a very strong three dimensional structure. A purpose designed machine applies the material onto the road followed by the application of calcined bauxite. The curing time is usually between 2 and 4 h. The excess aggregate is swept off the surface and the road can be opened to traffic.
- *Epoxy resin*. Such systems are a comparatively recent development. They have the advantages that they set very rapidly and adhere well to most surfaces, including concrete.
- *Polyurethane resin*. These resins are normally two part or three part binder systems with good adhesion to most surfaces. Some systems require a primer to be applied prior to application of the resin. They are normally applied by hand.
- *Acrylic resins*. These are fast setting two component systems that adhere well to most surfaces. They are transparent, which makes them ideal for pigmentation.
- *Thermoplastic rosin esters*. These are normally blended with calcined bauxite and heated, and hand screeded onto the road surface. The material cures rapidly and may be pigmented.

The UK's national road construction and maintenance specification recognises the benefits of these materials and permits the use of systems covered by highway authorities product approval scheme (HAPAS) certification (Highways Agency, 2008).

Figure 21.25 shows a photograph of a cold-applied high friction surface.

21.4 Grouted macadams

Grouted macadams have evolved from a specialist treatment to a mainstream surfacing option. They are proprietary products that do not require CE marking. In the UK, the Department for Transport is seeking to include grouted macadams in the new highways maintenance efficiency programme (HMEP – an initiative funded by the Department for Transport to support the highways sector to transform highway services) local roads standard maintenance specification (HMEP, 2014) in recognition of their growing popularity. Currently, agrément certification is used to distinguish between different variants and to ensure consistency of manufacture and installation.

Figure 21.25 A machine applied, cold, high friction surface. (Courtesy of EUROVIA Specialist Treatments)

Grouted macadams offer an innovative solution to problems with which conventional surfacing materials struggle to cope.

21.4.1 Development

Grouted macadams originated in France in the 1950s. The materials were designed from a simple idea: to alter the characteristics of a new running surface to meet a specific need. Their use in Europe is widespread but, although these products have had a presence in the UK since the 1960s, their use has been relatively limited there. However, that is now changing, and grouted macadams are beginning to emerge as a serious alternative to conventional materials.

21.4.2 Laying

There are two forms of grouted macadam

- cementitious grouted macadam
- asphaltic grouted macadam.

These are named according to the grout used within them, which is either cement or bitumen based. The cement grout adds strength and resistance to contaminants, whereas the bitumen grout provides an impervious seal.

The grouted macadam is laid using a two stage process. A design mix open graded bituminous 'receiving course' is laid by a conventional paving machine, generally to a depth of 40 mm, and compacted. The specified grout is then applied to the surface of the material and allowed to percolate into the voids to the required depth, providing the surface course with its chosen characteristics.

21.4.3 Cementitious grouted macadam

Cementitious grouted macadams are used in areas of high stress or where fuel, salt and leachate contamination is an issue. They are effectively a hybrid between asphalt and concrete. The most well known of these are Hardicrete, which was first awarded an agrément certificate in 1988 (certificate number 88/1969), and Hardipave, which was awarded a HAPAS certificate in 2006 (certificate number 06/H120).

Their use is widespread throughout the public and private sectors

- bus lanes, laybys and stations
- industrial yards
- coastal promenades
- roundabouts
- junctions
- waste and amenity
- ports and container terminals
- airfields.

The laying of cementitious grouted macadam is shown in Figure 21.26.

21.4.4 Asphaltic grouted macadam

Asphaltic grouted macadams use a bituminous grout instead of cement in order to seal the surface and improve flexibility. The sealed surface prevents the ingress of water, while the grout lying beneath the surface of the aggregate gives the product an increased bitumen content, which has a significant effect in slowing the rate of oxidation (discussed in section 20.1.1), and therefore extends product life. The earliest asphaltic grouted macadam, Milepave, gained HAPAS accreditation in 2006. Asphaltic grouted macadams have proved very successful when surfacing on top of concrete, by protecting the structure beneath. Preventing water ingress, they are designed to prevent erosion of the substrate and subsequent slab movement or breakage. In addition, the improved flexibility helps to delay crack formation or subsequent deterioration due to cracking.

The laying of asphaltic grouted macadam is shown in Figure 21.27.

Figure 21.26 Laying cementitious grouted macadam: (a) laying the receiving course; (b) application of cementitious grout; (c) cross section through the finished surface. (Courtesy of Miles Macadam Ltd.)

Figure 21.27 Laying asphaltic grouted macadam: (a) laying the receiving course; (b) application of asphalt grout; (c) finished overlay on concrete; (d) finished carriageway. (Courtesy of Miles Macadam Ltd.)

639

References

Asphalt Institute (2009) Asphalt in pavement preservation and maintenance. Asphalt Institute, Lexington, KY, USA, MS-16, 4th edn.

ASTM (American Society for Testing and Materials) (2009a) ASTM E1911-09ae1. Standard test method for measuring paved surface frictional properties using the dynamic friction tester. ASTM, West Conshohocken, PA, USA.

ASTM (2009b) ASTM D2419-09. Standard test method for sand equivalent value of soils and fine aggregate. ASTM, West Conshohocken, PA, USA.

ASTM (2010a) ASTM D5360-09. Standard practice for design and construction of bituminous surface treatments. ASTM, West Conshohocken, PA, USA.

ASTM (2010b) ASTM D6372-05 (Reapproved 2010). Standard practice for design, testing, and construction of micro-surfacing. ASTM, West Conshohocken, PA, USA.

ASTM (2011) ASTM D3910-11. Standard practices for design, testing, and construction of slurry seal. ASTM, West Conshohocken, PA, USA.

ASTM (2013) ASTM D2397/D2397M-13. Standard specification for cationic emulsified asphalt. ASTM, West Conshohocken, PA, USA.

Austroads (2006) *Update of the Austroads Sprayed Seal Design Method.* Austroads Inc., Sydney, Australia, AP-T60/06.

Austroads (2008) *Guide to Pavement Technology.* Part 4F: *Bituminous Binders.* Austroads Inc., Sydney, Australia, AGPT04F-08.

Austroads (2009a) *Guide to Pavement Technology.* Part 3: *Pavement Surfacings.* Austroads Inc., Sydney, Australia, AGPT03-09.

Austroads (2009b) *Guide to Pavement Technology.* Part 4K: *Seals.* Austroads Inc., Sydney, Australia, AGPT04K-09.

Austroads (2009c) *Guide to Pavement Technology.* Part 8: *Pavement Construction.* Austroads Inc., Sydney, Australia, AGPT08-09.

Austroads (2010) *Ball Penetration Test.* Austroads Inc., Sydney, Australia, AG:PT/T251.

Austroads (2013) *Update of Double/Double Design for Austroads Sprayed Seal Design Method.* Austroads Inc., Sydney, Australia, AP-T236-13.

BSI (British Standards Institution) (1989) BS 1707:1989. Specification for hot binder distributors for road surface dressing. BSI, London, UK.

BSI (2002a) BS EN 13043:2002. Aggregates for bituminous mixtures and surface treatments for roads, airfields and other trafficked areas. BSI, London, UK.

BSI (2002b) BS EN 12274-3:2002. Slurry surfacing. Test methods. Consistency. BSI, London, UK.

BSI (2002c) BS EN 12274-1:2002. Slurry surfacing. Test methods. Sampling for binder extraction. BSI, London, UK.

BSI (2002d) BS EN 12274-6:2002. Slurry surfacing. Test methods. Rate of application. BSI, London, UK.

BSI (2003a) BS EN 12274-4:2003. Slurry surfacing. Test methods. Determination of cohesion of the mix. BSI, London, UK.

BSI (2003b) BS EN 12274-5:2003. Slurry surfacing. Test methods. Determination of wearing. BSI, London, UK.

BSI (2003c) BS EN 12274-2:2003. Slurry surfacing. Test methods. Determination of residual binder content. BSI, London, UK.

BSI (2004) BS 598-112:2004. Sampling and examination of bituminous mixtures for roads and other paved areas – Part 112: Method for the use of road surface hardness probe. BSI, London, UK.

BSI (2005) BS EN 12274-8:2005. Slurry surfacing. Test methods. Visual assessment of defects. BSI, London, UK.

BSI (2006) BS EN 12274-7:2005. Slurry surfacing. Test methods. Shaking abrasion test. BSI, London, UK.

BSI (2008a) BS EN 13588:2008. Bitumen and bituminous binders. Determination of cohesion of bituminous binders with pendulum test. BSI, London, UK.

BSI (2008b) BS EN 12273:2008. Slurry surfacing. Requirements. BSI, London, UK.

BSI (2009) BS EN 1097-8:2009. Tests for mechanical and physical properties of aggregates. Determination of the polished stone value. BSI, London, UK.

BSI (2010) BS EN 13036-1:2010. Road and airfield surface characteristics. Test methods. Measurement of pavement surface macrotexture depth using a volumetric patch technique. BSI, London, UK.

BSI (2012) BS EN 933-8:2012. Tests for geometrical properties of aggregates. Part 8: Assessment of fines – Sand equivalent test. BSI, London, UK.

BSI (2013) BS EN 13808:2013. Bitumen and bituminous binders. Framework for specifying cationic bituminous emulsions. BSI, London, UK.

DfT (Department for Transport) (2013) *Reported Road Casualties Great Britain: 2012 Annual Report.* DfT, London, UK, p. 230.

DfT (2014) *Statistical Release – Road Conditions in England: 2013.* DfT, London, UK, Table RDC0320.

Dunford A (2010) *GripTester Trial – October 2009.* TRL, Crowthorne, UK, PPR497.

Energy Institute (2005) *Bitumen Safety Code, Model Code of Safe Practice in the Petroleum Industry*, 4th edn. Energy Institute, London, UK, Part 11.

Energy Institute (2011) *Tables 1.1–1.6 Product Specifications.* See https://www.energyinst.org/_uploads/documents/BitumenTables2011version1.pdf (accessed 01/10/2014).

Hatherly LW and Lamb DR (1970) *Accident Prevention in London by Road Surface Improvements.* Shell International Petroleum Company, London, UK.

Highways Agency (2008) *Manual of Contract Documents for Highway Works.* Volume 1, *Specification for Highway Works.* Series 0900, *Road Pavements – Bituminous Bound Materials.* The Stationery Office, London, UK, cl 924.

Highways Agency, Scottish Executive, Welsh Assembly Government and Department for Regional Development Northern Ireland (2006) *Design Manual for Roads and Bridges.* Volume 7, *Pavement Design and Maintenance.* Section 5, *Surfacing and Surfacing Materials*, Part 1, *Surfacing Materials for New and Maintenance Construction.* The Stationery Office, London, UK, HD 36/06.

HMEP (Highways Maintenance Efficiency Programme) (2014) *Guidance on the Standard Specification and Standard Details for Local Highway Maintenance.* See http://www.highwaysefficiency.org.uk/efficiency-resources/procurement-contracting-and-standardisation/guidance-on-the-standard-specification-and-standard-detailsfor-local-highway-maintenance.html (accessed 01/10/2014).

Holtrop W (2008) Sprayed sealing practice in Australia. *1st International Sprayed Sealing Conference.* Australian Road Research Board, Adelaide, Australia.

ISSA (International Slurry Surfacing Association) (2010a) *Recommended Performance Guidelines for Emulsified Asphalt Slurry Seal Surfaces*. ISSA, Annapolis, MD, USA, A105.

ISSA (2010b) *Recommended Performance Guidelines for Polymer Modified Micro Surfacing*. ISSA, Annapolis, MD, USA, A143.

ISSA (2012) *Wet Stripping Test for Cured Slurry Seal Mix*. ISSA, Annapolis, MD, USA, TB-114.

RIOH (Research Institute of Highway, Ministry of Transport) (2004) *Technical Specifications for Construction of Highway Asphalt Pavements*. China Communications Press, Beijing, China, JTG F40-2004.

RIOH (2006a) JTG D D50-2006. *Specifications for Design of Highway Asphalt Pavement*. China Communications Press, Beijing, China.

RIOH (2006b) *Technical Guidelines for Micro-Surfacing and Slurry Seal*. China Communications Press, Beijing, China.

Roberts C and Nicholls JC (2008) *Design Guide for Road Surface Dressing*. TRL, Crowthorne, UK, Road Note 39, 6th edn.

Roe PG, Tubey LW and West G (1988) *Surface Texture Measurements on Some British Roads*. TRL, Crowthorne, UK, Research Report 143.

Roe PG, Webster DC and West G (1991) *The Relation Between the Surface Texture of Roads and Accidents*. TRL, Crowthorne, UK, Research Report 296.

RSTA (Road Surface Treatments Association) (2014a) *RSTA Code of Practice for Surface Dressing 2014*, Parts 1–9. RSTA, Wolverhampton, UK.

RSTA (2014b) *Preparing Roads for Surface Dressing*. RSTA, Wolverhampton, UK, Guidance Note 2014.

Salt GF and Szatkowski WS (1973) *A Guide to Levels of Skidding Resistance for Roads*. TRL, Crowthorne, UK, Laboratory Report 510, Table 4.

Standards Australia (1996) AS 1160-1996. Bituminous emulsions for the construction and maintenance of pavements. Standards Australia, Sydney, Australia.

TxDOT (Texas Department of Transportation) (2010) *Seal Coat and Surface Treatment Manual*. TxDOT, Austin, TX, USA, Table 4-2.

Wood J, Janisch W and Gaillard S (2006) *Minnesota Seal Coat Handbook 2006*. Minnesota Department of Transportation, St. Paul, MN, USA, Table 4.6.

Chapter 22

Alternative asphalts

Frank Beer
Regional Technology
Manager, Shell
Deutschland Oil GmbH.
Germany

Jacques Colange
Global Technology
Development Executive, Société
des Pétroles Shell. France

Jayne Davies
Global Technology
Development Executive,
Shell International Petroleum
Co. Ltd. UK

**Dr Punith Veeralinga
Shivaprasad**
Application Advisor,
Shell Oil Products US. USA

Some asphalts are different from those commonly employed on highway construction or maintenance. These mixtures do not always comply with particular specifications but nevertheless are useful in helping to solve specific problems. Examples of these alternative asphalts discussed in this chapter are

- foamed asphalts
- warm mix asphalts
- asphalts made using Shell Thiopave® (also referred to as 'Thiopave asphalt'. THIOPAVE is a registered trade mark owned by Shell Brands International AG)
- recycling of asphalt pavements
- modification of bitumen by the addition of rubber.

In this chapter, these asphalts are described and explained.

22.1 Asphalts produced using foamed bitumen

Asphalts are extensively used for pavement construction, primarily due to their pressure dissipation and waterproofing properties. In order to mix bitumen with road building aggregates, the viscosity of the bitumen has to be reduced significantly. Traditionally, this was achieved by heating the bitumen and mixing it with heated aggregates, which act like a binding agent to produce an asphalt. Other methods of reducing the bitumen viscosity include dissolving the bitumen in solvents, the use of emulsifying agents and the use of foamed bitumen.

Foaming of bitumen is a means of temporarily reducing the binder viscosity and increasing the binder volume of a bitumen. This technique was first

The Shell Bitumen Handbook, Sixth Edition
ISBN 978-0-7277-5837-8
Shell International Petroleum Company Ltd: All rights reserved
http://dx.doi.org/10.1680/tsbh.58378.643

introduced at the end of the 1950s. Professor Csanyi at the Engineering Experiment Station of Iowa State University studied the potential of foamed bitumen in cold asphalt applications for full depth reclamation projects (Csanyi, 1957). At that time, steam was injected under pressure into hot bitumen through specially designed nozzles, yielding bitumen in the form of a foam. The process reduced the viscosity and surface energy in the foamed bitumen, enabling intimate coating when mixed with moist aggregate at ambient temperatures (Csanyi, 1957). This process was subsequently found to be impractical due to the complexity of the equipment and the difficulties in accurately metering the steam. Mobil Oil refined the foaming technology in the 1960s, and developed an expansion chamber in which cold water containing 1–5% by weight of bitumen is injected under pressure into hot bitumen to produce foam (Bowering and Martin, 1976; Maccarrone *et al.*, 1994, 1995).

Foamed bitumen is typically produced by adding small amounts of water (approximately 2–3% by weight of binder) to hot bitumen. This typically involves a process in which water is injected into the hot bitumen in an expansion chamber, resulting in foaming. The liquid bitumen thus expands by a factor of 15–20 times its original volume, causing foam, as shown in Figure 22.1 (Wirtgen Group, 2002).

The basis of the bitumen foaming process is as follows. The moment that a cold water droplet (at ambient temperature) makes contact with the bitumen at 170–180°C, the following chain of events occurs (Jenkins, 2000).

■ The bitumen exchanges energy with the surface of the water droplet, heating the droplet to a temperature of 100°C and cooling the bitumen.

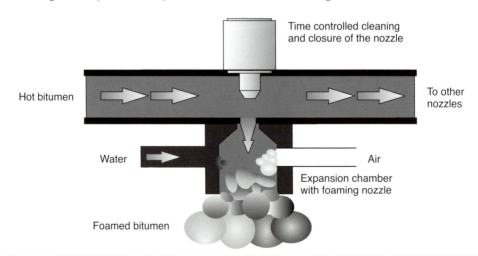

Figure 22.1 Mechanism of the bitumen foaming process

- The transferred energy of the bitumen exceeds the latent heat of steam, resulting in explosive expansion and the generation of steam. Steam bubbles are forced into the continuous phase of bitumen under pressure, in the expansion chamber.
- With emission from the spray nozzle, the encapsulated steam expands until a thin film of slightly cooler bitumen holds the bubble intact through its surface tension.
- During expansion, the surface tension of the bitumen film counteracts the ever diminishing steam pressure until a state of equilibrium is reached.
- Due to the low thermal conductivity of bitumen and water, the bubble can remain stable for a period of time, usually measurable in seconds.

This process occurs for a multitude of bitumen bubbles that occur, resulting in the material described as 'foamed bitumen'. The intensity and effectiveness of the foaming process is governed by the details of the operation, and is related to the applicable physical conditions such as the applied air pressure and bitumen temperature. The rheology of the foamed bitumen can vary and is difficult to predict. The physical properties of the foamed bitumen depend on various factors, including binder type, grade and modification, amount of water used, type of foaming technology used and temperature.

The physical properties of the bitumen are temporarily altered when the injected water comes into contact with the hot bitumen (at about 160–170°C) resulting in vapours that are trapped in thousands of tiny bitumen globules (Fu, 2009). The introduction of water can be effected by using a direct water injection system through a foaming nozzle, by adding water-bearing chemical additives such as zeolites, or by using moist aggregate (Chowdhury and Button, 2008; D'Angelo et al., 2008). The foam dissipates in less than a minute, and the bitumen retains its original physical properties. When the colloidal mass (the bitumen) cools at the ambient temperature, the steam in the bubbles condenses, resulting in bubbles that collapse and cause the foam eventually to 'break or decay' (Jenkins et al., 1999a). The collapse of foamed bitumen with time is referred to as 'decay', also described as 'breaking'.

22.1.1 Foam characterisation
The rapid evaporation of water produces a very large volume of foam that slowly reduces in volume with time. The foamed bitumens used in base layer stabilisation applications were typically characterised using the following three parameters (Jenkins, 2000)

- expansion ratio
- half life (lifetime)
- foam index.

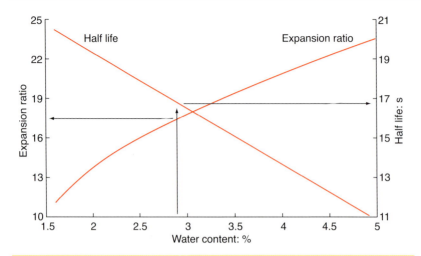

Figure 22.2 Example of the relationship between expansion ratio/water content and half life/water content of a bitumen

However, there is no test method available to measure these three parameters precisely. The expansion ratio of the foam is defined as the ratio between the maximum volume achieved during the foamed state and the final volume of the binder once the foam has dispersed (Wirtgen Group, 1998). The half life is defined as the time between the moment the foam achieves maximum volume and the time when it disperses to half of the maximum volume. Half life is measured in seconds, and it depends strongly on the binder/mixture temperature (i.e. higher temperature mixtures have a shorter half life) (Jenkins et al., 1999b). An example of the relationship between expansion ratio and water content, and also half life and water content, is shown in Figure 22.2.

Typically, for foamed bitumen the expansion ratio increases with foaming temperature and water content. In addition, the half life decreases with increasing foaming temperature and water content (Kim and Lee, 2006). Finally, a decrease in the half life causes the thinning of the foam film, as well as a reduction in the viscosity. As the viscosity decreases, the surface tension of the bitumen film decreases as well, and the steam pressure within the bubbles exceeds the surface tension of the bitumen and water bubbles, resulting in subsequent collapse of the bubbles (Hande, 2013).

The foam index is a measure of the area under the breaking or decay curve (i.e. the change in expansion ratio with time). It reflects the stored energy in the foam for a particular bitumen when foamed at a known temperature with water at a predetermined application rate (Jenkins, 2000). The decay curve defines the rate at which the foam collapses, and gives an indication of the time available for mixing. An important factor in foaming is the design of the nozzle and the injection pressure necessary to obtain an effective water droplet spray in contact with the hot bitumen. The foaming characteristics of specific bitumens are further influenced by the following.

- The temperature of the bitumen – for most bitumens, the foaming characteristics improve with higher temperatures.
- The expansion ratio – as more water is added, the expansion ratio increases, while the half life decreases. The water helps in creating the foam but the foam can collapse quickly due to the rapid escape of steam.
- Addition of compounds – some compounds (e.g. some silicones) can be effective anti-foaming agents. In contrast, other chemical compounds can increase the half life of the foam from seconds to minutes.

22.1.2 Foamed asphalts
'Foamed asphalt' refers to a mixture of mineral aggregates and foamed bitumen. Due to the foaming process, the surface area of the bitumen increases, resulting in a considerable reduction in its viscosity. This reduction in binder viscosity, together with the accompanying increase in volume, means that foamed bitumen effectively coats cold or wet aggregates. This contrasts with traditional asphalts, which can only be manufactured with hot and dry aggregates. Foamed asphalts can be produced in situ or in an asphalt production plant. Such plants can be fixed, free standing or mobile.

22.1.3 Mixing of foamed asphalt
In this section, typical examples of the use of foamed bitumen in the paving industry are discussed.

22.1.3.1 Cold mix asphalt
Foamed asphalts produced using cold or moist aggregates are widely used for stabilising granular materials. Foamed asphalt stabilisation can be effected through cold in-place recycling of traditional asphalts or by way of cold central plant recycling (Bonvallet, 2001; Brosseaud et al., 1997; Khweir et al., 2001). Foamed asphalt stabilised base is made up of combinations of recycled materials, including reclaimed asphalt pavement (RAP), recycled concrete and/or graded aggregate base with a foamed asphalt binder, to produce a partially stabilised base material. The foamed asphalt content is usually too low to coat all the aggregate particles fully contrary to what happens in conventional asphalts. During the mixing process, the bitumen 'bubbles' burst into tiny 'patches' that adhere, mainly, to the fine aggregate fraction. The resulting mortar of fine aggregate and bitumen binds the coarse aggregate particles after compaction, leading to increased

cohesion and stiffness of the aggregate assemblage (Schwartz and Khosravifar, 2013). The foaming characteristics, the moisture content and aggregate grading, together with adequate compaction, are key factors in maximising the performance of aggregates stabilised with foamed bitumen.

Mixtures with a low fines content will not mix well with foamed bitumen because the bitumen will not disperse properly. The expansion ratio of the foam for producing such cold mixtures should be high enough to coat as much of the aggregate as possible after the introduction of the foam into the mixer. The type and size of the mixer is important in ensuring the production of a homogeneous asphalt. The half life of the foam determines the distribution of the foam into the aggregate in these cold mixtures (Schwartz and Khosravifar, 2013). If the half life is too short, the bitumen reverts to a high viscosity liquid, with the risk of lumps of bitumen or pieces of mortar/mastic being formed (Kim et al., 2007). A longer half life may produce a mixture that has good workability when using cold or moist aggregates. For such mixtures, the lifetime of the bitumen foam can range from 20 s up to a few minutes. Long half lives are obtained by introducing additives into the mixture. Optimum values of expansion ratio and foam half life are dictated by the type of application. A key determinant is the residence time in the paddle mixer.

22.1.3.2 Warm mix asphalt

Warm mix asphalt technology originated in Europe. Warm mix asphalts are produced and placed at lower temperatures than is the case with conventional asphalts (Koenders et al., 2000). The use of foamed bitumen in asphalts manufactured using warm mix technology to construct pavements has increased (Prowell et al., 2007). However, there is no specification or test method to evaluate the quality of foam generated by different foaming techniques. Recently, an automated repeatable test procedure called the 'asphalt foam collapse test method' has been developed (Hande, 2013). This test can be used to measure the reduction in the height of foam bitumen over time by way of an image analysis process. The system permits calculation of parameters of the foam binder quality. In addition, two new parameters, bubble size distribution and surface area index, are introduced as quality parameters. It is reported that these quality parameters are promising candidates for evaluating the workability and the effectiveness of coatings, as well as the performance of asphalt pavements. In order to produce foamed asphalts, the bitumen has to be incorporated in the aggregates while still in its foamed state. The binder contents typically used are based on the mixture design, and are determined as the percentage (by weight) required for the mixture to have optimum properties. Some plant modifications are

(a)

(b)

(c)

Figure 22.3 Mixing process for producing foamed asphalts in the laboratory

necessary to incorporate these technologies. In addition, the mixture design process needs to be altered to include the steps of water injection and foaming action. Furthermore, the laboratory foam device can be used to determine the 'foamability' of the bitumen (Cazcliu *et al.*, 2008). Using this device, the asphalt mixture design can be carried out in the laboratory. The procedure adopted for the foaming process using laboratory foaming devices is very similar to that adopted in a large scale production plant facility. In the laboratory, foamed asphalts can be produced as shown in Figure 22.3. Foamed bitumen is poured over warm aggregates at a temperature of 100–135°C in the mixing bucket (Figure 22.3(a)), and the warm aggregates are mixed with the foamed bitumen using a bucket mixer (Figure 22.3(b)). The foamed warm mix asphalt produced after the mixing process is shown in Figure 22.3(c).

A major technical concern in using this technology is that, in mixtures produced at reduced temperatures, the moisture may not escape completely and may thus be trapped inside the mixture after compaction. Whether or not this occurs depends on the prevailing environmental conditions. The trapped moisture in the asphalt can cause decreased adhesion between the bitumen and the aggregates, and lower the cohesion of the bitumen itself, resulting in stripping, ravelling and other forms of pavement distress (Caro et al., 2008; Kiggundu and Roberts, 1988). It is reported that the rate of moisture dissipation in US high performance grade asphalts manufactured with stiffer binders was found to be slower than is the case in US low performance grade asphalts made with soft binders. Also, the size distribution of moisture bubbles in the binders varies with different bitumen grades (Kutay and Ozturk, 2012a). Furthermore, the results have revealed that the water content and air pressure have a significant influence on the expansion ratio, the half life and the flow index. Low water content and low pressure would result in foams with smaller bubbles than those in foams made with high water content and pressure. This would eventually affect the aggregate coating. Traditional mixing procedures that are adequate for the production of hot mixed and/or cold mixed asphalts have to be adapted in order to accommodate the different characteristics of foamed bitumen. Key parameters to consider are as follows.

- In mixing with hot bitumen or hot bitumen emulsions, the actual mixing has to ensure good coverage, resulting in increased mixing time and energy demand.
- Foam properties and mixing times must be set to ensure a homogeneous mixture.
- Good distribution of the bitumen in the mixture is obtained very quickly as a result of high expansion of the bitumen foam.
- Volume expansion must be such that the bitumen is distributed evenly in the asphalt.

22.1.4 Application and use

The use of foamed bitumen in base stabilisation and in situ recycling in base layers has attracted significant attention throughout the world (Schwartz and Khosravifar, 2013). The process is adaptable, allowing the treatment of virgin materials and recycled aggregates without heating and drying, thus saving energy compared with traditional hot mix production. Mixed material can be stockpiled for a period before being used, reducing the amount of waste. Although foamed bitumen is suitable for stabilisation and recycling work, the resultant mixtures still cannot be objectively regarded as being equivalent to traditional asphalts. In general terms, most cold products are currently inferior to traditional hot asphalt products.

If this is to change, three factors will influence the possibility of a move from traditional asphalts to cold mixed mixtures

- cost
- market demand
- performance.

In 2012, the US National Asphalt Pavement Association (NAPA) stated that the plant foaming of bitumen using water injection systems is a technology that enjoys widespread use in the USA for producing warm mix asphalt mixtures, with more than 88% of the market share as compared with other warm mix asphalt technologies used in various road projects (NAPA, 2012). The foaming process has gained popularity among asphalt producers due to the improved workability and compactibility of asphalts produced at lower temperatures. This technology can be used to add a new coating of bituminous binder to secondary aggregates, including RAP, creating a fresh asphalt and reducing demand for the extraction of virgin aggregate. When warm mix asphalt is used, there is a potential for the asphalt producer to save money on fuel costs, and the contractor is able to obtain better compaction (typically 1–2% higher in situ density) (Dale, 2012). Improved density translates into longer pavement life. Many paving contractors have reported that warm mix asphalts require lower compactive effort (fewer rollers/passes) than would be the case with traditional asphalts (Dale, 2012).

22.2 Warm mix asphalts

Warm mix asphalt technologies allow asphalt manufacturers to reduce production temperatures. In addition, warm mix asphalts can be compacted effectively at lower temperatures; typically, temperature reductions of 10–40°C have been reported (Bennert *et al.*, 2011; Estakhri *et al.*, 2010; Jenkins *et al.*, 1999b; Romier *et al.*, 2006). Such reductions in temperatures have the obvious benefit of reducing fuel consumption and decreasing the production of greenhouse gas emissions. World Bank estimates suggest that for every 10°C decrease in production temperature, savings of nearly 1 litre of fuel oil and 1 kg of carbon dioxide emissions are realised per tonne of asphalt produced (Krambeck, 2009). In addition, the engineering benefits include better compaction, the ability to haul over longer distances and extension of the paving season due to being able to lay asphalts at lower temperatures (Koenders *et al.*, 2000).

Numerous warm mix asphalt techniques have been developed with the goal of either reducing the effective viscosity of the binder or providing superior workability at lower temperatures than is the case with a traditional asphalt. Asphalt technologies can be classified on the basis of the amount of temperature reduction. Warm asphalt mixtures are separated from half-warm asphalt mixtures according to the mix temperature (D'Angelo *et al.*, 2008). There are many commercially available products that promise temperatures 10–40°C below those used with conventional asphalts, including some suggesting temperatures slightly above 100°C, with other technologies stating temperatures below the boiling point of water (Zaumanis, 2010). They are categorised as

- half-warm mix asphalt (65–100°C)
- warm mix asphalt (100–140°C)
- hot mix asphalt (above 140°C).

Another way to classify warm mix asphalts is based on the technology that is used to produce the asphalt (NAPA, 2012). Warm mix asphalt technologies are typically classified into three main categories according to the use of

- organic or wax additives
- chemical additives
- foaming techniques.

Foaming processes and organic or wax additives are used to achieve the temperature reduction by reducing the viscosity of the binder. These processes typically show a decrease in the viscosity of the binder above the melting point of the wax, making it possible to produce asphalts at lower temperatures (Hill, 2011). After crystallisation, these additives have the effect of increasing the stiffness of the binder and thus, in turn, improving the resistance of the asphalt to permanent deformation (Hanz et al., 2011). Chemical additives typically include a variety of chemical packages, which may be a combination of emulsification agents, surfactants, polymers and additives, to improve coating, mixture workability and compaction, as well as adhesion promoters (anti-stripping agents) (Hill, 2011). These chemical additives are typically used either in the form of an emulsion or are added to the bitumen during the asphalt production process and then mixed with hot or warm aggregates (D'Angelo et al., 2008). These materials affect the surface bonding between the asphalt binder and the aggregate, and are very likely to improve the fracture resistance of the mixture, and have the potential to improve the resistance to deformation (Xiao et al., 2012). Foaming processes and additives use water to foam the asphalt binder and reduce its viscosity prior to or during the mixing period. This particular category comprises the largest variety of warm mix asphalt methods available in the market, but there is the potential for such mixtures to show moisture damage and rutting due to the addition of water to the mixture during the foaming process (Hanz et al., 2011; Kvasnak et al., 2010; Xiao et al., 2011).

To date, more than 30 different warm mix asphalt technologies have been marketed and are available in North America. Depending on which process/technology is being used, the additional cost per tonne for warm mix asphalt with additives is now typically in the range 0–4% (Dale, 2012). At present, most of the US state departments of transportation and all the US Federal Lands Division have technical specifications and/or contractual language in place that would allow warm mix asphalt technology for use on federal aid or federal lands projects. Warm mix asphalt technology has gained popularity in most of US state departments of transportation and, from 2007, these organisations have road projects constructed using warm mix asphalt techniques. Some of the current challenges for implementing warm mix asphalt technologies are as follows.

- The technologies require changes to asphalt production plants for the incorporation of additives or foam into the mixture.
- Changes are needed in the asphalt mixture production operation.
- Greater quality control and adoption of best practices are needed for the mixture producers to achieve all the benefits of using warm mix asphalt technology, including fuel savings and lower emissions.

It is estimated that the use of warm mix asphalt in the paving industry in the USA will produce energy cost savings of US$3.6 billion by 2020 (NAPA, 2012).

Some warm mix asphalt techniques alter the physical and chemical properties of the asphalt that is produced, and this may cause changes in the long term behaviour of the material. Several researchers have evaluated the performance of warm mix asphalts with respect to various modes of pavement distress. One major concern is the increased susceptibility of warm mix asphalts to deformation (Vargas-Nordcbeck and Timm, 2012). It is possible that the asphalt binder in warm mix asphalt may not harden sufficiently at relatively lower production temperatures, and hence may develop higher post-construction densification or distortion under early age traffic. Conceptually, warm mix asphalt technology has the potential to be detrimental at early age and beneficial in the long term in-service performance. Another concern is that warm mix asphalts may have an increased propensity to moisture induced damage (Bonaquist, 2011; WMTWG, 2007; Xiao et al., 2009). In warm mix asphalts, aggregates are heated to relatively low production temperatures, and therefore may not thoroughly dry before they are mixed with the asphalt binder, thereby reducing the amount of bitumen absorbed in the aggregate. However, results from numerous field trials indicate that a variety of warm mix asphalt additives have required production temperatures 15–40°C lower than those used with traditional asphalts, with no detrimental effect on short term performance (EAPA, 2010). An examination of the field performance of a number of warm mix asphalt projects evaluated by the Texas Transportation Institute suggested that warm mix asphalt pavement test sections were found to be equivalent to comparable control sections made using conventional asphalts (Estakhri et al., 2010). X-ray computed tomography (a technology that uses computer processed X-rays to produce images of specific areas of the scanned object) applied to the laid warm mix asphalts has indicated that the in situ density of the mixture or air void distribution with depth in the material may be even more uniform than is the case with hot mix asphalts. The ability to produce asphalts using this technology has been proved through ongoing implementation in various paving projects across the USA and in Europe. This may eventually become the standard practice for asphalt production (WMTWG, 2007). In future, environmental aspects of projects will become even more important, and the use of warm mix asphalt may prove to be one means of achieving a lower carbon footprint (EAPA, 2010).

22.3 Shell Thiopave®, a sulfur-based asphalt modifier for enhanced asphalt road applications

Shell's technology using Shell Thiopave enables a proportion of the bitumen to be replaced in conventional asphalt. It is produced at lower temperature, equivalent to other warm mixtures, resulting in a modified asphalt with enhanced structural properties.

22.3.1 Introduction and background

The concept of using sulfur as a bitumen extender/modifier has been around for three-quarters of a century. Bencowitz and Boe (1938) demonstrated that sulfur is able to modify the bitumen properties, and sulfur-extended asphalts were used in the paving industry in the 1970s (Kandhal, 1982; Kennepohl et al., 1975). At that time, numerous projects around the world used hot liquid sulfur added to the bitumen, in an in-line blending process, prior to the introduction of the bitumen to the mixture. Pavement surveys have validated the performance of these sulfur-extended asphalts as, at the very least, comparable, and often improved, relative to conventional asphalts (Beatty et al., 1987). However, while hydrogen sulfide and sulfur dioxide emissions met the regulatory standards of the time, the odour and irritation were an ongoing source of worker concern.

A process to produce sulfur in pellet form was developed in the late 1990s, and this facilitated the handling of elemental sulfur in a solid form. The process was used to develop sulfur-extended asphalt modifier (SEAM), which is the previous Shell technology incorporating plasticisers in the sulfur pellets (Deme et al., 2004). This represented a major step forward because it enabled the sulfur compound to be added directly into the mixer while manufacturing the asphalt. SEAM significantly reduced emissions and odour compared with the addition of sulfur in liquid form. This technology has been developed further into Thiopave (Figure 22.4). Compared with conventional asphalts, the use of Thiopave enables asphalt production at lower temperatures (125–140°C), with a target temperature of 130°C, which reduces off-gassing while benefiting energy consumption at the hot mix plant.

22.3.2 Asphalt modification principle

22.3.2.1 Role of Thiopave

Thiopave comprises a range of sulfur-based and complementary non-sulfur-based additive components. Individual Thiopave components may differ in

Figure 22.4 Different forms of Thiopave pellets

type and appearance depending on the proposed usage. Although Thiopave pellets (i.e. pellets of Thiopave) act as an asphalt modifier, there are situations in which an additional workability additive is required. The workability additive improves the ability of the bitumen to coat the aggregates at lower production temperatures, and also lowers the temperature at which compaction can be effected. This also helps to limit the generation of hydrogen sulfide, reduces the energy used in production and serves as an adhesion promoter between the bitumen and the aggregate.

The addition of the pellets into the bitumen modifies the bitumen properties (Strickland *et al.*, 2007, 2008). Bitumen and sulfur combine at a temperature above the melting point of the pellets. Part of the sulfur is chemically combined with the bitumen and acts as an extender. This portion of sulfur is dissolved in the bitumen, modifying the bitumen properties by lowering the viscosity and increasing the ductility. Above a certain concentration, the remaining sulfur predominantly stays as free sulfur, which crystallises when the mixture cools. Depending on the quantity of pellets added, the crystallisation gives different levels of strengthening, with sulfur crystallisation acting as a structuring agent in the asphalt. An example of the structure of the modified asphalt obtained is shown in Figure 22.5.

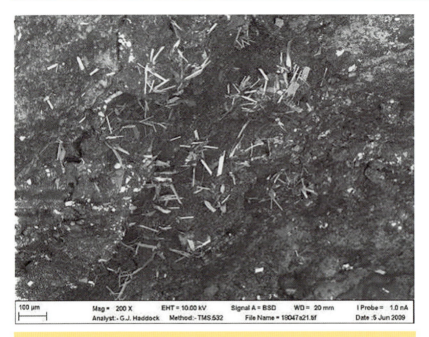

| 100 μm | Mag = 200 X | EHT = 10.00 kV | Signal A = BSD | WD = 20 mm | I Probe = 1.0 nA |
| | Analyst:- G.J. Haddock | Method:- TMS.532 | File Name = 18047a21.tif | | Date :5 Jun 2009 |

Figure 22.5 Scanning electron microscopy photograph of modified asphalt showing sulfur structure

22.3.2.2 Target applications

Thiopave is normally used in dense asphalts for binder courses and bases. However, in some countries it has been used in dense surface courses. It should not be used in tunnels, underground car parks or any places where there is a confined space, in case of fire. The modified asphalt is not prone to burning, and in this it is no different to conventional asphalt, but some sulfur compounds, mainly sulfur dioxide and hydrogen sulfide, may be emitted if the asphalt is exposed to a naked flame.

22.3.2.3 Design of Thiopave asphalt

The combination of Thiopave and bitumen results in a fraction of the bitumen being replaced by sulfur in the asphalt. Because the specific gravity of sulfur is virtually twice that of bitumen, in order to maintain the same volume of binder as in the conventional asphalt, the quantity of additive needed to replace the bitumen is almost doubled. In order to produce a paving mixture with enhanced structural properties, a bitumen/Thiopave mass ratio of 60w%/40w% (i.e. 60% bitumen/40% Thiopave by weight) is normally recommended, whereas for general applications the proportion should be limited to 70w%/30w% (i.e. 70% bitumen/30% Thiopave by weight) to retain maximum flexibility. However, these ratios are merely indicative, and the ideal proportion depends on the asphalt type, aggregate and bitumen quality. Therefore, it is recommended that an asphalt design using Thiopave be performed to verify the properties of a proposed mixture.

22.3.3 *Thiopave asphalt properties*
22.3.3.1 Workability and compaction
The blending of Thiopave and bitumen during manufacture of the asphalt results in a binder that has lower viscosity than the bitumen itself. Above its melting point (>120°C), Thiopave becomes liquid, with very low viscosity, and partly combines with the bitumen, coating the aggregates as a conventional binder would do but at a lower temperature.

Asphalt laying and compaction is carried out using the same equipment as would be employed with conventional asphalts, although the Thiopave asphalt is generally more workable. During paving, compaction starts at around 110°C although, as Thiopave includes a compaction additive, this results in compaction being effective at much lower temperatures. This technology has been accepted for use in warm mix asphalts in the USA (Powell and Taylor, 2012).

22.3.3.2 Water sensitivity
Asphalts made with Thiopave pellets generally exhibit the same resistance to water ingress as conventional asphalts. However, some specific studies have assessed the effect of water on both loose asphalts and compacted specimens, and found contradictory results. The immersion of loose asphalts in hot water (60°C) indicated no significant difference in binder film detachment for a period of up to 72 h, and indicated no difference in the effect of water saturation for either mixture type (Strickland *et al.*, 2007). However, with a particular limestone aggregate, the retained Marshall stability and the tensile strength ratio were found to have reduced by 10% for asphalts made with modified sulfur pellets compared with conventional asphalts. Typically, with these aggregates, the use of anti-stripping agents such as amine additives has been found to improve the water resistance properties significantly. There is no generic rule for water sensitivity with the new sulfur technology, and this demonstrates that when a new asphalt is designed it is always good engineering practice to examine its resistance to water ingress.

22.3.3.3 Deformation resistance
The specific structure formed in the binder and the air voids in the asphalt enhance the stiffness of the asphalt, without significantly impairing flexibility. As the nature of the crystallites does not change in the pavement temperature range, roads constructed using Thiopave asphalts exhibit superior resistance to deformation compared to those constructed using conventional asphalts. This behaviour was demonstrated in the rutting experiments carried out by

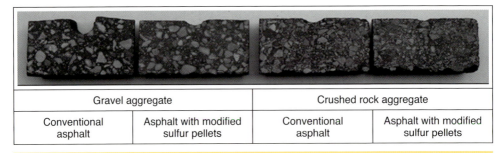

Gravel aggregate		Crushed rock aggregate	
Conventional asphalt	Asphalt with modified sulfur pellets	Conventional asphalt	Asphalt with modified sulfur pellets

Figure 22.6 Rutting behaviour of different asphalts at 58°C

Deme *et al.* (2004) utilising the asphalt pavement analyzer (APA – discussed in section 16.2.1), as shown in Figure 22.6, all made with a standard performance grade PG 58-28 bitumen (performance grade is the US framework classification for bitumen specifications). The conventional asphalts were much more susceptible to deformation than the same mixture when 40w% of the binder was Thiopave. As shown in Figure 22.6, gravel aggregate mixture, which shows the greatest deformation in a traditional asphalt, exhibits much less deformation when the asphalt is manufactured using modified sulfur pellets. In both cases, asphalt mixtures containing Thiopave demonstrate an improved resistance to deformation, and, to some extent, the inclusion of Thiopave has a positive impact when using lower quality aggregate sources.

22.3.3.4 Stiffness modulus
The stiffness modulus of an asphalt is generally increased when the concentration of Thiopave in the binder exceeds 25–30% by mass. This increase is significantly higher at elevated temperatures and long loading times than at low temperature and short loading times. This means that the modified asphalt is less sensitive to temperature and loading time than is the case with conventional asphalts. Figure 22.7 illustrates this with two asphalt master curves built at a reference temperature of 10°C using the principle of loading time or frequency–temperature superimposition. The asphalts compared are

- conventional asphalt made with a 70/100 pen grade bitumen
- the same asphalt formulation modified with Thiopave at 40% mass in the binder, achieving the same binder volume as for the asphalt above.

Master curves are commonly used when considering matters related to asphalt rheology. Figure 22.7 shows a representation of the stiffness of the two asphalts described above as a function of the temperature and loading time applied. Temperature and loading time are translated into a value of reduced frequency. Accordingly, Figure 22.7 enables the performance of

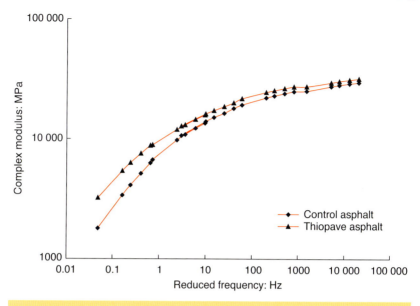

Figure 22.7 Comparison of the master curves of conventional and Thiopave asphalts

both asphalts to be compared against a key structural parameter (i.e. deformation). The left hand side of the figure represents high temperature and low speed of traffic (i.e. conditions in which the amount of deformation increases). The right hand side represents low temperature and high traffic speed. The higher the curve the less will be the amount of deformation. What Figure 22.7 demonstrates is that the mixture containing Thiopave is less susceptible to deformation, particularly when conditions are likely to lead to significant deformation.

22.3.3.5 Resistance to fatigue cracking

The resistance to fatigue cracking, as is the case for most asphalts, depends on the test conditions and the asphalt composition. Given similar binder volumes in a conventional and Thiopave asphalt, the fatigue resistance of the modified asphalt would be slightly lower in an imposed strain test because it has much higher stiffness, resulting in higher stresses being generated within the asphalt during fatigue testing. A decrease of 10% of the acceptable strain level of the asphalt has been observed in some cases, although some other studies have shown very similar behaviour in a two-point bending test at imposed strain (Cocurullo et al., 2014). Conversely, at imposed stress, the modified mixture always behaves better, with much higher allowable stresses. As an example, the lower fatigue resistance in the constant strain test mode can be overcome by increasing the amount of binder, as can be seen

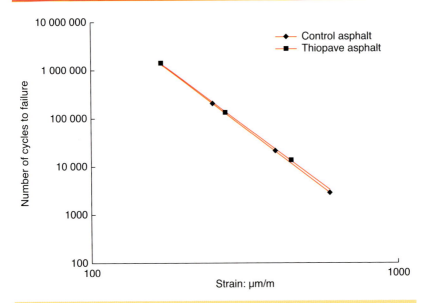

Figure 22.8 Fatigue test in the strain controlled mode at 20°C and 10 Hz

in Figure 22.8. The figure shows the fatigue lines as the test results of a four-point bending test carried out at 20°C and 10 Hz. The conventional asphalt had 4.6% bitumen PG 70-10 (this is a US asphalt manufactured with a bitumen PG 70-10). The modified asphalt used the same PG 70-10 bitumen but with 6.6% binder, of which 60% was bitumen and 40% was Thiopave.

22.3.3.6 Low temperature properties

The low temperature performance of asphalts made with Thiopave can be assessed using the thermal stress restrained specimen test (TSRST), in accordance with the AASHTO TP 10 test method (AASHTO, 1993), with some modifications to the specimen size. The temperature conditioning in the TSRST does not attempt to simulate the in situ temperature cycling condition, which causes low temperature fatigue cracking. Moreover, it monotonically decreases temperature until the specimen fails in tension (single event thermal cracking). However, some studies have shown that the low temperature behaviour of asphalt concrete pavements can be predicted using the TSRST (Bouldin *et al.*, 2000; Hannale *et al.*, 1994).

Similar studies were carried out at Shell laboratories in 2008 indicating that the presence of Thiopave in the binder does not detrimentally affect the low temperature performance of the mixture compared with hot mix asphalt, as shown in Figure 22.9. The development of the induced stress in the specimens, with the temperature decrease, was very similar for both types of asphalt, although the modified asphalt exhibited a slightly higher stress

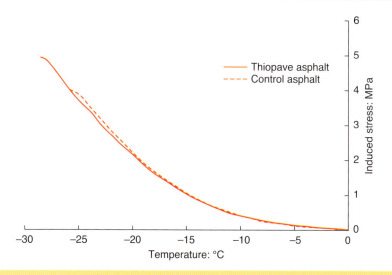

Figure 22.9 Thermal stress restrained specimen tests at a temperature
decrease of 10°C/h

and a lower fracture temperature. The conventional asphalt was made with
70/100 pen grade bitumen, and the sulfur modified mixture was made with
the same bitumen and 40% Thiopave pellets by mass. Both asphalts had the
same total volume of binder.

Figure 22.10 shows the fracture temperature and the induced stress for both a
traditional asphalt and Thiopave asphalt at 30% and 40% Thiopave pellets

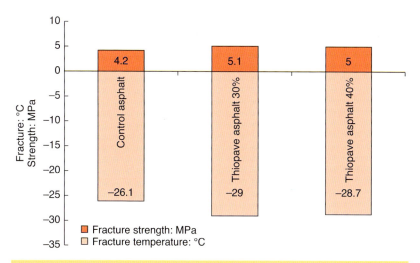

Figure 22.10 Fracture temperatures and induced stress for conventional hot
mix asphalt and two Thiopave asphalts

by mass in the binder (data represent the mean value of three tests per asphalt formulation). It can be seen that the sulfur modified specimens behave the same as, or even slightly better than, traditional asphalt in terms of cold fracture temperature and induced stress. This suggests that the bitumen dictates the cold temperature behaviour (Strickland *et al.*, 2008).

22.3.4 Road applications
22.3.4.1 Production of Thiopave asphalt
Thiopave is suitable for use in most batch and drum mix plants, with the exception of parallel flow drum plants, for which the drum temperature profile needs some investigation to ascertain their suitability. It is supplied as a solid pellet and added at ambient temperature towards the end of the asphalt manufacturing process. The high shear rate environment during mixing disperses Thiopave throughout the asphalt without the need to extend the mixing time. The target mixing temperature for asphalt containing Thiopave is 130°C, although the recommended range is 125–140°C. This represents a reduction of 20–40°C compared to asphalts made with unmodified bitumen.

Depending on the configuration of the asphalt plant, some modifications of the plant may be required to manage hazards associated with handling sulfur. These modifications are identified as a part of the plant assessment, which must be carried out for each individual plant.

22.3.4.2 Laying and compaction of Thiopave asphalts
Laying and compacting Thiopave asphalts is, in general, no different to that for a conventional asphalt. However, to avoid unwanted emissions during paving, the screed heater on the paver needs to be set to a maximum temperature of 130°C. Fuming typically occurs during rest breaks, when material under the screed may be heated for a prolonged period of time, but it may also occur due to localised heating if the screed heater is set at too high a temperature.

22.3.5 Road performance of asphalts containing Thiopave
More than 100 projects have been carried out worldwide using Thiopave. In most cases, the product brings benefits to the asphalt in terms of improved resistance to deformation and superior stiffness.

In China, in 2006, a 12 km long section of highway was built in Yunan Province (Getun Highway). The Shell technology was chosen to meet the strict specifications forming part of the Marshall mixture design, including a specified minimum level of resistance to deformation. Resistance to deformation was given particular attention because it is one of the main issues on this highway. The base was manufactured using 60% bitumen and

(a)

(b)

Figure 22.11 Views of the Getun Highway: (a) surface layer texture and (b) traffic

40% Thiopave pellets, while a 65%/35% ratio was used for the surface course. To date, the traffic has been assessed as 4500 vehicles/day, including around 400 heavy trucks. The value of a standard axle load used for pavement design in China is 100 kN. Regular inspections confirm that the section is performing well (Figure 22.11).

In Qatar, air temperatures around 50°C are not uncommon, which means that pavement rutting can be a significant problem on roads that carry high volumes of heavy commercial vehicles, a situation that is typical of construction related traffic in the region. The use of Thiopave was chosen as one of six technologies to meet these challenging conditions, and Thiopave was used in a 270 mm thick base (Figure 22.12). The road provides access to a sand processing plant carrying around 2000 heavy trucks per day. After three years of service, the section of the asphalt made with Thiopave is performing well, and no major distress has been observed. The road is still being monitored.

In India, in 2010, Thiopave was used on National Highway 3 near Nashik, Maharashtra (Figure 22.13). It was included in the dense bitumen macadam (DBM) as type I and II at 60 mm for each layer thickness (DBM type I and type II are standard asphalts in India). Three sections were paved

- Thiopave in both DBM layers
- Thiopave in DBM type I and conventional in DBM type II
- conventional asphalt in both layers.

Regular inspections of the road site are undertaken, including visual observations, coring of the layers and deflection measurements using a Benkelman beam. The fourth evaluation was carried out in March 2013 and confirmed the previous evaluations.

- The three sections are in good condition.

Figure 22.12 Laying and compacting Thiopave asphalt in Qatar in September 2010

■ The indirect tensile stiffness modulus tests carried out on cores taken from the road showed that the Thiopave asphalts have higher stiffness at all temperatures (20, 30 and 40°C) than the cores of conventional asphalt.

Figure 22.13 Laying Thiopave asphalt in India in 2010

Figure 22.14 NCAT pavement test track, Alabama, USA. (Courtesy of the NCAT pavement test track)

- The deflection on the full section of the Thiopave asphalt was reduced by 26% compared with the deflection found on the section constructed using conventional asphalt.
- The deflection measured on the hybrid section was 18% less than that on the section constructed using conventional asphalts.
- Overall, the Central Road Research Institute of India concluded that the results were encouraging, and that a further structure and functional evaluation would be sufficient to confirm the advantages of Thiopave asphalts over conventional asphalts.

In the USA, in 2009, Thiopave asphalt performance was assessed against conventional asphalt in a comparative full scale experimental test section at the National Center for Asphalt Technology (NCAT) pavement test track (Figure 22.14). The two sections were designed as follows

- surface course 32 mm thick
- binder course 70 mm thick
- base 76 mm thick over a 152 mm thick aggregate base.

The section of the Thiopave asphalt consisted of a binder course manufactured with 40% Thiopave by mass in the binder, a base manufactured with 30% Thiopave by mass in the binder, and a surface course manufactured with unmodified bitumen. The test sections were regularly monitored in terms of traffic, ride quality, rutting and fatigue cracking. At the conclusion of monitoring in 2012, 10.14 million equivalent single axle loads (ESALs) had been carried by both sections. It was observed that there was no cracking in any of the sections, and that the ride quality was excellent. The deformation performance was below the rutting failure threshold enforced at the test track. The two sections were left in place for further trafficking. In 2013, after an additional 5 million ESALs, the ride quality was still good for both sections. A small amount of cracking had begun to appear, although more cracking was found on the control section.

22.4 Recycling of asphalt pavements

Recycling in general has a long history and has always been an alternative source of materials when resources are scarce. There is proof from archaeological studies that, around 400 BC, humans had started to reuse broken tools and pottery (Wikipedia, 2014). The main reasons for the commercialisation of recycled asphalt are (Nordic Road Forum, 2012) to

- reduce the consumption of virgin materials
- cut energy use
- minimise air and water pollution.

Recycling has been defined as the 'use of waste materials for manufacturing new products'. RAP is by no means a waste material. On the contrary, it can be a high value component (Beer *et al.*, 2007). However, it is important to distinguish between recycling of asphalts containing bitumen and recycling of asphalts containing tar. The fumes of the latter are carcinogenic; hence recycling of asphalt containing tar can only be undertaken on cold material (e.g. in combination with an emulsion). The tar also has to be coated to avoid leaching and water pollution. This section applies only to recycling of asphalts containing bitumen, not tar.

For asphalts containing bitumen, the constraints as described above do not apply. The only limitation is the proportion of RAP that can be included in an asphalt. The dose depends on

- the layer in which RAP is to be included
- the type of road (motorway, country road etc.) in which RAP is to be included
- the moisture content of the RAP
- RAP quality
- RAP homogeneity
- RAP management
- asphalt plant capabilities
- local legislation/specification
- the contract with the road owner.

In general, the value of RAP increases with a higher proportion of bitumen. Accordingly, RAP originating in a surface course is likely to have a higher value than RAP originating from a base. Another relevant factor in

Usage of recycled asphalt

- Disposal site
  - expensive
  - in most cases neither ecological nor economically sensible
  - in some countries prohibited!
- Lower layers
  - resource friendly
  - economical
  - feasible with relatively simple techniques
- Top layers
  - resource friendly
  - economical
  - technically demanding
  - highest added value

Figure 22.15 Factors affecting the use of RAP

determining the value is whether it was used in a low trafficked or heavily trafficked road, with the latter probably having a higher value because it is likely that it contained a higher proportion of bitumen. As the constitution of RAP is variable, clearly the process of adding RAP to newly manufactured asphalts has, also, to be variable. Using RAP in the surface course of a heavily trafficked carriageway is technically demanding and, of course, comes at a cost (Riebesehl, 2007).

A precondition for recycling asphalt is that the performance of asphalt containing RAP is at least as good as the performance of virgin asphalt (EAPA, 2008). Figure 22.15 shows the factors that apply to the use of RAP in different layers of a pavement, with the greener areas of the arrow being the most financially beneficial (Federal Highways Administration, 2011; Riebesehl and Beer, 2006).

As the proportion of RAP in a mixture increases so too does the amount of labour required. This is because the higher the amount of RAP, the more effort is required in respect of quality control, screening, fractionating etc. Thus, it is relatively easy to include 20% RAP in a base but much more effort is required to include 40% RAP successfully in a surface course.

22.4.1 Production and storage of reclaimed asphalt
Reclaimed asphalt is obtained (Deutscher Asphaltverband, 2011)

- by milling asphalt
- by breaking and lifting up asphalt in slabs (demolition asphalt)

Figure 22.16 Stockpile of RAP

- from waste at asphalt production plants
- as returned unwanted loads from the laying site.

Demolition asphalt needs to be crushed and further processed before it can be used as reclaimed asphalt.

Milled asphalt results from milling out (originally described as 'planing') existing asphalt pavements, collecting the material as it is milled, and transporting it to an asphalt plant. At present, cold milling machines can process the full depth of a pavement in a single pass. However, in some cases, it may be worthwhile milling different layers separately. Very often, the upper layers contain aggregates or binders with special characteristics, such as high polished stone value aggregate or polymers in the binder.

RAP can be stockpiled (Figure 22.16). However, the quality and consistency of the stockpile has to be known. A disadvantage of stockpiling is that the RAP is not separated into different sizes. As a consequence, the amount of RAP that can be added to a mixture is limited.

In order to maximise the amount of RAP in an asphalt, the RAP should be separated into different sizes. Each size must be stored separately in a facility that prevents water getting into the RAP and thus limiting its use (Figure 22.17).

22.4.2 Quality and suitability

Quality control and consistency of the RAP are key factors in the successful recycling of asphalts. There are a variety of test methods used to determine

Figure 22.17 Storage of RAP under cover

the quality and suitability of RAP. Applicable local regulations must be obeyed. An unanswered question about RAP usage is how often an asphalt can be recycled. The answer is certainly not an infinite number of times. Research and practice demonstrate that RAP can be used more than once but then the dosage has to be limited.

The use of recycled asphalt shingles in asphalts is problematic. Usually these shingles contain special binders (oxidised grades, hard grades) that are not suitable for asphalts. Using such binders may result in premature failure of the pavement, as they can embrittle the mixture.

Asphalt containing tar as binder must not be used without special arrangements.

22.4.3 Recycling techniques
There are two main ways of recycling asphalt

- in an asphalt production plant
- in situ (FGSV, 2002).

22.4.3.1 Recycling in an asphalt production plant
This is discussed in detail in several sections in Chapter 14.

22.4.3.2 Recycling in situ
The main processes for recycling in situ are reshaping, repaving and remixing. A common feature of these techniques is that the existing pavement is heated in situ to make it flexible. In some processes, fresh material is mixed with the existing asphalt. Heating has to be carried out very carefully in order to avoid thermal ageing of the binder.

Reshape
In the reshape process (Figure 22.18), the top layer is carefully heated up, using infrared heaters, and scarified. The warm asphalt is then mixed, laid

In the USA, some states, including Arizona, California, Florida, Louisiana and Texas, have been predominantly using rubberised asphalt products for several decades. It was reported that, together, these states recycled over 35.6 million tyres in asphalt paving applications from 1995 to 2001 (Rubber Pavements Association, 2002). Approximately 1500 tyres are used to pave one lane mile of highway, an initiative that saves precious landfill capacity. The principal source of raw material for rubber modified binders is tyre rubber, which is a composite of natural rubber, synthetic rubber and carbon black (Chesner et al., 2005). Materials like natural rubber provide the elastic properties, while the synthetic rubber improves the thermal stability properties of the compound and the addition of carbon black improves the binder's durability.

Crumb rubber (CR), which may also be described as 'ground rubber' if it complies with a particular grading, is typically used to produce CR modified binders (CRMB). CR is produced by shredding and grinding reclaimed old scrap tyres from trucks, passenger cars or both, into very small particles (Kandhal, 2006). The tyre source affects the type of rubber present: truck tyres contain a greater proportion of natural latex than do passenger car tyres (Bandini, 2011). It should be noted that natural latex has been found to be more reactive with bitumen than ground tyre rubber. Some state agencies also specify the maximum ratio of car tyre rubber to truck tyre rubber used for binder modification. This is to ensure that an adequate amount of natural rubber is incorporated in the blended material. In the grinding process, most of the steel wires and reinforcing fibres or fluff from the waste recycled tyres are removed. The finer the CR is ground, the more it will react with bitumen; however, additional grinding adds to the cost of the CR material.

In the tyre recycling industry, the fine grinding technique used for scrap tyres is based on either the ambient method or the cryogenic method. The ambient method consists of grinding the rubber at or slightly above ambient temperature using a granulator or a cracker mill (Caltrans, 2014). Finer CR particles (i.e. <0.4 mm) can be produced using a micro-milling process. The CR particles produced using this grinding method are predominantly used for producing CRMB in the paving industry.

Cryogenic tyre grinding consists of using liquid nitrogen to freeze the scrap tyre until it becomes brittle, and then using a hammer mill to crack the frozen rubber into smaller particles. The resulting material is composed of clean, flat particles that generally have a smooth surface texture and a low surface area, and are considered to be less reactive with asphalt binders.

In the USA, most of the south eastern states have changed their bitumen specification to allow asphalt binder suppliers and mixture producers to use

(a)   (b)

Figure 22.21 Scanning electron microscopy images of a crumb rubber modified binders produced using: (a) particle size (−0.4 mm, ×100) and the ambient method; (b) particle size (−1.4 mm, ×100) and the cryogenic method

either ambient or cryogenic rubber for producing CRMB to meet the performance grading requirement (Willis *et al.*, 2012) (in the USA, bitumens are classified by their PG). Figure 22.21, shows scanning electron microscopy images of CRMB produced by ambient and cryogenic methods using different sizes of rubber particles. The CR is often sieved and separated into categories based on grading to meet the requirements of a particular application or agency. In the US market, three types of CR particle size are typically available

- type 1: 2 mm coarse
- type 2: 1.1–0.8 mm
- type 3: 0.6 mm.

The size designation indicates the first sieve with an upper range specification of 5–10% of material retained (Bandini, 2011). The use of CR in asphalts and pavement rehabilitation treatments such as fog seal, crack seal and chip seal can be achieved using two different processes: the 'wet process' and the 'dry process'. In addition, under the wet process, two types of production processes are used to produce rubber modified binders, one process requiring agitation and the other involving no agitation. Each of these is described below.

22.5.1 Wet process with agitation

In the wet process, the CR and other additives, such as extender oil, natural rubber and polymers (as required), are mixed thoroughly with the bitumen, which is maintained at temperatures in the range 180–225°C prior to manufacture of the asphalt. Extender oils such as naphthene or aromatic oil are typically used to promote the reaction between the binder and the CR. Extender oil assists with the dispersion by chemically suspending the rubber particles in the binder. In the US literature, this mixture is often referred to as 'asphalt rubber' (AR). This technology was developed in the early 1960s by Charles H. McDonald, while working as an engineer with the City of Phoenix, Arizona. The manufacture of AR can be done on site, and hence it is also called 'field blend asphalt rubber'. AR is currently being used in California, Arizona and Louisiana. California has been a major user of AR since the 1980s, with around 30% of all asphalts produced there incorporating AR.

For this technology, in order to meet the ASTM D6114/D6114M-09 specification (ASTM, 2009), the rubber content used should typically be at least 15% (by weight) of the total blend to provide acceptable properties in the material when laid. Use of higher rubber contents (18% and 22% by weight) is often reported or specified (Hanson *et al.*, 1996). When using such a high proportion of rubber in the asphalt, the rubber particles are typically agitated sufficiently in the hot bitumen until there is a reaction between the bitumen and rubber particles. This 'reaction' is not a chemical process but rather a dispersion process that includes the absorption of aromatic oils from the bitumen into the polymer chain of the rubber. The rubber particles swell to three to five times their original size as they absorb oils, which causes the viscosity of the AR modified binder to increase during the first hour or so (Kandhal, 2006). The rate of reaction is affected by the following properties

- the temperature of the binder (higher temperatures cause a quicker reaction)
- the surface characteristics of the rubber particle (rougher surfaces react quicker)
- the size of the rubber particle (smaller particles swell quicker but to a lesser extent)
- the period for which the rubber and bitumen are kept at the reaction temperature (a longer time gives a greater reaction).

During the blending process the rubber particles swell, and this changes the proportion of the ground rubber in the binder, which can partially break down. It is reported that the particle size reduction increases with mixing time, and decreases with increasing original rubber particle size (ARTS, 2010). After the 'reaction' process and associated swelling has stopped, the rotational viscosity of the blend levels off. The targeted viscosity at the job site for the product would be a minimum of 1.5 Pa·s at 190°C after the interaction period is completed. Because it possesses a short storage life, the AR modified binder has to be agitated constantly in order to maintain the rubber particles in suspension and ensure that they are distributed evenly in the storage tanks. Hence, it is preferable that mixing takes place adjacent to the production plant.

For AR modified binders, a minimum targeted rotational viscosity of 1.5 Pa·s at 190°C is necessary in order to increase binder contents significantly above those of conventional asphalts without excessive 'drain-down' (that portion of the bitumen that separates and flows downwards through the mixture) (Caltrans, 2014). If the AR modified binder is maintained at high temperature for a prolonged period of time (as little as 6 h), the rubber particles start to degrade, causing the viscosity of the binder to decrease. ASTM D6114/D6114M-09 (ASTM, 2009) specifies that, among other characteristics, the rubber particles used to produce AR modified binder should contain less than 0.75% moisture (by weight), no visible non-ferrous metal particles, no more than 0.01% ferrous metal particles (by weight) and should be free flowing.

22.5.2 Wet process – no agitation

Terminal blending is another type of wet process in which the CR is blended with the hot bitumen at the refinery or at the bitumen storage and distribution terminal (Bandini, 2011). In practice, terminal blended rubber modified binders are typically produced using rubber contents lower than 15% by weight. At present, this technology is gaining popularity with asphalt producers in the USA, as the process does not require agitation with paddles or augers when stored in tanks in order to keep the rubber particles in suspension. Previously, this was one of the major concerns in the industry when handling modified binders with higher rubber contents. Some of the methods or the processing conditions reported and in use to improve the storage stability of the CRMBs without agitation include (Perez-Lepe et al., 2003)

- the addition of substances to create a chemical stabilisation of the blends
- mixing rubberised binder blends in the presence of a low percentage of polymers
- the addition of a compatibiliser (see the definition below) to activate the CR, or with sulfur to improve cross-linking
- blowing oxygen gases through CR modified (CRM) binder blends
- pre-treatment of rubber particles with hydrogen peroxide (H_2O_2)
- lowering the storage temperature of CRM blends
- increasing the surface area of the rubber particles
- using fine CR (reducing particle size)
- high shear blending/mixing rate (see the note below)
- ensuring that the CR is fully digested into the binder, so that the resulting solubility for such binders is >97%.

In general, compatibilisers (a type of polymeric additive) are used in the production of CRM binders. Compatibilisers typically cause the CR to interact completely with the binder, thereby improving the rheological properties of the binder. The compatibiliser will typically consist of a reactive component such as one or more glycidyl groups. The effect of any of these reactive functional groups is beneficial in improving the solubility properties and also in aiding the dispersion of CR in bitumen.

The bitumen is heated in a tank to an elevated temperature, and CR is introduced into the tank and is digested into the bitumen. During this process, samples will be collected to test for solubility (i.e. to determine how many residual CR particles remain in the bitumen). This test is carried out during the manufacturing process to ensure that the rubber particles are completely digested in the binder. Most manufacturers use a high shear blending process to make sure that the ground tyre rubber is completely digested. The solubility of the finished product is generally above 97%.

Laboratory studies have shown that adding 10% rubber (passing 0.6 mm sieve) is equivalent to changing from a PG 67-22 binder to a PG 76-22 binder (Willis *et al.*, 2012). ('76' is the average 7 day maximum pavement design temperature (°C) and '22' is the minimum pavement design temperature (°C).) It is reported that the longer mixing process causes the CR particles to partially melt and separate, creating small CR 'chips' in the binder. These smaller CR 'chips' homogeneously mix with the binder to give a polymer-like structure, which improves their engineering properties (e.g. fatigue resistance) (Kutay and Ozturk, 2012b). Furthermore, increasing the rubber content in the binder would also increase the critical low temperature grade of the modified material. These research data suggested and provided support for US state agencies to move to a performance grade specification for CRMBs instead of specifying raw materials. With the development of performance related properties and the establishment of binder compatibility of rubber modified binders manufactured using the terminal blending process, many states have introduced new specifications for CRMBs, and these are often referred to as PG 76-22TR (tyre rubber) or PG 82-22RM (rubber modified).

In order to meet the PG specifications, terminal blended binder suppliers typically use a minimum CR content of 7% and the size of the rubber particle used would be smaller than 0.6 mm. In addition, they may also use 0.5–4% styrene–butadiene–styrene (SBS) polymer, depending on the performance grade required. Terminal blended products are provided by binder suppliers, and generally come with a certificate of compliance stating the material meets performance grade specifications. Terminal blended rubber modified binders are typically used in open graded and gap graded surface courses in Texas, Florida, Louisiana, New York, Arizona, California and Nevada (Asphalt Institute, 2008). The two main distinctions between the types of wet process methods are

- differences in the viscosities of the final product
- whether or not agitation is required to maintain relatively uniform distribution of the rubber particles in suspension in storage.

22.5.3 Dry process

CR modifier can also be used as a substitute for a percentage of the fine aggregate in an asphalt concrete grading rather than as part of the binder. This approach, called the 'dry process', was first developed in Sweden in the 1960s. In this process, the tyre rubber is ground to a size similar to that of some of the aggregate fractions, and introduced into the mixing zone (paddle mixer), clear of direct contact with any heating flames (Bandini, 2011). In this process, CR is introduced with the aggregate fraction before adding the binder. The CR acts as an 'elastic mineral aggregate' and replaces part of the aggregate fraction. Typically, the rubber percentage used in the dry process varies from 3% to 5% (by weight of the aggregate) (Hicks *et al.*, 2013). It should be noted that, while the bitumen in the dry process is considered to be unmodified, the rubber tends to absorb some of the lighter fractions from the bitumen during the mixing process. However, the properties of the resulting mixture would be modified (Amirkhanian, 2001). The designer must allow for this reaction to take place by slightly increasing the binder content, otherwise there will be a high risk of ravelling in service. The dry process can be used in dense, open graded and gap graded asphalts to accommodate the rubber particles in the aggregate grading, but cannot be used for cold mix applications such as chip seals and surface treatments. The mixture design should take into account the lower specific gravity of the CR compared with that of conventional aggregates (Caltrans, 2014).

The dry process can be implemented in batch asphalt plants and drum mix asphalt plants. The granulated or crumb rubber is usually packed and stored in sacks at the plant. Good control of the feeding of rubber and the temperature are critical in the mixing process, as these variables affect the performance of the resulting mixture and also the performance of the pavement into which the material is incorporated. There have been reports in the literature that batch plants require a dry mix cycle to ensure that the heated aggregate is mixed with the CR before applying the binder. For the mixing process, a hopper typically used for introducing RAP into a drum mixer can be used to introduce CR modifier in the same way (Colucci *et al.*, 1994). Due to the limited reaction time with this process, only the surface of the coarse rubber particle reacts with the binder, which creates a binder–rubber interface that bonds the two materials together. The applicable mixture design process includes a quantity of finer rubber particles for partial reaction with the binder, with a further quantity that acts, in effect, as coarse aggregate.

The fine rubber particles produce the swelling that occurs in the mixture, while the coarse CR particles retain their shape and rigidity to function as aggregate. Some of the major concerns and challenges when using this technology include the following.

- The dry process requires more CR modifier material per tonne of asphalt than does the wet process to achieve the desired properties. This has the effect of increasing the cost of the mixture compared with that of material resulting from the wet process (Colucci *et al.*, 1994).
- The additional binder content needed to produce sufficient rubber–binder reaction on the surface of the rubber particles also carries with it a higher cost.
- These mixtures also appear to be more temperature sensitive, as vertical expansion and extensive cracking have been observed in laboratory studies following 6 h of conditioning at 60°C (Rahman, 2004).
- Visual inspection studies have shown that plucking of rubber particles from the asphalt specimens after moisture conditioning is predominant, especially for mixtures having a rubber content of 5% or more.
- Surface ravelling may also occur due to a lack of bonding between the rubber and binder, meaning that the surface aggregate is not retained.

The above are some of the major reasons why wet process technology is favoured over dry process systems.

22.5.4 Applications of CRMB

The USA and India are currently the leading users of CR for the modification of asphalt. China, Canada and some European countries are also now taking a serious look at this technology. Although initially limited to trial roads, the use of CRMB now covers a wide range of applications, including

- anti-reflective cracking systems in asphalts
- thin overlay systems on concrete or asphalt, including acting as a stress absorbing membrane interlayer
- stress absorbing membranes
- waterproofing treatment of pavements, typically using chip seal technology beneath the top layer or on the base of the pavement
- bridge deck waterproofing
- noise reduction pavement
- anti-icing pavement (Way *et al.*, 2010).

Thus CRMB can be found in and around the surfacing courses (i.e. the binder course and surface course of new pavements). Typical concentrations of rubber particles in bitumen are in the range 7–15%. High rubber particle content (>20%) allows the formation of a physical network that partially prevents the sedimentation of the particles. In all cases, when dispersed into bitumen, the CR particles highly influence its rheology and increase its viscosity, which is then sometimes detrimental to the workability.

22.5.5 Limitations of CRMB

Asphalt modification with ground tyre rubber provides the obvious environmental benefit of using scrap tyres. The initial costs of this technology are high and this is still an issue, but the results of life cycle cost analysis indicate that the use of CRMB is, in many cases, an economical option (Hicks *et al.*, 2013). Notwithstanding, aspects of CRMB technology pose a number of concerns

- a lack of availability of rubber tyre recycling processing facilities in the vicinity
- the cost of such facilities
- a shortage of quality CR
- the need to establish a Superpave mixture design procedure (an asphalt design procedure commonly used in the USA)
- inexperience in the asphalt industry of using this technology
- binder performance grading of AR modified binder
- mixtures are highly temperature sensitive
- compaction issues
- weather restrictions
- limited paving window
- higher mixing temperatures (180–195°C) are required, and these cause the generation of odours and fumes
- the fact that storing the CRMB at high temperature might accelerate the process of particle sedimentation to the bottom of the tank or truck, which results in inconsistent supply quality (Presti, 2013).

22.5.6 Technology gaining popularity in the USA

While environmental stewardship is important, some state agencies and/or contractors are investigating using CRMBs instead of polymers such as SBS in binder modification. If CRMBs can perform such that they are perceived to be equivalent to polymer modified binders, state agencies and contractors will have additional options in the event of another polymer shortage such as occurred in 2008 in the asphalt industry in the USA. Many state agencies are considering using or increasing the use of CR in paving applications, and are looking to other state agencies that have strong recycling programmes for guidance on best practice when handling and using recycling materials. Presently, many asphalt producers are looking at increasing the range of rubberised asphalt paving applications. They hope to do so by using different techniques, including the use of additives in the mixtures, which will enable a reduction in paving temperatures (by using warm mix technologies) (Punith *et al.*, 2012). This could allow mixtures to be placed at night and in cooler climates, or pelletised asphalt rubber (a patented technology) to be added to the heated aggregates and mixed thoroughly to produce polymerised asphalts (Amirkhanian and Kelly, 2012). These pellets are less than 20 mm in size and contain some form of stiffener (e.g. hydrated lime) and other chemicals.

Recent efforts have focused on using bio binders consisting of bio oil that has reacted with CR. Preliminary laboratory results have shown that such binders can produce a binder that is comparable to binders derived from crude petroleum (Peralta *et al.*, 2012).

Shell Bitumen developed a patented technology (Shell Mexphalte RM) to improve the homogeneity and the stability of CRMB during transport and storage at the asphalt plant. It also limited the maximum mixing and laying temperature to 170°C, thus avoiding the odours and other emissions from rubber particles.

At present, CR serves as an asphalt modifier that can ease the asphalt industry's dependence on the supply of polymers. Studies conducted at NCAT have shown that the use of CR in speciality mixtures such as open graded friction courses and stone mastic asphalts (described in the USA as 'stone matrix asphalt') can allow contractors to prevent drain down (that portion of the bitumen that separates and flows downwards through the mixture) without the use of additional fibres such as cellulose fibres. In addition to removing additional fibres from such mixtures, using CRMBs to modify the mixture can also eliminate the need to use polymer modified binders to avoid rutting. Two year field studies carried out at the NCAT test track, showed that the test sections made of CR modified mixes performed in a manner that is similar to SBS modified mixes in every aspect of pavement performance (West *et al.*, 2012). Based on this research, many US states allow the use of CR as an alternative to traditional PG 76-22 for heavily trafficked roads.

References

AASHTO (American Association of State Highway and Transportation Officials) (1993) TP 10-93. Standard test method for thermal stress restrained specimen tensile strength. AASHTO, Washington, DC, USA.

ASTM (American Society for Testing and Materials) (2009) ASTM D6114/ D6114M-09. Standard specification for asphalt-rubber binder. *Annual Book of ASTM Standards*. ASTM, West Conshohocken, PA, USA.

Amirkhanian SN (2001) *Utilization of Crumb Rubber in Asphaltic Concrete Mixtures – South Carolina Experience*. Department of Civil Engineering, Clemson University, Clemson, SC, USA.

Amirkhanian SN and Kelly S (2012) Development of polymerized asphalt rubber pelleted binder for HMA mixtures. *5th Asphalt Rubber Conference, Munich, Germany*.

ARTS (Asphalt Rubber Technology Service) (2010) *Chemical Analysis of Interaction of Crumb Rubber Modifier and Asphalt Binder*. South Carolina Department of Health and Environmental Control (DHEC) and ARTS, Clemson University, Clemson, SC, USA.

Asphalt Institute (2008) Terminal blended rubberized asphalt goes mainstream – now PG graded, asphalt. *Magazine of the Asphalt Institute*. See http://www.asphaltmagazine.com/news/detail.dot?id=4c868096-abe6-4109-8387-1995098c4f4a (accessed 01/10/2014).

Bandini P (2011) *Rubberized Asphalt Concrete Pavements in New Mexico, Market Feasibility and Performance Assessment*. Department of Civil Engineering, New Mexico State University, Las Cruces, NM, USA.

Beatty TL, Dunn K, Harrigan ET, Stuart K and Weber H (1987) Field evaluation of sulfur-extended asphalt pavements. *Journal of the Transportation Research Board* **1115**: 161–170.

Beer F, Damm KW, Nölting M and Riebesehl G (2007) Ein dynamischer Wachstumsmarkt. *Asphalt* **42(4)**: 8–15 (in German).

Bencowitz I and Boe ES (1938) Effect of sulfur upon some of the properties of asphalts. *Proceedings of the American Society for Testing and Materials* **39(2)**: 539–550.

Bennert T, Ali M and Robert S (2011) *Influence of Production Temperature and Aggregate Moisture Content on the Performance of Warm Mix Asphalt*. Transportation Research Board, Washington, DC, USA, TRR 2208, pp. 97–107.

Bonaquist R (2011) *Mix Design Practices for Warm Mix Asphalt*. TRB, Washington, DC, USA, NCHRP Report 691.

Bonvallet J (2001) Fabrication industrielle et contrôlée d'enrobes à froid avec des mousses de bitumen. *XVI World Congress of the International Road Federation (IRF), Paris, France*, Paper 0027 (in French).

Bouldin MG, Dongre R, Rowe GM, Sharrock MJ and Anderson DA (2000) Predicting thermal cracking of pavements from binder properties: theoretical basis and field validation. *Proceedings of the Association of Asphalt Paving Technologists* **69**: 455–496.

Bowering RH and Martin CL (1976) Foamed bitumen production and application of mixtures evaluation and performance of pavements. *Proceedings of the Association of Asphalt Paving Technologists* **45**: 453–477.

Brosseaud Y, Gramsammer JC, Kerzreho JP, Goacolou H and Le Bourlot F (1997) Experimentation (premiere partie) de la Grave-Mousse sur le manege de fatigue. *Revue Générale des Routes et Aérodromes* **752**: 61–70 (in French).

Caltrans (2014) *Asphalt Rubber Usage Guide*. California Department of Transportation, Sacramento, CA, USA. See http://www.dot.ca.gov/hq/maint/Pavement/Offices/Pavement_Engineering/PDF/Asphalt-Rubber-Usage-Guide.pdf (accessed 01/10/2014).

Caro S, Masad E, Bhasin A and Little DN (2008) Moisture susceptibility of asphalt mixtures, Part 1: Mechanisms. *International Journal of Pavement Engineering* **9(2)**: 81–98.

Cazcliu B, Peticila M, Guieysse B et al. (2008) Effect of process parameters on foam bitumen-based road material production at ambient temperature. *International Journal of Road Materials and Pavement Design* **9(3)**: 499–523.

Chesner WH, Collins RJ and MacKay MH (2005) *User Guidelines for Waste and By-product Materials in Pavement Construction*. Federal Highway Administration, Research and Development, Turner-Fairbank Highway Research Center, Washington, DC, USA.

Chowdhury A and Button JW (2008) *A Review of Warm Mix Asphalt*. Texas Transportation Institute, College Station, TX, USA, Report SWUTC/08/473700-00080-1.

Cocurullo A, Grenfell JRA, Yusoff NI, Maggiore C and Airey GD (2014) *Dynamic Mechanical and Fatigue Analysis of Thiopave Asphalt Mixtures*. Nottingham Transportation Engineering Centre, Department of Civil Engineering, University of Nottingham, Nottingham, UK, NTEC Report 14131.

Colucci B, Colucci AJ, Prieto JV and Dayton D (1994) *Feasibility of Using Crumb Rubber Modifier in Hot Mix Asphalt Pavement Applications in Puerto Rico*. Puerto Rico Transportation Technology Transfer Center. University of Puerto Rico, Mayaguez, Puerto Rico.

Csanyi LH (1957) *Foamed Asphalt in Bituminous Paving Mixtures*. Highway Research Board, Washington, DC, USA, Bulletin 160, pp. 108–122.

Dale AR (2012) *Warm Mix Asphalt*. TEEX – Texas A&M University, College Station, TX, USA, Lone Star Roads, Issue 1.

D'Angelo J *et al.* (2008) *Warm Mix Asphalt: European Practice*. Federal Highways Administration, US Department of Transportation, Washington, DC, USA, FHWA-PL-08-007.

Deme I, Kennedy B and Keenan K (2004) Performance properties of semi-rigid sulfur-extended asphalt mixes. *Proceedings of the 3rd Eurasphalt and Eurobitume Congress, Vienna, Austria*, Papers Technical Sessions, Book 1, Paper 365, pp. 835–849.

Deutscher Asphaltverband (2011) *Asphalt. Recycling of Asphalt*. Deutscher Asphaltverband, Bonn, Germany.

Estakhri C, Button J and Alvarez AE (2010) *Field and Laboratory Investigation of Warm Mix Asphalt in Texas*. Texas Transportation Institute, College Station, TX, USA, Report FHWA/TX-10/0-5597-2.

EAPA (European Asphalt Pavement Association) (2008) Arguments to stimulate the government to promote asphalt reuse and recycling. EAPA, Brussels.

EAPA (2010) *The Use of Warm Mix Asphalt*. EAPA, Brussels, Belgium, EAPA Position Paper.

Federal Highways Administration (2011) *Reclaimed Asphalt Pavements in Asphalt Mixtures: State of the Practice*. US Department of Transportation, Washington, DC, USA, FHWA-HRT-11-021.

FGSV (Forschungsgesellschaft für Straßen- und Verkehrswesen) (2002) *Merkblatt für das Rückformen von Asphaltschichten*. FGSV, Cologne, Germany, FGSV 786/1 (in German).

Fu P (2009) *Micromechanics for Foamed Asphalt Stabilized Materials*. PhD Thesis. University of California, Davis, CA, USA.

Gillen S (2007) *Preliminary Summary Report on Ground Tire Rubber (GTR) Asphalt Pavement Demonstration Project*. Illinois Tollway, Downers Grove, IL, USA, Report RR-06-9092.

Hande IO (2013) *Quantification of Quality of Foamed Warm Mix Asphalt Binders and Mixtures*. PhD Dissertation. Michigan State University, East Lansing, MI, USA.

Hannale K, Vinson TS and Huayang Zeng (1994) *Low-Temperature Cracking: Field Validation of the Thermal Stress Restrained Specimen Test*. Department of Civil Engineering, Oregon State University, Strategic Highway Research Program, National Council, Washington, DC, USA, SHRP-A-401.

Hanson DI, Epps JA and Hicks RJ (1996) *Construction Guidelines for Crumb Rubber Modified Hot Mix Asphalt*. Federal Highway Administration, Washington, DC, USA, Report DTFH 61-94-C-00035.

Hanz A, Mahmoud E and Bahia H (2011) Impacts of WMA production temperatures on binder aging and mixture flow number. *Journal of Asphalt Paving Technology* **80**: 459–490.

Hicks RJ, Tighe S, Tabib S and Cheng D (2013) *Rubber Modified Asphalt – Technical Manual*. Ontario Tire Stewardship, Toronto, Ontario, Canada.

Hill B (2011) *Performance Evaluation of Warm Mix Asphalt Mixtures Incorporating Reclaimed Asphalt Pavement*. MSc Thesis, University of Illinois at Urbana-Champaign, IL, USA.

Jenkins KJ (2000) *Mix Design Considerations for Cold and Half-Warm Bituminous Mixes with Emphasis on Foamed Bitumen*. PhD Dissertation, University of Stellenbosch, Cape Town, South Africa.

Jenkins KJ, De Groot J and Van De Ven MFC (1999a) Characterization of foamed bitumen. *7th Conference on Asphalt Pavements for Southern Africa, CAPSA '99*, Victoria Falls, Zimbabwe.

Jenkins KJ, De Groot J, Van De Ven MFC and Molenaar AAA (1999b) Half-warm foamed bitumen treatment, a new process. *7th Conference on Asphalt Pavements for Southern Africa, CAPSA '99*, Victoria Falls, Zimbabwe.

Kandhal PS (1982) Evaluation of sulfur extended asphalt binders in bituminous paving mixtures. *Proceedings of the Association of Asphalt Paving Technologists* **51**: 189–221.

Kandhal PS (2006) Quality control requirements for using crumb rubber modified bitumen (CRMB) in bituminous mixtures. *Journal of the Indian Roads Congress* **67(1)**: 99–104.

Kennepohl GJA, Logan A and Bean DC (1975) Conventional paving mixes with sulfur–asphalt binders. *Proceedings of the Association of Asphalt Paving Technologists* **44**: 485–518.

Khweir K, Fordyce D and McCabe G (2001) Aspects influencing the performance of foamed bitumen. *Asphalt Yearbook*. Institute of Asphalt Technology, Edinburgh, UK, pp. 27–34.

Kiggundu BM and Roberts FL (1988) *Stripping in HMA Mixtures: State-of-the-art and Critical Review of Test Methods*. National Center for Asphalt Technology (NCAT), Auburn University, Auburn, AL, USA, Report 88-02.

Kim Y and Lee H (2006) Development of mix design procedure for cold in-place recycling with foamed asphalt. *Journal of Materials in Civil Engineering* **18(1)**: 116–124.

Kim Y, Lee H and Heitzman M (2007) Validation of new mix design procedure for cold in-place recycling with foamed asphalt. *Journal of Materials in Civil Engineering* **19(11)**: 1000–1010.

Koenders BG, Stoker DA, Bowen C *et al.* (2000) Innovative process in asphalt production and application to obtain lower operating temperatures. *Proceedings of the 2nd Eurasphalt and Eurobitume Congress, Barcelona, Spain*, Book 2, Session 3, pp. 830–840.

Krambeck H (2009) *Draft GHG Emissions Calculator for Asphalt Production*. The World Bank Group, Washington, DC, USA.

Kutay M and Ozturk H (2012a) Investigation of moisture dissipation in foam-based warm mix asphalt using synchrotron-based X-ray microtomography. *Journal of Materials in Civil Engineering* **24(6)**: 674–683.

Kutay ME and Ozturk HI (2012b) Internal structure characteristics of crumb rubber modified asphalt binders: an analysis using 3D X-ray microtomography imaging. *5th Asphalt Rubber Conference, Munich, Germany*.

Kvasnak A, Taylor A, Signore JM and Bukhari SA (2010) *Evaluation of Gencor Green Machine, Ultrafoam GX*. National Center for Asphalt Technology (NCAT), Auburn, AL, USA, NCAT Report 10-03.

Maccarrone S, Holleran G, Leonard DJ and Dip SH (1994) Pavement recycling using foamed bitumen. *Proceedings of the 17th ARRB Conference, Vermont South, Victoria, Australia* **17(3)**: 349–365.

Maccarrone S, Holleran G and Key A (1995) Cold asphalt systems as an alternative to hot mix. *9th AAPA International Asphalt Conference, Paradise, Queensland, Australia*, Session 7a.

NAPA (National Asphalt Pavement Association) (2012) *Annual Asphalt Pavement Industry Survey on Recycled Materials and Warm-Mix Asphalt Usage: 2009–2012*. NAPA, Lanham, MD, USA, Information Series No. 138.

Nordic Road Forum (NVF) (2012) *Sustainable Asphalt*. See http://www.eapa.org/userfiles/2/Publications/NVF_en.pdf (accessed 01/10/2014).

Oliver J (1999) *The Use of Recycled Crumb Rubber*. ARRB Transport Research, Vermont South, Victoria, Australia, Technical Note 10.

Peralta J, Williams CR, Rover M and Silva MRDH (2012) *Development of Rubber-Modified Fractionated Bio-Oil for Use as Noncrude Petroleum Binder in Flexible Pavements*. Transportation Research Board, Washington, DC, USA, TRB Circular Number E-C165, pp. 23–36.

Perez-Lepe FJ, Martinez-Boza, FJ *et al.* (2003) Influence of the processing conditions on the rheological behavior of polymer-modified bitumen. *Fuel* **82**: 1339–1348.

Powell RB and Taylor AJ (2012) *Design, construction and performance of sulfur-modified mix in the WMA Certification Program at the NCAT Pavement Test Track*. National Center for Asphalt Technology, Auburn University, Auburn, AL, USA, NCAT Report 12-01.

Presti DL (2013) Recycled tyre rubber modified bitumens for road asphalt mixtures: A literature review. *Construction and Building Materials* **49**: 863–881.

Prowell BD, Hurley GC and Crews E (2007) Field performance of warm mix asphalt at national center for asphalt technology test track. *Journal of the Transportation Research Board* **1998**: 96–102.

Punith VS, Xiao F and Amirkhanian SN (2012) Evaluation of moisture sensitivity of stone matrix asphalt mixtures using polymerized warm mix asphalt technologies. *International Journal of Pavement Engineering* **13(2)**: 152–165.

Rahman M (2004) *Characterisation of Dry Process Crumb Rubber Modified Asphalt Mixtures*. PhD Thesis, School of Civil Engineering, University of Nottingham, Nottingham, UK.

Riebesehl G (2007) Wohin mit dem Ausbauasphalt. *Asphalt* **42(4)**: 35–40 (in German).

Riebesehl G and Beer F (2006) Ausbauasphalt in Asphaltschichten mit hohen Belastungen. *Asphalt* **41(6)**: 11–18 (in German).

Romier A, Audeon M, David J, Martineau Y and Olard F (2006) Low-energy asphalt with performance of hot-mix asphalt. *Transportation Research Record* **1962**: 101–112.

Rubber Pavements Association (2002) *Air Quality Issues and Best Management Practices with the Production of Asphalt–Rubber Asphaltic Concrete*. Rubber Pavements Association, Tempe, AZ, USA. See http://www.asphaltrubber.org/ari/Emissions/RPA_Environmental_Issues_With_AR_2002.pdf (accessed 01/10/2014).

Schwartz CW and Khosravifar S (2013) *Design and Evaluation of Foamed Asphalt Base Materials*. Maryland Department of Transportation, Baltimore, MD, USA, Research Report MD-13-SP909B4E.

Strickland D, Colange J, Martin M and Deme I (2007) Performance properties of sulfur extended asphalt mixtures with SEAM. *23rd World Road Congress Conference (PIARC)*, Paris, France.

Strickland D, Colange J, Shaw P and Pugh N (2008) Study of the low-temperature properties of sulfur extended asphalt mixtures. *Proceedings of the 53rd Annual Conference of the Canadian Technical Association (CTAA), Saskatoon, Canada* (Goodman S (ed.)). Polyscience Publications, Quebec, Canada, pp. 105–118.

Vargas-Nordcbeck A and Timm DH (2012) Rutting characterization of warm mix asphalt and high RAP mixtures. *International Journal of Road Materials and Pavement Design* **13(Suppl 1)**: 1–20.

Way GB, Kaloush KE and Biligiri KP (2010) *Asphalt–Rubber Standard Practice Guide* (Draft). See http://www.rubberpavements.org/Library_Information/Draft_RPA_AR_Std_Practice_Guide_Sept_14_2010_VA1.pdf (accessed 01/10/2014).

West R, Timm D, Willis R *et al.* (2012) *Phase IV NCAT Pavement Test Track Findings*. National Center for Asphalt Technology, Auburn University, Auburn, AL, USA, Report 10.

Wikipedia (2014) http://en.wikipedia.org/wiki/Recycling#Origins (accessed 01/10/2014).

Willis JR, Plemons C, Turner P, Rodezno C and Mitchell T (2012) *Effect of Ground Tire Rubber Particle Size and Grinding Method on Asphalt Binder Properties*. Auburn University, Auburn, AL, USA, NCAT Report 12-09.

Wirtgen Group (1998) Foamed bitumen – the innovative technology for road construction. In *Wirtgen Cold Recycling Manual*. Wirtgen GmbH, Windhagen, Germany.

Wirtgen Group (2002) *Foamed Bitumen – The Innovative Binding Agent for Road Construction*. Wirtgen GmbH, Windhagen, Germany.

WMTWG (Warm Mix Technical Working Group) (2007) *WMA Technical Working Group Meeting*, Day 1. Federal Highways Administration, Baltimore, MD,

USA. See http://www.warmmixasphalt.com/wmatwg07.aspx (accessed 01/10/2014).

WRAP (Waste & Resources Action Programme) (2008) *A Review of the Use of Crumb Rubber Modified Asphalt Worldwide.* WRAP, Banbury, UK, Project TYR032-001.

Xiao F, Jordon J and Amirkhanian SN (2009) Laboratory investigation of moisture damage in warm-mix asphalt containing moist aggregate. *Journal of Transportation Research* **2126**: 115–124.

Xiao F, Punith VS, Putman B and Amirkhanian SN (2011) Utilization of foaming technology in warm mix asphalt mixtures containing moist aggregates. *Journal of Materials in Civil Engineering Division, ASCE* **23(9)**: 1328–1337.

Xiao F, Punith VS and Amirkhanian SN (2012) Effects of non-foaming WMA additives on asphalt binder properties at high performance temperatures. *Fuel* **94**: 144–155.

Zaumanis M (2010) *Warm Mix Asphalt Investigations.* MSc Thesis, Department of Civil Engineering, Technical University of Denmark, Kongens Lyngby, Denmark.

Chapter 23

Certification of bitumens and asphalts

David O'Farrell
Principal Consultant, PTS Ltd. UK

This chapter examines the certification of bitumen and asphalts (in the context of construction products) in Europe, where this practice is well established, although other similar systems are used elsewhere.

Certification is the term used to describe the procedures used to confirm defined characteristics of a product or process, usually undertaken by an organisation that is independent of the producer. This is referred to as a third party certification scheme, and is the most common type of certification. The third party is referred to as a certification body.

Other types of certification can exist in which a producer self-certifies products under its own quality system, or when the purchaser is involved in assessing the product (a second party scheme).

Certification is usually applied to either a management system or a product produced to a stated specification. Management systems are certified to BS EN ISO 9001:2008 (BSI, 2008a), which can be supplemented by additional requirements, such as those found in the national highway sector schemes. Product certification is a declaration of the properties or performance of the product in the terms defined by the appropriate standard, and a declaration of compliance with a stated standard.

Initially, certification was adopted voluntarily by some producers as a means of improving customer confidence in their products. Gradually, clients in the construction sector began to include quality management and certification requirements in their specifications, and this has continued until it is now a requirement of most specifications. In recent years, product certification has become a regulatory requirement for some products.

Certification is the responsibility of the producer, who selects a certification body with the relevant experience to certify the product or management

The Shell Bitumen Handbook, Sixth Edition
ISBN 978-0-7277-5837-8
Shell International Petroleum Company Ltd: All rights reserved
http://dx.doi.org/10.1680/tsbh.58378.687

system. The producer will need to develop, document and implement a system prior to the initial assessment by the certification body.

The duration of the assessment and subsequent surveillance arrangements will depend on the scope of the quality system and the size of the organisation being assessed. It is usually a staged process involving examination of the system documentation and observing the implementation of the quality system.

Once the certification body has awarded the certification, it carries out routine surveillance, usually on an annual basis with reassessment every 3 years.

23.1 The regulatory framework

In the construction industry, the principal regulatory requirements for certification are found in the Construction Products Regulation 2011 (EC, 2011), which was implemented in the UK on 1 July 2013. This Regulation superseded the Construction Products Directive (EC, 1989) and the accompanying UK legislation.

The Construction Products Regulation builds upon the earlier Construction Products Directive and aims to break down technical barriers to trade in construction products within the European economic area.

The Regulation stipulates that construction works as a whole and in their separate parts must be fit for their intended use, taking into account, in particular, the health and safety of persons involved throughout the life cycle of the works.

It also requires that, subject to normal maintenance, construction works must satisfy defined basic requirements for construction works for an economically reasonable working life.

These basic requirements are found in Annex 1 of the Regulation

- mechanical resistance and stability
- safety in case of fire
- hygiene, health and the environment
- safety and accessibility in use
- protection against noise
- energy economy and heat retention
- sustainable use of natural resources.

These basic requirements apply to the works as a whole, not individual products. The characteristics of the products themselves must include at least one essential characteristic that relates to the basic requirements above. A product may have several essential characteristics relating to more than one of the basic requirements.

The Construction Products Regulation (EC, 2011) provides for four main elements

- a system of harmonised technical specifications
- an agreed system of conformity assessment for each product family
- a framework of notified bodies
- CE marking of products.

23.1.1 Technical specifications

The requirements for technical specifications are discussed in detail later in section 23.2.

23.1.2 Conformity assessment

The methods of conformity assessment are set out in Annex V of the Regulation, and six systems of assessment and verification of the constancy of performance (AVCP) are described with requirements for certification with differing levels of involvement of external bodies.

All the harmonised standards require that a system of factory production control (FPC) is put in place by the producer. An FPC system is similar to a BS EN ISO 9001:2008 system but includes requirements for inspection, sampling and testing during production. For the higher levels of AVCP, the FPC system must be certified against the relevant product standard, and some producers also choose to have their system certified for products with the lower levels of AVCP.

Asphalts, bitumens and surface treatments require AVCP system 2+. In the UK, aggregates with a polished stone value (PSV) of 58 or greater also require system 2+, and those with a PSV less than 58 require AVCP system 4.

The differences between these two systems of AVCP are shown in Table 23.1.

Table 23.1 Tasks and responsibilities for AVCP systems 2+ and 4

	AVCP system	
	2+	4
Tasks for the producer		
Factory production control	Yes	Yes
Further testing of samples taken at a factory according to a prescribed test plan	Yes	No
Declaration of performance (initial type testing)	Yes	Yes
Tasks for third party notified accreditation body		
Certification of factory production control	Yes	No
Surveillance of factory production control	Yes	No

23.1.3 *Notified bodies*

Notified bodies are the product certification bodies, FPC certification bodies and testing laboratories, which are considered to be competent to carry out the conformity assessment tasks described in the Construction Products Regulation (EC, 2011). These organisations are first approved by their respective member states to carry out certain designated tasks, and then notified to the European Commission (EC) and other member states.

In the UK, the notification of bodies under the Construction Products Regulation is carried out by the Department for Communities and Local Government, which is the UK notifying authority. It delegates the assessment and ongoing surveillance of notified bodies to the United Kingdom Accreditation Service (UKAS).

Notified bodies are required to participate in the group of notified bodies, with their European counterparts, to discuss practical implementation matters in order to achieve a consistent approach to the tasks.

Once a harmonised technical specification is available for its product, a manufacturer required to use a notified body for assessment can approach any such body in the European economic area that has been notified for the particular harmonised technical specification and task, according to the appropriate conformity assessment procedure. It does not have to use a body operating in the same country as the place of manufacture or where the product is to be used.

Details of notified bodies and their designated tasks can be found on the New Approach Notified and Designated Organisations Information System (Nando) website (Nando, 2014).

23.1.4 *CE marking*

CE marking on a product is the manufacturer's declaration that the product complies with the basic requirements of all the directives or regulations that apply to it. It also indicates to the appropriate bodies that the product may be legally offered for sale in its country.

The CE mark can only be applied to products within the scope of a harmonised European standard or a European technical assessment. It indicates that a product is consistent with its declaration of performance (DoP) as made by the manufacturer. The requirements for the DoP are given in the relevant product standard and are normally based on those properties measured for FPC.

The mark is shown in Figure 23.1 and is required to be fixed to the product itself. Where this is not practical, it should be on the accompanying documentation. In addition, the documentation supplied with the product should include the manufacturer's DoP. Typical examples of the CE mark certificate can be found in Annex ZA of a harmonised standard.

For those products that require notified body involvement in the CE marking (e.g. those subject to AVCP 2+), the number of the notified body must also be shown beneath the CE symbol.

23.1.5 Implications for specifiers and purchasers

In addition to the requirements for the manufacture and labelling of products, the Construction Products Regulation (EC, 2011) also contains requirements that relate to public bodies, their specifications and procurement practices.

Articles 8.4 and 8.5 of the Construction Products Regulation place obligations on member states to ensure that the use of construction products bearing the CE mark shall not be impeded by rules imposed by public bodies or private bodies acting as a public undertaking.

Those acting as such a body, either in a monopoly position or under a public mandate, should not specify the performance of products other than in accordance with the basic requirements covered by the harmonised section of the harmonised European standards or European technical assessment under which the CE mark is applied.

Article 8.6 requires that methods used by the member states in their requirements for construction works, as well as other national rules in relation to the essential characteristics of construction products, shall be in accordance with harmonised standards.

Public bodies also have other requirements with which they have to comply when specifying and procuring materials, and these are given in both European directives and UK legislation to implement European directives. At the time of writing (May 2014), a revision to the public procurement Directive (EC, 2004) is due to be published, and the relevant UK regulations will be updated accordingly in due course.

23.2 Standards

Certification is usually undertaken in relation to a standard. A standard is a document that provides requirements, specifications, guidelines or characteristics that can be used consistently to ensure that materials, products, processes and services are fit for their purpose.

Standards themselves are used by producers on a voluntary basis but there are circumstances in which their use is mandatory, if required by legislation.

While some clients or industries may develop their own standards, the majority of standards used are developed and published by the recognised standards bodies at national, European or international level.

23.2.1 National standards

Most countries have their own national standards body. The British Standards Institution (BSI) fulfils this role in the UK, with DIN and AFNOR, for example, operating in Germany and France, respectively (AFNOR, 2014; DIN, 2014). They work with government, industry and consumers to develop standards at national, European and international levels. They represent their country as members of the International Organization for Standardization (ISO) and the European Committee for Standardization (CEN).

British standards are developed through a committee structure, with representation from producers, consumers and government (local and national). Additional expertise can be co-opted as necessary. Other national standards bodies operate in a similar manner.

The committees are responsible for providing the national representatives to participate in European or international standards work, and also provide a national response when required in the development of these European or international standards.

National standards can be used as the basis for certification of products and systems.

Further information about the BSI and other national standards bodies, their processes and the standards they produce can be found on their websites.

23.2.2 European standards

CEN is an association that brings together the national standardisation bodies of 33 European countries. It is one of three European standardisation organisations (together with the European Committee for Electrotechnical Standardization (CENELEC) and the European Telecommunications Standards Institute (ETSI)) that have been officially recognised by the European Union and by the European Free Trade Association (EFTA) as being responsible for developing and defining voluntary standards at European level.

CEN operates with a structure of technical committees, each having responsibility for a defined technical area. The committees are subdivided into working groups and task groups.

The development of a European standard can be initiated by a proposal from any interested party, although most standardisation work is proposed through the national standardisation bodies.

If a standard is required to support legislation, the European standardisation organisations will receive a mandate from the EC or the secretariat of EFTA. This is a request to develop and adopt European standards in support of European policies and legislation.

Once CEN accepts a proposal to develop a standard, all member countries must put any national activity within the scope of the project on hold. This means that they do not initiate new projects, nor revise existing standards at national level. This is referred to as 'standstill'.

The new standard is developed by the technical committees and working groups, and then submitted for public comment through the CEN enquiry process. Once this is completed and comments resolved, a final version is drafted, which is then submitted to the CEN national members for a weighted formal vote.

When approved, the standard is announced in the *Official Journal of the European Union* (OJEC, 2014) and subsequently published. It must be adopted and given the status of a national standard in all member countries within 2 years, and any national standards that would conflict with it must be withdrawn.

When a new standard is published, the national standards body is able to provide a national annex. Such an annex provides the opportunity for specific guidance to be given about the application of the standard in that country.

In the UK, if the BSI committee considers that more extensive guidance is required, then this is made available in the form of a BSI published document (PD). For example, PD 6691:2010 (BSI, 2010b) gives guidance on the use of BS EN 13108 Bituminous mixtures – Material specifications (BSI, 2006b, 2006c).

More information about CEN and European standards can be found on the CEN website (CEN, 2014).

23.2.3 Harmonised European standards

A harmonised standard is a European standard produced in response to a mandate to develop a European standard that provides solutions for compliance with a legal provision. The mandate provides guidelines with which these standards must comply to meet the basic requirements or other provisions of relevant EU harmonisation legislation.

Compliance with harmonised standards provides a presumption of conformity with the corresponding requirements of harmonisation legislation. Manufacturers, suppliers or conformity assessment bodies can use harmonised standards to demonstrate that products, services or processes comply with relevant EU legislation.

CE marking can only be applied to products manufactured to a harmonised standard.

Harmonised standards contain an Annex ZA, which is the part of the standard that is used for the purpose of CE marking of construction products, and transforms all or part of a European product standard into a harmonised European product standard.

The Annex ZA identifies those clauses of the standard (or refers to clauses in another standard) that cover the essential characteristics included in the mandate to which it responds, considered in the answer to the mandates prepared by the technical committee and accepted by the commission.

It also gives details of the AVCP system(s) to which the product needs to be submitted before the manufacturer is able to draw up the DoP and to affix the CE marking.

23.2.4 *International standards*
The ISO develops standards for international use. At the time of writing (May 2014), 161 countries are ISO members.

ISO standards are voluntary standards and do not inhibit any development or revision of national standards. In some instances they have been adopted and implemented as European standards (e.g. BS EN ISO 9001:2008 (BSI, 2008a)), and in other instances as British standards, when there is no conflict with European standards.

More information about ISO and international standards can be found on the ISO website (ISO, 2014).

23.3 Certification for products and systems outside the scope of CEN harmonised standards
There are several options available to enable products, systems and services outside the scope of CEN harmonised standards to gain certification. These include

- European technical assessments
- national highway sector schemes (NHSS)
- highways authorities product approval scheme (HAPAS)
- BSI Kitemark.

23.3.1 European technical assessments

The European technical assessment is defined in the Construction Products Regulation (EC, 2011) as a document providing information on the assessment of the performance of a construction product, in relation to its essential characteristics.

It is considered to be a harmonised technical specification for the purposes of the Regulation, and consequently provides a way for the manufacturer to CE mark a product in the following situations

- the product is not or not fully covered by any harmonised technical specification such as European assessment documents (EADs) or harmonised European standards
- no mandate has yet been issued by the EC for an appropriate standard
- the product is covered by a EAD.

European technical assessments are carried out by technical approval bodies, which are members of the European Organisation for Technical Assessment (EOTA), using EADs. New EADs, where required, are developed by EOTA and technical approval bodies in conjunction with manufacturers. The EC is informed of the development work and comments on the final draft. After approval, publication of the EAD is announced in the OJEU.

The EOTA website contains further information about the procedures and available documents (EOTA, 2014).

23.3.2 National highways sector schemes

NHSS are developed in partnership with all sides of the highways industry, and their purpose is to interpret the international quality management standard (BS EN ISO 9001:2008 (BSI, 2008a)) for specific activities within the UK highways sector. These include

- NHSS 13: The Supply and Application of Surface Treatments to Road Surfaces (UKAS, 2013)
- NHSS 14: The Quality Management of the Production of Asphalt Mixes (UKAS, 2014b)
- NHSS 15: The Supply of Paving Bitumens (UKAS, 2014c)
- NHSS 16: The Laying of Asphalt Mixes (UKAS, 2014d).

Each scheme is managed by technical advisory committees with industry and client representation. Training organisations, certification bodies and UKAS are also represented.

Consensus is reached on key elements of the schemes such as

- minimum levels of services or products
- minimum standards of workmanship

- testing
- training and competency of the workforce
- training and competency of auditors used by certification bodies.

Once an organisation has been successfully assessed, its BS EN ISO 9001:2008 registration is endorsed with the details of the relevant schemes. The organisation is then required to arrange for its details to be included on the schedule of registered suppliers for the NHSS, maintained by LANTRA. This is available for viewing on the NHSS schedule of suppliers website (LANTRA, 2014).

The use of registered suppliers, where an NHSS exists, is a mandatory requirement of the *Specification for Highway Works* (SHW), and is referenced by other specifiers (Highways Agency, 2014).

NHSS are mostly used in the UK but any competent organisation registered to BS EN ISO 9001:2008 can apply for assessment. At the time of writing, only UK certification bodies are able to provide assessments.

Since the introduction of NHSS in the mid-1990s, some or parts of the activities covered have become the subject of harmonised standards or are affected by the requirements of the Construction Products Regulation (EC, 2011). These schemes have been either withdrawn or amended to remove any potential conflicts, and now provide specific guidance on elements such as training and competency.

Details of the current schemes and copies of the scheme requirements can be found on the UKAS website (UKAS, 2014a).

23.3.3 *The highway authorities product approval scheme*
The highway authorities product approval scheme (HAPAS) was established in 1995 at the request of specifiers and producers to provide a national approval scheme for products that were not within the scope of existing standards or were innovatory.

It was intended to rationalise the ad hoc processes that many clients had instigated for proprietary products, and to provide a form of approval that would be transferrable between clients.

HAPAS is operated by the British Board of Agrément, and consists of assessment processes for a number of different product sectors. The scheme has operated principally in the UK but is open to any supplier that wishes to supply in the UK. It has been used on a very limited basis by specifiers in other countries.

The process for each product sector is described in guidelines produced by a specialist advisory group drawn from all sides of the highways industry.

These guidelines follow a standard approach to assessment, which typically includes

- Stage 1 – assessment of applicant's data
- Stage 2 – assessment of quality assurance/factory production control
- Stage 3 – laboratory testing
- Stage 4 – system installation trial
- Stage 5 – system performance trial (if required)
- Stage 6 – certification.

Once certification has been achieved by an organisation, it is subject to annual surveillance and periodic reassessment. Changes made to the design of the product may also necessitate a reassessment, depending on the scale of the change.

HAPAS certification has been a requirement of the SHW for products such as thin surface course systems and high friction surfacing, but the introduction of the Construction Products Regulation (EC, 2011) allows alternative forms of certification to comply with the requirements in Series 100 of SHW.

Further details of HAPAS can be found on the British Board of Agrément website (British Board of Agrément, 2014).

23.3.4 *BSI Kitemark*
The BSI Kitemark (BSI, 2014b) is a registered product certification scheme developed many years ago and operated by the BSI in its role as a certification body. Although it originated in the UK, it is now recognised worldwide.

Certification of conformity with standards is provided through Kitemark schemes to many industry sectors, including the construction industry. The schemes involve an initial assessment of conformity to the relevant standard and an assessment of the quality management system operated by the supplier.

A programme of ongoing surveillance is undertaken, including routine testing of the product or service and assessment of production quality controls.

This type of scheme can provide improved customer confidence when products are not within the scope of CE marking.

The Kitemark scheme is fully described on the BSI website (BSI, 2014a).

23.4 Requirements for product certification
Product certification to meet the requirements of harmonised standards consists of two key elements

- a DoP
- implementation of FPC and certification if AVCP 2+ applies to the product.

The preparation of the DoP is the responsibility of the producer, and it must be renewed at least every 5 years. The tests on which it is based must also have been undertaken within the preceding 5 years. The tests required for the DoP are detailed in Annex ZA of the product standard, and are usually drawn from those tests included within the FPC regime.

Previously, under the Construction Products Directive, the DoP was referred to as the initial type test (ITT).

The testing schedules and frequencies are also included in the product standards.

For those products to which AVCP 2+ applies, the producer will need to involve a notified body in the initial inspection, certification and annual surveillance of the FPC. A reassessment of the FPC is undertaken every 3 years.

Some more details about the different product types are given below. References are given to the UK version of the published European standard.

23.4.1 Bitumen
Bitumen for paving applications is certified to the requirements of one of the following standards

- BS EN 12591:2009 Bitumen and bituminous binders – Specifications for paving grade bitumens (BSI, 2009)
- BS EN 13808:2013 Bitumen and bituminous binders – Framework for specifying cationic bituminous emulsions (BSI, 2013a)
- BS EN 13924:2006 Bitumen and bituminous binders – Specifications for hard paving grade bitumens (BSI, 2006d)
- BS EN 14023:2010 Bitumen and bituminous binders – Specification framework for polymer modified bitumens (BSI, 2010a)
- BS EN 15322:2013 Bitumen and bituminous binders – Framework for specifying cut-back and fluxed bituminous binders (BSI, 2013b).

All these standards contain sections describing FPC, and an Annex ZA describing AVCP requirements and CE marking.

AVCP system 2+ applies to all these products.

23.4.2 Aggregate
Aggregate requirements for road surfacing are given in BS EN 13043: 2002, Aggregates for bituminous mixtures and surface treatments for roads, airfields and other trafficked areas (BSI, 2002).

FPC and CE marking requirements are included but this standard contains two systems of AVCP, the one that applies being dependent on the properties of the aggregate.

The products are categorised in accordance with their PSV, and AVCP 2+ applies to those aggregates with a PSV of 58 or greater. AVCP 4 applies to the remainder. This is because, in the UK, coarse aggregates for use in skid-resistant surfacings (PSV $\geq$ 58) have been defined as high specification aggregates with an influence on road safety.

23.4.3 *Asphalt*

The specifications for asphalts are found in the BS EN 13108 series of standards, which are currently being reviewed and updated. The requirements for FPC and ITT (the previous term for a DoP) for all the mixtures are grouped together as

- BS EN 13108-20:2006 Bituminous mixtures – Material specifications – Part 20: Type testing (BSI, 2006b)
- BS EN 13108-21:2006 Bituminous mixtures – Material specifications – Part 21: Factory production control (BSI, 2006c).

Part 20 of the standard identifies which test methods are used for each of the different types of asphalt, and Part 21 gives the requirements for the frequency of testing. Testing is undertaken on the constituent materials and also on the manufactured asphalt.

AVCP 2+ applies to asphalts, except where they are used in situations subject to reaction to fire regulations. Where this situation applies, the level of AVCP may be higher, and will need to be determined from Annex ZA.2, taking into account the implications of any national legislation.

The scope of BS EN 13108 is the manufacture of asphalts, and the CE marking applied relates to the material as it leaves the production facility, not the installed material.

23.4.4 *Surface treatments*

Both surface dressing and slurry surfacing are now within the scope of harmonised standards, and can therefore be CE marked. The relevant standards are

- BS EN 12271:2006 Surface dressing – requirements (BSI, 2006a)
- BS EN 12273:2008 Slurry surfacing – requirements (BSI, 2008b).

Each of these standards contains requirements for FPC, including the required tests and frequencies. The testing requirements include the constituent materials, the equipment used and a method of visual assessment of defects after 1 year in service.

The declaration of performance is treated differently than in most other standards, as these products cannot be sampled from a production plant prior to installation and tested in a laboratory in the conventional way. Instead, the producer carries out a type approval installation trial (TAIT).

The trial is undertaken on a representative site at least 200 m long, selected by the producer. The installation is monitored by the producer and all the data required by the FPC are recorded. The design details, including the traffic volumes defined by road category, are also recorded. The performance is then assessed by the producer after 1 year in service using the visual assessment method in either BS EN 12272-2:2003 (BSI, 2003) or BS EN 12274-8:2005 (BSI, 2005) as appropriate and measurement of the macrotexture is carried out.

The results from these tests are recorded on a TAIT report, which provides the performance categories for the CE mark certificate.

AVCP 2+ is applied to both types of surface treatment.

References

AFNOR (2014) French national standards organisation. See http://www.afnor.org/en (accessed 16/10/2014).

British Board of Agrément (2014) HAPAS Certificates. See http://www.bbacerts.co.uk/product-approval/hapas (accessed 16/10/2014).

BSI (British Standards Institution) (2002) BS EN 13043:2002. Aggregates for bituminous mixtures and surface treatments for roads, airfields and other trafficked areas. BSI, London, UK.

BSI (2003) BS EN 12272-2:2003. Surface dressing. Test methods. Visual assessment of defects. BSI, London, UK.

BSI (2005) BS EN 12274-8:2005. Slurry surfacing – Test methods. Visual assessment of defects. BSI, London, UK.

BSI (2006a) BS EN 12271:2006. Surface dressing – requirements. BSI, London, UK.

BSI (2006b) BS EN 13108-20:2006. Bituminous mixtures – Material specifications – Part 20: Type testing. BSI, London, UK.

BSI (2006c) BS EN 13108-21:2006. Bituminous mixtures – Material specifications – Part 21: Factory production control. BSI, London, UK.

BSI (2006d) BS EN 13924:2006. Bitumen and bituminous binders – Specifications for hard paving grade bitumens. BSI, London, UK.

BSI (2008a) BS EN ISO 9001:2008. Quality management systems. Requirements. BSI, London, UK.

BSI (2008b) BS EN 12273:2008. Slurry surfacing – requirements. BSI, London, UK.

BSI (2009) BS EN 12591:2009. Bitumen and bituminous binders – Specifications for paving grade bitumens. BSI, London, UK.

BSI (2010a) BS EN 14023:2010. Bitumen and bituminous binders – Specification framework for polymer modified bitumens. BSI, London, UK.

BSI (2010b) PD 6691:2010. Guidance on the use of BS EN 13108 Bituminous mixtures – Material specifications. BSI, London, UK.

BSI (2013a) BS EN 13808:2013. Bitumen and bituminous binders – Framework for specifying cationic bituminous emulsions. BSI, London, UK.

BSI (2013b) BS EN 15322:2013. Bitumen and bituminous binders – Framework for specifying cut-back and fluxed bituminous binders. BSI, London, UK.

BSI (2014a) See http://www.bsigroup.com/en-GB (accessed 16/10/2014).

BSI (2014b) BSI Kitemark. See http://www.bsigroup.com/en-GB/our-services/product-certification/kitemark (accessed 16/10/2014).

CEN (European Committee for Standardization) (2014) See http://www.cen.eu/Pages/default.aspx (accessed 16/10/ 2014).

DIN (Deutsches Institut für Normung) (2014) German national standards organisation. See http://www.din.de/cmd?level=tpl-home&contextid=din (accessed 16/10/2014).

EC (European Commission) (1989) Directive 89/106/EEC of the European Parliament and of the Council of 21st December 1988 on the approximation of laws, regulations and administrative provisions of the Member states relating to construction products. EC, Brussels, Belgium.

EC (2004) Directive 2004/18/EC of the European Parliament and of the Council of 31st March 2004 on the coordination of procedures for the award of public works contracts, public supply contracts and public service contracts. EC, Brussels, Belgium.

EC (2011) Regulation (EU) No. 305/2011 of the European Parliament and of the Council of 9th March 2011 laying down harmonised conditions for the marketing of construction products and repealing Council Directive 89/106/EEC. EC, Brussels, Belgium.

EOTA (European Organisation for Technical Assessment) (2014) See http://www.eota.eu/en-GB/content/home/2 (accessed 16/10/2014).

Highways Agency (2014) *The Manual of Contract Documents for Highway Works*. Volume 1 – *Specification for Highway Works*. See http://www.dft.gov.uk/ha/standards/mchw/vol1/index.htm (accessed 16/10/2014).

ISO (International Organization for Standardization) (2014) See http://www.iso.org/iso/home.htm (accessed 16/10/2014).

LANTRA (2014) NHSS Schedule of suppliers. See http://www.lantra-awards.co.uk/Schedule-of-Suppliers.aspx (accessed 16/10/2014).

Nando (New Approach Notified and Designated Organisations Information System) (2014) See http://ec.europa.eu/enterprise/newapproach/nando (accessed 08/10/2014).

OJEU (Official Journal of the European Union) (2014) See http://eur-lex.europa.eu/JOIndex.do?ihmlang=en (accessed 16/10/2014).

UKAS (United Kingdom Accreditation Service) (2013) NHSS 13: The Supply and Application of Surface Treatments to Road Surfaces. UKAS, Feltham, UK.

UKAS (2014a) Publications relating to sector schemes for quality management in highway works. See http://www.ukas.com/Technical-Information/Publications-and-Tech-Articles/Publications/PubsForCBAccred.asp (accessed 16/10/2014).

UKAS (2014b) NHSS 14: The Quality Management of the Production of Asphalt Mixes. UKAS, Feltham, UK.

UKAS (2014c) NHSS 15: The Supply of Paving Bitumens. UKAS, Feltham, UK.

UKAS (2014d) NHSS 16: The Laying of Asphalt Mixes. UKAS, Feltham, UK.

Chapter 24

Other important uses of bitumens and asphalts

Dr Sherry Guo
Senior Application Specialist,
Shell Oil Products US. USA

Xie Yongqing
Application Advisor,
Shell China Ltd. China

Xu Liting
Senior Application Specialist,
Shell China Ltd. China

Andrew Wayira
Global Product Portfolio
Specialist, Shell Markets
(Middle East) Ltd. UAE

Jeyan Vasudevan
Senior Application Specialist,
Shell Malaysia Trading Sdn
Bhd. Malaysia

The majority of bitumen and asphalt usage is in road construction. However, bitumens and asphalts are also used in a range of other important applications. This chapter considers some of these applications

- bitumen in roofing applications
- bituminous adhesives for pavement markers application
- airfield pavements
- conventional and high speed railway applications
- bridges
- recreational areas
- motor racing tracks including Formula 1 tracks
- vehicle testing circuits
- hydraulic applications
- coloured surface courses and surface treatments
- asphalt kerbs.

The UK's Mineral Products Association (MPA) has published a number of useful application guides, including one for decorative and coloured surfacing

The Shell Bitumen Handbook, Sixth Edition
ISBN 978-0-7277-5837-8
Shell International Petroleum Company Ltd: All rights reserved
http://dx.doi.org/10.1680/tsbh.58378.703

(MPA, 2009a), asphalt in construction of games and sport areas (MPA, 2009b), miscellaneous uses of asphalt (MPA, 2009c) and airfields (MPA, 2009d). They can all be found on the MPA's website (MPA, 2014). Additionally, two books published by Shell provide substantial detail on industrial (Morgan and Mulder, 1995) and hydraulic applications (Scönian, 1999).

24.1 Bitumen in roofing applications
The bitumen industry has three primary areas for its products

- paving
- roofing
- industrial applications.

Roofing products consume approximately 10–15% of the market share of asphalts and bitumens. The roofing sector of the bitumen market is divided between residential steep-slope shingles and commercial modified waterproof membranes. This section addresses bitumen roofing applications in both residential and industrial construction.

24.1.1 Roofing asphalt flux manufacture
Roofing flux is a vacuum residue produced during the distillation of crude oil. It is called roofing flux because it is used to produce harder grades of oxidised bitumen (via a bitumen blowing unit) for use in roofing products. Over 2000 different crudes are produced worldwide, and they are all different in terms of both their physical and chemical properties. The most critical aspect of roofing flux is the source of the crude oil; less than 20% of crudes make acceptable material. The main sources are Middle East and North and South American crudes.

24.1.2 Atmospheric distillation
Crude oil is a very complex mixture of hydrocarbons differing in molecular weights and boiling ranges. Manufacturing of residue involves several steps of physical separation and chemical treatment. The first step is fractional atmospheric distillation, where the crude is heated to temperatures in the range 350–380°C. The column bottom stream is called atmospheric or 'long' residue. It requires further vacuum flashing before it can be used as roofing flux. The process is illustrated in Figure 2.1.

24.1.3 Vacuum flashing
The atmospheric residue needs further distillation at reduced pressure in a high vacuum unit. Typical process conditions are 10–100 mmHg pressure with a temperature in the range 350–425°C. The vacuum column removes the non-condensable gas that enters the vacuum unit with the feed. The process is illustrated in Figure 2.2.

To prevent cracking of the heavy components, the residence time at temperatures above 350°C has to be as short as possible. The bottom stream obtained by vacuum distillation of the long residue is called the vacuum or 'short' residue (also known as vacuum tower bottom (VTB)). This VTB residue can be used as roofing flux if the crude source is suitable.

24.1.4 Roofing products

The two general categories of roofing products are steep slope and low slope. Steep slope roofings are coverings installed on slopes exceeding 14°, and mostly occur on residential properties. Low slope roofing includes waterproofing systems installed on slopes that are 14° or less. This type of roofing is usually used for commercial buildings. Steep slope roofing is the dominant product, representing over 75% of the total roofing market. Data on the usage of the available types are shown in Table 24.1 (ARMA et al., 2011). As can be seen from Table 24.1, in the USA most steep slope roofing consists of shingles, an example of which is shown in Figure 24.1. Note that built-up systems are those where several layers of roof felt are laminated together with bitumen.

Asphalt shingles are composed of the following components

- fibreglass felt
- coating asphalt (to provide weather resistance and adhesion)
- mineral filler
- mineral granules.

Table 24.1 North American and European bitumen roofing production by roof type (2006) (© 2011 Asphalt Roofing Manufacturers Association)

	Steep slope roofing: $m^2 \times 10^6$	Low slope roofing: $m^2 \times 10^6$
North America		
Shingles	3403	0
Built-up roofing	0	298
Bitumen membranes and underlay	1418	0
Polymer modified bitumen roofing	0	309
Total	4821	607
Market share	89%	11%
Europe		
Shingles	40	0
Built-up roofing	0	2
Bitumen membranes and underlay	14	273
Polymer modified bitumen roofing	0	610
Total	54	885
Market share	6%	94%

Figure 24.1 Shingles on a roof

24.1.5 Roofing coating specification

The American Society for Testing and Materials (ASTM) produces standard specifications for coating bitumens used in the roofing market in ASTM D312-00 (ASTM, 2000) (Table 24.2).

24.1.6 Roofing shingles specification

Standard specifications for asphalt shingles made from glass felt and surfaced with mineral granules are given in ASTM D3462/D3462M-10a (ASTM, 2010d). The document requires that several properties are met by

Table 24.2 Physical requirements for bitumen in roofing (ASTM D312-00 (ASTM, 2000)). (© 2011 Asphalt Roofing Manufacturers Association)

Test	Minimum/maximum			
	Type I	Type II	Type III	Type IV
Softening point: °F	135/151	158/176	185/205	210/225
Softening point: °C	57/66	70/80	85/96	99/107
Penetration (25°C,100 g, 5 s): dmm	18/60	18/40	15/35	12/25
Penetration (0°C, 200 g, 60 s): dmm	3+	6+	6+	6+
Penetration (46°C, 50 g, 5 s): dmm	90/180	100	90	75
Ductility (25°C, 5 cm/min): cm/min	10.0+	3.0+	2.5+	1.5+
Solubility in TCE: %	99.0	99.0	99.0	99.0
Flash point: °C	288+	288+	288+	288+

TCE, trichloroethylene.

Table 24.3 Standard specification for roofing shingles (ASTM D3462/D3462M-10a (ASTM, 2010d))

Property	Min.	Max.
Loss of volatile matter: %	–	1.5
Sliding of granular surfacing: in.	–	1/16
Tear strength: g	16.7	–
Fastener pull-through resistance (73°F): lbf		
Average of single-layer specimens	20	–
Average of multi-layer specimens	30	–
Fastener pull-through resistances (32°F): lbf		
Average of single-layer specimens	23	–
Average of multi-layer specimens	40	–
Wind resistance	Class A	–
Fire resistance	Class A	–
Penetration of bitumen (without mineral stabiliser, 77°F): dmm	15	–
Pliability (73°F, all directions)	4 of 5 shall pass	
Asphalt softening point (without mineral stabiliser, R&B): °F		
Unmodified	190	235
Polymer modified	190	320
Weight of displaced granules: g	–	1.0

R&B, ring and ball test.

bitumen shingles, as shown in Table 24.3. As can be seen in Table 24.3, two principal properties – penetration and softening point – are required that relate solely to the coating bitumen used in the shingle manufacture. These two properties provide some indication of the stiffness of the bitumen at intermediate and high temperatures.

24.1.7 Roofing flux specification

There is no standard specification for roofing flux, either from the ASTM or from the industry itself. Thus, flux specifications vary between producers, manufacturers and suppliers. Roofing industry companies have different requirements for flux quality depending on their specific needs.

The most critical criteria for roofing flux, as related to roofing manufacture, are the properties of the crude oil used in making the roofing flux, as the end product is the oxidised coating material resulting from the air blowing process. The chemistry of the crude oil contributes to the resulting performance. Even if a flux meets the required physical properties, it does not follow that it has the required coating properties. Therefore, in most cases, a new flux has to pass an approval process before it can be used in the manufacture of roofing products.

As there is no well established relationship between the flux and the coating performance, the flux requirement (not a specification) can only come from a pre-approved crude oil resource or from a pre-approved supply. Assay of the crude oil does not indicate whether it will result in a good roofing flux or a

Table 24.4 Physical properties of four bitumens typically used in North American roofing made from crude source A (ARMA et al., 2011)

Test and units	Test method[a]	Straight run bitumen		Oxidised bitumen	
		Polymer modified bitumen base	Roofing flux	Shingle saturate	Shingle coating
Softening point: °C	ASTM D36	41	41	54	101
Penetration (25°C): dmm	ASTM D5	170+	220+	64	18
Viscosity (60°C): poise	ASTM D4402	140	140	355	13 000
Loss on RTFO (163°C): %	ASTM D2872	0.1880	0.0418	−0.0554	0.0043
Flash point (COC): °C	ASTM D92	313	313	320	320
Stain index	ASTM D2746				b
Weathering	ASTM D4798				b

[a] ASTM D36/D36M-14e1 (ASTM, 2014b); ASTM D5 (ASTM, 2006); ASTM D4402/D4402M-13 (ASTM, 2013b); ASTM D2872-12e1 (ASTM, 2012c); ASTM D92 (ASTM, 2012b).
[b] Varies by manufacturer

good shingle coating. Roofing manufacturers need good oxidised coating performance rather than good flux properties.

However, there are some basic requirements that can be used for flux: for example, for paving grade bitumens the flash point (Cleveland open cup (COC) method) needs to be a minimum of 550°F (250°C)); and, for reasons of safety, some roofing manufacturers require the closed cup flash point (Pensky–Martens closed cup (PMCC) test) to be a minimum of 450°F (232°C). For a roofing flux made from a pre-approved crude oil, the viscosity range can be used to specify the properties. This can be the absolute viscosity at 60°C, or the kinematic viscosity at 135°C. The stain index is also critical, because it represents the volatile oil content and relates to the discolouration performance of the coating. The requirement for the stain index varies between different manufacturers (Tables 24.4 and 24.5).

Table 24.5 Physical properties of four bitumens typically used in North American roofing made from crude source B (ARMA et al., 2011)

Test and units	Test method[a]	Straight run bitumen		Oxidised bitumen	
		Polymer modified bitumen base	Roofing flux	Shingle saturate	Shingle Coating
Softening point: °C	ASTM D36	42	37	54	100
Penetration (25°C): dmm	ASTM D5	170+	300+	53	17
Viscosity (60°C): poise	ASTM D4402	150	100	310	10 000+
Loss on RTFO (163°C): %	ASTM D2872	0.066	0.165	−0.0730	−0.0208
Flash point (COC): °C	ASTM D92	320	318	335	330
Stain index	ASTM D2746				b
Weathering	ASTM D4798				b

[a] ASTM D36/D36M-14e1 (ASTM, 2014b); ASTM D5 (ASTM, 2006); ASTM D4402/D4402M-13 (ASTM, 2013b); ASTM D2872-12e1 (ASTM, 2012c); ASTM D92 (ASTM, 2012b).
[b] Varies by manufacturer

24.1.8 Product approval

Shingle manufacturers provide guarantees on their products with periods ranging from 15 years' to a lifetime warranty. The durability of the coating (filled and unfilled) is critical. The accelerated xenon arc weathering test (ASTM, 2011) is the traditional method used to evaluate the long term ageing performance. The coating panel must show less than 10% of failure after a 2000 h cycle of exposure. The time needed to pass the weathering test is usually 3–4 months.

The coating material also needs to meet certain softening point/penetration relationship criteria, without the inclusion of any additives. Once the coating has passed the testing phases, a plant trial is carried out to evaluate the material's performance in the process. A trial is typically one to three truck loads (approximately 25–75 tonnes). Following a trial, it usually takes around a month to complete the product evaluation.

24.1.9 Air blowing of roofing flux

Oxidation requires a complex series of reaction processes in order to produce the bitumen coating product from a roofing flux. The blowing process can be carried out in batch or continuous mode. The batch production mode is the one most commonly used in the USA. The reaction process introduces oxygen into the system; the temperature is typically in the range 480–500°F (218–260°C). A series of chemical thermal reactions takes place during production. There are mainly two types of reaction

- dehydrogenation, which introduces oxygen to hydrogen at the molecular level
- polymerisation, which converts most aromatic compounds to a condensed asphaltene structure.

It has been found that all the oxygen taken up by the bitumen can be accounted for by the formation of hydroxyl, carbonyl, acid and ester groups (ARMA *et al.*, 2011).

The air blowing time depends on the properties of the flux, and the final oxidised product requirements. As a general rule, the softer the flux is at the start, the longer it will take to achieve the final product. This is illustrated in Figure 24.2: the higher the softening point to be achieved, the longer the time that is needed.

In terms of its chemical composition, the flux changes during the oxidation process, the proportion of saturates tends to remain the same, some aromatics convert to resins, some resins convert to asphaltenes and the asphaltenes themselves remain unchanged. The net result is that there is an increase in asphaltene content and a dramatic drop in aromatic content (Table 24.6).

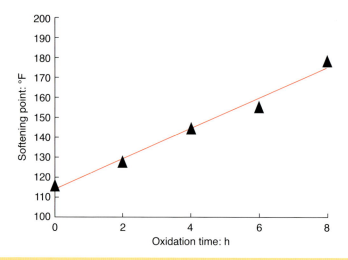

Figure 24.2 Softening point/oxidation time chart

After air blowing (Figure 2.4), the end product (typically called 'bitumen coating' in Europe and 'asphalt coating' in the USA) exhibits decreased penetration, an increased softening point, higher viscosity, improved weathering resistance and good temperature susceptibility.

24.1.10 Asphalt low slope roofing products

Low slope roofs have slopes of 14° or less. Most low slope roofs are used in commercial or industrial buildings. Bitumen is used in two different low slope roof systems: polymer modified bitumen membranes and built-up roofing (BUR) products. In the USA, these two systems account for 35–40% of the low slope market. The total share of low slope bitumen systems is even higher in western Europe, where polymer modified bitumen systems dominate the low slope bitumen market (Figure 24.3).

24.1.11 Polymer modified membrane systems

Polymer modified bitumen membranes are often used in commercial roofing, including low slope roofing applications.

The most common base bitumens are European paving grades 160/220 and 100/150 pen, depending on the market and customer requirements.

Table 24.6 Typical changes in SARA between the flux and the coating

	Roofing flux	Oxidised roofing flux
Saturates: %	15.0	14.5
Aromatics: %	38.5	27.2
Resins: %	35.5	27.2
Asphaltenes: %	11.0	31.1

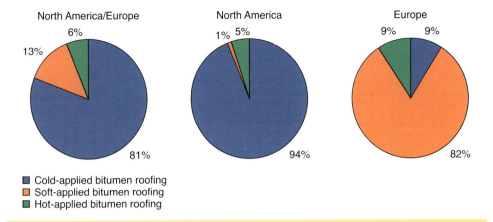

Figure 24.3 North American/European bitumen roofing production by application temperature (ARMA *et al.*, 2011) (© 2011 Asphalt Roofing Manufacturers Association)

Polymers typically used in this application are atactic polypropylene (APP) and styrene–butadiene–styrene (SBS). Different polymers require different bases. In most instances, a base with the highest saturate content would be suitable with an APP, whereas a base with a higher aromatic content would be appropriate for use with an SBS application.

An APP modified bitumen membrane felt typically contains

- bitumen 52–63%
- APP 15–25%
- filler 20–30%,

while an SBS modified membrane felt typically contains

- bitumen 60–70%
- SBS 5–15%
- filler 20–35%.

Specifications for APP modified bituminous sheet materials are given in ASTM D6223-02 (ASTM, 2009e), and specifications for SBS modified bituminous sheet materials using glass fibre reinforcement can be found in D6163-00 (ASTM, 2008) (Table 24.7).

24.1.12 *Future trends for roofing bitumen shingles and coating evaluation*

Although rheological testing of paving grade bitumens has become much more widespread over the last two decades as a result of the Strategic Highway Research Program (SHRP), the conventional empirical tests (e.g. penetration, ring and ball (R&B) softening point and viscosity) remain the

Table 24.7 Polymer modified membrane roofing felt requirements

Source information for APP roofing felt requirements[a]	UEAtc[b]	Manufacturer			
		A	B	C – low performance	C – high performance
Country		France	France	UK	UK
Penetration (25°C, 0.1 mm; ASTM D5)				20–30	20–30
Penetration (50°C, 0.1 mm; ASTM D5)			<80		
Penetration (60°C, 0.1 mm; ASTM D5)				90–120	70–100
Dynamic viscosity (180°C; ASTM D2170): cP				4000–6000	3500–5500
Softening point (R&B; ASTM D36): °C	120	140–155	150	>150	>150
Flexibility at low temperature (ASTM C711-14): °C	−5	−10	−20	−5	−10
Durability requirement – ageing 6 months at 70°C					
Softening point (R&B; ASTM D36): °C	110	130	150		
Flexibility at low temperature (without damage): °C	0	0	−20		

R&B, ring and ball test.
[a] ASTM D5 (ASTM, 2006); ASTM D2170/D2170M (ASTM, 2010b); ASTM D36/D36M-14e1 (ASTM, 2014b); ASTM C711-14 (ASTM, 2014a).
[b] UEAtc: European Union for technical approval in construction.

dominant tests and requirements for oxidised coating and BUR products in the bitumen roofing industry in North America. The rapid development of the dynamic shear rheometer (DSR), including the 4 mm parallel plate measurement for low temperature testing and the master curve application for paving grade bitumen, and the belief that the results can be linked to bitumen performance, has created an interest in employing these techniques for use in the roofing industry. More detailed information about rheological testing of bitumen is given in Chapter 7.

Due to proprietary issues in the roofing industry, some individual roofing companies have undertaken research based on rheological properties using rotational rheology tests. This is an attempt to link the rheological properties of the bitumen to the performance of coatings and shingles, and characterisation of roofing fluxes. However, little has been done to produce a unified or standard specification for bitumen fluxes or coatings. For the last 10 years, the Asphalt Institute has, in cooperation with the roofing industry, been very actively involved in research based on rheological properties (Asphalt Institute, 2014). The purpose of this research is to explore the possibility of developing a new standard based on the use of rheological parameters as a measure of oxidised bitumen coating, which could be linked to shingle performance with a view to replacing the conventional empirical testing. For example, a useful bitumen relationship has been developed based on the complex modulus (G^*) and penetration (for more details see Chapter 7)

$$\log(G^*_{T=25°C, \ f=0.4 \ Hz}) = 2.923 - 1.9 \log(\text{penetration})$$

Table 24.8 Relationship between G*2.5 and penetration at 25°C

Penetration: dmm	>20	16–20	11–15	≤10
Average G*2.5 at 25°C: pa	3.37E6	4.42E6	6.98E6	1.44E7
Average penetration at 25°C: dmm	21	18	13	9
Number of samples	4	9	7	2

The current consensus is to use the rotational rheology test to characterise the oxidised bitumen coating material at intermediate and higher temperatures

■ DSR at 25°C, 8 mm plate, 1% strain at a frequency of 2.5 rad/s
■ DSR temperature sweep tests from 90°C to 110°C in 10°C increments.

Similar studies of predicting ageing behaviour using rheological testing have been carried out on paving grade binders (Farrar *et al.*, 2013; Mike, 2013; Qin *et al.*, 2014), and these suggest that a master curve can be generated on the basis of the temperature sweep test.

The rheological parameter ($G*2.5$ – the complex modulus multiplied by 2.5) was chosen to create a correlation with conventional physical properties such as penetration and softening point (Mike, 2013) (Table 24.8). The data suggest that there is a linear relationship in the penetration range 13–25 with a variability of 13.5%. Figure 24.4 shows the relationship between $G*$ (complex modulus) and the softening point for oxidised coating materials.

The relationship from all data is

■ log $G*2.5 = -1.6281*$log(pen) $+ 8.6582$
Penetration $= 12$ dmm

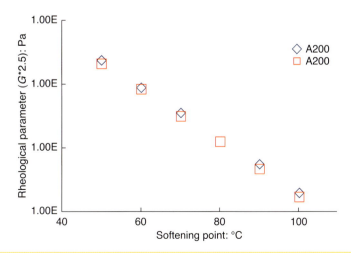

Figure 24.4 Relationship between the rheological parameter $G*2.5$ (Pa) and the softening point (°C)

- $G^*2.5 = 7.96E6$ Pa
  Penetration $= 25$ dmm
- $G^*2.5 = 2.41E6$ Pa
  Proposed specification values:
  $2.50E6 \leq G^*2.5(25°C) \leq 8.00E6$
- Testing variability for T 315 (PAV DSR)
  Single operator d2s% $= 13.8\%$.

The proposed use of the rheological parameter ($G^*2.5$) criterion to replace the softening point and penetration tests is as follows.

- Proposed procedure to replace softening point
  - perform a temperature sweep using a DSR
  - 25 mm parallel plate, 10% strain, 1 rad/s
  - start at 90°C, increase to 110°C in 10°C increments
  - plot η^*(1 rad/s) versus temperature on a semi-logarithmic graph
  - calculate T_c when η^*(1 rad/s) $= 1200$ Pa·s.
- Proposed procedure to replace penetration at 25°C
  - perform single point temperature test using a DSR
  - 8 mm parallel plate, 1% strain, 2.5 rad/s, 25°C
  - determine $G^*2.5$ and compare with the recommended specification value.

Work is underway to evaluate the proposed criteria further.

24.2 Bituminous adhesive for use with pavement markers

Flexible bituminous pavement marker adhesive is a hot-melt thermoplastic bituminous material used for bonding raised pavement markers and recessed pavement markers to the pavement. There are two types of bituminous marker adhesive

- standard bituminous marker adhesive, which consists of an asphalt base with a homogeneously mixed mineral filler
- flexible bituminous marker adhesive, which consists of a highly polymer modified asphalt – this type normally contains ground tyre rubber (GTR) as one modifier.

Both adhesives must be suitable for bonding ceramic and plastic markers to cementitious, asphalt concrete and surface dressed road surfaces, and be applicable when road surface and marker temperatures are in the range 4–71°C. Prior to application, adhesives are heated in either an air or an oil-jacketed melter up to a temperature of around 220°C, and must be able to tolerate these elevated temperatures without deteriorating.

24.2.1 Manufacture of bituminous marker adhesive

Standard bituminous marker adhesive (BMA) can be made from a relatively soft bitumen base made harder through light distillation, an oxidation process

Table 24.9 Adhesive properties

Property	Material		Standard adhesive	
	Test method[a]		Min.	Max.
Softening point: °C	ASTM D36		93	
Penetration (100 g, 5 s, 25°C, 0.1 mm)	ASTM D5		10	20
Flow (5 h, 70°C (158°F)): mm				5
Heat stability flow (5 h, 70°C (158°F)): mm				5
Viscosity (10 rpm, 204°C (400°F)): Pa·s				7.5
Flash point (COC): °C	ASTM D92		288	
Ductility (5 cm/min, 25°C): cm	ASTM D113			
Flexibility (1 in. mandrel, 90° bend, 10 s)	ASTM D3111			

[a] ASTM D36/D36M-14e1 (ASTM, 2014b); ASTM D5 (ASTM, 2006); ASTM D92 (ASTM, 2012b); ASTM D113 (ASTM, 1999); ASTM D3111 (ASTM, 2010c).

or the addition of a hard component. Alternatively, a hard bitumen base could be employed at the outset, modified with additives and filler. More often than not, the formula and production process are the intellectual property of the manufacturer (DMS, 2008).

Flexible BMA is a polymer modified marker adhesive. It is normally a GTR-based modified product. In order to achieve other requirements such as ductility, other elastic polymers may be needed in the formulation. Companies that make these products in the USA include

- Crafco (Crafco, 2008)
- Martin Asphalt (Martin Asphalt, 2014)
- Henry.

Each company produces both standard and flexible BMA in their own brands. Both standard and flexible BMA need fillers added to them to comply with the specific standards listed in Tables 24.9 and 24.10. The filler content can be varied, depending on the neat base. Normally, the filler content is in the range 40–75%.

Table 24.10 Properties of flexible bituminous marker adhesives

Specification	Test method[a]	Requirements	Typical results
Softening point: °C	ASTM D36	93 min.	102
Penetration (100 g, 5 s, 25°C, 0.1 mm)	ASTM D5	30 max.	22
Rotational viscosity (191°C): Pa·s	ASTM D4402	2–6	3.8
Flow (70°C): mm	ASTM D5329	5 max.	0
Flexibility (−7°C)	ASTM D3111	No breaks or cracks	pass
Ductility (25°C, 5 cm/min): cm	ASTM D113	15 min.	30
Ductility (4°C, 1 cm/min): cm	ASTM D113	5 min.	10
Flash point: °C	ASTM D92	288 min.	315

[a] ASTM D36/D36M-14e1 (ASTM, 2014b); ASTM D5 (ASTM, 2006); ASTM D4402/D4402M-13 (ASTM, 2013b); ASTM D5329-09 (ASTM, 2009d); ASTM D3111 (ASTM, 2010c); ASTM D113 (ASTM, 1999); ASTM D92 (ASTM, 2012b).

Table 24.11 Filler-free bitumen properties

Property	Minimum	Maximum	Test method[a]
Penetration (100 g, 5 s, 25°C): dmm	25		ASTM D5
Viscosity (135°C): Pa s	1.2		ASTM D2171
Viscosity ratio (135°C)		2.2	ASTM D1754

[a] ASTM D5 (ASTM, 2006); ASTM D2171 (ASTM, 2010a); ASTM D1754 (ASTM, 2009b).

24.2.2 Requirements and specifications for bituminous marker adhesives

24.2.2.1 Adhesive properties

The adhesive must be smooth and homogeneous and contain no visible particles, and must comply with the requirements shown in Table 24.9.

24.2.2.2 Properties of base bitumen

This applies to standard bituminous adhesive only. The filler-free bitumen, obtained from the extraction and Abson recovery process, must have the properties shown in Table 24.11.

24.3 Airfield pavements

Uninterrupted flow of air traffic and safe traffic conditions are essential for all airport owners and their customers. The combination of the large number of airports and the increase in air travel encouraged by low cost flights means that construction works are ongoing at many airports. This is necessary for reasons of expansion and maintenance to keep the taxiways and runways in a serviceable condition.

In airports, both asphalt pavements and concrete pavements are used. Consideration of a number of factors related to cost, performance and environment will assist in explaining the increasing use of asphalt pavements on airfields.

In matters of cost, flexible pavements have the following characteristics.

- Initial construction costs for flexible pavements are generally lower than those for concrete pavements.
- Less time is required for curing before the facility can be opened to traffic.
- Although, in general, maintenance costs for asphalt pavements can be slightly higher than those for concrete pavements, repair of flexible pavements is far easier and quicker than repair of concrete pavements, meaning less downtime for airport operation, a facet that may completely reverse the maintenance cost argument.
- Whole life costs of an asphalt pavement are generally lower than those of a concrete pavement.

In relation to performance and environmental issues, asphalt pavements can exhibit the following characteristics in comparison to concrete pavements

- lower noise levels
- less cracking, and thus less ravelling or loss of aggregate, which is an important safety aspect for aircraft
- lower spray and aquaplaning risks, particularly with negatively textured asphalts
- higher friction when wet
- vastly superior surface regularity
- recycling of existing asphalt in new construction is easy and already common practice
- asphalt pavements require about 20% less energy to poduce and construct than other pavements (EAPA, 2014).

The use of flexible pavements manufactured with polymer modified bitumen has become much more common in airfield works. Traditional asphalt mixture design and the design of flexible pavements do not take account of the improved characteristics of asphalts that are modified in this manner.

Many UK airports now use open graded mixtures for the surface course. This material, known as the 'friction course', allows water to permeate through the layer, thus maximising the frictional characteristics of the running surface. An impervious binder course, placed below the surface course, and laid to appropriate falls, causes the rainwater to flow away from the pavement. On airfields carrying aircraft that are relatively light, the use of the standard road mixtures may well be quite adequate.

24.3.1 Design of flexible airfield pavements

Airfield and highway flexible pavements have many similarities but also some significant differences. The key distinctions between highways and airfields are the types and frequencies of loads that are experienced when the pavement is in service. Airfield pavements tend to experience far fewer load repetitions over their design lives than do highway pavements. For highways, the number of heavy goods vehicles is the primary characteristic used to specify pavement structure and materials. This spectrum of traffic has been quantified by the use of standard axles in Europe, and equivalent single axle loads (ESALs) in the USA and other areas where US practice is followed as the controlling factor for both pavement design and selection of asphalts. For airfields, the pavement is generally designed and specified based on a particular design of aircraft. Depending on whether the airport is a small general aviation airport or a large commercial airport, the design aircraft can be as small as a Cessna Skyhawk, which has a gross weight of approximately 1.36 tonnes, or an Airbus A380, which has a maximum take off weight of 590 tonnes (Burns Cooley Dennis, Inc., 2009).

Another factor related to loads is tyre pressure. Small aircraft can have tyre pressures similar to cars, while some military fighter jets can have tyre pressures in excess of 20 bar.

Another difference between airfield and highway pavements is the traffic patterns. For highways, the traffic is generally channelised and falls within narrow wheel tracks on the road. Traffic patterns on airfields can vary from channelised moving (taxiways) to channelised stacked (ends of runways or taxiways) to evenly distributed and random (aprons) to occasional (runway edges) to almost never (shoulders and blast pads).

The number of loads or passes on airfields is, in general, much lower than that found on roads carrying high volumes of traffic but the loads and tyre pressures on aircraft can be significantly higher. This has important consequences for the pavement design and the choice of materials. Figure 24.5 illustrates the significant difference in loading and tyre pressures on roads and airfields.

The high loads on airfields will produce critical stresses and strains in the subgrade and lower layers of the pavements if not properly designed. The extent to which critical stresses and strains are produced in the upper layers of the pavement is dependent on the wheel configuration of the aircraft.

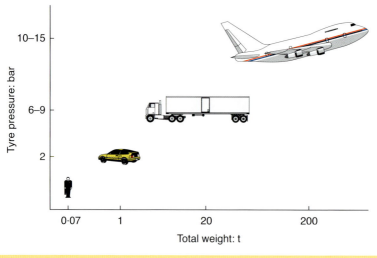

Figure 24.5 Loading and tyre pressures on road and airfield

The traffic situation and loading severity varies greatly at different locations in an airfield, with the ends of runways and taxiways being subject to the greatest concentrations of traffic. In particular, the design of taxiways that are used for departing traffic is critical. Runways are more critical with respect to cracking and adhesion but have a lesser priority from a design perspective.

24.3.1.1 Design methods

The International Civil Aviation Organization (ICAO) classifies the bearing capacity of airfield pavements in terms of a 'pavement classification number' (PCN). Similarly, the severity of loading from aircraft is expressed in terms of an 'aircraft classification number' (ACN). Several design procedures can be used to select materials, layer thickness, mixture type etc. that will provide a pavement structure with PCN values that exceed the ACN values required by the ICAO.

Some of the most commonly used design procedures for flexible pavements on military and civilian airfields are

- FAA method (Federal Aviation Administration, USA)
- LCN method (UK load classification number system)
- the French method (ICAO French practice)
- Canadian Ministry of Transportation method
- Corps of Engineers method (CBR method)
- Asphalt Institute method
- Shell Pavement Design Method (SPDM).

Most of these systems are empirical, being based on the assessment of the California bearing ratio (CBR). Exceptions are the Asphalt Institute method and SPDM, which are analytical pavement design methods based on calculations and criteria for stresses and strains in the various pavement layers.

These design systems recommend a minimum thickness of asphalt pavement. The methods listed above have few or no specific requirements for the asphalt layers. For this reason, theoretical models for pavement responses are often used as a supplementary tool to optimise the pavement construction and selection of materials. In relation thereto, analytical software such as BISAR is often used. BISAR was designed to calculate and analyse the stresses and strains within the layers of a pavement, and this program was used in drawing up the SPDM design charts that were issued in 1978. More information on SPDM can be found in sections 13.9 and 13.10.

24.3.2 Requirements for airfield pavements

Airfield pavements must perform certain functions in line with agreed international standards. The consequences of a pavement that performs poorly may be dangerous and very costly. Detached pieces of aggregate may be ingested by jet engines or strike aircraft moving at speed, causing serious and possibly fatal foreign object damage (FOD). As a consequence, maintenance on airfield pavements is often focused on minimising the risk of FOD.

The most important requirements for asphalt pavements on airfields are adequate

- wet friction for landing aircraft
- resistance to ravelling and cracking to minimise the risk of FOD
- surface regularity
- surface water drainage
- resistance to rutting and fatigue
- pavement bearing capacity
- resistance to deterioration due to spilled fuel (especially on military airfields), de-icing chemicals and other chemicals
- sufficient bonding strength between asphalt layers to avoid slippage cracking.

The asphalt/bitumen solutions used to meet specific airport requirements will vary substantially from location to location depending on several factors such as climate, traffic intensity, types of aircraft etc. Asphalts offer great flexibility in their composition and physical characteristics, and may be specially designed to meet the needs at individual airports. Bitumen, obviously a very important component in relation to the properties of the asphalt, can be selected in different stiffness grades, being specially designed to meet climatic challenges, good resistance to damage resulting from fuel spillage, and improved rutting, low temperature cracking and fatigue properties. Conventional bitumen cannot always meet these demands. Thus, the use of polymer modified bitumen is, in most cases, necessary to meet the specific needs of an airport pavement.

Asphalt concrete with specification requirements for Marshall stability, flow and void content has normally been used for airfields in the past. The surfaces of these pavements were usually grooved to provide sufficient macrotexture and to create drainage channels, leading to improved frictional characteristics. However, the introduction of stone mastic asphalt and porous asphalt on airfields has resulted in improved wet friction properties. Grooves, which often lead to premature ravelling, are unnecessary when stone mastic asphalt or porous asphalt is used as the surface course. The use of an optimised and adequate asphalt composition may offer improved pavement service life. Polymer modified bitumens based on SBS are generally used in dense asphalts, stone mastic asphalt and porous asphalts (e.g. Shell Cariphalte® PG76 and Shell Cariphalte® PG82). These bitumens provide the best possible aggregate adhesion to prevent cracking and ravelling, while at the same time providing the pavement with an improved anti-rutting performance.

The development of new asphalts has been followed by improvements in mechanical tests for bitumen and asphalts, as well as in specifications and mixture design procedures. Specifications for some recent airport works have been based on various sets of special binder specifications (Dubai Airport in the UAE and Muscat Airport in Oman, Kuala Lumpur in Malaysia, Gardermoen Airport in Norway, Schiphol Airport in the Netherlands, Arlanda Airport in Sweden and various airports in Greenland). At Dubai Airport, Muscat Airport and Schiphol Airport, specifications for rutting and crack resistance were used based on the wheel tracking test, dynamic creep test and indirect tensile test, while the binder for Gardermoen Airport was specified using SHRP specifications. Superpave Mix Design was implemented at Muscat Airport. These developments in tests and test criteria for airports have led to new asphalt and bitumen solutions that will further encourage clients to choose asphalts for airfield applications.

The nature of the loading on airfield pavements may produce critical shear stresses in the interface between asphalt layers, and result in slippage cracking if the bonding strength is insufficient. For main airfields with severe traffic, it is advisable to use bond coats (bond coats are always manufactured with polymer modified bitumen) with an application rate resulting in 300 g/m^2 of residual binder.

Some airports at which Shell Bitumen supplied the bitumen (as at 2014) are listed in Table 24.12.

24.4 Conventional and high speed railway applications
24.4.1 Conventional railway
In the USA, Japan and many European countries, asphalts have been tested and used in the construction of railway track beds for many years. One of the earliest known applications dates back to 1894, when an asphalt base, approximately 150 mm thick surfaced with a 30 mm thick surface course, was laid as a tramway bed in Visalia, California. In Japan, asphalt has been used in railway construction since the late 1950s, and it has been the subject of numerous tests and commercial applications in Europe.

In railway construction, asphalts are used in one of four ways (Beecken, 1986)

- to stabilise the aggregate ballast
- as an asphalt layer below the aggregate ballast
- as a direct base for the track sleepers
- as a direct base for the rails.

Table 24.12 Some airports at which Shell Bitumen supplied the bitumen (as at 2014)

Country	Name of job/city or state	Product group	Year
Germany	Heringsdorf (public)	Cariphalte	1995
Germany	Parchim (public)	Cariphalte	1996
Norway	Bardufoss Airport	Cariphalte	1996
Germany	Spangdahlem (military)	Cariphalte	1998
Northern Ireland	St Angelo Airport	Cariphalte	1999
Germany	Baden-Baden (public)	Cariphalte	2000
UK	East Midlands Airport	Cariphalte	2000
France	Merignac Airport/Bordeaux	Mexphalte Fuelsafe	2001
France	Toulouse Blagnac Airport/Toulouse	Mexphalte Fuelsafe	2001
Germany	Hamburg (public)	Cariphalte	2001
Philippines	Diosdado Macapagal International Airport	Cariphalte	2001
Brunei	Brunei Airport	Cariphalte	2002
Philippines	Subic Bay International Airport/Zambales	Cariphalte	2002
South Africa	Cape Town Airport	Cariphalte	2002
Argentina	Ezeiza International Airport	Cariphalte	2003
Germany	Berlin (public)	Cariphalte	2003
Greece	Alexandroupoli Airport	Cariphalte	2003
Australia	Sydney Airport Maintenance/NSW	Cariphalte Fuelsafe	2004
Belgium	Zaventem Airport/Brussels	Cariphalte	2004
Germany	Fraport Airport/Frankfurt (public)	Cariphalte	2004
Germany	Finkenwerder Hamburg (private Airbus)	Cariphalte	2004
Thailand	Bangkok International Airport	Cariphalte Fuelsafe	2004
Thailand	Udorn Thaini Airport	Cariphalte	2004
UK	Heathrow Airport	Cariphalte Fuelsafe	2005
Cambodia	Siem Reap Airport	Cariphalte Fuelsafe	2006
Cambodia	Phnom Penh Airport	Cariphalte Fuelsafe	2008
Scotland	Edinburgh Airport	Cariphalte	2008
UAE	Al Bateen Air Base/Abu Dhabi	Cariphalte PG76-22	2010
UAE	Qusaiwera Air Base/Abu Dhabi	Cariphalte Fuelsafe	2012
UAE	Sweihan Air Base/Abu Dhabi	Cariphalte PG76-22	2013
Netherlands	Schiphol Airport	Cariphalte	1993–2004
Norway	Gardermoen Airport repairs	Cariphalte	2000–2004
Greenland	Thule (military)	Cariphalte	2004–2005
UAE	Dubai International Airport	Cariphalte, Cariphalte Fuelsafe	2005–2013
Hong Kong	HKIA	Cariphalte Fuelsafe	2006–2007
Singapore	Changi Airport	Cariphalte	2006–2007
UAE	Abu Dhabi International Airport	Cariphalte PG76-10	2007–2010
UAE	Jebel Ali/Al Maktoum International Airport	Cariphalte, Cariphalte Fuelsafe	2007–2013
Sri Lanka	Hambantota International Airport	Cariphalte PG76-22	2011–2013
Oman	Muscat International Airport	Cariphalte PG76-22	2011–2014
Oman	Salalah International Airport	Cariphalte PG76-22	2012–2014

24.4.2 Cement asphalt mortar used in high speed railway application

Recent years have seen moves towards improving the capacity of railways and the speeds of trains. China, in particular, has invested heavily in high

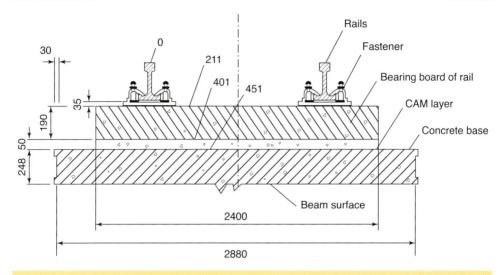

Figure 24.6 Cross section of ballastless slab track. (Liu and Liu, 2009)

speed railway technology, having carried out a number of major projects in the main cities.

Traditional rail tracks tend to require significant maintenance, and often have weak subgrades and drainage deficiencies. High speed railways overcome this by using a simple but advanced structure consisting of a continuous slab of concrete base (like a highway structure) with the rails supported on a bearing board, which in turn are supported on a layer of cement asphalt mortar. Such a system, shown in Figure 24.6, allows for adjustment of levels while buffering the high speed heavy loading. Figure 24.7 shows a

Figure 24.7 Ballastless high speed track in China

photograph of such a system, a ballastless high speed track in China (Liu and Liu, 2009; Ministry of Railways, 2008).

24.4.2.1 Composition of cement asphalt mortar

Cement asphalt mortar consists of a bitumen emulsion, fine aggregate and additives. Additives include defoamers, superplasticisers and aluminium powder, which are added to give better control of the performance of the emulsion in the field. It is very common to use slow setting modified bitumen emulsion, which will tolerate longer application times without final curing. Both cationic and anionic modified bitumen emulsions have been used to make the mortar in high speed railway projects. Examples where modified bitumen emulsion systems have been used are two high speed railway projects in China: the Harbin–Dalian railway, in which a cationic modified emulsion base was employed, and the Beijing–Shanghai railway, in which an anionic modified emulsion base was employed.

24.4.2.2 Materials specification

Polymer modified cationic bitumen emulsion

The properties of the cationic polymer modified bitumen emulsion plays an important role in successful cement asphalt mortar performance. Typical requirements are shown in Table 24.13.

Table 24.13 Polymer modified cationic bitumen emulsion

Test item	Unit	Test method[a]	Specification
Sieve test (1.18 mm)	%	ASTM D244	<0.1
Appearance	–	–	Brown, smooth, no impurity
Particle charge	–	ASTM D244	positive
Viscosity (Engela, 25°C)	s	ASTM D244	5–15
Storage stability (1 day, 25°C)	%	ASTM D244	<1.0
Storage stability (5 days, 25°C)	%	ASTM D244	<5.0
Miscibility	%	ASTM D244	<1.0
Discharge temperature	°C	–	5–35
Low temperature storage stability (1 day, −5°C)	%	ASTM D244	No coagulation
Residual properties			
Residual content	%wt	ASTM D244	58–63
Penetration (25°C, 100 g, 5 s)	0.1 mm	ASTM D5	60–120
Softening point	°C	ASTM D36	≥42
Solubility	%	ASTM D244	≥97
Ductility (15°C)	cm	ASTM D113	≥50
Ductility (5°C)	cm	ASTM D113	≥20

[a] ASTM D36/D36M-14e1 (ASTM, 2014b); ASTM D5 (ASTM, 2006); ASTM D113 (ASTM, 1999); ASTM D244-09 (ASTM, 2009c).

Table 24.14 Polymer modified anionic bitumen emulsion

Test item	Unit	Test method[a]	Specification
Sieve test (1.18 mm)	%	ASTM D244	<0.1
Appearance	–	–	Brown, smooth, no impurity
Particle charge	–	ASTM D244	negative
Viscosity (Saybolt Furol, 25°C)	s	ASTM D244	>20
Storage stability (1 day, 25°C)	%	ASTM D244	<1.0
Storage stability (5 days, 25°C)	%	ASTM D244	<5.0
Brookfield viscosity (40°C)	MPa s	ASTM D2171	≤1000
Miscibility		ASTM D244	Pass
Discharge temperature	°C	–	5–35
Low temperature storage stability (1 day, −5°C)	%	ASTM D244	No coagulation
Average particle size	μm		≤7
Mode	μm		≤5
Residual properties			
Residual content	% wt	ASTM D244	≥60
Penetration (25°C, 100 g, 5 s)	0.1 mm	ASTM D5	40–120
Softening point	°C	ASTM D36	≥42
Solubility	%	ASTM D244	≥97
Ductility (25°C)	cm	ASTM D113	≥100
Ductility (5°C)	cm	ASTM D113	≥20

[a] ASTM D36/D36M-14e1 (ASTM, 2014b); ASTM D5 (ASTM, 2006); ASTM D113 (ASTM, 1999); ASTM D2171 (ASTM, 2010a); ASTM D244-09 (ASTM, 2009c).

Polymer modified anionic bitumen emulsion
Typical requirements for polymer modified anionic bitumen emulsion are shown in Table 24.14.

Important requirements for the emulsified bitumen include binder content, cement miscibility, mixing temperature, water blending ratio, mixing time and workability.

Fine aggregate quality requirements
The source of the fine aggregate ranges from river gravel to crushed rock. The maximum particle size of aggregate is 1.18 mm. The finished aggregate should meet the quality requirements shown in Table 24.15.

24.4.2.3 Cement asphalt mortar performance requirements
The workability of cement asphalt mortar is difficult to control for the following reasons

- it has a relatively high bitumen content
- the working temperature
- the mixing conditions
- the temperature of the mixture.

Table 24.18 Shell Cariphalte Racetrack® technology for racing circuit projects

Country	Circuit	Year	Remarks
China	Zhuhai	1996	
Germany	Oschersleben	1996	
Malaysia	Sepang International Racing Circuit (SIC)	1998	Formula 1 circuit
Australia	Willowbank Raceway/QLD	1999	
Philippines	Carmona Race Track/Carmona, Cavite	1999	
Germany	Sachsenring	2000	
Germany	Eurospeedway Laustiz (test track)	2000	
Philippines	Batangas Racing Circuit/Batangas	2000	
Brasil	Interlagos	2002	Formula 1 circuit
Germany	Hockenheim Grand Prix	2002	Formula 1 circuit
Italy	Fiorano	2002	Ferrari test track
Germany	Nurburgring	2003	
Bahrain	Bahrain Formula 1 track	2003	Formula 1 circuit
Qatar	Losail International Racetrack	2004	
Malaysia	Sepang International Racing Circuit (SIC)	2007	Formula 1 circuit – resurfacing
Singapore	Singapore Marina Bay	2008	Formula 1 circuit/normal road between races
UAE	Abu Dhabi Yas Marina	2009	Formula 1 circuit

Table 24.18 lists a number of projects where Cariphalte Racetrack has been used.

24.8 Vehicle testing circuits

Many vehicle testing circuits have steep banked curves, and special techniques have been developed by specialist contractors to ensure that the finished pavement is fully compacted, meets the surface tolerances and has a high degree of surface regularity. Figure 24.8 shows a surfacing being laid on the Rockingham Speedway track in England. On this project, considerable skill and expertise was required to lay and compact materials to the required specification under the prevailing conditions.

24.9 Hydraulic applications

The waterproofing properties of bitumens and asphalts have been recognised since ancient times. The ancient civilisations of Ur, Egypt, Babylon, Assyria and the Indus all used the naturally occurring surface seepages of bitumen for waterproofing and building. The reservoir dam at Mohenjo Daro in the Indus basin is particularly well preserved, and demonstrates that, in this field, asphalt can claim a life of 5000 years (Menzies, 1988).

Bitumens and asphalts have been used effectively for a range of hydraulic applications such as canal lining, reservoir lining, sea wall construction, coastal groynes, dam construction and the lining of leisure lakes. In the UK, dams at Dungonnell, Colliford, Winscar, Marchlyn and Sulby incorporate bitumens. A further example of the use of bitumens in hydraulic applications is shown in Figure 24.9, which shows a photograph of the

Figure 24.8 Surfacing the Rockingham Speedway track. (Courtesy of Colas Ltd.)

Figure 24.9 Bituminous grouting on the face of the Megget Dam

Megget Dam in the south of Scotland. Reservoirs at Shotton and Towey in Wales and Leamington in England also make significant use of bitumens for their waterproofing properties.

The two principal properties that make bitumen ideal for hydraulic applications are its impermeability and the fact that it is chemically inert. In combination with suitable aggregates in mixtures that are specifically designed for particular applications, bitumen can impart these properties to the structures in which the asphalt is used. In many of these structures it is essential to have impermeability under pressure, and asphalts with void contents below 3% can successfully contain water column pressures up to 200 m.

Reservoir and canal embankments often have a slope of 1 : 1.75 but can be as steep as 1 : 1.25. Thus, asphalts used in such applications must have sufficient stability to be laid and compacted on such slopes without cracking during application and in service. The material must be flexible enough to accommodate differential settlement of the substrate.

One great advantage of asphalt is that it can be laid on the dam or reservoir face in a continuous manner, thus eliminating joints. Any joint constitutes a discontinuity and, at the high pressures that exist at the base of dams and reservoirs, a joint provides the opportunity for the water to find an outlet. *The Shell Bitumen Hydraulic Engineering Handbook* provides substantial detail on hydraulic applications (Sconian, 1999).

24.10 Coloured surface courses and surface treatments
The appearance of traditional asphalts, especially when finished with the normal contrasting white lines, is generally pleasing to the eye. However, there are some locations where a specific colour is desired (Table 24.19), and there are a number of ways in which this can be achieved

- incorporation of coloured pigments in the mixture during manufacture
- application of suitably coloured chippings to the asphalt during laying
- application of a coloured surface treatment after laying
- use of a conventional bitumen with a coloured aggregate
- use of a suitably coloured aggregate with a translucent binder.

An example of the application of coloured asphalt in a recreational park is shown in Figure 24.10.

24.10.1 Application of coloured chippings
Hot rolled asphalt and fine graded asphalt concrete (previously called fine cold asphalt) can have decorative coloured chippings rolled into the surface during the compaction phase. Hot rolled asphalt is suitable for most traffic conditions but fine graded asphalt concrete is only appropriate for lightly trafficked applications such as private drives and pedestrian areas.

Table 24.19 Reasons for and applications where the use of a coloured pavement surface may be advantageous

Purpose of providing coloured surface	Examples of applications
Increasing safety	Road crossings, obstacles, bus stops, turning pre-selections, zone 30 km, tunnels
Controlling and guiding traffic	Footpath, cycle lane, bus lane, parking area
Improve visibility/safety, save energy	Tunnels, in front of schools, fire department exits
Optical illusion (make objects bigger)	Football area
Increase attractiveness	Pleasure areas, park lanes, pleasure grounds
Increasing attractiveness by avoiding contrast	Historical sites, natural or historical environment, sport areas, zoos, prestige sites, square, footpaths
Improving comfort	Cycling, walking, skating
Improving comfort or performance, reduce pavement warming	Footpath – reduced heat radiation Heavily trafficked road – reduced rutting
Improved appearance	Using light chippings in footpaths

To provide a decorative finish to both these types of asphalt, pigmented bitumen coated or clean resin coated chippings can be applied during laying. The bitumen or resin coating promotes adhesion to the surface. The use of uncoated chippings is not recommended, as they are liable to become detached from the surface. However, in the case of fine graded

Figure 24.10 Example of coloured asphalt used in a recreational park

macadam that is laid on areas subject to little traffic, a light application of uncoated chippings (e.g. white limestone chippings) produces a very attractive finish if they are uniformly applied and properly embedded. Many footways in Scotland are finished with a 30 mm thick layer of hot rolled asphalt sand carpet (i.e. hot rolled asphalt surface course without the coarse aggregate) into which white limestone chippings are added. Decorative chippings cannot be successfully rolled into the surface of high stone content mixtures such as dense asphalt concretes or high stone content hot rolled asphalts.

24.10.2 Application of coloured surface treatments

There are three surface treatments that can be applied to an asphalt surface course to provide a decorative finish

- pigmented slurry seals
- surface dressings
- application of a coloured paint.

Pigmented slurry seals are available as proprietary products in a range of colours. However, as they are applied in a very thin coat (approximately 3 mm thick) they are only suitable for pedestrian and lightly trafficked areas.

Surface dressings are suitable for most categories of road application (Roberts and Nicholls, 2008) but are less suitable for pedestrian situations. The final colour of the surface dressing will be that of the aggregate used.

Proprietary coloured paints are available for painting on top of conventional black surface courses. These are only suitable for pedestrian and games areas (e.g. tennis courts, street painting).

24.10.3 Coloured aggregate bound by a conventional bitumen

When conventional bitumen is used, the depth of colour achieved is dependent on

- the colour of the aggregate itself
- the thickness of the binder film on the aggregate
- the rate at which the binder on the road surface is exposed by traffic.

In medium and heavily trafficked situations, the natural aggregate colour will start showing through fairly quickly, with a year being typical. In lightly trafficked situations, where coloured surfacings are often specified, it may take significantly longer (3 years or more) for the aggregate colour to become apparent.

24.10.4 Coloured aggregate bound by a translucent binder

Several proprietary asphalt concretes are available in which the binder is a clear resin rather than bitumen. A range of coloured surfacings can be manufactured by selecting appropriately coloured aggregates. The major

Figure 24.11 An extruded asphalt kerb on a UK motorway (Paving Expert, 2014)

advantage of such a system is that the colour is obtained immediately. For more information on this subject see section 8.3.2.

24.11 Asphalt kerbs

Extruded asphalt kerbs are very simple to construct. They are produced by extruding a hot rolled asphalt through specialised equipment directly onto a binder course or surface course, which has been treated with a tack coat or bond coat. Asphalt kerbs are unsuitable for the edge of carriageways carrying heavy traffic, as they may be damaged if hit by vehicles, particularly commercial vehicles. However, they do provide a convenient means of producing a raised edge on other paved areas (e.g. perimeter kerbing to drives or car parks). In the UK, BS 5931:1980 (BSI, 2014) gives guidance on the use of asphalt for producing kerbs in situ.

An extruded asphalt kerb on a UK motorway is shown in Figure 24.11.

References

Asphalt Institute (2014) Use of a DSR to evaluate effect of aging on roofing membrane. AI spring meeting, San Antonio, TX, USA.

Asphalt Paving Association of Iowa (2014) *Designs for Recreational Uses*. See https://www.apai.net/cmdocs/apai/designguide/Chapter_6B.pdf (accessed 16/10/2014).

ARMA, BWA, NRCA and RCMA (Asphalt Roofing Manufacturers Association, Bitumen Waterproofing Association, National Roofing Contractors Association and Roof Coatings Manufacturers Association (2011) *The Bitumen Roofing Industry – A Global Perspective: Production, Use, Properties, Specifications and Occupational Exposure.* ARMA/BWA/NRCA/RCMA, Washington, DC/Maumee, OH/Rosemont, IL/Washington, DC, USA.

ASTM (American Society for Testing and Materials) (1999) D113. Standard test method for ductility of bituminous materials. ASTM, West Conshohocken, PA, USA.

ASTM (2000) ASTM D312-00 (2006). Standard specification for asphalt used in roofing. ASTM, West Conshohocken, PA, USA.

ASTM (2006) D5. Standard test method for penetration of bituminous materials. ASTM, West Conshohocken, PA, USA.

ASTM (2007a) C289. Standard test method for potential alkali–silica reactivity of aggregates (chemical method). ASTM, West Conshohocken, PA, USA.

ASTM (2007b) D2746-07. Standard test method for staining tendency of asphalt (stain index). ASTM, West Conshohocken, PA, USA.

ASTM (2008) D6163-00. Standard specification for SBS modified bituminous sheet materials using glass fibre reinforcements. ASTM, West Conshohocken, PA, USA.

ASTM (2009a) C29. Standard test method for bulk density ('unit weight') and voids in aggregate. ASTM, West Conshohocken, PA, USA.

ASTM (2009b) D1754. Standard test method for effect of heat and air on asphaltic materials (thin-film oven test). ASTM, West Conshohocken, PA, USA.

ASTM (2009c) D244-09. Standard test methods and practices for emulsified asphalts. ASTM, West Conshohocken, PA, USA.

ASTM (2009d) D5329-09. Standard test methods for sealants and fillers, hot-applied, for joints and cracks in asphaltic and Portland cement concrete pavements. ASTM, West Conshohocken, PA, USA.

ASTM (2009e) D6223-02. Standard specification for APP modified bituminous sheet materials using a combination of polyester and glass fibre reinforcements. ASTM, West Conshohocken, PA, USA.

ASTM (2010a) D2171. Standard test method for viscosity of asphalts by vacuum capillary viscometer. ASTM, West Conshohocken, PA, USA.

ASTM (2010b) D2170/D2170M-10. Standard test method for kinematic viscosity of asphalts (bitumens). ASTM, West Conshohocken, PA, USA.

ASTM (2010c) D3111. Standard test method for flexibility determination of hot-melt adhesives by mandrel bend test method. ASTM, West Conshohocken, PA, USA.

ASTM (2010d) D3462/D3462M-10a. Standard specification for asphalt shingles made from glass felt and surfaced with mineral granules. ASTM, West Conshohocken, PA, USA.

ASTM (2011) D4798/D4798M-11. Standard practice for accelerated weathering test conditions and procedures for bituminous materials (xenon-arc method). ASTM, West Conshohocken, PA, USA.

ASTM (2012a) C127. Standard test method for density, relative density (specific gravity), and absorption of coarse aggregate. ASTM, West Conshohocken, PA, USA.

ASTM (2012b) D92. Standard test method for flash and fire points by Cleveland open cup tester. ASTM, West Conshohocken, PA, USA.

ASTM (2012c) D2872-12e1. Standard test method for effect of heat and air on a moving film of asphalt (rolling thin-film oven test). ASTM, West Conshohocken, PA, USA.

ASTM (2013a) C117-13. Standard test method for materials finer than 75-μm (No. 200) sieve in mineral aggregates by washing. ASTM, West Conshohocken, PA, USA.

ASTM (2013b) D4402/D4402M-13. Standard test method for viscosity determination of asphalt at elevated temperatures using a rotational viscometer. ASTM, West Conshohocken, PA, USA.

ASTM (2014a) C711-14. Standard test method for low-temperature flexibility and tenacity of one-part, elastomeric, solvent-release type sealants. ASTM, West Conshohocken, PA, USA.

ASTM (2014b) D36/D36M-14e1. Standard test method for softening point of bitumen (ring-and-ball apparatus). ASTM, West Conshohocken, PA, USA.

BSI (British Standards Institution) (2014) BS 5931:1980 Revision 2014. Code of practice for machine laid in situ edge details for paved areas. BSI, London, UK.

Beecken G (1986) Asphaltic base course for railway track construction. *Shell Bitumen Review* **61**: 10–21.

Burns Cooley Dennis, Inc. (2009) *Implementation of Superpave Mix Design for Airfield Pavements*. Volume 1, *Research Results for Airfield Asphalt Pavement Technology Program (AAPTP) Project 04-03*. Burns Cooley Dennis, Inc., Ridgeland, MS, USA.

Crafco (2008) *Hot-Applied Bituminous Marker Adhesive*. Crafco Inc., Chandler, AZ, USA. Product data sheet.

DMS (Departmental Materials Specification) (2008) DMS 6130. Bituminous adhesive for pavement markers. Florida Department of Transportation, Tallahassee, FL, USA.

EAPA (European Asphalt Pavement Association) (2013) *Asphalt Pavements on Bridges Decks*. EAPA, Brussels, Belgium. Position paper.

EAPA (2014) See http://www.asphaltadvantages.com/advantages/sustainability/6 (accessed 09/11/2014).

European Union (1995) European patent application No. 828 893 (claiming a priority date of 26.5.1995; title: *Impact Absorbing Macadam*).

Farrar MJ, Turner TF, Planche JP, Schabron JP and Hansberger MP (2013) *Evolution of the Crossover Modulus with Oxidative Aging: A Method to Estimate the Change in Viscoelastic Properties of an Asphalt Binder with Time and Depth on the Road*. Transport Research Board, Washington, DC, USA, Report 13-3059.

Formula 1 (2014) *Formula One Circuits – Layer by Layer*. See http://www.formula1.com/news/features/2006/11/5333.html (accessed 16/10/2014).

Green EL and Tolonen WJ (1977) The Chemical and Physical Properties of Asphalt–Rubber Mixtures. Arizona Department of Transportation, Phoenix, AZ, USA.

Heitzman M (1992) Design and construction of asphalt paving materials with crumb rubber modifier. *Transportation Research Record* **1339**: 1–8.

Liu S and Liu D (2009) *Q&A for Cement Asphalt Mortar in CRTS TYPE High Speed Railway with Ballastless Track*. China Railway Press, Beijing, China.

Martin Asphalt (2014) *Everything Asphalt*. See http://www.themartincompanies.com/sites/themartincompanies.com/files/asphalt/pds/Evergrip%20Flexible%20GTR%20Bituminous%20Marker%20Adhesive%20Tech%20Sheet%20California%20DOT.pdf (accessed 16/10/2014).

Menzies I (1988) Waterproofing with asphalt. *Water and Waste Treatment* **31**: 20–21.

Mike A (2013) *Rheology*. Asphalt Institute, Lexington, KY, USA. Spring meeting.

Ministry of Railways (2008) *2008-74. Technical Requirements for Cement Asphalt Mortar in CRTS TYPE High Speed Railway with Ballastless Track (Draft)*. China Railway Press, Beijing, China.

Morgan P and Mulder A (1995) *The Shell Bitumen Industrial Handbook*. Shell International Petroleum Company Ltd, London, UK.

MPA (Mineral Products Association) (2009a) Decorative and coloured finishes for asphalt surfacings. *Asphalt Applications 4*. MPA, London, UK.

MPA (2009b) Use of asphalt in the construction of games and sport areas. *Asphalt Applications 7*. MPA, London, UK.

MPA (2009c) Miscellaneous uses of asphalts. *Asphalt Applications 9*. MPA, London, UK.

MPA (2009d) Airfield uses of asphalts. *Asphalt Applications 10*. MPA, London, UK.

MPA (2014) See http://www.mineralproducts.org/prod_asp02.htm (accessed 16/10/14).

Paving Expert (2014) *Road Kerbs and Channels*. See http://www.pavingexpert.com/edging5.htm (accessed 16/10/2014).

Qin Q, Schabron JF, Boysen RB and Farrar MJ (2014) Field aging effect on chemistry and rheology of asphalt binders and predications for field aging. *Fuel* **121**: 86–94.

Roberts C and Nicholls JC (2008) *Design Guide for Road Surface Dressing*. TRL, Crowthorne, UK, Road Note 39, 6th edn..

Scönian E (1999) *The Shell Bitumen Hydraulic Engineering Handbook*. Shell International Petroleum Company Ltd, London, UK.

Takallou HB and Hicks RG (1988) Development of improved mix and construction guidelines for rubber modified asphalt pavements. *Transport Research Record* **1171**: 113–120.

Appendix 1
Physical constants of bitumens

A1.1 Specific gravity

The specific gravity of a bitumen is primarily dependent on the grade of the bitumen and the temperature (Pfeiffer, 1950). Typical values of specific gravity for a range of grades of bitumen are given in Table A1.1, and the effect of temperature on specific gravity is detailed in Table A1.2. The conversion from volume of bitumen in litres to weight in tonnes at various temperatures and values of specific gravity is detailed in Table A1.3.

Table A1.1 Typical specific gravities of selected bitumens at 25°C	
Grade of bitumen	Typical specific gravity at 25°C
Penetration grades	
160/220	1.015–1.025
100/150	1.020–1.030
70/100	1.020–1.030
50/70	1.020–1.030
40/60	1.025–1.035
30/45	1.025–1.035
20/30	1.030–1.040
Oxidised grades	
75/30	1.015–1.030
85/25	1.020–1.035
85/40	1.010–1.025
95/25	1.015–1.030
105/35	1.000–1.015
115/15	1.020–1.035
Hard grades	
H 80/90	1.045–1.055
H 110/120	1.055–1.065
Cut-back grades	
50 s	0.992–1.002
100 s	0.995–1.005
200 s	0.997–1.007

The Shell Bitumen Handbook, Sixth Edition
ISBN 978-0-7277-5837-8
Shell International Petroleum Company Ltd: All rights reserved
http://dx.doi.org/10.1680/tsbh.58378.739

Table A1.2 Typical specific gravities of bitumen at various temperatures

Temperature: °C	Specific gravity at 25°C					
	1.00	1.01	1.02	1.03	1.04	1.05
15.5	1.006	1.016	1.026	1.036	1.046	1.056
25	1.000	1.010	1.020	1.030	1.040	1.050
45	0.988	0.998	1.008	1.018	1.028	1.038
60	0.979	0.989	0.999	1.009	1.019	1.029
90	0.961	0.971	0.981	0.991	1.001	1.011
100	0.955	0.965	0.975	0.985	0.995	1.005
110	0.949	0.959	0.969	0.979	0.989	0.999
120	0.943	0.953	0.963	0.973	0.983	0.993
130	0.937	0.947	0.957	0.967	0.977	0.987
140	0.931	0.941	0.951	0.961	0.971	0.981
150	0.925	0.935	0.945	0.955	0.965	0.975
160	0.919	0.929	0.939	0.949	0.959	0.969
170	0.913	0.923	0.933	0.943	0.953	0.963
180	0.907	0.917	0.927	0.937	0.947	0.957
190	0.901	0.911	0.921	0.931	0.941	0.951
200	0.895	0.905	0.915	0.925	0.935	0.945

Table A1.3 Conversion factors linking volume to weight for bitumen at various temperatures and specific gravities

Temperature: °C	Specific gravity at 25°C					
	1.00 l/tonne	1.01 l/tonne	1.02 l/tonne	1.03 l/tonne	1.04 l/tonne	1.05 l/tonne
25	995	984	973	963	953	943
45	1010	999	988	978	968	958
60	1020	1009	998	988	978	968
90	1041	1030	1019	1009	999	989
100	1047	1036	1026	1015	1005	995
110	1054	1043	1032	1022	1011	1001
120	1060	1049	1038	1028	1017	1007
130	1067	1056	1045	1034	1024	1013
140	1074	1063	1052	1041	1030	1019
150	1081	1070	1058	1047	1036	1026
160	1088	1076	1065	1054	1043	1032
170	1095	1083	1072	1060	1049	1038
180	1103	1091	1079	1067	1056	1045
190	1110	1098	1086	1074	1063	1052
200	1117	1105	1093	1082	1070	1058

A1.2 Coefficient of thermal expansion

The coefficient of thermal expansion of bitumen is effectively independent of the grade, and is $6.1 \times 10^{-4}/°C$.

A1.3 Coefficient of thermal contraction

The coefficient of thermal contraction (Jones *et al.*, 1968) of bitumen is effectively independent of the grade, and is $3.45 \times 10^{-4}/°C$.

A1.4 Electrical properties
Electrical resistance

Bitumen has a very high resistance and is, therefore, an ideal insulating material. Hard grades of bitumen have slightly higher resistance than soft grades. However, these differences are insignificant. The resistance of all grades of bitumen falls with increasing temperature, as shown in Table A1.4.

The influence of fillers on electrical resistance is negligible unless conductive fillers such as graphite, coke or metal powders are used in significant quantities.

Table A1.4 Relationship between temperature and electrical resistance

Temperature: °C	Resistance: Ω/cm
30	10^{14}
50	10^{13}
80	10^{12}

Dielectric strength

The dielectric strength is measured in kilovolts per millimetre, and depends on the conditions of measurement and the shape of the electrodes. Hard grades of bitumen tend to have a higher dielectric strength than soft grades. The dielectric strength of all grades decreases with increasing temperature, as shown in Table A1.5.

Table A1.5 Relationship between temperature and dielectric strength

Temperature: °C	Dielectric strength (flat electrodes): kV/mm
20	20–30
50	10
60	5

Dielectric constant (permittivity)

The dielectric constant (or permittivity) of bitumen is about 2.7 at 25°C, rising to about 3.0 at 100°C. The dielectric losses in bitumen rise with increasing temperature but fall with increasing frequency. The Transport and Road Research Laboratory showed that the rate at which bitumen weathers under the combined effects of oxygen, rain, oil deposition from traffic and ultraviolet component of sunlight is related to the dielectric constant of the bitumen (Green, 1977).

A1.5 Thermal properties
Specific heat

The specific heat of bitumen is dependent on both the grade of the bitumen and its temperature. Values of specific heat vary from 1675 to 1800 J/kg/°K at

0°C. The specific heat is increased by 1.67–2.51 J/kg for every 1°C increase in temperature.

Thermal conductivity

Bitumen is a good thermal insulating material, and typically has a thermal conductivity of 0.15–0.17 W/m/°K.

References

Green EH (1977) *An Acceptance Test for Bitumen for Rolled Asphalt Wearing Courses*. Transport and Road Research Laboratory, Crowthorne, UK. Laboratory Report *777*.

Jones GM, Darter MI and Littlefield G (1968) Thermal expansion–contraction of asphaltic concrete. *Journal of the Association of Asphalt Paving Technologists* **37**: 56–97.

Pfeiffer JPH (ed.) (1950) *The Properties of Asphaltic Bitumen*, vol. 4, *Elsevier's Polymer Series*. Elsevier Science, Amsterdam, The Netherlands.

Appendix 2
Conversion factors for viscosities

The viscosity of bitumen is usually measured using either capillary or cup viscometers (for more detail, see section 6.3). In capillary viscometers, viscosity is determined by timing the flow of bitumen through a glass tube at a given temperature. The product of flow time and a calibration factor gives the kinematic viscosity in centistokes (cSt) or mm^2/s. In cup viscometers, the time is recorded in seconds for a standard volume of bitumen to flow out through the orifice in the bottom of the cup. The values given in Table A2.1 enable conversion of a cup viscosity to a kinematic viscosity for a number of different types of cup viscometer, and vice versa.

Table A2.1 Conversion factors for viscosities

Known viscosity	To obtain unknown viscosity multiply by:						
	Kinematic: mm^2/s	Redwood I: s	Redwood II: s	Saybolt Universal: s	Saybolt Furol: s	Engler: °E	Standard tar viscometer: s
Kinematic: mm^2/s	–	4.05	0.405	4.58	0.458	0.132	0.0025
Redwood I: s	0.247	–	0.1	1.13	0.113	0.0326	–
Redwood II: s	2.47	10	–	11.3	1.13	0.326	0.0062
Saybolt Universal: s	0.218	0.885	0.0885	–	0.1	0.0287	–
Saybolt Furol: s	2.18	8.85	0.885	10	–	0.287	0.0054
Engler: °E	7.58	30.7	3.07	34.81	3.48	–	–
Standard tar viscometer (10 mm cup): s	400	–	162	–	183	528	–

Appendix 3
Blending charts and formulae

Bitumen can be blended with a wide variety of crude oil-based fractions for different applications. Volatile light fractions (e.g. white spirit) are used when rapid curing is required. Less volatile fractions, such as cut-back kerosene, are used for the manufacture of cut-back bitumens. Figures A3.1 and A3.2 show the effect of adding cut-back kerosene and diesel, respectively, to 70/100, 160/220 and 250/330 pen bitumen. Similarly, Figure A3.3

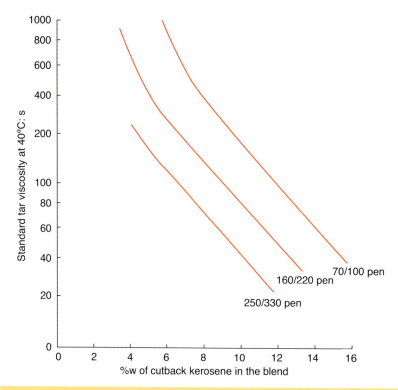

Figure A3.1 Typical blending curves for 70/100, 160/220 and 250/330 pen bitumen and cut-back kerosene

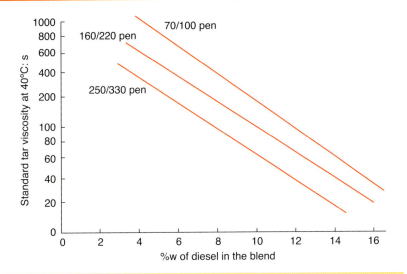

Figure A3.2 Typical blending curves for 70/100, 160/220 and 250/330 pen bitumen and diesel

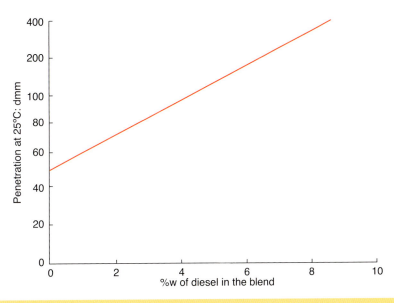

Figure A3.3 Typical blending curve for 40/60 pen bitumen and diesel

shows the effect on penetration when diesel is added to 40/60 pen bitumen.

Bitumens are miscible with each other in all proportions. The penetration and softening point of a blend of two bitumens can be estimated using the blending charts shown in Figures A3.4 and A3.5. These charts should only be used for blends of bitumen of the same type (i.e. those having the same penetration index). They should not be used for blends of oxidised and penetration bitumens. As an alternative to these charts, the following formulae can be used to estimate the penetration and softening point of a blend

$$\log P = \frac{A \log P_a + B \log P_b}{100}$$

where P is the penetration of the final blend, P_a is the penetration of component a, P_b is the penetration of component b, A is the percentage of component a in the blend, and B is the percentage of component b in the blend

$$S = \frac{AS_a + BS_b}{100}$$

where S is the softening point of the final blend, S_a is the softening point of component a, S_b is the softening point of component b, A is the percentage of component a in the blend, and B is the percentage of component b in the blend.

How to use the blending charts – Figures A3.4 and A3.5

For Figure A3.4, mark the measured penetration value (at 25°C) for the two bitumens you wish to blend (bitumens A and B) on the left hand and right hand y axes of the graph. Draw a straight line joining these two points – this is the blend line. Select the desired penetration of the final product you wish to produce and draw a horizontal line from the left hand y axis to the point at which it intersects the blend line. Draw a vertical line down from the point of intersection to the x axis and read the percentage of bitumen A and the percentage of bitumen B from the x axis. This is the estimated blend ratio required to produce bitumen of the desired penetration from the two components.

An example. Bitumen A has penetration of 100 dmm and bitumen B has a penetration of 20 dmm. To produce a bitumen with a penetration of 50 dmm from these two components would require a blend of 60% of bitumen A and 40% of bitumen B.

For Figure A3.5 the process is exactly the same but softening point values are used.

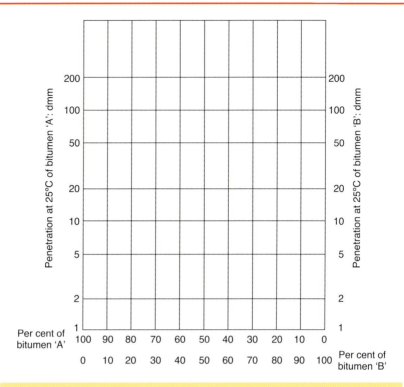

Figure A3.4 Chart for estimating the penetration of a blend of two bitumens

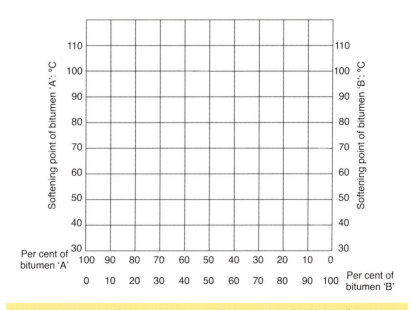

Figure A3.5 Chart for estimating the softening point of a blend of two bitumens

Appendix 4
Calculation of bitumen film thickness in an asphalt

It is possible to estimate the surface area of a sample of aggregate by coating it with an oil and measuring the quantity of oil required for complete coating. Alternatively, the surface area of an aggregate can be calculated by assuming a specific aggregate particle shape. Hveem (ASTM, 1992) calculated surface area factors, assuming a spherical particle shape and a specific gravity of 2.65. Typical surface area factors are shown in Table A4.1.

Table A4.1 Typical surface area factors

Sieve size: mm	Surface area factor: m^2/kg
0.075	32.77
0.150	12.29
0.300	6.14
0.600	2.87
1.18	1.64
2.36	0.82
>4.75	0.41

The surface area of the aggregate is calculated by multiplying the total mass, expressed as a percentage passing each sieve size, by the appropriate surface area factor, and adding the resultant products together. All the factors must be used, and if different sieves are used different factors are necessary. The theoretical bitumen film thickness is then calculated as

$$T = \frac{b}{100 - b} \times \frac{1}{\rho_b} \times \frac{1}{SAF}$$

where T is the bitumen film thickness (mm), ρ_b is the density of bitumen (kg/m^3), SAF is the surface area factor (m^2/kg), and b is the bitumen content (%).

Another method, developed in France (Norme Française, 1991), gives an approximation of the binder film thickness using the formula

$$T = \frac{b}{a \times \sqrt[5]{\Sigma}}$$

where T is the bitumen film thickness (mm), b is the bitumen content (% by total mass of the mixture), and a is a correction coefficient taking into account the density of the aggregate. The coefficient a is given by

$$a = \frac{2650}{SG_a}$$

where SG_a is the density of the aggregate (kg/m^3). Σ is the specific surface area of the aggregate, and is given by

$$\Sigma = 0.25G + 2.3S + 12s + 135f$$

where G is the proportion by mass of aggregate over 6.3 mm, S is the proportion by mass of aggregate between 6.3 and 3.15 mm, s is the proportion by mass of aggregate between 3.15 and 0.80 mm, and f is the proportion by mass of aggregate smaller than 0.80 mm.

References

ASTM (American Society for Testing and Materials) (1992) D1560–09a. Standard test methods for resistance to deformation and cohesion of bituminous mixtures by means of Hveem apparatus. ASTM, West Conshohocken, PA, USA.

Norme Française (1991) NF P 98-140 *Couches d'assis: Enrobé à élevé*. Norme Française, Paris, France (in French).

Appendix 5
Bitumen product standards used across the world

The specification systems used in bitumen product standards can be broadly described as either empirical or performance based. Those based on empirical testing such as penetration, softening point and viscosity have been used for many decades in a variety of climates. The performance based specification system (PG grading) used in the USA is well established in product standards.

Derivations, combinations and local adaptations of these standards are used depending on the regulatory environment, climate and historical factors. Even when regional product standards are used (e.g. the harmonised norms in Europe), differences still exist between countries.

Penetration and the ring and ball softening point test are used as the main characteristics in product standards throughout Europe, many parts of Asia (including China), Africa and Argentina. Viscosity based standards are used more commonly in Australia and New Zealand, and for the softer grades in Europe and South America. India is the latest country to move to viscosity grading. The VG grading used in India is slightly different from the Australian C grade, the VG grades having a higher requirement at 60°C.

In Argentina, although penetration and viscosity standards are valid, viscosity specifications are becoming more widely used for both soft and hard grades. In Columbia, penetration grades are used, and in other countries in Latin America viscosity standards are most common.

China is unique in that it employs ductility (at both normal temperature and low temperature) for conventional bitumen specifications.

The PG grading system is the primary method for specifying bitumen in the USA and Canada. Here, individual regulators and specifiers determine the grades to be used on state roads. Other countries, such as China, use PG grading for premium binders and for some large construction projects such as airfields, motor sport circuits, heavy duty expressways or highways exposed to extreme weather conditions. However, other performance based systems are (at the time of writing, May 2014) under development in many countries.

The map in Figure A5.1 provides a summary of the type of specification systems used throughout the world.

Penetration/softening point used in:
Europe
Many parts of Asia
Africa
Argentina
Brazil

Viscosity used in:
Australia
New Zealand
Europe (soft grades)
South America: Chile, Mexico and Argentina
India

PG grading used in:
USA and Canada
Other locations: mainly for premium binders (e.g. China)

Figure A5.1 The type of specification systems used throughout the world

Table A5.1 lists the main bitumen product standards used in a selection of countries.

Table A5.1 The main bitumen product standards used in various countries

Region/country	Regulatory body	Main product standards for bitumen used in asphalt
Member countries of the European Union (EU)	European Committee for Standardization (CEN)	Published in each country in the local language: EN 12591:2009. Bitumen and bituminous binders. Specifications for paving grade bitumens EN 14023:2010. Bitumen and bituminous binders. Specification framework for polymer modified bitumens EN 13924:2006. Bitumen and bituminous binders. Specifications for hard paving grade bitumens

Table A5.1 Continued		
Region/country	Regulatory body	Main product standards for bitumen used in asphalt
USA and Canada	State department for transportation/ local agency	AASHTO M 320-10. Standard specification for performance-graded asphalt binder. (American Association of State and Highway Transportation Officials, Washington, DC, USA) State and agency-specific standards
China	(Approval by) Ministry of Transport of The People's Republic of China (Edited by) Research Institute of Highway Ministry of Transport	JTG F40-2004. Technical specification for construction of highway asphalt pavements. (China Communications Press, Beijing, China)
India	Bureau of Indian Standards (BIS)	IS 73:2013. Paving bitumen – Specification IS 15462:2004. Polymer and rubber modified bitumen – Specification IS 8887:2004. Bitumen emulsion for roads (cationic type)
	Indian Roads Congress (IRC)	IRC SP 53:2010. Guidelines on use of modified bitumen in road construction
Australia	Australia Standard Austroad Test Method and Specification	AS2008 – 2013. Bitumen for pavements AGPT/T190 2014. Specification framework for polymer modified binders
Thailand	Department of Highways Standard Specification for Highway Construction (DH-T)	American Society for Testing and Materials (ASTM) and American Association of State and Highway Transportation Officials (AASHTO) for standard testing procedures
Malaysia	Standards & Industrial Research Institute of Malaysia (SIRIM)	MS 124:1996. Specification for penetration grade of bitumen for use in pavement construction
	Jabatan Kerja Raya (JKR)	JKR/SPJ/2008-S4. Standard specification for road works
Philippines	Department of Public Works & Highways (DPWH)	Standard specifications for highways, bridges and airports
South Africa	South African Bureau of Standards (SABS)	SANS 4001-BT1:2012. Civil engineering test methods: Penetration grade bitumen
Argentina	IRAM (Argentinian Rationalisation Materials Institute)	IRAM 6604:1993. Asfalto de petróleo para pavimentación. Para uso en mezclas asfálticas y tratamientos superficiales [Petroleum asphalt paving. For use in asphalt mixtures and surface treatments] IRAM 6835:2002. Asfaltos para uso vial – Clasificados por viscosidad – Requisitos [Asphalts for pavements – Classified by viscosity – Requirements] IRAM 6596:2002. Asfaltos modificados con polímeros para uso vial. Clasificación y requisitos [Polymer modified bitumen for pavements. Classification and requirements]

Useful links for more information

Asphalt Institute. *US State Binder Specification Database*. http://www.asphaltinstitute. org/public/engineering/state_binder_specs/index.dot (accessed 20/10/2014).

Austroads. Specifications & Test Methods. http://www.austroads.com.au/road-operations/asset-management/resources/specifications-test-methods (accessed 20/10/2014).

Eurobitume. Types of Bitumen, Bituminous Products & Binders. http://www.eurobitume. eu/bitumen/types-bitumen (accessed 20/10/2014).

IRAM (Instituto Argentino de Normalización y Certificación). http://www.iram.org.ar (accessed 20/10/2014).

Kerajann Malaysia and Jabatan Kerja Raya Malaysia. *Standard Specification for Road Works*. Section 4 – *Flexible Pavement*. http://jkrmarang.terengganu.gov. my/maxc2020/appshare/widget/mn_img/76668file/Arizan/Standard% 20Specification%20For%20Road%20Works(Flexible%20Pavement).pdf (accessed 20/10/2014).

Appendix 6
Asphalt product standards used across the world

A wide variety of asphalt product standards exist across different countries and regions and construction sectors. These are controlled, reviewed and amended by approved regulatory bodies.

Table A6.1 lists the main asphalt standards used in a selection of countries.

Table A6.1 The main asphalt standards used in various countries		
Region/country	Regulatory body	Main standards for asphalt in road construction
Member countries of the European Union (EU)	European Committee for Standardization (CEN)	EN 13108 series (there are several parts in the series specifying various types of asphalt)
USA and Canada	State department for transport/ local agency	State and agency specific standards
China	(Approval by) Ministry of Transport of The People's Republic of China (Edited by) Research Institute of Highway, Ministry of Transport	JTG F40-2004. Technical specification for construction of highway asphalt pavements
India	Indian Roads Congress (IRC)	MORT&H:05. Specifications for road and bridge works, 5th revision (2013)
Argentina	Direccion Nacional de Vialidad	Specifications for road construction (1998) Asphalt Commission recommendations are applied for specific public and private contracts
Australia	Austroad Test Method and Specification	AGPT/T190 2014. Specification framework for polymer modified binders
Thailand	Department of Highways Standard Specification for Highway Construction (DH-T)	ASTM and AASHTO for Standard Testing Procedures

Table A6.1 Continued		
Region/country	Regulatory body	Main standards for asphalt in road construction
Malaysia	Standards & Industrial Research Institute of Malaysia (SIRIM)	MS 124:1996. Specification for penetration grade of bitumen for use in pavement construction
	Jabatan Kerja Raya (JKR)	JKR/SPJ/2008-S4. Standard specification for road works
Philippines	Department of Public Works & Highways (DPWH)	Standard specifications for highways, bridges and airports

Appendix 7
Reporting asphalt compositional analysis results

The composition of an asphalt is a basic factor that critically impacts its properties such as stiffness, fatigue resistance and durability etc. This explains why the choice of individual constituents and their proportions in the asphalt mixture design (see Chapter 12) is so important. Compositional analyses of asphalts are taken routinely by asphalt producers in order to verify their compliance with applicable standards. Clients often also carry out compositional analyses to give confidence that the asphalts used in their projects meet specified requirements.

The requirements of the asphalt composition (i.e. type and size distribution of the aggregates, and the type and content of the bitumen) have been covered in the relevant sections of this book. This appendix describes how the results of the asphalt compositional analysis are reported in different parts of the world.

In general, there are two aspects to asphalt compositional analysis: aggregate particle size distribution and bitumen content. The former, also known as the aggregate 'grading' or 'gradation', is typically determined by a sieve analysis that involves the use of a set of increasingly finer sieves to separate the aggregate into gradually smaller particle sizes. The mass of material retained by each sieve is then determined as a proportion of the total aggregate weight and reported as the cumulative quantity (by percentage) passing each sieve.

Bitumen content is most commonly reported as a percentage by mass of the total mixture but it can also be reported as a percentage by mass of the dry aggregate only. Both methods are acceptable provided that the method to be employed is clear, in order to avoid any misunderstanding. The choice of method depends largely on the road authorities or agencies concerned, and some allow both methods to be used. Table A7.1 indicates which method is used in different parts of the world.

Table A7.1 Method of reporting bitumen content by country

Bitumen content reported as a percentage of the total mixture	Bitumen content reported as a percentage of the dry aggregate
North America (as per Superpave mix design)	Argentina
Argentina	Chile
Chile	Colombia
Colombia	China
Europe (as per EN 13108)	Thailand
South Africa	Vietnam
Middle East	
India	
China	
Hong Kong	
Vietnam	
Brunei	
Indonesia	
Malaysia	
Philippines	
Singapore	
Australia	

The two percentages can be easily converted as follows

$$P_{b(agg)} = \frac{100 \times P_b}{100 - P_b}$$

or

$$P_b = \frac{100 \times P_{b(agg)}}{100 + P_{b(agg)}}$$

where $P_{b(agg)}$ is the bitumen content as a percentage by mass of the dry aggregates, and P_b is the bitumen content as a percentage by mass of the mixture.

In a small number of specifications, it is required that the effective bitumen content is reported rather than the total bitumen content. This will involve additional effort to determine and deduct the binder quantity that is absorbed by the aggregates, in accordance with the following formula

$$P_{be} = P_b - \left(\frac{P_{ba}}{100}\right) P_s$$

where P_{be} is the effective bitumen content as a percentage by mass of the total mixture, P_b is the bitumen content as a percentage by mass of the total mixture, P_{ba} is the absorbed bitumen as a percentage by mass of the dry aggregate, and P_s is the aggregate content as a percentage by mass of the total mixture.

In the asphalt mixture design process, the volumetric properties of the mixture and its components need to be determined. These, in turn, require the determination of the specific gravities of the constituent ingredients in addition to their masses. The percentage of bitumen by volume of mixture is given by

$$P_{b(vol)} = \left(\frac{G_{mb}}{G_b}\right) P_b$$

where $P_{b(vol)}$ is the bitumen content as a percentage by volume of the total mixture, P_b is the bitumen content as a percentage by mass of the total mixture, G_{mb} is the bulk specific gravity of the mixture, and G_b is the specific gravity of the bitumen.

However, for purposes of asphalt mixture design, the parameters that are commonly used are the volume of air voids, voids in the mineral aggregate and voids filled with bitumen (see Chapter 12). Hence, for the convenience of asphalt production control, the easier method of measuring and reporting the compositional analysis by the total mass is used.

Appendix 8
Volumetrics symbols

USA		UK/EU[a]	
Description	Symbol	Description	Symbol
Volume of voids in mineral aggregate	V_{ma}	Volume of voids content in the mineral aggregate	VMA
Bulk volume of compacted mix	V_{mb}		
Voidless volume of paving mix	V_{mm}		
Volume of voids filled with asphalt	V_{fa}	Volume of voids in the mineral aggregate filled with binder	VFB
Volume of air voids	V_a	Volume of air voids	V_m
Volume of asphalt binder	V_b	Volume of binder	V_{br}
Volume of absorbed asphalt binder	V_{ba}		
Volume of mineral aggregate (by bulk specific gravity)	V_{sb}	Volume of aggregate	V_a or VIM
Volume of mineral aggregate (by effective specific gravity)	V_{se}		
Total mass of asphalt mixture	M_{mix}		
Mass of asphalt binder	M_b		
Mass of effective asphalt binder	M_{be}		
Mass of aggregate	M_{agg}		
Mass of air$=0$	M_{air}		

[a] BS EN 12697-8:2003. Bituminous mixture – Test methods for hot mix asphalt – Part 8: Determination of air void characteristics of bituminous specimens. (This standard expresses volume as a percentage by volume.)

Index

Note: Page numbers in *italics* reference figures and tables separate from the corresponding text.

2L-PA *see* double layered porous asphalt
2010 World Emulsion Congress 186

AAMAS *see* asphalt–aggregate mixture analysis
    system
AASHO *see* American Association of State
    Highway Officials
AASHTO *see* American Association of State
    Highway and Transportation Officials
AAV *see* aggregate abrasion value
abrasion resistance, aggregates 238, 240
absorption of dust, surface dressings 619
AC *see* asphalt concrete
accelerated xenon arc weathering test 709
acid–base theory 551
acids, emulsions 204, 206
ACN *see* aircraft classification numbers
acrylic resins, high friction surfaces 636
AC-TL *see* asphalt concrete for very thin layers
additives
    adhesion 568–9
    alternative asphalts 652–3
    asphalt production plants 403–4
    oxidation 60
    polymer modified bitumen 150–1, *152–3,*
        173
    slurry surfacing 626
adhesion 549–71
    adsorption 568–9
    disbonding mechanisms 553–7
    emulsion components *188*
    improvement 567–9
    measuring and assessment 557–67
    moisture damage 549, 557–67
    polymer modified bitumen 151, *152*
    quality checks 73–5
    surface dressings 607, 609–10
    thermodynamic principles 550–2
adhesive failure, binder properties 537
adhesive power, surface dressings 607
adhesives, pavement markers 714–16
adsorption
    adhesion/moisture damage 563–4, 568–9
    elemental analysis 49–50
Advanced Asphalt Technologies, 2011 297,
    304
ageing 59, 586–8
    binder properties 513, 519–20, 537
    durability testing 465
    hardening *577, 578, 579, 580*
    indices 60, *61, 577, 578*
    quality checks 75
    hot mixture storage test 77–8
    PAV test *77, 78*
    RTFOT test *75, 76, 77*
    structural ageing 79
Agency for Toxic Substances and Disease
    Registry 37
aggregate abrasion value (AAV) 238, 240,
    241, 596, 598, 603
aggregates 217–59
    adhesion
        improvement 567–9
        measuring and assessment 557–67
        moisture damage 557–67
        surface energy measurements 558–9
    all-in 217, 232–9
    asphalt mixture design
        apparent specific gravity 302
        bulk specific gravity 301–2
        coarse aggregate angularity 333–4
        Corps of Engineers method 316–23
        densification curves 333
        effective specific gravity 302
        fine aggregate angularity 333–4
        grading 304–11, *313*
        Marshall method 316–323
        Superpave 325–42
        voids in the mineral aggregate 300–1
    bitumen–aggregate adhesion 552–3, *554*
    certification 698–9
    coarse 217, 231–49
        basalt 'sonnenbrand' 246–7
        bituminous binders 245–6
        broken/crushed surfaces 234–5
        chemical properties 247, *248*
        durability 243–4
        fragmentation 236
        geometrical properties 232–5
        lightweight contaminators 247
        physical properties 236–43, 247, *248,*
            249
        polishing 236–8, *239*
        property data sheets 249
        shape of 234
        thermal shock 244–5
        wear 240–1
    cold mix plants 404–6
    coloured surface course 168–171, 734–5
    continuous asphalt plants 394–402
    creep properties 487, *489*
    filler 217, 250–5
        chemical requirements 254
        consistency 254–5

aggregates (*continued*)
  delta ring and ball 253–4
  dry compacted filler 253
  geometrical requirements 252
  grading 252
  harmful fines 252
  particle density 252
  physical requirements 252–4
  properties 252–4
  stiffening properties 252–3
  testing 252–4
  water content 252
fine 231–49
  basalt 'sonnenbrand' 246–7
  bituminous binders 245–6
  chemical properties 247, 248, 249
  durability 243–4
  geometrical properties 232–5
  physical properties 236–43, 247, 248, 249
  property data sheets 249
  thermal shock 244–5
hardening 576–8
heater batch mixing plants 390–3
high speed railways 722–4
origin 218–24
oxidation/chemical changes 60
processing 224–31
  crushing 226–31
  extraction 224–6
  screening 231
quality 255–6
Shell Thiopave 658
simulative measurements 561–4
slurry surfacing 624–5, 628
surface energy measurements 560
type 218–24
agriculture, use of bitumen 6, 7
AH-70 heavy traffic bitumen 290
AH-90 heavy traffic bitumen 290
air blowing
  chemical composition & physical properties 58–9
  chemistry 23–4
  short residues 20–4
air blowing of roofing flux 709–10
aircraft classification numbers (ACN) 719
airfield pavements 716–21, 722
  requirements 719–21, 722
airfields, design methods 719
airfield slurries 623–4
air-rectified bitumen 22
air voids (VIM)
  asphalt mixture design 304, 317
  volumetrics 298–9
aliphatic compounds 47
Al-Khateeb model 485–6
all-in aggregate 217, 232–9
alternative asphalts 643–86
  foamed bitumen 643–51
  recycling 666–71
    production 667–8, 669
    quality 668–9
    storage 667–8, 669
    suitability 668–9
    techniques 669–71

Shell Thiopave 654–6
  background 654
  compaction 657, 662, 663, 664
  deformation resistance 657–8
  design 656–7
  fatigue cracking 659–60
  laying 662, 663, 664
  low temperature properties 660–2
  pellet form 654, 655
  production 662
  properties 657–62
  road applications 662
  road performance 662–6
  role 654–5, 656
  stiffness moduli 658–9
  target applications 656
  water sensitivity 657
  workability 657
warm mix asphalt 651–4
altitude, surface dressing 610
aluminates, polymer modified bitumen 171
ambient rubber, alternative asphalts 672–3
American Association of State Highway Officials (AASHO)
  flexible pavement design 362, 366, 371
American Association of State Highway and Transportation Officials (AASHTO)
  see also Mechanistic–Empirical Design Guide
  asphalt mixture design 296, 303, 308, 332, 334–40
  dynamic tests 455–6
  flexible pavement design 366–8
  flexible pavement layers 284, 285, 286, 287
  immersion mechanical tests 564, 565–6
  immersion tests 562
  M 320 127, 134, 285, 336, 752
  M 323 284, 335, 336, 337, 338, 339–40
  MP 19-10 138
  PP 61 455–6
  R 35 339
  repeated load test 450
  rheology 120, 134, 138
  Shell Thiopave 660
  T 166 303
  T 182 562
  T 209 332
  T 228 303
  T 275 303
  T 283 340, 465, 564, 565–6
  T 304 334
  T 312 337–8
  T 322 287
  T 324 287, 459
  T 340 287, 459
  TP 10 660
  TP 70 138
  TP 79 455–6
  wheel tracking 459
  Witczak equation 483–5
American Society for Testing and Materials (ASTM) 87, 98
  alternative asphalts 674, 675
  asphalt mixture design 302–3, 304, 334

binder properties 532
bitumen recovery 471
C29 726
C117 726
C127 726
C289 726
C711 712
capillary viscometers 96, 98
cohesion 114
cohesion testing 115
D5 88, 708, 712, 715, 716, 724, 725,
    727
D36/D36M 89–90, 708, 712, 715, 724,
    725
D70 303
D92 708, 715
D113 114, 176–7, 715, 724, 725
D244 198, 208, 724, 725
D312 706
D1664 562
D1754 586, 716
D1876-08 560
D2041 302, 304, 321
D2170/D2170M 96, 712
D2171 98, 716, 725
D2397 208, 214, 625
D2419 624
D2746 708
D2872 586, 708
D3111 715
D3381 516
D3462/D3462M 706–7
D3625 563
D3910 626, 627, 629
D4402 98, 708, 715
D4798/D4798M 708
D5329 715
D5360 594
D5404 471
D5801 115
D5821 334
D6084 175
D6114/D6114M 674, 675
D6163 711, 712
D6223 711
D6372 626–7, 628, 629
D6648 134–5
D6723 83
D6862 560
D6925 464
D6927 468
D6931 469
D7173 174
D7405 532
D7460 452, 454–5
direct tension test 83, 85
durability 586
dynamic tests 454–5
E1911 600
emulsions 198, 208
flexural bending test methods 452
immersion tests 562, 563
Marshall test 468
pavement marker adhesive 717
peel tests 560

penetration test 88
polymer modified bitumen 174–7
rheology 134–5
roofing 706–7, 711, 712
rotational viscometers 98
slurry surfacing 624, 625, 626–7, 628, 629
softening point tests 89–90
surface dressing 593–4
temperature susceptibility 92
ammonium compounds 193, 224
ammonium nitrate fuel oil 224
amplitude, compaction 438
AMPT see Asphalt Mixture Performance Tester
analytical pavement design 368–9
angle of attack 419, 423, 426
angularity of fine aggregate 235
anhydrides, polymer modified bitumens 159
anionic emulsified cement asphalt mortar
    requirements 725–6, 727
anionic emulsions 188, 190
    chemical components 193
    slurry surfacing 625
anionic modified emulsion base, high speed
    railways 724, 727
anionic soaps 187
anionic surfactants 186
anti-oxidants 60
anti-oxidants, polymer modified bitumens 151,
    152
anti-rutting performance, Shell Cariphalte 179
anti-stripping agents 568–9
APA see Asphalt Pavement Analyzer
APP see atactic polypropylene
apparent specific gravity G_{sa} asphalt mixture
    design 302
aqueous film-forming foam extinguishers 35
aqueous phases, emulsions 198–9, 201, 204
'archipelago'-type asphaltenes 56
aromatics
    broad chemical composition 59
    constitution/rheology/structure relationships
        57
    elemental analysis 49
    oxidation/chemical changes 60–1
    polymer modified bitumen 157, 158, 164
    solubility 50, 51, 52
    structure 55
ARRB see Australian Road Research Board
Arrhenius equation 133
artificial ageing 587
asbuton see Buton asphalt
asphalt–aggregate mixture analysis system
    (AAMAS) 325–6
asphalt concrete (AC) 262, 268–9, 276
    binder properties, fretting 537–8
    stone to stone contact 534–5
    terminology 6
    test data charts 99, 101
asphalt concrete for very thin layers (AC-TL) 262,
    278
Asphalt Emulsion Manufacturers Association 211
Asphalt Institute 10, 36, 135, 213, 296, 311,
    316, 318–9, 323, 324–5, 482, 494,
    496, 593, 712, 719
asphalt mixture design 266, 312–14, 323–5

Asphalt Institute (*continued*)
  bitumen emulsions 213
  definitions 4
  fatigue deformation 494, 496
  stiffness 482
asphaltenes
  air blowing 20, 24
  broad chemical composition 59
  constitution/rheology/structure relationships
    57–8
  elemental analysis 49
  oxidation/chemical changes 60–1
  polymer modified bitumens 158
  solubility 50, 52–3
  structure 55–6
asphalt flux manufacture 704
asphaltic grouted macadam 638, *639*
Asphalt Mixture Performance Tester (AMPT) 450,
  451, 455–6
Asphalt Pavement Analyzer (APA) 459–60
asphalt rubber (AR) 673–5
asphalts
  aggregates 217–59
    all-in 232–9
    coarse 217, 231–49
    filler 217, 250–5
    fine 231–49
    origin 218–24
    processing 224–31
    quality 255–6
    type 218–24
  airfield pavements 716–21, *722*
  binder properties 503–47
  bond coats 415–16
  bridges 726–728
  certification 687–701
  chipping of hot rolled 434–5
  coloured surface courses 168–171, 732–5
  compaction 413, 437–44
  compositional analyses 758
  conventional railways 721
  definitions 4
  delivery 413, *414*, 417–18
  disbonding mechanisms 553–7
    blistering 555–6
    detachment 554
    displacement 554
    film rupture 554–5
    hydraulic scouring 556
    pitting 555–6
    pore pressure 556–7
  discharge 417–18
  durability 573–89
    ageing tests 586–8
    hardening 574–5, 576–86
  emulsions 209–13, 214
  film thickness calculation *748–9*
  flexible pavement 261–93
    China 289–90
    design 373, 374–5, *374*
    Europe 261–4
    France 277–82
    Germany 282–4
    India 287–9
    United Kingdom 264–77

USA 284–7
gussasphalt 435–6
high speed railways 721, 722–4,
hot-on-hot paving 427, *428*
hydraulic applications 730–2
joints 430–4
kerbs *735*, 735
laboratory testing 447–77
  bitumen recovery 470–4
  empirical 449, 467–70
  fundamental 449–59
  simulative 449, 459–67
layer thickness 429–30
laying
  by hand 427–8, 443
  machine 424–7, 441–3
  weather 423
low slope roofing products 710
manufacture
  changes 59
  durability quality checks 77–8
mixtures
  Asphalt Institute design method 324–5
  California design method 296, 314–6
  Corps of Engineers design method 296,
    316–24
  design 295–345
  grading 304–11, *312*, *313*, 334–5
  history 295–7
  Hubbard Field method 296, 312–4
  Hveem design method 296, 314–6
  Marshall design method 296, 316–24
  methods 311–42
  quotations 341–2
  Smith design method 324–5
  Superpave 296, 325–42
  volumetrics 297–304
moisture damage 549–71
mortar 722–726, *726*, *727*
motor racing tracks 729–730
opening to traffic 443
other important uses 703–36
pavement marker adhesive 714–7
pavers 418–36
polymer modified bitumen 65, 150–3, *152*,
  164, 168–71
preparatory works 413–15
production plants 379–412
  additive systems 403–4
  capacities 402–3, *404*
  cold mix plants 404–6
  control 407–10
  sampling 410–12
  types 380–402
  warm mix plants 406–7
product standards 754–5
properties 479–501
  cracking failure theories 496–8, *499*
  deformation 486–91
  fatigue characteristics 491–6
  stiffness 479–86
railways 723–7, *728*
recreational areas 728–729
roofing applications 704–14
simulative measurements 564–7

Superpave binder specifications 82–3, *84*
surface regularity 428, 429–30
surface treatments 732–4
terminology 6, 13
transportation 413
vehicle testing circuits 730
weather 423, 443
asphalt for ultra-thin layers (AUTL) 262
Assessment and Verification of the Constancy of Performance (AVCP) 689, 694, 698–9
ASTM *see* American Society for Testing and Materials
atactic polypropylene (APP) 710–11
atmospheric distillation 16, *17*, 58, 704
Australia
ageing 587
AS 1160-1996 604–5
Austroads 594
hardening 585–6
surface dressing 594, 596, 604–5
Australian Road Research Board (ARRB) 585–6, 587
AUTL *see* asphalt for ultra-thin layers
AVCP *see* Assessment and Verification of the Constancy of Performance
averaging beams 429, *430*
aviation fuel 16, *19*, 20

bag filters, conventional batch mixing plants 383–4, 385
Bailey method, packing theory 307–9
ball-and-plug-type valves 36
ballastless high speed railway tracks *723*, *724*
ball penetration tests, surface dressing 596
BANDS (bitumen and asphalt nomographs developed by Shell) 369
basalt, 'sonnenbrand' 246–7
base bitumen, marker adhesive 716
base concrete, high speed railways *723*, 726, *727*
base course
flexible pavement
asphalts 263, 265–6, 282–3
design 348, *349*, 351–2, 355, 358–9
hardening 584–5
PMB manufacture *177*
basic surface dressing 592–3
batch emulsion plants 194, *195*
batch heater plants 381, 390–4
batch mixing plants 380, 381–94
continuous 394–402
conventional 381–90
heater 381, 390–4
batch processes, PMB manufacture 178
BBME *see* bétons bitumineux à module élevé
BBR *see* bending beam rheometers
BBSG *see* béton bitumineux semi-grenus
BBTM *see* béton bitumineux très mince
bending beam rheometers (BBR)
low temperature linear elastic regions 134–5
quality and specifications 66, 71, *72*, 83, 85
bending beam tests, low temperature cracking 466, *467*
beneficial filler aggregates 250

béton bitumineux drainant 278
see also porous asphalt
béton bitumineux semi-grenus (BBSG) 278, 279, 282
béton bitumineux très mince (BBTM) 262, 278
béton bitumineux ultra mince 278
bétons bitumineux à module élevé (BBME) 279
Bible 3
binder course
airfields 717
flexible pavement
asphalts 263, 266–7, 282–3, 288
design 348–50, 352–3, 359–60
binders 503–47
see also bituminous binders; CR modified binders
aggregates 245–6
asphalt mixture design 319, 323, 336, 338–40
bearing capacity 507
coloured surface course 170–1, 732–3
compaction 509–11, *512*
during construction 508–11, *512*
dynamic shear modulus 483–6
emulsions 206–7
engineering requirements 504, *505*
fatigue cracking 507
fretting resistance 508
handling temperatures 620–1
industrial 171–2
low temperature cracking 508
mechanical properties 507
mixing 508–9, *510*, *512*
pigmentable 168–71
in service performance 511–42
cracking 511–27
deformation 527–35
fatigue cracking 522–3
fatting up 535
fretting 535–40
load-associated surface cracking 517–20
low temperature cracking 513–17
multiple stress creep recovery 532–3
ravelling 540–1
reflective cracking 523–7
skid resistance 541–2
softening point 533–5
surface cracking models 520–2
thermal cracking 513–14
thermal fatigue 517–20, 522
zero shear viscosity concepts 532
specifications 120, 127
surface dressing 604–10, 619
transport 508–9, *510*, *512*
Bingham, Eugene C. 120
BISAR *see* Bitumen Stress Analysis in Roads
Bitufresh *see* Shell Bitufresh
bitumen, terminology 13
bitumen absorption (P_{ba}) 300
bitumen adhesion 549–71
bitumen and asphalt nomographs developed by Shell (BANDS) 369
bitumen content (P_b)
see also thickness
asphalt mixture design 295–342

coating evaluations 711–14
coating specifications, roofing 706
coating test 561–3
cobalt aluminate/oxide 171
COC *see* Cleveland open cup method
coefficients of thermal contraction 740
coefficients of thermal expansion 740
cohesion 114–16
  ductility 114
  force ductility 114
  quality checks 72–3, *74*
  slurry surfacing 626–7
  tenacity 114–15
  thermodynamic principles 550–2
  toughness 114–15
  Vialit cohesion 115–16
cohesive failure, binder properties 537
cohesive strength, surface dressings 607
coked chippings 539–40
coke deposits, health hazards 38
cold-applied joint paint 431
cold in-place recycling 212
cold milling 414–15
cold mix asphalt 647–8
cold mix plants 404–6
cold plant mixes 212–13
cold RAP feeds 386–7, *388*
Cole–Cole diagrams 131
colloidal models 54–6
colloid mill operating conditions 204
coloured asphalt 168–71
coloured surface courses 732–4
column chromatography, elemental analysis
  49–50
combination mixing plants 399, 401
combination rollers 440
combination screed 421
combustion 38
Comité Européen de Normalisation (CEN)
  see also European Standards
  binder properties 506
  BSI specifications 81
  bitumen specification 81
  certification 693
  flexible pavement layers 264, 279, *280–1*
  surface dressing 593
co-modifiers, polymer modified bitumens 159
Compactasphalt systems 427, *428*
compaction 413, 437–44
  binder properties 509–11, *512*
  gyratory compaction 463–4, *465*
  Shell Thiopave 657, 662, 663, 664
complex colloidal models 55–6
complex modulus 128–33, 138–45
complex shear modulus *84*, 128, 130
complex viscosity 129–30
composite pavements, binder properties 523–4
compositional analyses 756–8
composition of bitumen life cycle 58–61
compression crushers 228, *229*, 230
computed tomography 235, 653–6
computer controlled spray bars 614
concrete curing 214
conductivity, thermal 7442
cone crushers 230–1

consensus properties, asphalt mixture design
  333–4
constitution 47–54, 57–8
  elemental analysis 48–50
  in-service changes 59–60
  other classification methods 53–4
  solubility 50–3
Construction Products Directive (CPD) 81–2,
  688, 698
Construction Products Regulation (CPR) 81, 255,
  688–9, 690, 691, 695, 696, 697
contact angle measurements 557, *558*
contaminators, coarse aggregate 247
continuous air blowing process 20, 22, *22*,
  23–4
continuous cold mix asphalt plants 404, 405
continuous emulsion plant 194, *195*
continuously graded asphalt, grading envelopes
  269–70
continuously graded surface course mixtures,
  hardening 582
continuous mixing plants 380, 394–402
continuous PMB manufacture 178
contraction, thermal coefficients 740
control points, asphalt mixture design 334, *335*,
  337
control systems, asphalt production plants
  407–10
convection losses, asphalt compaction 443
conventional asphalt batch plants 381–90
conventional bitumen emulsions, surface dressing
  604–6
conventional pavers 418–19
conventional railways 721
conversion factors for viscosities 743
convertible mixing plants 399, 401
cooling times, asphalt compaction 442
core pairs, asphalt compaction 444
coring, binder properties 523–4, *525*
Corps of Engineers design method 296,
  316–24
  see also Marshall methods
counter flow drum mix plants 397–402, 403
CPD *see* Construction Products Directive
CR *see* crumb rubber
cracking 141, 466–7
  binder properties 506, 507, 511–27
  failure theories 496–8, *499*
  flexible pavement design 351–2
  Fraass breaking point test 91
  Shell Thiopave 659–60
crawler tracked pavers 418
creep
  aggregate shape 487, *489*
  asphalt deformation 486, *488*
  binder properties 528–9, 532–3
  bitumen content 486–7, *489*
  direct tension tests 83, 85
  Marshall stability 528–9
  rheology 124–6, 130, 137–8
  stiffness moduli 107
CRMB *see* CR modified binders
cross-linking materials, polymer modified
  bitumens 159, 162
Cross model 123

crude coal tar 12–13
crude oil 5, 15–20
crumb rubber (CR)
  alternative asphalts 672–3, 675–80
  polymer modified bitumen 162–3
crumb rubber modified bitumen (CRMBs)
  alternative asphalts 672–3, 675–6
    applications 678–9
    limitations 679
  flexible pavement layers 288–9
  USA technology 679–80
crushed rock aggregate, Shell Thiopave 658
crushing of aggregates 226–31
  compression crusher 228, 229, 230
  cone crusher 230–1
  crusher types 228–31
  gyratory crusher 228, 229
  impact crusher 230
  jaw crusher 228, 229
  scalping 227–8
  simplified crushing process flow 228
crushing of chippings, surface dressing 619
cryogenic rubber, alternative asphalts 672–3
cryogenic tyre grinding, alternative asphalts 672
cubical aggregate particle shapes 234
cubic close packing, asphalt mixture design 305
cumulative deformation, flexible pavement
    design 351–2
cumulative strain curves 450, 451
cup viscometers 94–6
curing
  concrete 214
  hardening 584–5
cut-back bitumens 6, 9
  cup viscometers 94–5, 743
  health hazards 39
  specifications 65
  surface dressings 608–9, 620–1
cutting wheels 431–2
cycloaliphatic compounds 47

dashpot visco-elasticity 125
data interpretation in asphalt mixture design
    321–3
DBM see dense bitumen macadam
Deborah number (De) 123–4
decay curves, foamed bitumen 646–7
declaration of performance (DoP) 690, 694,
    697–8
  see also Initial Type Test
definition of bitumen 4–5
deformation 486–91
  see also fatigue
  alternative asphalts 653
  binder properties 527–35
  flexible pavement design 351–2, 357
  MEPDG rutting model 490–1
  Shell method 489–90
  VESYS method 490
deformation resistance, Shell Thiopave 657–8
dehydrogenation, roofing flux 709
delamination, slurry surfacing 631–2, 635
delayed elastic response 123
delivery 24–7, 413, 414
  see also transportation

de Lorentweiler Tunnel 169
Delphi method 333–4
delta ring and ball 253–4
de-mixing/phase separation, polymer modified
    bitumens 157, 159, 174
dense asphalt concrete 583
dense bitumen macadam (DBM)
  flexible pavement 268–72
    design 356–7, 366
    grading limits 288
    India 288
  Shell Thiopave 663–5
dense graded flexible pavement 285–6, 287
densification curves, asphalt mixture design
    331–2, 333
density
  asphalt mixture design 321
  emulsions 205–6
Department of Transportation (DOT)
  batch mixing plants 389–90
  flexible pavement layers 284, 285–7
depth of penetration, surface dressing 595–6,
    597
design traffic, flexible pavement design 361–3
detachment disbonding mechanisms 554
dewetting surface energy 558
di-block precursors, polymer modified bitumens
    154
dielectric constants 741
dielectric strength 741
diesel, crude oil distillation 16, 19, 20
dilatant liquids 122
DIO Specification 40 251
dip sampling 35
direct tension tests 83, 85
direct water jet extinguishers 34
disbonding mechanisms
  blistering 555–6
  displacement 554
  hydraulic scouring 556
  pitting 555–6
  pore pressure 556–7
distillation 5, 16, 17
  bitumen recovery 470–1
  chemical composition & physical properties
    58–9
  crude oil 15–20
distillation, atmospheric 704
distortion energy failure theory 498
DoP see declaration of performance
double chip techniques, surface dressings 606
double drum counter flow plants 399, 400,
    401–2, 403
double layered porous asphalt (2L-PA) 262,
    263
double surface dressings 611–12, 613
drag marks, slurry surfacing 631–2, 633
dried aggregates, counter flow drum mix plants
    399, 401
drinking water 43–4
droplet tests, exudation 79
drum mix feed conveyors, continuous mixing
    plants 394–5
drum mix plants 394–402, 403
dry chemical powder extinguishers 35

dry compacted filler aggregate 253
dryers, RAP feeds 388, *389*
drying capacities, asphalt production plants 402
drying zones, continuous mixing plants 395, 398, 399, *400*
dry processes
    alternative asphalts 677–8
    asphalt production plants 403
    polymer modified bitumens 162–3
DSR *see* dynamic shear rheometer
dual phase modified emulsions 206, 207, *207*
dual RAP feeds, conventional batch mixing plants 387
ductile bitumens, direct tension tests 83, 85
ductility
    cohesion 114
    metals 123
    polymer modified bitumens 175–6
    rheology quality checks 72, *73*
dump trucks 224, 226
Duntilland Quarry Scotland 224, *225*
durability 573–89
    aggregates 243–4
    quality checks 75, *76*, 77–9
    test 587
dust absorption, surface dressings 619
dynamic compaction 437
dynamic display screens, asphalt production plants 407, *408*
dynamic flow 127–30
dynamic friction tester 600, *601*
dynamic immersion tests 563
dynamic shear modulus of asphalt binders at glassy state 485–6
dynamic shear modulus of binders 484
dynamic shear rheometer (DSR) 127, 138–41, *139*, 140–1
    binder properties 507, 520
    quality and specifications 69–70, 83, 85
    roofing 713
dynamic stability (DS)
    binder properties 533–4
    wheel tracking tests 462–3
dynamic tests 452–6
dynamic viscosity 130

EADs *see* European Assessment Documents
EAPA *see* European Asphalt Pavement Association
EC *see* European Commission
echelon paving 433, *434*
effective bitumen content (P_{be}) 299–300
effective specific gravity G_{se} 302
EFTA *see* European Free Trade Association
EI *see* Energy Institute
elasticity 105–9
    recovery tests 175, *176*
    rheology
        deformation and flow 120, 121, 123–30
        low temperature linear elastic regions 134–5
    polymer modified bitumens 143
elastomers 142, 151, *152*, 153–9
electrical properties of bitumens 741
electrical resistance 741

elemental analysis of bitumen 48–50
elongated aggregate particle shapes 234
elongation at break 109, 111, *112*, 113
EME *see* enrobé à module élevé
EME2 *see* enrobé à module élevé class 2
empirical design, flexible pavement design 368
empirical laboratory tests 449, 467–70
    indirect tensile strength tests 469–70
    Marshall tests 467, 468–9
empirical testing 141–2
    see also penetration test; ring and ball test; softening point; viscosity
    roofing 711–12
emulsified cement asphalt mortar requirements 725–6, *726*, *727*
emulsion base, high speed railways *724*, *725*, 724, 725
emulsions 10, 185–216
    breaking 189, 196, 200–3, 206–7, 208–10, 415, 598, 605–6, 608, 618, 623, 626
    classification 208–9
    components 187–93
    consumption 186, *187*
    cup viscometers 94–5
    further information 214
    handling 194, 196
    history 186–7
    joints 431
    manufacture 193–4, *195*, 204–5, 207–8, *209*
    polymer modified bitumen 206–8
    properties 196–203
        adhesivity 203
        breaking of 200–3
        effects of 205–6
        modification 203–8
        stability 196–7
        viscosity 197–200
    slurry surfacing 625
    specifications 208–9
    storage 194, 196
    surface dressings 604–6, 608
    uses 209–14
        miscellaneous 213–14
        new developments 214
        roads 209–13
        trends 214
energy curves, quality checks 73, *74*
Energy Institute (EI) 34, 87, 89, 91
engineering properties 105–14
    elongation at break 109, 111, *112*, 113
    fatigue strength 113–14
    stiffness moduli 105–9, *110*
enrobé à module élevé (EME) 278–9
enrobé à module élevé class 2 (EME2) 268–73, 353, 355–6, 365, 366
environment 10, 11, 15–27, 29, 42–4
    emulsions 201
    life cycle assessment 42–3
    potable water lining 43–4
Environmental Protection Act of 1990 212
Environmental Protection Agency (2008) 36–7
EOTA *see* European Organisation for Technical Assessment

epoxy resins 609, 636
equivalent single axle load (ESAL) *318*, *328*, 333, *337*, *338*, *340*, 665–6, 717
equiviscous temperature 508–9, 510
erosion control 6, *7–8*
ESAL *see* equivalent single axle load
esters
   high friction surfaces 636
   polymer modified bitumen 159
ethylene–butyl acrylate random copolymers 161
ethylene–vinyl acetate (EVA) 159–62, 206
ETSI *see* European Telecommunications
   Standards Institute
EU chemical limits for drinking water 43–4
Eurobitume 4, 6, *7–9*, 9–11
   binder properties 506–7, 508
   life cycle assessment 43
Europe, flexible pavement layers 261–4
European Asphalt Pavement Association (EAPA)
   11, 728
European Assessment Documents (EADs) 695
European bitumen roofing production 710, *711*
European Commission (EC), certification 690,
   693
European Committee for Electrotechnical
   Standardization (CENELEC) 692
European Committee for Standardisation *see*
   Comité Européen de Normalisation
European Free Trade Association (EFTA) 692,
   693
European Organisation for Technical Assessment
   (EOTA) 695
European Standards
   see also British Standards
   ageing tests 586
   aggregates 217–18, 222–3, 232–8, 240,
      242–56
   asphalt production plants 380, 410–12
   binder properties 508
   bitumen recovery 471
   capillary viscometers 96, 98
   certification 687, 689, 692–4, 696, 698,
      699, 700
   cohesion tests 114–15
   crack propagation 466–7
   cup viscometers 94–5
   durability 465, 586
   dynamic tests 452, 453–4
   emulsions 186, 198, 200, 208–9, *210*
   EN 58 36
   EN 196-2 254
   EN 196-6 255
   EN 196-21 *248*, 254
   EN 459-1 250–1
   EN 459-2 254
   EN 536 380
   EN 932-3 222, *223*, 247
   EN 933-1 232, 233
   EN 933-3 234
   EN 933-4 234
   EN 933-5 235
   EN 933-6 235
   EN 933-8 624
   EN 933-9 233, *248*, 252
   EN 933-10 252

EN 1097-1 240, *248*
EN 1097-2 236, 247, *249*
EN 1097-3 242, 255
EN 1097-4 253
EN 1097-5 *248*, 252
EN 1097-6 242, *248*, *249*
EN 1097-7 252, 254
EN 1097-8 236–40, *248*, *249*, 598
EN 1097-9 242
EN 1367-1 244
EN 1367-2 244, *248*, *249*
EN 1367-3 247
EN 1367-5 244–5
EN 1426 67–68, 88–9
EN 1427 67–68, 89–90
EN 1744-1 247, *248*, 254, *249*
EN 1744-4 254
EN 12271 699
EN 12272-2 700
EN 12273 629, 699
EN 12274-1 to 12274-8 629
EN 12274-1 629
EN 12274-2 629
EN 12274-3 626, 629
EN 12274-4 626–7, 629
EN 12274-5 627, 629
EN 12274-6 629
EN 12274-7 628, 629
EN 12274-8 629, 700
EN 12591 65–9, 82, 89, 253, 268, 269,
   508, 586, 698, 751
EN 12592 69
EN 12593 67–68, 91
EN 12595 67–68, 69, 96
EN 12596 98
EN 12597 186
EN 12607-1 67–68
EN 12607-2 69
EN 12697 411–2
EN 12697-3 471
EN 12697-4 471
EN 12697-8 759
EN 12697-11 246, 563
EN 12697-22 459
EN 12697-23 469, 480
EN 12697-24 *452*, 453–5, 459
EN 12697-25 450
EN 12697-26 *452*, 458, 565
EN 12697-33 464, *465*
EN 12697-34 476
EN 12697-44 466–7
EN 12697-45 465, 565–6
EN 12697-46 465–6
EN 12846-1 94–5, 198
EN 12846-2 94–5
EN 12848 200
EN 13036-1 444, 602
EN 13043 217–18, 232–3, 242–7,
   250–6, 411, 596, *598*, 698
EN 13075-1 200
EN 13108 264, 268, 282, 389–90, 393,
   401, 411, 693, 699, 754, 757
EN 13108-1 264, 268, 269, *280–1*
EN 13108-2 264, *280–1*
EN 13108-3 264

European Standards (continued)
  EN 13108-4 250, 264, 269
  EN 13108-5 264
  EN 13108-6 264
  EN 13108-7 264, 281
  EN 13108-8 223, 264
  EN 13108-20 264, 699
  EN 13108-21 264, 410–11, 699
  EN 13179-1 253
  EN 13179-2 254
  EN 13302 98, 198
  EN 13304 65
  EN 13305 65
  EN 13398 69, 175
  EN 13399 174
  EN 13587 69, 114–15
  EN 13588 115, 608
  EN 13589 69, 114, 176–7
  EN 13703 69, 114, 115, 177
  EN 13808 200, 208, 210, 604, 608,
      625, 698
  EN 13924 65, 82, 89, 269, 698, 751
  EN 13924-2 166
  EN 14023 65, 66, 82, 177, 698, 751
  EN 14769 77
  EN 14770 69
  EN 14771 66
  EN 15322 65, 698
  EN ISO 2719 68, 69
  EN ISO 9001 687, 689, 694, 695–6
  flexible pavement layers 264, 268, 269,
      272–4, 279, 280–1, 282
  flexural bending test methods 452
  Fraass breaking point test 91
  gyratory compaction 464, 465
  immersion mechanical tests 565–6
  immersion tests 563
  indirect tensile fatigue tests 459
  indirect tensile stiffness modulus 458
  low temperature cracking 465–6
  marking 82
  Marshall test 468
  penetration test 88, 89
  polymer modified bitumen 174, 175–7
  repeated load test 450
  rotational viscometers 98
  saturation ageing tensile stiffness 565–6
  slurry surfacing 624–7, 629
  softening point test 89–90
  surface dressings 596, 608
  valves 36
  wheel tracking tests 459
European Telecommunications Standards Institute
    (ETSI) 692
EVA see ethylene–vinyl acetate
evaporative hardening, quality checks 75, 76,
    77–8
excavators 224, 226
excessive binder application 619
exhaust systems
  continuous mixing plants 395–6
  control 409
  conventional batch mixing plants 383–4
  heater batch mixing plants 392
expansion

alternative asphalts 645–6, 647, 648
  thermal coefficients 740
extensional flow 121
extinguishers 34–5
extraction of aggregates 224–6
extruded asphalt kerbs 735, 735
exudative hardening 79, 574, 575
eye burns 42

FAA see fine aggregate angularity
factory production control (FPC) 255–6, 689,
    690, 697–9, 700
failure
  binder properties 537
  cracking 496–8, 499
  flexible pavement design 351–2
  Octahedral shearing stress theory 498, 499
  surface dressings 618–20
fatigue
  binder properties 517–20, 522
  cracking 141, 466–7
    binder properties 506, 507, 513–14,
        522–3
    flexible pavement design 351–2
    Shell Thiopave 659–60
  deformation
    Asphalt Institute 496
    asphalts 491–6
    MEPDG 496
    predictions 493–6
    Shell method 494, 495
  dynamic tests 452
  flexible pavement design 351–2, 369
  indirect tensile tests 458–9
  strength 113–14
fatting up, binder properties 535
fatty acids 193
Federal Highway Administration (FHWA)
  binder properties 530, 533
  flexible pavement layers 284, 285, 286
  repeated load tests 450
felts 172
ferrous oxide 171
fibreglass felt 705
fibres, polymer modified bitumen 151, 152
filler
  aggregates 217, 250–5
    calcium carbonate content 251, 254
    calcium hydroxide content 251, 254
    chemical requirements 254
    consistency 254–5
    delta ring and ball 253–4
    dry compacted filler 253
    geometrical requirements 252
    grading 252
    harmful fines 252
    particle density 252
    physical requirements 252–4
    properties 252–4
    stiffening properties 252–3
    testing 252–4
    water content 252
    water solubility 254
    water susceptibility 354
  bituminous marker adhesive 714, 716

polymer modified bitumen 151, *153*
silos
    asphalt production plants 409–10
    conventional batch mixing plants 385
slurry surfacing 625, 628
filler, mineral 705
film rupture disbonding mechanism 554–5
film thickness
    see also thickness
    calculations 748–9
    hardening 583–4
final mixture designs 323–4
fine aggregates 217, 231–49
    angularity 235, 333–4
    basalt 'sonnenbrand' 246–7
    bituminous binders 245–6
    chemical properties 247, *248*, 249
    durability 243–4
    geometrical properties 232–5
    high speed railways 725
    packing theory 307–8, *309*
    physical properties 236–43, 247, *248*,
        249
    property data sheets 249
    thermal shock 244–5
finishers 435–6
finite element modelling, surface cracking
    520–1
fire-fighting and fire prevention 34–5
first aid, skin burns 39, 42
Fischer–Tropsch synthetic waxes 163–4
fixed width screed 422
flaky aggregate particle shapes 234
flask methods 563
flexible bituminous marker adhesive 714, *715*
flexible pavement 261–93, 347–77
    analytical 368–9
    background 351–3
    China 289–90
    elements of 356–60
        foundation 357–8
        pavement 358–60
        subgrade 356–7
    empirical 368
    Europe 261–4
    France 277–82
    future 374–5
    Germany 282–4
    India 287–9
    Shell Bitumen 369–72
    SPDM-PC examples 372–4
    stages involved 360–6
        design traffic 361–3
        foundation 363
        pavement 363–6
    stiffness 353–6
    United Kingdom 264–77
    United States of America 284–7, 366–8
flexible pavements, airfields 718, *717–9*
flexural bending test methods 452
Flintkote *see* Shell Flintkote
flocculation, emulsions 197
flooring 172
flowcharts, Superpave 329
flow content, airfield pavements 720

flow diagrams
    continuous mixing plants 394, *395*, 398
    conventional batch mixing plants 382, *383*,
        386–7, 388, *389*
    heater batch mixing plants 390–1, *392*
flow, Marshall 468
flow rates, emulsions 199
flow regions, polymer modified rheology 143
flow studies *see* rheology
fluorescence microscopy, polymer modified
    bitumens 154–6
flushing up, binder properties 535
flush surface, slurry surfacing 631–2, 634
flux, roofing 703–6, 707–8
fluxed bitumens 6, 9
    specifications 65
    surface dressings 608–9
fluxing, PMB manufacture 177
flux oils 164
foamed asphalts 647
foamed bitumen 406
    alternative asphalts 643–51
        application 650–1
        characterisation 645–7
        mixing 647–50
        use 650–1
foam extinguishers 35
foam indices 645, 646–7
foaming techniques 652
FOD *see* foreign object damage
fog seals 211
footway slurry 622
force ductility 114
foreign object damage (FOD) 720
Formula 1 racing tracks 729–30
foundations to flexible pavements 265, 348,
    349, 357–8, 363
four-point bending tests 452, 454, *455*
FPC *see* factory production control
Fraass breaking point test 91
fractional distillation, crude oil 15–20
fracture temperatures, Shell Thiopave 661–2
fragmentation, aggregates 236
France, flexible pavement layers 277–82
freeze–thaw cycles, aggregates 243–4
French Central Laboratory of Roads and Bridges
    (LCPC) 507
French Institute of Science and Technology for
    Transport, Spatial Planning, Development
    and Networks 507
French rutting tester (FRT) 459–60, 462
frequency, compaction 438
fretting 508, 535–40
friction course, airfields 717
FRT *see* French rutting tester
fuel oil, crude oil distillation 16, *19*, 20
fully blown bitumens *21*, 23
fume odour neutralisation 173
fundamental bitumen testing 141–2
fundamental laboratory tests 449–59
    dynamic tests 452–6
    indirect tensile tests 456–9
    repeated load tests 450, *451*
    static creep tests 449–50
furnaces 16

gap graded asphalt, grading envelope 269–70
gap graded surface course mixtures, hardening 582
gas oil 16, *19*, 20
gasoline 16, *19*, 20
gas–liquid chromatography (TBP–GLC) 77
G$_b$ see specific gravity of a bitumen
gel-type bitumens 54–5, *157*
geometrical properties of aggregates 232–5
Germany, flexible pavement layers 282–4
Gershkoff formulae 141–2
Gilsonite 12
glassy regions, polymer modified rheology 143
glycerine, softening point test 89–90
G$_{mb}$ see bulk specific gravity of an asphalt
G$_{mm}$ see maximum specific gravity of the loose mixed material
goniometers 558
Good–Van Oss–Chaudhury theory 551–2
grades of bitumen 6
    see also specifications
grading
    aggregates 232–3
    asphalt mixture design 304–11, *312*, *313*, 334–5
    asphalts in China 290
    bituminous concrete 288
    dense bitumen macadam 288
    filler aggregate 252
grading envelopes
    continuously graded asphalt 269–70
    gap graded asphalt 269–70
    slurry surfacing 624–5
granites 247, *248*, *249*, 552–3
granules, mineral 705
grave bitume 279
grave émulsion 212–13
gravel 224, *225*, 658
gravity 301–2, *303*
    physical constants 739–40
GripTester 598, *599*, *600*
grit application 443
gritstone aggregate specimens 240
ground tyre rubber (GTR) 714, *715*
grouted macadams 636–8, *639*
grouting, hydraulic applications 732–3
G$_{sa}$ see apparent specific gravity
G$_{sb}$ see bulk specific gravity
G$_{se}$ see effective specific gravity
GTR see ground tyre rubber
gussasphalt 272, 435–6
gyratory compaction 460, 462, 463–4, *465*
gyratory crushers 228, *229*

HA 41/90 *248*
half life, alternative asphalts 646
half-warm emulsion technology 214
half-warm mix asphalt 651–2
Hamburg wheel tracking device (HWTD) 179, 459–60, 462
hand laying 427–8, 443
handling 15, 24–7, 29–36
    emulsions 194, 196
    polymer modified bitumens 177–9
handling temperatures, surface dressings 620–1

HAPAS see Highway Authorities Product Approval Scheme
hard bitumens 10
    specifications 70
    terminology 13
    vapour emissions 37–8
hardening 574–5, 576–86
    aggregates 576–8
    bulk storage 576, *577*
    evaporative 75, 76, 77–8
    exudative, quality checks 79
    hot storage 578–9
    laying 578–9
    mixing 576–86
    pavement 579–86
    roads 579–86
    in service 576–86
    storage 576–86
hard grade paving bitumen specifications 82
hard industrial bitumen specifications 65
hardness, surface dressings 595–6, *597*
hard paving grade bitumen specifications 65
hard penetration grade bitumen specifications 66
hard rock deposits *225*, *226*
harmful fines, filler aggregate 252
hazards 36–9
    combustion 38
    elevated temperature 36
    hydrogen sulfide 38, *40–1*
    skin contact 39
    vapour emissions 36–8
    water 39
HBMs see hydraulically bound mixtures
HCs see hydrocarbons
HDM see heavy duty macadam
head of material 425–6
head parts of emulsifying agents 189–91
health 10–11, 29, 36–42
    first aid & skin burns 39, 42
    hazards 36–9
heated aggregates, counter flow drum mix plants *399*, *401*
heated tanks, conventional batch mixing plants 384–5
heaters, batch mixing plants 381, 390–4
heating of emulsions 197
heating zones
    continuous batch mixing plants 395
    continuous mixing plants 398
heavily trafficked roads 214
heavy duty macadam (HDM) 268–73, 353, 355–6, 366
heavy traffic bitumens 290
hexagonal close packing 305
high compaction screed 421
high friction surfaces 609, 635–6, *637*
high level RAP dryers 388, *389*
highly modified bitumen A15E 168
high performance surface dressing 214
high shear mills, PMB manufacture 177–8
high speed railways 722–6,
high stone content hot rolled asphalt 277
high temperature viscous regions 135–6

Highway Authorities Product Approval Scheme (HAPAS)
  asphaltic grouted macadam 638
  cementitious grouted macadam 638
  flexible pavement layers 274–5
  high friction surfaces 636
Highways Agency
  flexible pavement design 356, 372–3
  flexible pavement layers 272–5
Highways Authorities Product Approval Scheme (HAPAS) 694, 696–7
Highways Maintenance Efficiency Programme (HMEP) 636
Hirsch model 485
history of bitumen research 1, 2
HMAs see hot mix asphalts
Hooke's law, rheology 121, 127
horizontal action impact crushers 230
horizontal heated tanks, conventional batch mixing plants 384–5
hot mix asphalts (HMAs) 651–2
  emulsions 212
  flexible pavement layers 284
  fracture temperatures 661–2
  induced stresses 661–2
  terminology 6, 13
hot mixture storage tests 77–8
hot-on-hot paving 427, 428
hot RAP feeds, mixing plants 388, 389
hot rolled asphalts (HRAs) 262–3, 268–74
  binder properties 508–9, 510, 539, 540
  chipping of 434–5
  flexible pavement design 356–7, 360
  high stone content 277
  typical thickness 583
hot stone bins, asphalt production plants 386, 387, 409–10
hot stone elevators, conventional batch mixing plants 384
hot storage, hardening 578–9
Hubbard Field method, asphalt mixtures 296, 312–14
humped grading 334–5
Hveem design method, asphalt mixtures 296, 314–16
HWTD see Hamburg Wheel Tracking Device
hydrated lime 568
hydraulically bound mixtures (HBMs) 267–8, 348, 349
hydraulic applications 730–2
  polymer modified bitumens 172
hydraulic scouring 556
hydraulics and erosion control, use of bitumen 6, 7–8
hydrocarbons (HCs) 15
hydrodynamic size 47
hydrodynamic volumes 53–4
hydrogen sulfide 38, 40–1
hydrophilic surfactants 186
hygiene 30, 621

IAA see impact absorbing asphalt
IAN 73/06 see Interim Advice Note 73/06
IAN 156/12 see Interim Advice Note 156/12

IARC see International Agency for Research on Cancer
ICAO see International Civil Aviation Organization
IFSTTAR see Institut Français des Sciences et Technologies des Transports, de l'Aménagement et des Réseaux
igneous aggregate
  fragmentation 236
  micro-Deval test 241
igneous rock 219–21, 224, 225, 226
immersion test 561–3
  mechanical 564–6
  trafficking 566–7
impact absorbing asphalt (IAA) 729
impact compactors 468
impact crushers 230
India
  flexible pavement layers 287–9
  Shell Thiopave 663
India Roads Congress (IRC) 288–9
indirect tensile fatigue tests (ITFT) 458–9
indirect tensile stiffness moduli (ITSM) 458, 565
indirect tensile strength tests 469–70
indirect tensile tests 456–9
induced damage see moisture damage
industrial applications, modified binders 171–2
industrial bitumen specifications 65
  see also hard industrial bitumens; oxidised bitumens
industrial gas oil 16, 19, 20
industrial paving, use of bitumen 6, 7
industrial use of bitumen 6, 8–9
infrastructure applications, polymer modified bitumens 172
Initial Type Test (ITT) 256, 698, 699
  certification 699
InLine Paving system 427
in-service changes to bitumen constitution 59–60
in service hardening 576, 579–86
in situ penetration index 528
in situ recycling 669–71
in situ void content 444
Institute of Science and Technology for Transport, Spatial Planning, Development and Networks 507
Institut Français des Sciences et Technologies des Transports, de l'Aménagement et des Réseaux (IFSTTAR) 507
Instron pull off tests 561, 562
intelligent rollers 441
Interim Advice Note (IAN) 73/06 265, 356, 358, 361, 373
Interim Advice Note (IAN) 156/12 238
International Agency for Research on Cancer (IARC) 10, 37
International Civil Aviation Organization (ICAO) 720
International Organization for Standardization (ISO) 11, 43, 692, 694
International Slurry Surfacing Association (ISSA) 624–5, 626, 628–9
International Union of Laboratories and Experts in Construction Materials, Systems and Structures 507

International Union of Pure and Applied Chemistry 185
inverted double surface dressing 611–12, 613
ions, emulsions 201–2, 205
IRC *see* India Roads Congress
iron oxide 38
'island' structures 55–6
ISO *see* International Organization for Standardization
isochronal plots 130, 133, 144–5
isothermal plots 131
ISSA *see* International Slurry Surfacing Association
ITFT *see* indirect tensile fatigue tests
ITSM *see* indirect tensile stiffness moduli
ITT *see* initial type testing

jaw crushers 228, *229*
joints 415, 424, 430–4

Kelvin elements 125
Kepler conjecture 305
kerbs, asphalt *735, 735*
kerosene 16, *19, 20*, 178–9
kinematic viscosity test equipment 96, *97*
Kitemark 697

Laboratoire Central des Ponts et Chaussées (LCPC) 459
laboratory testing
  asphalt mixture design 319–20, 330–3
  bitumen recovery 470–4
  empirical 449, 467–70
    indirect tensile strength test 469–70
    Marshall test 467, 468–9
  foamed asphalt mixing 649
  fundamental 449–59
    dynamic tests 452–6
    indirect tensile tests 456–9
    repeated load tests 450, *451*
    static creep test 449–50
  simulative 449, 459–67
    crack propagation 466–7
    durability testing 465
    gyratory compaction 463–4, *465*
    low temperature cracking 465–6, *467*
    wheel tracking 459–63
lake asphalt 4, 11–12, 151, *152*
laser texture measurements 602, *603*
latex
  emulsions 206, 207–8
  polymer modified bitumens *152*
latitude, surface dressings 610
law of mixture parallel models 485–6
layer coefficients, flexible pavement design 366
layered elastic models, flexible pavement design 369, 371
layer thickness 429–30
laying
  asphaltic grouted macadam 638, *639*
  binder properties 509–11, *512*
  by hand 427–8, 443
  cementitious grouted macadam 638, *639*
  grouted macadam 637–8, *639*
  hardening 578–9
  Shell Thiopave 662, 663, 664

LCA *see* life cycle assessment
LCPC *see* Laboratoire Central des Ponts et Chaussées
lean base asphalt compositions 297
life cycle assessment (LCA) 42–3
life cycles, chemical composition and physical properties 58–61
lightweight contaminators, coarse aggregate 247
lignosulfonate 191, *192*
lime
  adhesion 568
  filler aggregates 250–1
limestone filler aggregate 254
limits of proportionality 121
limpet pull off tests 561
linear aliphatic compounds 47
linear visco-elastic models 125
lipophilic surfactants 186
liquids, rheology and behaviour 121–30
liquid–vapour and liquid interfaces, thermodynamic principles 550–1
load-associated cracking 368, 513, 517–20
loaded wheel tests, slurry surfacing 627–8
loading-induced pavement stresses, binder properties 520–1
loading jigs, Marshall test 468
loading time
  alternative asphalts 658–9
  mixture stiffness 486, *487*
  stiffness modulus 107, *108*
long residue 16, *18*
long-life pavements 350–1
long term ageing, binder properties 513, 537
Lorentweiler Tunnel 169
lorries 413, *414*
Los Angeles tests 236
Los Angeles coefficient *598*
loss-in-weight systems 396
loss modulus *see* shear loss modulus
Lottman procedure 564–5
low shear mixing 177–8
low slope roofing products 705, 710
low temperature cracking 465–6, *467*, 506, 508, 513–17
low temperature linear elastic regions 134–5
low temperature properties, Shell Thiopave 660–2
lubricating oil 16, *19, 20*

MA *see* mastic asphalt
machine laying 424–7, 441–3
macrotexture 444, 541, 596, 598–604, 625, 700, 720
magnesium sulfate test 244
maltenes
  air blowing 20
  elemental analysis 49
  polymer modified bitumen *157*
  structure 55
manufacture 15–20
  aggregates 217, 222–3
  asphalt 59
  emulsions 193–4, *195*, 204–5, 207–8, *209*
  polymer modified bitumens 177–9

Marshall flow *322*, 468
Marshall method
  asphalt mixtures 296, 316–24
  flexible pavement layers 289–90
Marshall quotient 468
Marshall stability 317–8, 468
  airfield pavements 720
  binder properties 529–30, *531*
  immersion 564, 565
Marshall test 80–1, 467, 468–9, 564
master curves 131–40, 453, 455–6, 486,
    488, 516, 520, 658–9, 712–3
mastic asphalt, bridges 727
mastic asphalt (MA) 262, 263, 282–3
mastic flooring 172
material safety data sheets (MSDSs) 10, 29, 34,
    36
material segregation 426–7
materials specifications, high speed railways
    724, *726*, 727
material transfer vehicles 417–18
maximum density lines, asphalt mixture design
    334, *335*
maximum distortion energy failure theory 498
maximum principal stress–strain theory 497
maximum shearing stress theory 497
maximum specific gravity of the loose mixed
    material (G_{mm}) 302, 304
Maxwell elements 125
mean summer SCRIM coefficient (MSSC) 600
mechanical properties 87–118
  bitumen test data chart 99–105
    blown class bitumens 102, *104*
    straight line class bitumens *100*, 102, *104*,
      105
    waxy class bitumens 102, *104*, 105
  cohesion 114–16
    ductility 114
    force ductility 114
    tenacity 114–15
    toughness 114–15
    Vialit cohesion 115–16
  engineering properties 105–14
    elongation at break 109, 111, *112*, 113
    fatigue strength 113–14
    stiffness moduli 105–9, *110*
  routine testing procedures 87–91
    Fraass breaking point test 91
    penetration tests 87, 88–9
    softening point test 87, 89–91
    temperature susceptibility 92–3, *94*, *95*
  viscosity 93–9
    capillary viscometers 96, *97*, 98
    cup viscometers 94–6
    rotational viscometers 98–9
    test data charts 99, *101*
mechanical testing
  asphalt mixture design 320–1
  immersion 564–6
Mechanistic–Empirical Flexible Pavement Design
    (MnPave) 368
Mechanistic–Empirical Pavement Design Guide
    (MEPDG)
  deformation 490–1
  fatigue prediction 496

pavement design 367–8
  stiffness 483–5
medical care, skin burns 39, 42
medium temperature asphalt (MTA) 277
Megget Dam *731*, *732*
melt flow index (MFI) 161
MEPDG *see* Mechanistic–Empirical Pavement
    Design Guide
metallic elements 48–9
metals, oxidation/chemical changes 60
Mexphalte C *see* Shell Mexphalte C
Mexphalte RM *see* Shell Mexphalte RM
MFI *see* melt flow index
micelles, emulsions 201–2
microcalorimetric measurements 58
micro-Deval test 240, 241, *305*
micropaver, slurry surfacing 629, *630*
microscopy, polymer modified bitumens 154–6
microsurfacing *188*, 207, 209, 214, *241*,
    621–635
microtexture 541, 596, 600–2, 613

mill flow rate 199
mill operating conditions 204
mill viscosity 200
milling 223, 414–15, 423, 429, 667–8
mineral filler 705
  slurry surfacing 625
mineral granules 705
mineralogy, rock types 219–20
Mineral Products Association (MPA) 539, 540,
    703–4
Ministry of Road Transport and Highways
    (MoRT&H) 288
mixed material silos 385–6
mixing
  asphalt production plants 402–3
  binder properties 508–9, *510*, *512*
  hardening 576–86
mixing zones
  continuous batch mixing plants 398
  continuous mixing plants 395, 399, *400*,
    401
mixture design, slurry surfacing 626–9
mixture parallel models 485–6
mixture volumetrics, asphalt mixture design
    297–304, 310–11, *313*
MnPave *see* Mechanistic–Empirical Flexible
    Pavement Design
Model Code of Safe Practice (Energy Institute,
    2005) 34
modern surface dressing sprayers 614, *615*
modified bitumen 9
  see also polymer modified bitumen
modified Lottman procedures 564
moisture 536–8, 539, 549, 557–67
molecular weight, ethylene–vinyl acetate 161
MoRT&H *see* Ministry of Road Transport and
    Highways
mortar 722–7, *726*, 727
mosaics, surface dressing 612
motor racing tracks 729–30
MPA *see* Mineral Products Association
MSCR *see* multiple stress creep recovery
MSDSs *see* material safety data sheets

MSSC see mean summer SCRIM coefficient
MTA see medium temperature asphalt
multigrade bitumen 166–8
multi-layered elastic analysis 369, 371
Multiphalte see Shell Multiphalte
multiple hot stone bin plant 386, *387*, 393
multiple stress creep recovery (MSCR) 126, 532–3
Mu-Meter 600

nano-aggregates 56
naphtha 16
National Asphalt Pavement Association (NAPA) 285, *286*, 651, 653
National Center for Asphalt Technology (NCAT)
    alternative asphalts 665, 680
    wheel tracking tests 459
National Cooperative Highway Research Program (NCHRP) 450, 483
National Highway Sector Schemes (NHSS) 694, 695–6
natural aggregate 217
natural asphalts 4, 11–12, 151, *152*
natural rubber latex 206, *207*
NCAT see National Center for Asphalt Technology
NCHRP see National Cooperative Highway Research Program
needle penetration tests, specifications 87, 88–9
neutron-scattering measurements 58
Newbold Quarry 224, *225*
Newtonian behaviour, rheology 122–3, 135–6
New Zealand road transport, surface dressing 594
n-heptane 58
NHSS see National Highway Sector Schemes
NMAS see nominal maximum aggregate size
    see also nominal stone size
no flow indicators 394, 409
nominal maximum aggregate size (NMAS)
    flexible pavement layers 284, 285, *286*
    packing theory 308, 309
    Superpave 334, 335, *336*, 337
nominal stone size (NSS)
    see also nominal maximum aggregate size
    flexible pavement layers 285, *286*
nomographs
    elongation at break 111, *112*
    fatigue deformation 494, *495*
    flexible pavement design 369, *370*
    penetration index 93, *94–5*
    stiffness moduli 109, *110*
non-ionic emulsions 188, 191, 193
non-ionic surfactants 186
non-Newtonian behaviour, rheology 57, 122–3, 135–6, 197, 484
North American/European bitumen roofing production 710, *711*
notified bodies, regulatory frameworks 690
Nottingham asphalt testers 565
NSS see nominal stone size

occupational exposure 10–11, 29–30, 31–4, 37–8, 40–1

occupational exposure levels (OELs)
    hydrogen sulfide 38, *40–1*
    PPE 29–30, *31–4*
Octahedral shearing stress theory 498, *499*
odour neutralisation 173
Official Journal of the European Union 693
oil–aggregate mixtures 314
oils
    distillation 16, *19*, 20
    exudative hardening 574, *575*
    polymer modified bitumen 164, 178–9
olefinic polymers 160–1
optimum binder content 323
organic additives 652
origins of bitumen 3–4
oscillation 127–30, 138–41, 437–9, *440*
out-of-phase component 129, 130
overlapping screed plates 433
oxidation
    air blowing 20
    chemical changes 60–1
    constitution/in-service changes 59–60
    hardening 574, 575, 580, 582–3
    quality checks 75, *76*, 77–8, 81
oxidation time charts, roofing flux 709, *710*
oxides, polymer modified bitumens 171
oxidised bitumen 10, *22*, 23, 24, 55, 60–1
    specifications 65, 69–70
    terminology 13
    vapour emissions 37–8
oxidised coatings, roofing 711–12
oxygen, hardening 583

PA see phosphoric acid; porous asphalt
packing theory 304–9
PACs see polycyclic aromatic compounds
pad coats 611, *612*
paddle mixers 390, *391*, 392–3, 405
PAHs see polycyclic aromatic hydrocarbons
paints 734
paraffinic flux oils 164
paraffin wax 164
parking areas 169, *170*
parks 732, *733*
particle density 242, 252, 254
particle rounding 241
particle size distribution
    aggregates 232–3
    emulsions 201, 203–5
patching 428, 443
pavement
    see also flexible pavement
    binder properties 520–1
    fatigue cracking 141
    hardening 579–86
    polymer modified bitumen 150–3
pavement classification numbers (PCN) 719
pavement marker adhesive 714–16
pavements, airfields 716–721, *722*
pavers 359, 418–23, 629, *630*
paving
    speed 426
    surface course 629, *630*
    use of bitumen 6, 9

paving grade bitumens 5
  binder properties 508
  specifications 66, *67, 68*, 82
PAV *see* pressure ageing vessel
P_b *see* bitumen content
P_{ba} *see* bitumen absorption
P_{be} *see* effective bitumen content
PCN *see* pavement classification numbers
PCS *see* primary control sieve
peel tests 560–1
pendulum tests 115–16
penetration grade bitumens 5
  binder properties 504
  C320 bitumen 168
  cohesion curves 116
  durability quality checks 75
  Qualagon tests 79–80
  reduced-frequency master curves 139–40
  specifications 65, 66, *67, 68*
  stiffness modulus 107, *108*, 109
  test data chart 99, *100*, 102, *103*
penetration index (PI) 90–1, 92–3, *94–5*
  binder properties 505, 508–9, 511, *512*,
    528, *529, 530*
  elongation at break 111
  stiffness modulus 107
penetration test
  binder properties 517
  rheology 119, 137
  specification 87, 88–9
penetration-viscosity numbers (PVNs) 505
Pensky–Martens closed cup (PMCC) test 708
percentage of air voids (V_a) 483–4
percentage of crushed and broken surfaces in
    coarse aggregate 234–5
Performance Graded (PG) Bitumen
    Specifications 120, 127, 285
performance related tests, US asphalt
    specifications 286, *287*
permittivity 741
peroxides 159
perpetual pavements *see* long-life pavements
personal computer (PC)
  asphalt production plants 407–8
  flexible pavement design 372–4
personal hygiene 30
personal protective equipment (PPE) 29–30, 36,
    621
petroleum pitches 4
PG *see* Performance Graded Bitumen
    Specifications
phase angle isochrones 144–5
phase separation, polymer modified bitumen
    157, 159, 174
phosphoric acid (PA) 165–6
photo-oxidation, hardening 580, 582
physical constants of bitumens 739–42
physical hardening 574, *575*
physical properties of bitumen life cycle 58–61
physio-mechanical adsorption,
    bitumen–aggregate adhesion 553
pigmentable binders 168–71
PI *see* penetration index
pitting, disbonding mechanisms 555–6
planing *see* milling

plasticity
  polymer modified rheology 142
  rheology 121
plastomers 142, 159–62
plateau regions, polymer modified rheology 143
PLCs *see* programmable logic controllers
PMB *see* polymer modified bitumen
PMCC *see* Pensky–Martens closed cup test
pneumatic tyred roller (PTR) 438, *439*
polar aromatics 157
polar functional groups 60
polarity
  see also anionic...; cationic...
  emulsions 189–91, *192*
polished stone value (PSV)
  certification 689
  polishing resistance 236–8, *239*
  surface treatments 596, 598, 602–4, 609
polychlorophene 206, *207*
polycyclic aromatic compounds (PACs) 36
polycyclic aromatic hydrocarbons (PAHs) 36–7
polyethylene 159–60, 161–2
polymer modified bitumen (PMB) 9, 149–80
  airfields 717
  asphalt pavements 150–3
  CE marking 82
  cohesion 73, *74*, 116
  durability quality checks 75, 77, 78
  emulsions 206–8
  flexible pavement layers 288–9
  handling 177–9
  high speed railways 724, *724, 725*
  industrial applications 166–73
  manufacture 177–9
  modification of bitumens 153–66
    plastomers 159–62
    reactive chemistry 165–6
    rubbers 162–3
    thermoplastic elastomers 153–9
    viscosity modifiers 163–4
  other products 166–73
    industrial binders 171–2
    multigrade bitumen 166–8
    odour neutralisation 173
    synthetic pigmentable binders 168–71
  performance for road applications 179–80
  properties 174–7
  related test methods 174–7
  rheology 142–5
  road applications 166–73
  specifications 65, 66, 69
  storage 177–9
  surface dressings 606–8
  terminology 13
polymer modified emulsion, joints 431
polymer modified membrane systems, roofing
    710–11
polymerisation, roofing flux 709
polymers
  see also rubber
  polyolefins 160–1
  thermoplastics 151, *152*, 153–9, 636
  thermosetting 151, *152*
polyphosphoric acid (PPA) 159, 165–6
polypropylene 159–60

polystyrene 142, 153–4, 159–60
polyurethane resins 636
polyvinyl chloride 159–60
PONOS 369
poor joints, slurry surfacing 631–2
pore pressure 556–7
porous asphalt (PA)
    bond coats 210
    flexible pavement 262–3, 276–8, 282–3,
        348, 349
    typical thickness 584
porphyrins 48–9
portable extinguishers 35
potable water lining 43–4
power law model 122, 136
PPA see polyphosphoric acid
PPE see personal protective equipment
precoated chippings 273, 377, 539–40, 610,
    618
pre-graded aggregates 390–1, 405
preparations of bitumen 6
pressure ageing vessel (PAV) 77, 78, 587–8
pressure–stress dissipation 374–5
primary control sieve (PCS)
    packing theory 307–8, 309
    Superpave 337
primary crushers 226, 227
prime coats 211
primers 726–7
principal chippings 612
principal strain theory 497
principal stress theory 497
processing of bitumen 59
product approvals, roofing 709
production plants 379–412
    additive systems 403–4
    capacities 402–3, 404
    cold mix plants 404–6
    control 407–10
    sampling 410–12
    types 380–402
    warm mix plants 406–7
product standards 750–3, 754–5
programmable logic controllers (PLCs) 407, 409
property data sheets 249
protective coats 214
protective layers, bridges 726–7
pseudoplastic liquid 122
PSV see polished stone value
PTR see pneumatic tyred roller
pull off test 561, 562
pumping of emulsions 197
pumping temperatures 27
punching shear type tests 313–14
purchasers, regulatory frameworks 691
PVN see penetration-viscosity number
pyrophoric material 38

Qatar, Shell Thiopave 663, 664
Qualagon tests 79–81
quality control 65, 70–81
    adhesion 73–5
    aggregates 255–6
    alternative asphalts 668–9
    asphalt compaction 444

cohesion 72–3, 74
durability 75, 76, 77–9
Qualagon tests 79–81
rheology 70–1, 72
quartzites, adhesion 552–3
quaternary ammonium compounds 193
quick setting cationic emulsions 625
quotients, Marshall 468

racing tracks 729–30
racked in surface dressings 611–13
railways 721–6
    use of bitumen 6, 9
Rankine maximum principal stress theory 497
RAP see reclaimed asphalt pavement
ravelling
    binder properties 540–1
    slurry surfacing 634
REA see Road Emulsion Association
REACH see Registration, Evaluation,
        Authorisation and Restriction of Chemicals
reactive chemistry, polymer modified bitumens
        165–6
reactive polymers, polymer modified bitumens
        151, 153
reclaimed asphalt pavement (RAP)
    alternative asphalts 666–9, 673, 677
    cold mix plants 405
    continuous mixing plants 396, 397–9
    conventional batch mixing plants 386–90
    flexible pavement layers 282
recovery tests, rheology 137–8
recreation, use of bitumen 6, 9
recreational areas 728–9
recreational parks 732, 733
recycled aggregates 217, 223–4
recycled polymer modified bitumen 151, 152
recycling, alternative asphalts 666–71
reduced-frequency master curve 139–40
reduced temperature susceptibility 607
reduced-time/frequency scale 131
refined Trinidad lake asphalt 11–12
reflective cracking 523–7
Registration, Evaluation, Authorisation and
        Restriction of Chemicals (REACH) 6, 7–9
rejuvenation agents 212
relaxation test 124, 126, 127, 466
remix processes, recycling 670
removal of bitumen and burns 42
repairs, hand laying 428
repave processes, recycling 670, 671
repeated load axial test (RLAT) 450
repeated load tests 450, 451
research
    see also laboratory testing
    aggregates 235
    history 1, 2
reshape processes, recycling 669–70
residual binder, emulsions 206–7
resins
    broad chemical composition 59
    constitution/rheology/structure relationships
        57–8
    elemental analysis 49
    high friction surfaces 636

oxidation/chemical changes 60–1
solubility 50, 51
structure 55
resistance, electrical 741
resistance to abrasion 238, 240
resistance to crack propagation 466–7
resistance to deformation 657–8
resistance to fatigue 452
resistance to fretting 508
resistance to low temperature cracking 508
resistance to polishing 236–8, *239*
resistance to thawing 243–4
restricted zones, asphalt mixture design 334,
    335
retained Marshall stability test 564
retained stiffness test 565
Réunion Internationale des Laboratoires et Experts
    des Matériaux (RILEM) 507
reverse flow viscometer 96, *97*
rheology 47, 57–8, 119–48
    data representation 130–4
        master curves 131–4
        simple graphs 130–1
    deformation and flow 120–34
    elongation at break 109, 111
    empirical bitumen testing 141–2
    fundamental bitumen testing 141–2
    high temperature viscous regions 135–6
    low temperature linear elastic region 134–5
    polymer modified bitumen 142–5
    quality checks 70–1, *72*
    visco-elastic region 137–41
rhyolites, adhesion 552–3
rideability *see* surface regularity
rigid pavement 347–8
rip joints 424
ripper attachment 224, *225*, *226*
RLAT *see* repeated load axial test
road applications
    polymer modified bitumens 166–73
        binders 171–2
        multigrade bitumens 166–8
        synthetic pigmentable binders 168–71
    Shell Thiopave 662
Road Emulsion Association (REA) 213
Road Engineering and Environmental Division of
    the Highways Agency 274
roads
    see also Bitumen Stress Analysis in Roads;
        Highways...
    emulsions 209–13
        bond coats 210–11
        cold in-place recycling 212
        cold plant mixes 212–13
        fog seals 211
        prime coats 211
        soil stabilisation 211–12
        tack coats 210
    hardening 579–86
    Shell Thiopave 662–6
    surface dressing 595–6, *597*
rock
    aggregates
        classification 222
        petrographic examination 222

asphalt 12, 164
    cycles/formation 219
    types
        aggregates 219–22
        igneous rock 219–21
        metamorphic rock 222
        sedimentary rock 221
Rockingham Speedway track 730
roller compactors 464, *465*
rollers
    compaction 438–40
    slurry surfacing 631
    surface dressings 616–17
rolling bottle method 246, 563
rolling thin-film oven test (RTFOT) 59, 75, *76*,
    77, 141, 174, 586–8
roofing
    applications 704–14
    coating 706
    felts 172
    flux 707–10
    future trends 711–14
    products 705, *706*
    shingles 705, 706–7, *706*, 711–14
rosin esters, high friction surfaces 636
rotary evaporators 471
rotating dryers, conventional batch mixing plants
    382–3
rotating heater drums, heater batch mixing plants
    392
rotational viscometer 83, 98–9, 136
rounding, particle 241
routine testing procedures 87–91
    Fraass breaking point test 91
    penetration test 87, 88–9
    softening point test 87, 89–91
RTFOT *see* rolling thin-film oven test
rubber
    see also crumb rubber...
    alternative asphalts 671–3
    asphalt production plants 403–4
    binders 676
    emulsions 206, *207*
    polymer modified bitumens 158, 162–3
    steel drummed vibratory rollers 616–17
    thermoplastics 158
rut filling, surface course 629, *630*
rutting 490–1
    see also deformation
    binder properties 506, 527–35
    flexible pavement design 350–1, 356–7
    Shell Thiopave 658

SA *see* soft asphalt
Safepave 274
safety
    delivery 25
    handling 10, 29–36
        fire-fighting and prevention 34–5
        personal hygiene 30
        PPE 29–30
        sampling 35–6
safety data sheet (SDS) 10
Saint Venant maximum principal strain theory
    497

sample valves 35–6
sand deposits 224, *225*
sandwich surface dressing 605, *611*, 613–14
SARA *see* saturates, aromatics, resins, asphaltenes
SATS *see* saturation ageing tensile stiffness
saturates
  broad chemical composition 59
  constitution/rheology/structure relationships 57
  elemental analysis 49
  oxidation/chemical changes 60
  polymer modified bitumens 157
  solubility 50, 51
  structure 55
saturates, aromatics, resins, asphaltenes (SARA) *50, 57, 60, 61, 709, 710*
saturation ageing tensile stiffness (SATS) 246, 251, *465*, 565–6
SAVLs *see* specified apparent viscosity levels
SBE *see* styrene–butadiene elastomer
SBS *see* styrene–butadiene–styrene
scalping, aggregates 227–8
scalping screens, continuous mixing plants 394
scanning electron microscopy 655, *656*
SCB *see* semi-circular bending test methods
scouring, hydraulic 556
scratch course 629, *630*
screed 419–23, 433
screening, aggregate processing 231
screw-driven plunger type sample valves 35–6
SCRIM *see* sideways-force coefficient routine investigation machine
SDS *see* safety data sheet
sealing 211, 585–6
sealing layers, bridges726–728
SEAM *see* sulfur-extended asphalt modifier
sedimentary rock 221
segregation, material 426–7
semi-blowing of bitumens 22, 23
semi-circular bending (SCB) test methods 466–7
semi-coarse asphalt concrete *see* béton bitumineux semi-grenu
sensor measured texture depth (SMTD) 602, *603*
settlement of emulsions 196–7
severe fretting, binder properties 538
SFC *see* sideways-force coefficient
SGC *see* Superpave gyratory compactor
shape of coarse aggregate 234
shear flow 121
shearing stress theory 497, 498, *499*
shear loss moduli 129
shear moduli 106, 484–6
  see also complex shear moduli
shear rate dependency/susceptibility 136
shear storage modulus 128–9
shear thickening/thinning liquids 122
sheet asphalt, mixture design 312–13
Shell Bitufresh 2, 173
Shell Bitumen
  airfield pavements *721, 722*
  flexible pavement design 369–72
  rheology 120

Shell Cariphalte *73, 116*, 180, 374, *720–21*
  performance for road applications 179–80
  storage 178
Shell Cariphalte Fuelsafe 180, *722*
Shell Cariphalte Racetrack 180, *729*
Shell Caritop 172
Shell Flintkote 2, 172, 196
Shell method
  deformation 489–90, 494, *495*
  stiffness predictions 480–2
Shell Mexphalte C 2, 169–71
Shell Mexphalte Fuelsafe *723*
Shell Mexphalte RM 2, 163, 680
Shell Multiphalte *2*, 166–8
Shell Pavement Design Manual (SPDM) 371, 372–4
Shell Pavement Design Method on a Personal Computer (SPDM-PC) 372–4
Shell Research, exudative hardening 79
Shell sliding plate viscometers 587
Shell Thiopave 654–66
  compaction 657, 662, 663, 664
  deformation resistance 657–8
  design 656–7
  fatigue cracking 659–60
  laying 662, 663, 664
  low temperature properties 660–2
  production 662
  properties 657–62
  road applications 662
  road performance 662–6
  stiffness modulus 658–9
  target applications 656
  water sensitivity 657
  workability 657
Shell Tixophalte 171–2
shift factor 131–2
shingles 705, *706–7, 706*, 711–14
short residues
  air blowing 20–4
  crude oil distillation 16, *19*, 20
short term ageing, binder properties 513, 537, 576
SHRP *see* Strategic Highway Research Program
shuttle buggies 417
SHW *see* Specification for Highway Works
sideways-force coefficient routine investigation machine (SCRIM) 598, *599*, 600
sideways-force coefficient (SFC) 598, 600
silica (SiO_2) content of aggregates 245–6
Silly Putty 123
silos
  asphalt production plants 385–6, 409–10
  hardening 578–9
simple colloidal models 54–5
simulative laboratory tests 449, 459–67
  crack propagation 466–7
  durability testing 465
  gyratory compaction 463–4, *465*
  low temperature cracking 465–6, *467*
  wheel tracking 459–63
single phase modified emulsions 206, *207*
single point viscosity measurements 198
single surface dressings 611–12
site geometry, surface dressings 610

size exclusion chromatography (SEC) 53–4
skid resistance
  binder properties 541–2
  surface dressings 598, *599*, 600–4
skin contact and safety 39, 42
skinning 606
slab compactors 464, *465*
sliding plate rheometer 138, *139*
sliding plate viscometer 587
slip layers 214
slope pavers 418–19
slot jets 614
slow setting emulsions 211–12
slurry seals 734
slurry surfacing 621–35
  applications 622–3
  certification 699
  construction 629–31
  defects 631–5
    bleeding 631–2, 634
    delamination 631–2, 635
    drag marks 631–2, *633*
    flush surface 631–2, 634
    poor joints 631–2
    ravelling 631–2, 634
    washboarding 631–2, *633*
  materials 623–6
  mixture design 626–9
  specifications 626–9
small angle neutron scattering measurements 58
SMA *see* stone mastic asphalt
Smith method, asphalt mixture design 324–5
SMTD *see* sensor measured texture depth
soaps 187
sodium chlorides, emulsion components *188*
soft asphalt (SA) 262, *263*
softening points
  binder properties 508–9, *510*, 533–5
  durability quality checks 77–8
  hardening 576, *577*
  roofing flux 709, *710*
  specifications 87, 89–91
  visco-elastic rheology 137
soft paving grade bitumen specifications 66, 69
soil stabilisation 211–12
solids, rheology and behaviour 121, 123–30
solid–liquid and vapour interfaces,
    thermodynamic principles 550–1
'sol'-type bitumens 54, *157*
solubility, bitumen classification 50–3
solvents
  elemental analysis 49
  emulsion components *188*
'sonnenbrand', basalt 246–7
soundness, aggregate 244
SPDM *see* Shell Pavement Design Manual
SPDM-PC *see* Shell Pavement Design Method on
    a Personal Computer
specialist slope pavers 418–19
specialty asphalts, flexible pavement layers
    286–7
Specification for Highway Works (SHW)
    217–18
  aggregates 217–18, 246
  filler 251

flexible pavement layers 275
joints 430
HAPAS certification 697
registered suppliers 696
surface courses 360
surface regularity 266
thin surface course systems 349, 353
Type 1 unbound mixture 355
specifications
  see also British Standards; European
      Standards
  asphalt compaction 444
  binders 120, 127
  bituminous marker adhesive 716–17
  breaking point test 91
  CEN 81–2
  emulsions 208–9
  Fraass breaking point test 91
  hard bitumens 70
    industrial 65
    paving grade 65, 82
  hard grade paving bitumen 82
  hard industrial bitumen 65
  hard paving grade bitumen 65
  hard penetration grade bitumens 66
  high speed railways 724, *724*, 725, *725*
  industrial bitumens 65
  oxidised bitumens 65, 69–70
  paving grade bitumens 66, *67*, *68*
  penetration grade bitumens 65, 66, *67*, *68*
  penetration tests 87, 88–9
  polymer modified bitumens 65, 66, 69
    CE marking 82
    cohesion *73*, *74*
    durability *75*, *77*, *78*
    quality checks 65, 70–81
      adhesion 73–5
      cohesion 72–3, *74*
      durability 75, *76*, 77–9
      Qualagon tests 79–81
      rheology 70–1, *72*
  roofing coating 706
  roofing flux 707–8
  roofing shingles 706–7
  routine testing 87–91
  SHRP 82–5
  slurry surfacing 626–9
  softening point test 87, 89–91
  soft paving grade bitumens 66, 69
  Superpave 82–5
  surface dressings 594
  viscosity grade bitumens 66, 69
specific gravity 739–40
  asphalt mixture design 302–3
specific gravity of a bitumen (G_b) 303
specific heat 741–2
specified apparent viscosity level (SAVL) 587
specifiers, regulatory framework 691
specifying bitumens 65–86
splitt mastic asphalt 272
sprayed seals 585–6
sprayers, surface dressings 614, *615*, *616*
spring visco-elasticity 125
S-shape curves, polymer modified bitumens
    156–7

stability, Marshall 468, 529–30, *531*, 564, 565
stability values, asphalt mixture design 313
standard bituminous marker adhesive 714, *715*
static compaction 437
static creep test 449–50
static immersion test 562–3
steady state creep measurements 137
steam injection extinguishers 35
steel drummed vibratory rollers 616–17, *617*
steel slags 247
steel structures, bridges 728
steric hardening *574*, *575*
steric stabilisers 186
stiffness 479–86
    Al-Khateeb model 485–6
    Asphalt Institute 482
    dynamic tests 452
    filler aggregate 252–3
    flexible pavement design 353–6
    Hirsch model 485
    loading time–temperature curves 486, *487*
    MEPDG 483–5
    retained stiffness test 565
    Shell method 480–2
    Witczak equation 483–5
stiffness modulus 105–9
    breaking strength 113–14
    fatigue strength 113–14
    flexible pavement design 369, *370*
    indirect tensile test 456, *457*, 458
    prediction 109, *110*
    Shell Thiopave 658–9
Stokes' law 196–7
stone mastic asphalt (SMA) 210
    asphalt mixture design 297
    flexible pavement design 359–60
    flexible pavement layers 262, 271–3, 274, 286–7
    generic 277
    Germany 282–4
    typical thickness 584
stone-to-stone contact, asphalt concrete 262, 534–5
storage 15, 24–7
    emulsions 194, 196
    hardening 576–86
    polymer modified bitumens 174–5, 177–9
    reclaimed asphalt 667–8, *669*
    stability test 174–5
storage modulus 128–9
straight run bitumen 5, 11
    terminology 13
    vapour emissions 37–8
strain controlled testing 127
strain theory 497
Strategic Highway Research Program (SHRP) 60
    adsorption tests 563–4
    asphalt mixture design 296, 325–7, 333–4
    binder properties 505–6, 530
    bitumen specifications 82–5
    durability 587–8
    elemental analysis 50
    rheology 120
    roofing 711–12

Strategic Transportation Research Study (STRS) 325
stress criteria, typical fatigue lines 491–2, *492*
stress distribution, indirect tensile tests 456, *457*, 458
stresses of wheel loads 447, *448*
stress theory 497
stress–strain relationships
    creep tests 124–6
    flexible pavement design 369, 371
    low temperature cracking 465–6
    oscillatory test 127–8
    relaxation test 126, *127*
stripping 167, 233, 245–6, 275, 305, 538, 541, 552–3, 562–3, 564, 566–7, 568–9, 595, 620, 626, 650
STRS *see* Strategic Transportation Research Study
structural ageing, quality checks 79
structure of bitumens 47, 54–6, 57–8
    complex colloidal models 55–6
    simple colloidal models 54–5
study of flow *see* rheology
styrene–butadiene elastomers (SBEs) 153–9
styrene–butadiene–styrene (SBS)
    alternative asphalts 676
    emulsions 206, *207*
    polymer modified bitumens 153–9, *160*
    polymer modified rheology 142, 143–5
    roofing *708*, 710–11
styrene–isoprene elastomers 153
subbase, flexible pavement design 355
subgrade, flexible pavement 265, 356–7
substrates, bitumen–substrate tests 560–1, *562*
sulfonates 193
sulfur
    alternative asphalts 654–66
    polymer modified bitumens 159, *160*
sulfur extended asphalt modifier (SEAM) 654, *655*
superficial fretting 538
Superpave gyratory compactor (SGC) 330–2, 337–8
Superpave Mix Design, airfield pavements 721
Superpave (Superior Performing Asphalt Pavement)
    asphalt mixture design 296, 325–42
    binder properties 530
    bitumen specifications 82–5
    durability 587–8
    flexible pavement layers 285, *286*
    mixture classifications 285, *286*
    repeated load test 450
surface abrasion, aggregates 238, 240
surface activity, slurry surfacing 624
surface course
    asphalt mixture design 297
    binder properties 508–9, 510, 538–9, *540*
    chipping of hot rolled asphalt 434–5
    flexible pavement
        design 348, *349*, 350, 352–3, 360
        Europe 261–3
        Germany 282–3
        India 288
        United Kingdom 267, 273–7
    grit application 443

hardening 579–84
rut filling and paving 629, *630*
volumetric patch test 444
surface cracking, binder properties 517–22
surface dressings 214, 591–621
 adhesion 607, 609–10
 altitude 610
 binders 604–9
 bitumen hardening 585
 breaking of emulsion 598, 605–6, 608,
  616, 618
 certification 699
 chipping 596, 598
 coloured surface course 734
 design 593–610
 equipment 614–18
  chipping spreaders 615–16
  rollers 616–17
  sprayers 614, *615, 616*
  sweepers 617
  traffic control and aftercare 617–18
 existing road surface condition/hardness
  595–6, *597*
 failure types 618–20
 hardening 585–6
 influential factors 594–610
 latitude 610
 local circumstances 610
 performance 594–610
 safety 620–1
 season 619–20
 site geometry 610
 size 596, 598
 skid resistance 598, *599*, 600–4
 texture 598, *599*, 600–4
 time of year 610
 traffic volumes and speeds 594–5
 types 610–14
  double 611–12, *613*
  inverted double 611–12, *613*
  pad coats 611, *612*
  racked-in 611–13
  sandwich 611–12, *613*–14
  single 611–12
surface factor, asphalt mixture design 315
surface free energy
 moisture damage 557–60
 thermodynamic principles 550–2
surface-initiated reflection cracking 523–4,
 525, 526–7
surface layer texture, Shell Thiopave 662–3
surface oxidation, hardening 582–3
surface regularity 428, 429–30
surface texture, surface dressings 598, *599*,
 600–4
surface treatments 591–642
 certification 699–701
 coloured surface courses 732–734
 grouted macadams 636–8, *639*
 high friction surfaces 635–6, *637*
 microsurfacing 622, *623*, 626
 slurry 622–35
 surface dressing 591–621
surfacing, vehicle testing circuits 730
surfacing layers, bridges 726–728

surfactants 186
sweepers, surface dressings 617
swell tests 316
synthetic pigmentable binders 168–71
synthetic waxes 163–4

tack (adhesive power) 607
tack coats 210
tail parts of emulsifying agents 189–90
TAIT *see* type approval installation trials
tamping screed 420
tandem rollers 438–9, *440*
tanks, storage 25–6
tar 12–13
TBP–GLC *see* true boiling point gas–liquid
 chromatography
TCT *see* tensile creep test
technical committees (TCs) 82
temperature 24–7
 see also thermal…
 adhesion, thermodynamic principles 550–2
 alternative asphalts 658–9
 binder properties 508–9, *510*, 533–4
 coefficients of thermal contraction 740
 coefficients of thermal expansion 740
 cohesion 550–2
 emulsions 204
 hardening *576, 577*
 hazards 36
 low temperature cracking 465–6, *467*, 506,
  508, 513–17
 low temperature linear elastic regions 134–5
 mixture stiffness 486, *487*
 penetration index 92–3, *94, 95*
 polymer modified bitumens 164
 rheology
  high temperature viscous regions 135–6
  low temperature linear elastic regions
   134–5
  time–temperature superposition principles
   131–2, 137–9
  visco-elastic regions 137–41
 Shell Thiopave 658–9, 660–2
 stiffness modulus 107, *108*
 surface dressings 595–6, *597*, 607, 620–1
 surface free energy 550–2
 thermal properties of bitumens 741–2
tenacity 114–15
tensile creep test 466
tensile strength, binder properties 515
tension–compression fatigue test 507–8
test data chart (BTDC) 99–105, 167
testing circuits, vehicle 730
texture, surface dressings 598, *599*, 600–4
TFOT *see* thin-film oven test
thawing resistance 243–4
The Bitumen Industry – A Global Perspective 4
thermal conductivity 742
thermal cracking, binder properties 513–14
thermal fatigue, binder properties 517–20, 522
thermally induced stress contours 526–7
thermal shock 244–5
thermal stress restrained specimen test (TSRST)
 466, 516, 660–1
thermodynamically meta-stable emulsions 186

thermodynamic principles 550–2
thermoplastic polymers 151, *152*, 153–9, 636
thermoplastic rubber (TR) 158
thermoset binders 609
thermosetting, surface dressings 609–10
thermosetting polymers 151, *152*
thick carriageway slurries 623
thickness
    asphalts 429–30
    calculations 748–9
    flexible pavement 282, *283*, 350–1,
        363–5, 373, *374*
    hardening 583–4
thin bonded overlay mixtures 287
thin carriageway slurries 622
thin-film oven test (TFOT) 59, 586
thin overlay mixture (TOM) classification 287
thin surface course systems (TSCS) 149, 151,
    174, 180, 210, 238, *239*, 241, 265,
    267, 272, 273–6, 349, 353, 404, 508,
    567–8, 697, 728
Thiopave *see* Shell Thiopave
third party certification schemes 687
three component acid–base theory 551
three phase bitumen–aggregate–water system
    555
three-point bending tests 454
three-point flexural bending test 452
three wheeled rollers 438, *439*
tiles 172
time–temperature superposition principles
    131–2, 137–9
titanium oxide 171
Tixophalte *see* Shell Tixophalte
TL Asphalt-StB (FSGV, 2013a) standard 282,
    283
TOM *see* thin overlay mixture classification
top down cracking 521–2
toughness 114–15
tourniquets, skin burns 42
TR *see* thermoplastic rubber
traffic
    airfields 719
    asphalt laying 443
    asphalt mixture design 313
    flexible pavement design 361–3
    immersion tests 566–7
    Shell Thiopave 662–3
    surface dressings 594–5, 617–18
transient flow 124–6, *127*
transition regions, polymer modified rheology
    143
translucent binders 732
transport
    binder properties 508–9, *510*, *512*
    emulsions 197
    hardening 578–9
Transportation Research Board (TRB), asphalt
    mixture design 325–6, 335, 340–1
Transport Research Laboratory (TRL)
    curing 584–5
    flexible pavement 157, 274–5, 350–2,
        355–7, 363
    hardening 584–5
    long-life pavements 350–1

surface dressing 595–6, *597*, 600–3, *603*,
    604, 611–12
thicker asphalt pavements 513
trapezoidal two-point bending tests 453
TRB *see* Transportation Research Board
Tresca maximum shearing stress theory 497
trial mixes, slurry surfacing 626
Trinidad épuré 11–12
TRL *see* Transport Research Laboratory
true boiling point gas–liquid chromatography
    (TBP–GLC) 77
trunk roads, surface course materials 275–6
TSCS *see* thin surface course systems
TSRST *see* thermal stress restrained specimen test
Tunnel de Lorentweiler 169
Tunnicliff and Root procedures 564–5
two-point bending tests 452, 453
type approval installation trials (TAIT) 699–700
tyre loading-induced pavement stresses 520–1
tyre rubber 676

UCTST *see* uniaxial cyclic tension stress test
UKAS *see* United Kingdom Accreditation Service
UK Environmental Protection Act of 1990 212
UL-M (ultra-mince) mixtures 274
ultraviolet fluorescence microscopy 154–6
ultraviolet radiation, hardening 580, 582
uniaxial cyclic tension stress test (UCTST) 466
uniaxial tension stress test (UTST) 466
United Kingdom Accreditation Service (UKAS)
    690
United Kingdom (UK)
    see also British Standards...
    asphalt kerbs 735
    asphalt production plants 411
    flexible pavement 264–77, 359–60, 362–3
    grouted macadam 636
    surface dressing 593, 598, *599*, 614
    wheel tracking tests 459–60
United States of America (USA)
    see also American Association of State Highway
        and Transportation Officials; American
        Society for Testing and Materials; Strategic
        Highway Research Program
    alternative asphalts 665, 672–3, 679–80
    emulsions 208
    flexible pavement design 366–8
    flexible pavement layers 284–7
    Shell Thiopave 665–6
    slurry surfacing 624–5, 626, 627–9
    surface dressing 593–4, 596
    wheel tracking tests 459
US Agency for Toxic Substances and Disease
    Registry 37
US Environmental Protection Agency (2008)
    36–7
uses of bitumen 5–10, 43–4
US National Asphalt Pavement Association
    (NAPA) 651, 653
US Superpave specifications 587–8
UTST *see* uniaxial tension stress test
U tube reverse flow viscometers 96, *97*

vacuum capillary viscometers 135–6
vacuum distillation 16, *18*, 58

vacuum flashing 704–5
valves, sample 35–6
Van der Poel nomographs
   elongation at break 111, *112*
   flexible pavement design 369, *370*
   stiffness moduli 109, *110*
vapour emissions
   binder properties 536–7
   fire-fighting and prevention 35
   health hazards 36–8
vapour–liquid and solid interfaces,
   thermodynamic principles 550–1
variable speed belt feeders 382, 390–1, 394,
   405, 409
variable width screed 422
vehicle testing circuit 732
verification of the constancy of performance
   (AVCP) 689, 694, 698
vertical heated tanks 384–5
VESYS method, deformation 490
VFB *see* voids filled with bitumen
Vialit cohesion 115–16
vibrating grizzly 228
vibrating screed 420–1
vibrating screens 384
vibratory rollers 616–17
VIM *see* air voids
vinyl acetate content in ethylene–vinyl acetate
   161–2
visco-elastic region rheology 137–41
viscometers 83, 587
viscosity 93–9
   capillary viscometers 96, *97*, 98
   conversion factors 743
   cup viscometers 94–6
   emulsions 197–200, 205
   polymer modified bitumens 151, *153*, 163–4
   rheology
      deformation and flow 120, 121–30
      high temperature viscous regions 135–6
      polymer modified bitumens 143
   rotational viscometers 98–9
   structure 55
   test data charts 99, *101*
viscosity grade bitumen specifications 65, 66,
   69
viscosity–temperature relationships
   see also penetration index
   polymer modified bitumens 164
viscous flow, polymer modified bitumens 150
V_{ma} *see* voids in mineral aggregate
void content
   airfield pavements 720
   asphalt compaction 444
   hardening 580, *581*
voids analyses 321
voids of dry compacted filler aggregate 253
voids filled with bitumen (VFB)
   asphalt mixture design 301
   stiffness 483–4
void size, packing theory 308, 309
voids in mineral aggregate (V_{ma})
   asphalt mixture design 300–1, 304,
      310–11, *313*, 317, 341
   stiffness 483–4

voids over-filled with binder 305
volatilisation, hardening 574, 575
volume
   blast furnace/steel slags 247
   classification methods 53–4
volumetric belt feeders 394
volumetric patch test 444
volumetrics
   asphalt mixture design 297–304, 310–11,
      312, 313, 322
   symbols 759
von Mises maximum distortion energy failure
   theory 498, *499*
vulcanisation 159
vulcanised rubber 162

warm emulsion based mixtures 214
warm mix asphalt 648–50, 651–4
warm mix plants 406–7
Warm Mix Technical Working Group
   (WMTWG), alternative asphalts 653–4
washboarding, slurry surfacing 631–2, 633
water
   aggregates in asphalts 242, 243
   alternative asphalts 644–6
   binder properties 536–8
   boiling water test 563
   disbonding mechanisms 555, 556–7
   drinking water 43–4
   emulsion components *188*
   filler aggregate 252, 254
   fire fighting and prevention 34–5
   health hazards 39
   potable lining 43–4
   Shell Thiopave 657
   slurry surfacing 626, 627–8
   softening point test 89–90
   warm mix plants 406–7
   waterproofing 211
   work of adhesion 559–60
waterproofing layers, bridges 727, 728
Waterways Experiment Station (WES) 317
waxes 9, 69, *153*, 163–4, 406, 652
wear of coarse aggregates 240–1
weather
   asphalt laying 423, 443
   binder properties 536, 539, *540*
weigh hoppers
   conventional batch mixing plants 385
   heater batch mixing plants 391–2, 393
Weissenberg number (Wi) 124
WES *see* Waterways Experiment Station
wet processes
   alternative asphalts 673–7
   polymer modified bitumens 162
wet stripping, slurry surfacing 626, 627
wetting surface energy 558
wet track abrasion, slurry surfacing 627–8
WGs *see* working groups
wheeled pavers 418, *419*
wheel track cracking 517, *518*
wheel tracking devices 459–60, 566
wheel tracking tests 459–63
   binder properties 529–30, *531*
   Shell Cariphalte 179

Wi *see* Weissenberg number
Williams, Landel and Ferry (WLF) equation
    132–3
wind speed, asphalt compaction 443
Witczak equation, stiffness 483–5
WLF *see* Williams, Landel and Ferry equation
WMTWG *see* Warm Mix Technical Working
    Group
wood-derived emulsifiers 191, *192*

work of adhesion calculations 557–60
working groups (WGs) 82
World Emulsion Congress 186

xenon arc weathering test 709
X-ray computed tomography 235, 653–4

zero shear viscosity concepts 532
zwitterionic surfactants 186